BIOMATERIALS IN REGENERATIVE MEDICINE

Edited by **Leszek A. Dobrzański**

Biomaterials in Regenerative Medicine
http://dx.doi.org/10.5772/66233
Edited by Leszek A. Dobrzański

Contributors

Amirkianoosh Kiani, Candace Colpitts, Anna Dobrzańska-Danikiewicz, Xingguo Cheng, Yasemin Budama Kilinc, Rabia Cakir Koc, Sandra Amado, Pedro Morouço, Paula Pascoal-Faria, Nuno Alves, Carmen Mortellaro, Massimo Del Fabbro, Ewa Zuba-Surma, Malgorzata Sekula, Leandra Ernst Kerche Silva, Mohammad Hosein Amirzade-Iranaq, Neide Pena Coto, Zoia Duriagina, Roman Holyaka, Tetiana Tepla, Volodymyr Kulyk, Peter Arras, Elena Eyngorn, Ildiko Peter, Ladislau Matekovits, mario rosso, Tsanka Dikova, Mônica Fernandes Gomes, Marcello V. Mergulhão, Dr. Maurício David Martins Das Neves, Carlos Eduardo Podestá, Lech B. Dobrzanski, Ralf Pörtner, Hao-Hsiang Hsu, Katharina Schimek, Uwe Marx, Leszek A. Dobrzański

Published by InTech
Janeza Trdine 9, 51000 Rijeka, Croatia

Publishing Process Manager Dajana Pemac
Technical Editor SPi Global
Cover InTech Design team

First published February, 2018
Printed in Croatia
Legal deposit, Croatia: National and University Library in Zagreb

Additional hard copies can be obtained from orders@intechopen.com

Biomaterials in Regenerative Medicine, Edited by Leszek A. Dobrzański
p. cm.
Print ISBN 978-953-51-3776-4
Online ISBN 978-953-51-3777-1

For EU product safety concerns:
IN TECH d.o.o., Prolaz Marije Krucifikse Kozulić 3, 51000 Rijeka, Croatia,
info@intechopen.com or visit our website at intechopen.com.

Meet the editor

Prof. Leszek Adam Dobrzanski, Hon. Prof., Dr. HC (multi), DSc, PhD, MSc, Eng., born in 1947, is a full professor of Materials Engineering, Manufacturing Engineering, Nanotechnology, Medical and Dental Engineering, Management and Organisation. He is one of the 55 most frequently cited contemporary Polish scientists of all disciplines. He is the supervisory board chairman, project manager, principal investigator and director of the Science Centre in the Medical and Dental Engineering Centre for Research, Design and Production ASKLEPIOS in Gliwice, Poland. He is an honorary professor of the 'Lviv Polytechnic' National University in Lvov, Ukraine, and a doctor honoris causa of the Universities in Ruse, Bulgaria; in Miskolc, Hungary; and in Khmelnitsky, Ukraine. He is a fellow of the Materials Science Committee of the Polish Academy of Sciences (PAS) and the president of the Metallic Materials Section of this Committee. He is the vice president and a fellow of the Academy of Engineers in Poland. He is a foreign fellow of the Ukrainian Academy of Engineering Sciences and the Slovak Academy of Engineering Sciences. He is the president of the World Academy of Materials and Manufacturing Engineering and president of the International Association of Computational Materials Science and Surface Engineering. He is the editor in chief of the *Journal of Achievements in Materials and Manufacturing Engineering, Archives of Materials Science and Engineering* and *Open Access Library*. He organised ca. 50 international conferences in Poland and was the member of the scientific committees of the next 100 conferences around the world. He advised 60 finished PhD degree theses and ca. 1000 MSc and BSc degree theses and is the author of as many as 2400 publications, more than 50 books, 40 chapters and 50 patents. He has received a lot of awards, including Gold Medals of William Johnson, Albert Schweitzer, Ferdinand Martinengo, Tadeusz Sendzimir, Rudolf Clausius and Wieslaw Chladek et al., and the orders and the crosses, including Commander's Crosses of Rebirth of Poland and 'Merite de L'Invention' by the Kingdom of Belgium. And, he has also won ca. 80 awards on International Innovation Fairs around the world.

Contents

Preface

The book *Biomaterials in Regenerative Medicine* is my successive book published in my personal, academic career and second one prepared with InTechOpen from Rijeka, Croatia. The topic is very familiar to me because I am interested in it as it is the main area of my scientific interest during the last few years. I have already published many studies, as well as books, monographs and numerous scientific papers on this topic. This is also one of the main objects of the company's interest in the Medical and Dental Engineering Centre for Research, Design and Production ASKLEPIOS in Gliwice, Poland.

Severe cases of diseases and different types of accidents at work, transport accidents, travel accidents and sports accidents are known in each family. People want to spend their time actively, so they practise sports for leisure, but it turns out that in this way they expose themselves to severe health complications and accidents. People tend to live longer and longer, and it is often the reason for the wear of human joints. Many people suffer from extensive burns and persistent ulcerations, often threatening directly one's life. People lose some or all of their teeth, or extensive losses are caused by tooth decay, natural wear or accidents. Sick teeth cause complications for many organs, heart and joints. Intervention medicine and surgery save lives in many cases. However, the price for saving one's life is sometimes extensive mutilations caused by the amputation of different organs, limbs and bones and skin losses. Naturally, patients are offered prostheses, even implants. A range of mechanical devices, screws, fixation plates, nails and glues is required to fixate such medical devices to a body, and the implanted device, although it serves the patient, is artificial. More advanced action concerns the implantation of living cells and even entire organs or their restoration using tissue engineering methods. Regenerative medicine and dentistry methods gain wider and wider applications.

A concept has come into the life of specific 3D spatial scaffolds containing numerous pores into which living cells may grow, and such meshes after some time may disappear in an organism or may stay there permanently provided they are not harmful. They have proposed indirect solutions also; rigid ones involve high strength and transmission of high mechanical loads but may also be elastic and thin, when a very light dressing supplying living cells is applied to an extensive wound, e.g. on the skin, in a way ensuring their fast fusion with the defected surface of the body. We have proposed implant scaffolds, i.e., rigid devices composed of a solid metal core and a surface or porous transition zone into which living cells may grow. The interior of such openings, extending along the entire part of the material, needs to be covered from the inside with a very thin coating, which can be accepted by living cells so that they can develop in such conditions and penetrate such openings deep inside. After performing suitable radiographs with a method similar to computer tomography, a shape of such a device—accurately adapted to a patient's anatomical features and the loss—can be designed. Cells can be cultured in a laboratory on the so-prepared metal, and such an engineering and biological material can be introduced during surgery, in a single procedure, into a patient's body, ensuring the fusion of cells from such implant surface with the own cells of the patient's body.

It also applies to a removed tooth that can be replaced with a suitably shaped implant with the very own cells of a patient's bone cultured in a laboratory but also to an artificial oesophagus and prostheses of some blood cells, although the endings connecting with post-operative stumps will only be made as rigid here. When burns and skin losses are treated, a multilayer engineering and biological plaster with cultured skin cells has to be provided outside and a layer that will disappear without detaching the outer cover carrying such cells, which is shaped like a mat with very thin fibres of special polymers, has to be provided inside. It is difficult to make such living cells to develop on such surfaces, especially inside such small openings. If such basic aspects are resolved, appropriate implants with the patient's own cells can be produced in the future on demand, with the shape and dimension accommodated to the individual size of the damaged and removed part of a given patient's body. It will create for many people better conditions of recovery and of returning to normal life after experiencing severe operations, sometimes as the only life-saving solution.

The examples given indicate that it also requires continuous research and progress in the implementation of the ever-new biomaterials being applied for this purpose. It is what this book deals with, developed by scientists and practitioners from many countries around the world. The book presents a theoretical introduction on multi-aspect bibliographic analysis of the synergy of technical, biological and medical sciences with reference to materials and technologies used for medical and dental implantable devices and on metallic biomaterials. It describes 16 case studies provided in each of the chapters, comprehensively describing the authors' accomplishments of numerous teams from different countries across the world in advanced research areas relating to the biomaterials applied in regenerative medicine and dentistry.

I have a great honour to thank all my co-authors and co-operators from InTechOpen for their efforts in preparing this book. I hope that the detailed information collected in the book, largely deriving from own and original research and R&D works pursued by the authors, will be beneficial for the readers to develop their knowledge and harmonise specific information concerning these topics and will convince the application of the advanced types of prosthesis, implants, scaffolds and implant-scaffolds including personalised ones. The book will be useful for the scientists, PhD degree students, as well as the engineers and medical practitioners. I wish the readers enjoy reading this book and hope it serves them in solving real engineering and medical problems.

Prof. Leszek A. Dobrzański, Hon. Prof., M Dr HC
Director of Science Centre,
Medical and Dental Engineering Centre for Research, Design and Production ASKLEPIOS,
Gliwice, Poland

Chapter 1

Introductory Chapter: Multi-Aspect Bibliographic Analysis of the Synergy of Technical, Biological and Medical Sciences Concerning Materials and Technologies Used for Medical and Dental Implantable Devices

Leszek A. Dobrzański

Additional information is available at the end of the chapter

http://dx.doi.org/10.5772/intechopen.73094

1. General role of the regenerative medicine and dentistry

In many scientific centres, the World intensive research is under way related to the significant development of science in the field of Materials Engineering in connection with cell biology, thus expanding the new groups of advanced materials and technologies finding their application in regenerative medicine and dentistry. The author has already published on this topic studies [1–4] as well as books [5–9], monographs [10–24], scientific papers [25–76], patents [77–84] awarded in international fairs and exhibitions of research, invention and innovation [85–102]. In this area, several of author's own scientific and research projects have also been realised [103–107]; some are in progress [108], and others are planned for implementation [109]. Generally, one of the main reasons of this activity is the dynamic growth of post-injury defects, post-resection defects, as well as those originating from the operative treatment of cancerous tumours, whereas the number of cases is systematically increasing, or inflammation processes and as a result of other disorders of the human population, including the consequences of tooth decay, in particular due to local and systemic complications. Surgical treatment, often saving human lives, causes the necessity to replace losses in such organs or tissues, including in the dental system, to prevent biological and social degradation of patients and to restore their living functions. A growing number of road accidents and severe injuries of more and more people frequently requires a surgical intervention with replacement or supplementation of losses in organs or tissues and numerous interventions in the stomatognathic system. For example, the data of the Association for Improving Safety of Road Traffic reveal that about 1.7 million people suffer injuries in the EU every year. The growing number of sports accidents and bodily injuries frequently requires a surgical intervention with supplementation of lost tissues, also in the dental system. For example, according to EU IDB catalogue, annually, an average 6.1 million people in the EU are treated in hospitals for sports injuries. The number of people aged over 65 will have systematically grown, e.g., in EU by 70% to 2050. Systematically proceeding population ageing, which considerably increases the number of patients requiring surgical intervention with supplementation of losses in organs or tissues, often in the stomatognathic system.

One of the fundamental tasks globally, considered also to be one of the European Union's priorities, is to improve the society's condition of health, medical care and health safety. The numerous tasks related to this topic include the prevention of health risks, early recognition of diseases, rapid and effective implementation of medical procedures, comprehensive and continued therapies leading to an improved health condition and the improved health condition and improved quality of the society's life [1]. For example, the European Health Strategy regards health-related aspects to be a central focus of the Community's policy and proposes a programme of actions for citizens, by recognising their right to own health and healthcare and through the promotion of the ageing society's health, through the protection of citizens against risks for their health and life and through supporting dynamic health care systems and new technologies, related to technical support for medicine, including dentistry. The idea is to protect against serious health risks, especially such as civilisational diseases, pandemics and bioterrorism and to support research, especially such applying advanced technologies, to ensure the fullest prophylactics of diseases and safe treatment of patients, taking into account the relationships between health and economic well-being. A significant and costly problem of modern medicine is the necessity to replace or supplement organs or tissues, in particular in orthopaedics and traumatology and maxillofacial surgery and restorative dentistry, to prevent the biological and social degradation of patients and to restore their living functions [1]. It also applies mostly to the elimination of consequences of tooth decay, considered to be an exceptionally burdensome civilisational disease, also one of the costliest ones, in particular, due to local and systemic complications [110–125].

A relatively new branch of medicine is regenerative medicine and dentistry. The achievements of modern implantology depend not only on the knowledge and experience of medicals, but it also requires the application of advanced engineering problems, both in the field of engineering design, material engineering, nanotechnology and material technology. These are very responsible research, and the most avant-garde trends concern the offering of personalised medical devices manufactured according to individual anatomical features of the patient, according to complex and advanced original technologies and ever newer biomaterials used for implantable devices. These issues are the subject of many years of scientific interest of the Author, and they belong to the group of the most avant-garde, so far relatively little known, but extremely promising technical problems for use in regenerative medicine and dentistry. This study presents the results of previous studies and own research, derived from previously published original own work done by the Author with a team of co-workers [1–19]. As in all other cases, in material engineering and material science, in order to satisfy the functional functions of implantable devices as well as all other products, it is necessary to design and apply engineering materials that, subjected to the appropriate technological processes of shaping the geometric form, and especially the structure, will ensure appropriate physico-chemical properties of the material. This book presents exhaustively the achievements of numerous teams from different countries of the world, inscribing itself in the discussed European Health Strategy, in advanced research areas related to biomaterials used in regenerative medicine and dentistry. In the overall analysis of the issue concerning biologically active cellular structures and a substrate with an engineering composite material matrix used for scaffolds and for newly developed implantable devices applied in regenerative medicine and dentistry, further in this description all the aspects are analysed separately, because the lack of holistic approach and general references is in the literature, apart from own works.

2. Importance of regenerative medicine and dentistry and tissue engineering

The development of regenerative medicine, whose technical aspects are covered by the book with the first reports dating back to 1992 [126], is a relatively new field of medicine. Implantable biomedical devices encompass numerous solutions eliminating various dysfunctions of a human organism and are currently aggregately considered to be medical bionic implants. The development of regenerative medicine, whose technical aspects are covered by the book, previously described in author's theoretical study [1], started with the first reports dating back to 1992 [126], as a relatively new field of medicine. Implantable biomedical devices encompass numerous solutions eliminating various dysfunctions of a human organism and are currently aggregately considered to be medical bionic implants. Bionics is understood as production and investigation of biological systems to prepare and implement artificial engineering systems which can restore the lost functions of biological systems [127, 128]. Autographs, allografts or metal devices or such made of other engineering materials are primarily the current methods of organ and tissue replacement employ [129]. The purpose of the regenerative medicine is treatment—by replacing old and sick cells with young cells, also using tissue engineering methods and cell-based therapies or organism regeneration using a gene therapy. It raises numerous new challenges, notably in counteracting the symptoms and consequences of diseases, and even their causes [130–133]. The application of therapies based on living cells in medicine is a relatively new concept. The first successful allogeneic transplantation of human haematopoietic stem cells (HSC) was seen as late as in 1968 [134] (29), then the cells were used for other therapeutic applications [135, 136]. The therapies based on living cells were intensively developed in the 1990s [137], in particular for skin and cartilage implants. Tissue engineering provides technical support for regenerative medicine and is an interdisciplinary field employing the principles of engineering and life sciences for development of biological substitutes, for restoration, maintenance and improvement of functions of tissues or entire organs [138, 139]. Tissue engineering, introduced in 1985 [140], as a field of technical sciences, utilises medical knowledge and materials engineering methods to develop biological materials capable of restoring, maintaining or improving the functions of particular tissues or organs [138] and to produce their functional substitutes [141, 142]. Tissue engineering is based on understanding the principles of tissue growth and on applying this for functional production of a replacement tissue for clinical use [143].

An overview of the present situation points to a diversity of the currently available therapeutic methods based on cells, undergoing the phase of clinical studies [144]. The global market of cell-based therapies boasts a revenue of more than a billion USD per year [145]. Therapeutic strategies include direct transplantation of the desired type of cells collected using biopsy or such originating from cultures of stem cells, both in the autologous and allogeneic system. Multipotent and self-renewing stem cells (MSCs), depending on the tissue development stage, can be grouped into adult and embryonic cells [136, 146, 147]. Embryonic cells exist in umbilical cord blood [148, 149], in the placenta [150], in amniotic fluid [151] or deciduous teeth pulp in infants, termed as multipotent stromal cells (MSCs). The clinical application is most beneficial, especially of autologous cells [136], as they do not cause an immunological response and do not require immunosuppressive treatment [151–154], and, nevertheless, such application is limited [155]. There is a much greater potential for somatic and especially haematopoietic stem cells (HSC) from bone marrow stromal [156–158], supporting bone marrow stromal cells (BMSCs) as a standard [136, 158]. Adult stem cells also occur, in particular, in synovial fluid, tendons, skeletal muscles [158, 159] and adult muscles, fat tissue—ASCs (adipose-derived stem cells ASCs) [158, 159], the stroma of cornea [149], peripheral blood, nervous tissue and dermis and have an ability of transformation into multiple tissues.

The efficiency of cell-based therapies depends on the preservation of their viability after implantation [160, 161], to prevent the ischaemia of tissues and necrosis [162–167]. Stem cells originating from bone marrow and fatty tissue may be used for breeding mesenchymal cells and tissues, in adipocytes, chondrocytes, osteoblasts and skeletal myocytes and can be used for producing tissues, e.g., muscles fat, gristle and bones [168–171]. Stringent safety requirements must be considered in cell-based therapies because raw materials of the animal origin are used, which poses a potential threat of transmitting a pathogen to a recipient or of immunological complications [172] and as post-production cleaning is required [173]. Progress in this domain seen since 2004 requires further detailed considerations concerning the mechanism of *in vivo* therapeutic activity, to facilitate the development and optimisation and for the development of automated processes, with improved efficiency and with quality control and with reference standards established [174–177]. About 100 companies specialised in this area are operating at the American and European market. For better characterisation, the conditions of cell-based therapies must be realised more intensive the basic research. The introduction of the new clinical regenerative procedure having no competition requires the development of breeding techniques of human stem cells. However, the adult stem cells could be used to a limited extent [178]. The fabrication of the majority of therapeutically meaningful cell types (except the mesenchymal stem cells) has not yet been mastered at a technologically satisfactory scale. However, it should be emphasised that the outcomes achieved to date in this area be promising. Therefore, a wide commitment of the scientific environment is generated for the elaboration and explanation of the phenomena accompanying the growth of tissue structures in conditions allowing their industrial production. The purpose is the development of the adequately organised therapeutic processes, introduction appropriate engineering materials and obtaining the technological processes for them, including nanotechnology. A very important cognitive task is the explanation the interaction between the surface structure of engineering materials forming the substrate and the tissue structures deposited onto it and explication the role of a substrate for culturing tissue structures.

Opposite to pure therapies, in which stem cells are injected directly into peripheral circulation or located in particular tissues, in numerous clinical cases it is necessary to use stem cells' carriers to transport them, and especially scaffolds for their three-dimensional grouping in a particular place of an organism, and such research is being constantly developed [179–188]. It should be noted that the notion of scaffolds is quite comprehensive, as it may not only refer to engineered extracellular matrices 'scaffolds', but also to rigid microporous materials into which osteoblasts may grow, and also to microporous mats made of polymer nanofibres, into which living cells are growing and may be used as specific plasters, to treat for example burns or to reconstruct large fragments of skin [7]. The latest publications [5, 6] also note the relevance and need of developing such therapeutic methods for the dental system. A microscopic, porous structure of scaffolds is required, enabling the diffusion of nutrients and metabolism products through them. Scaffolds, including also bone scaffolds, must enable the adhesion and migration of cells and facilitate their development to form a three-dimensional tissue structure in conditions simulating a natural micro-environment [187, 189]. Scaffolds should exhibit adequate mechanical properties for ensuring an appropriate environment of development for cells and ensure the mechanical preservation of living tissues in a three-dimensional structure. The task of scaffolds is to ensure also the necessary conditions of cells growth by the creation of new blood vessels [190–193].

Two separate aspects need to be considered: a porous structure, especially the size of pores and manufacturing technologies of appropriate porous materials, and the selection of engineering materials of which scaffolds are produced. Both aspects are of primary importance for the undertaking's success, and hence each of them has been analysed based on the state-of-the-art.

3. Designing of geometric properties of porous materials

Porous materials have been used for not too long in orthopaedic procedures to replace damaged bones, and porous scaffolds are geometrically similar to natural hard tissues made up of a skeleton penetrated by interlinked pores [194], thus solid metals are unsuitable because they are, by nature, non-permeable and prevent the adhesion and proliferation of living cells. Porous materials, mainly metals, are implanted to repair damaged bones having a critical size and in the majority of cases are used as carrying devices [195]. It is vital that an implant has properties similar to a recipient bone and the surrounding tissue [196]. A human bone has a hierarchical structure with three major anatomic cavities of different sizes which are Haversian canals (50 μm) [197], osteocytic lacunae (few micrometres) [198–201] and canaliculi (<1 μm) [201, 202]. All the three cavities play an important role in the mechanical integrity of the bones and the processes reconstruction [203]. A porous structure ensures appropriate space for the transport of nutrients and the growth of living cells [204, 205]. A bone elasticity modulus, much smaller than for non-noble metals, is regarded to be one of the major problems in the construction of implants because, as a result, the screening of stresses would often lead to implant damages. Porous metals can duplicate bone properties if their structures are digitally designed and are produced using advanced manufacturing technologies [205]. A bone, despite its anisotropy [206], is being replaced by porous engineering materials with similar rigidity, making them efficient for transmitting loads and for alleviating the effect of stress screening and for regeneration of bone tissue in the damaged place [195]. The following is therefore required: (1) biocompatibility; (2) appropriate surface for adhesion, proliferation and differentiation of cells; (3) a very porous structure with a network of open pores for the growth of cells and for the transfer of nutrients and metabolic wastes; (4) mechanical properties adjusted to the requirements of the surrounding cells to reduce or eliminate excessive stresses and meeting anatomic requirements to avoid a mechanical damage [207–211]. The key features for the design of porous metallic implants include careful selection of porosity, size of pores and mutual connections between pores, aimed at achieving satisfactory clinical results. Such constructional features have a large effect on the mechanical properties and biological activity of scaffolds [195]. Bone regeneration *in vivo* porous implants consists of recruitment and penetration of cells from the surrounding bone tissue and vascularisation [212]. Higher porosity may facilitate such processes and enable bone growth [213, 214]. The influence of the pore size on the growth of cells, e.g., bone cells, is, however, still controversial [195, 215]. Pore sizes of artificially produced scaffolds should be adapted to the specific cell type. They should be small, because then they ensure more space for cells' growth, e.g., bone tissue [195], but small enough to prevent the sealing of pores in a scaffold [187]. They should also be large enough to prevent blood clots [216], enable the migration of cells and ensure conditions to fill up scaffold pores by the reconstructed cells, guaranteeing neovascularisation [217]. For this reason, the upper suggested limit is 400 μm [215], because it is no longer beneficial to increase it. It was found, though, that for pores sized 600–900 μm, the growth of bone cells is higher than for the pores sized 300 μm [195]. The permeability of a porous structure of a scaffold ensures the transport of cells, nutrients and growth factors as a result of blood flow, and a blood pressure gradient depends on the pore size and has also influence on vascularisation and accelerates the osseointegration process [195], although this requires further studies. It is thought that the optimum pore size for cells growth, and especially mineralised bone growth, is 100–400 μm [195]. The aspect is still important, because although increased porosity, pore size and mutual connection of pores are the key factors, which are having a large impact on facilitated bone growth and the transport of cells and nutrients, evidently ameliorating the quality of biological processes, nevertheless they may considerably reduce rigidity and strength of scaffolds [195]. Although the shape and size of pores can be adjusted by changing the conditions of even traditional fabrication processes, e.g., casting or powders metallurgy; however, only a randomly organised porous structure can then be obtained, which is not fully open or is not open at all [218]. To produce porous scaffolds as well as other medical implants, including dental ones, additive technologies are used most often in combination with prior CAD/CAM, as highly competitive against traditional manufacturing methods, such as casting or machining [219–224]. Such technologies can be applied for various engineering materials, not only metals and alloys which are prepared, respectively, as powder or liquid, rolled material or thin fibres. Additive technologies have been widely used for fabricating diverse, customised elements applied in medicine, in particular, scaffolds with required porosity and strength with living cells implanted into an organism [225–227], models of implants and dental bridges [228–230], implants of individualised implants of the upper jaw bone, hip joint and skull fragments [231–238]. Considering the additive technologies applied most widely, the following have found their application for scaffold manufacturing, in implantology and prosthetics, i.e., electron beam melting (EBM) [222, 239–243], and also 3D printing for production of indirect models, although selective laser sintering/selective laser melting (SLS/SLM) and its technological variants offers broadest opportunities [220, 222, 244–253], which was noted in discussing each group of materials. SLS/SLM techniques permit to produce a structure with open pores, e.g., with a lattice structure promoting osseointegration, while maintaining different external shapes of the whole implant [254]. The determination of optimum geometric features of scaffolds as a result of computer-aided design in conjunction with the optimisation of technological conditions of the applied additive technology, and broad experimental verification in the engineering and biological aspect, is of key importance.

4. Selection of materials of implantable devices in regenerative medicine and dentistry

The selection of appropriate materials for application in the regenerative medicine and dentistry, in conjunction with the optimisation of technological conditions of fabrication of implantable devices, including porous scaffolds and implant-scaffolds, are the important aspects of the analysed problems, previously just described in author's own theoretical study [2]. It occurs as a synergy of classical prosthetics/implantation of bone and organ post-injury or post-resection losses together with the methods of tissue engineering in the connection (interface) zone of bone or organ stumps with prosthetic elements/implants. It calls for the use of porous and high-strength non-graded metal and/or composite and/or polymer materials (which is strongly, but not exclusively, dependent on the specificity of the clinical application) together with using at the same time biodegradable materials for tissue scaffolds. One of the solutions strives to achieve bioactive connections, as most advantageous regarding bond strength, which is formed between bone tissue and implants/scaffolds made or coated with bioactive materials, considerably improving the stability and durability of connection, especially for porous scaffolds/implants. Another acceptable approach is a very durable biological connection characteristic for porous implants/scaffolds whereby the bone tissue is growing through the material pores and is mechanically "anchored" in the bone. Porous resorbable bioglass may be used for scaffold fabrication [255], e.g., from the CaO-SiO_2-P_2O_5 system, Hench bioglass [256], produced both, with classical melting methods and with sol-gel methods, and also bioglass from the SiO_2-Al_2O_3 system endowed with silver, due to their biocompatibility [257, 258] and bacteriocidity, and with pore walls coated with hydroxy carbonate apatite (HCA) [259], ensuring enhanced activity of osteoblasts [260] and expression of genes connected with bones [261]. The formation methods of porous structures from ceramic materials, in particular such as aluminium oxide, zirconium oxide, calcium carbonate, hydroxyapatite (HA), titanium oxide, include casting of sections from mass containing a fine-grained ceramic material with additives facilitating foaming and then material sintering, and also the use of other methods, e.g., an organic matrix and lyophilisation of ceramic slip [255]. The basic bioactive ceramic materials used for scaffolds is calcium phosphate (CaPs), as the main component of bone, and in hydroxyapatite (HA), β-tricalcium phosphate (β-TCP) or a mixture of HA and b-TCP, known as biphasic calcium phosphate [262–264]. A classical solution in the domain of ceramics, are porous scaffolds/corundum implants completely biocompatible, growing through the fully valuable bone tissue. They possess mechanical strength sufficient for many types of clinical procedures and ensuring freedom of manipulation during a surgical procedure. After growing through, have an appropriate modulus of elasticity, which ensures their good interworking with the bone and also allow for sterilisation with any method. Both, bioactive and biodegradable polymers can be employed [193, 265, 266], including in particular natural polymers such as: alginate, chitosan, collagen, fibrin, hyaluronic acid and silk, and used, e.g., for bone reconstruction [267], as well as synthetic ones, such as poly(lactic acid) (PLA), poly(glycolic acid) (PGA), and polycaprolactone (PCL) and poly(propylene fumarate) (PPF) with high compressive strength, comparable to this of a cortical layer of bone [266]. Some of them, such as poly(lactic acid) (PLA), poly(glycolic acid) (PGA) may cause negative tissue reactions [268]. Composite materials satisfy mechanical and physiological requirements, e.g., CaP-polymer scaffolds, interconnected tricalcium phosphate (TCP) scaffolds coated inside pores with polycaprolactone (PCL) [269], hydroxyapatite HA/poly(ester-urethane)(PU) [270] or a nanocomposite of collagen and Bioglass [271].

Metallic materials represent one of the largest groups of engineering materials used for this purpose and comprise, among others, titanium, tantalum, niobium and their alloys, and in dentistry also cobalt-matrix alloys and alloys of noble metals, for which, SLS/SLM technologies were especially applied [272, 273], irrespective of stainless steels often used until now. The SLS/SLM technique has been used successfully for production of complex porous and cellular structures made of austenite stainless steels [218]. One of the main grades of stainless austenitic steel which can be used for medical purposes is nickel, now thought to be one of the main allergens [274], with 17% of adults [275] and 8% of children [276] being sensitised to it, approx. 50–60 million people altogether in the EU. Apart from many disorders [277, 278], this chemical element causes rejection of orthopaedic gradients [279], and dental implants [280] and, for this reason, Directive 94/27/EC [281] was put into force prohibiting the use of nickel and materials containing it for prosthetic and implantological purposes. Co-matrix alloys such as Co-Cr-Mo, Co-Cr-W and cast Co-Cr-W-Mo alloys, for machining or manufactured by powder metallurgy methods, for which the SLS/SLM technology was successfully employed [272, 273, 282–289], can be seen as the basic classical materials for dental prosthetic restorations [290–296], despite their high density, which is more and more often considered as counter-indicative for their application for such purposes.

Porous metal materials, though not biodegradable, are used for scaffolds, mainly Ti and Ta [297], also after treatment of pores' surface [298]. Therefore, interest in light metals and their alloys have been on the rise. Ti and its alloys represent engineering materials which are particularly suitable for use in additive technologies, e.g., for selective laser sintering [5, 6, 10, 77–81, 85–91, 299–305]. Porous Ti [297] can be utilised for non-biodegradable scaffolds, including such after treatment of the pore surface, applied primarily due to relatively high compressive strength and fatigue strength [193, 306]. It was proven that structures of porous Ti6Al4V are also effective in aiding the growth of cells and bone tissue [307–311]. Ti and its alloys with Al, Nb and Ta and Ti alloys with Al, V and Nb, well tolerated by a human organism, are the metallic materials more and more often used these days for joint prostheses and for various implants, also intramedullary wires and for prosthetic restorations and dental implants [33, 312–322]. When used for dental crowns, though, they have a significant disadvantage, consisting of porcelain reactions with titanium oxide, causing bruising and colour darkening, which - for aesthetic reasons - practically eliminates such prosthetic restorations, but does not exclude their use for a root part of the implant [323]. Titanium-matrix materials do not cause allergic reactions and are stainless and feature high strength and hardness, and also thermal conductivity several times lower than traditional prosthetic materials [324]. Titanium is a very thrombogenic material [240]. The biocompatibility, especially thrombocompatibility, of Ti can be enhanced by introducing alloy elements [325]. Some publications [326–334] provide limited information on the toxic activity of V as an alloy element in the Ti6Al4V alloy, because it was found that V could be regarded as a potentially toxic factor [335, 336], which is, however, true only for a considerable concentration of V in a body, due to development of its undesired immunological response, when the freed V ions migrate from the material surface to a soft tissue, binding with proteins [337, 338]. Some research on cells implies that Ti6Al4V exhibits high cytotoxicity [307–311, 339]; there are reports that V may cause sterile abscesses, and Al may cause scarring, while Ti, Zr, Nb and Ta show good biocompatibility [340]. Some publications argue that a risk associated with an unfavourable effect of Ti alloys can be limited by replacing V, as in, e.g., Ti6Al4V alloy, by Nb, e.g., in Ti6Al7Nb alloy, which would show better properties, e.g., corrosion resistance and bioavailability [334, 341–346]. It was revealed, however, by direct comparison in the same conditions that differences between Nb and V in Ti alloys are not too high [334], and even that Ti6Al4V alloys exhibit better properties than Ti6Al7Nb alloy [334, 347–349], such as thrombocompatibility, more intensive antibacterial activity and resistance to colonisation of Gram-positive bacteria, although worse for colonisation of Gram-negative bacteria [334]. Other alloys, with a higher concentration of Nb, are however employed successfully, e.g., Ti24Nb4Zr8Sn, including those manufactured by selective laser sintering, with an elasticity modulus better adapted to a bone than Ti6Al4V alloy, which prevents bone resorption and does not cause implant loosening in use [350]. Although porous Ti6Al4V was comprehensively studied, the potential release of toxic ions has led to a search for alternative Ti alloys, including, among others, Ti24Nb4Zr8Sn, Ti7.5Mo and Ti40Nb, with comparable mechanical properties as their counterparts manufactured traditionally [350–353].

The concept of the synergic use, for this purpose, of the existing achievements in tissue engineering in the scope of selection of materials and scaffold fabrication technologies, in materials engineering and production engineering in the scope of design and manufacture of prostheses/implants with different engineering materials, and in surgery and regenerative medicine in the scope of prosthetics/implantation in the treatment of the above-mentioned civilisational diseases and their effects have been outlined in the earlier works and projects by the author [1–109].

Regeneration in a natural condition is forcing the removal of an artificial scaffold [354–356]. The topic of scaffolds and biodegradable implants, including porous ones, both made of polymers [193], despite doubts of Mg-Ca alloys [357], and of composite materials [269–271], as well as a concept of separating the redundant pieces of cell-based products after finishing a therapy performed with their use [7], were thoroughly analysed in the own works [5, 6, 10]. Relatively high compressive strength and fatigue strength [193, 306] are primarily the reason for the application of Mg and its alloys [358–360]. Introducing pure Mg with interconnected porosity onto bearing plates, bolts and networks made by rapid pressure assisted densification methods, such as rapid hot pressing/Spark Plasma Sintering (SPS) is an innovative approach [361]. Due to Mg biocompatibility, and the manufacturing technology ensuring good mechanical strength and recovery of the bone pure Mg gives an optimum solution. Mg with Ca, Zn and Mn alloys can be used, to reduce the rate of *in vivo* corrosion and prevent necrosis and the blocking of blood flow. A human organism well tolerates these alloys [362, 363] mainly with coatings adapted to bioresorbable implants [364–367]. Reinforcements of magnesium MMCs usually include, notably, HA [368–373], FA [374], calcium polyphosphate [375], and calcium [374] and well affect biocompatibility. Mg and its alloys can be used for non-biodegradable scaffolds [372]. Except for several studies, the application of Mg as a biomaterial has not won popularity as late as until the end of the 1990s, because this pure metal cannot ensure appropriate mechanical properties or corrosion resistance in orthopaedic uses [376, 377]. Such popularity, though, has risen exponentially since then [378], owing to major improvements in Mg production [378] and various techniques elaborated, including the use of Mg alloys, substrate surface treatment or coating technologies [379–381]. Generally speaking, such alloys contain Al or rare-earth elements (REE) [382–384], although there are reports about studies over additives of non-toxic elements, such as Ca, Mn, Zn and Zr, and even Li, Cd, Sn, Sr., Si, Ag and Bi [378]. Al is a common additive for Mg alloys, as it is conceded that it improves strength and resistance to Mg alloys' corrosion Mg (258), although Al is shown in numerous pathological conditions in humans [385–388]. The additives of rare-earth elements are used for increasing strength, plasticity and wear resistance of Mg alloys and their corrosion resistance [383] in environments with a high content of chlorides in connection with a passivation layer rich in oxygen [389, 390]. The influence of such elements on the physiological system is unknown, [378], although it was found that they possess both, anti-carcinogenic and anti-coagulation properties [391], and when used as vascular stents without side effects [392], and also without La and Ce, do not have an effect on cytotoxicity and have a positive effect on cellular life [393], although quite the opposite was also found, that at least some REEs are highly toxic [393, 394]. The value of the research carried out in this scope may be limited if it turns out that such elements are too toxic in use for biomaterials [377] and this requires further systematic studies. Magnesium synthesised by SLM is closely adapted to a human bone [395]. This technology was employed for producing complex porous/cellular structures of magnesium alloys [350, 396–398], although the results of such studies are normally not available in the available literature [254]. A selective laser sintered Mg2Mn alloy is predisposed to use for bone implants [254]. The use of Mn, as a component of Mg alloys, consists of the improvement of corrosion resistance and may increase the plasticity of Mg alloys [383, 399]. Zn improves the strength of Mg alloys [399, 400] and their corrosive resistance [401]. However, its influence on increased cytotoxicity was identified [402, 403]. Magnesium and its alloys feature a high potential for orthopaedic uses because it has proven to be fully bioresorbable, their mechanical properties are adapted to bones and do not cause an inflammatory response; moreover, they are osteoconductive, supportive to bone growth and play a positive role in the binding of cells [404]. The application of biomaterials made of Mg highly increases a risk of hypomagnesia and, probably, of the excessively stored and circulating Mg [378]. Corrosion analysis and Mg concentration monitoring in serum must be an important aspect of Mg-based biomaterials' assessment [378]. Due to an effect of diverse factors such as pH, concentration and type of ions, adsorption of proteins on orthopaedic implants and the biochemical activity of the surrounding cells in the presence of body fluids [405, 406], further research into biomedical uses of Mg is indispensable. Powders are serially manufactured for, in particular, SLS/SLM [407].

Third-generation scaffolds made of CaP, Si-TCP/HA [408] and collagen hydrogel [409] are osteoinductive and allow to create a new bone and also its biomineralisation. Substitute scaffolds of bones are often administered with medicines, including gentamycin, vancomycin, alendronate, methotrexate and ibuprofen [410, 411] and with growth factors and transcription factors [265, 412, 413].

5. Selection of technologies of implantable devices in regenerative medicine and dentistry

The issue was previously considered in the Author's own review work [2], and some of the information compiled there was used here. The traditional, and also the oldest fabrication technologies of scaffolds with a porous structure differentiate the method of emulsifying/lyophilisation [414], to thermally induced phase separation (TIPS) [415], solvent casting & particulate leaching (SCPL), where solvent residues may have an adverse effect on cellular structures [416]. The aforementioned classical methods being unable to control accurately a general shape of the scaffold as well as the size, shape, distribution and interconnections of pores. Nevertheless, these methods have not been completely abandoned. They are used as a modern method in tissue engineering replication technologies with micro-/nanopatterned-surfaces [417–419]. The master moulds are produced using a hard or soft material, for the reason of mould rigidity. The structures with small feature resolution and micro/nanofabricated moulds, including for hot embossing (also known as nanoimprint lithography) and soft lithography (micro-casting) for achieving patterns with dimensions about of 5 nm [420–423] can be cast using synthetic and natural biodegradable polymers [424, 425]. Currently used methods do not require moulds for fabrication of scaffolds (solid freeform fabrication SFF) made not only polymer materials and hydrogels but also ceramic and metallic materials [193, 426–429]. The particular fabrication methods find wide application for the processing both the mentioned biocompatible engineering materials and biological materials [430]. The particular layers of powder are sprayed with an adequate biocompatible binding agent, e.g., for merging powder to fabricate scaffold from collagen [431], and a 25% acrylic acid solution in a mixture of water with glycerine [432] using the three-dimensional printing method (3DP) [433]. This method is also used for the integration of hydroxyapatite used for bone regeneration, and an aqueous citric acid solution is used for integration of ceramics based on calcium phosphate [434]. A method of three-dimensional printing hot wax droplets [435] could be used for manufacture a replica of the scaffold surface, e.g., bone and gristle substitutes fabricated with the SFF method. The limitations of the method originate from wax impurities with biologically incompatible solvents [436], which are not exhibited by new generation materials such as BioBuild and BioSupport dissolving in ethanol or water [436]. The stereolithography method permits to shape three-dimensional form of liquid polymer [437], in particular using poly(propylene fumarate) (PPF) [438, 439], poly(ethylene glycol) (PEG) [440, 441]. Polymer materials without solvents, including poly(ε-caprolactone) PCL [429, 442], poly(ethylene glycol)–poly(ε-caprolactone)–poly(lactide), PEG-PCL-PLA [442, 443] acrylonitrile-butadiene-styrene (ABS) and hydroxyapatite-poly(ε-caprolactone) HA-PCL [442, 444] are used in fused deposition modelling (FDM) [445]. The particular layers are placed from a computer controller and using computer-aided design (CAD) methods. Selective laser sintering (SLS) is similar to 3D printing. The process is starting with uniform spreading of a thin layer of powder onto the surface and then followed by the merging of powder grains as a result of sintering with the neighbouring grains with partial pre-melting. The next layers are manufactured subsequently according to the same method until the full dimensions of the manufactured element are achieved. The manufactured element, including the scaffolds, shows the assumed constructional features. This technique, used commonly for additive manufacturing of products, includes scaffolds and implants for the dental purpose, from metallic and ceramic materials [5, 6, 10, 446]. This technology was also utilised for scaffolds preparation [426] from biodegradable polymers, e.g., polyether polymer, poly (vinyl alcohol), polycaprolactone [447] and poly(L-lactic acid) [448], and also hydroxyapatite [449] and from composites composed of some of such polymers and hydroxyapatite [448, 450, 451].

Nanofibrous scaffolds are manufactured by electrospinning, and the so obtained nanofibres with the diameter of 5 nm to over 1 mm are continuous and randomly interconnected [452, 453]. Due to the character of electrospinning, fibres are arranged in an orderly manner or are oriented randomly [454]. They have a large specific surface area, exhibit high porosity, the small size of pores and small density [453] and their structure is similar to the extracellular matrix (ECM). Natural and engineering materials can be used as a material, including, in particular, collagen, gelatin and chitosan [453]. Scaffolds are fabricated using of non-covalent interactions for spontaneous fabrication of a three-dimensional structure in response to the activity of environmental factors [455]. The ability of peptides and nucleic acids, to self-organisation, is utilised for scaffolds fabrication. Such types of scaffolds were used, e.g., for regeneration of nervous tissue to stop bleeding and repair infatuated myocardia, as well as in medical products for slow release of a medicine [456, 457] and for DNA, where the branched DNA particles are hybridising with each other in the presence of ligases in hydrogel [458]. The scaffold fabrication method employing self-organisable nanofibres is one of few allowing to produce biocomponents with their properties similar to the natural extracellular matrix (ECM), and scaffolds containing hydrogel, made using such technology, employ more advantageous toxicological properties and higher biocompatibility than traditional materials. Conventional hydrogels are particularly useful for three-dimensional placement of cells [459]. Hydrogels used in tissue engineering should have low viscosity before injection and should be gelling fast in the physiological environment of the tissue, and the most important is gelling (sol-gel transition) by cross-linking, which may take place when producing them *in vitro* and *in vivo* during injection. Physical cross-linking is used in particular in the case of poly(N-isopropylacrylamide) (poly(NIPAAM)), which may be used in tissue engineering after introducing acrylic acid (AAc) or PEG [460, 461] or biodegradable polymers, including such as chitosan, gelation, hyaluronic acid and dextran [462–466] to block copolymers, such as poly(ethylene oxide) PEO-PPO-PEO (Pluronic), poly(lactide-co-glycolide) PLGA-PEG-PLGA, PEG-PLLA-PEG, polycaprolactone PCL-PEG-PCL and PEG-PCL-PEG [467–471], and also agarose (a polysaccharide polymer material, extracted from seaweed as one of the two principal components of agar) [459], as thermo-sensitive systems [472], to avoid the use of potentially cytotoxic ultraviolet radiation. Poly(NIPAAM) and block copolymer hydrogels may undergo cross-linking as a consequence of temperature and pH acting at the same time, as in the case of acrylates [473, 474], such as 2-(dimethylamino)ethyl-methacrylate (DMAEMA) or 2-(diethylaminoethyl) methyl methacrylate. Self-assembling peptides hydrogels, including such containing peptide amphiphiles (PAs), can form nanofibres [475, 476] used for three-dimensional formation of tissue cultures [476–480].

Chemical cross-linking hydrogels having convalescent bonds include photo-cross-linkable poly(ethylene glycol)-diacrylate (PEGDA), poly(ethylene glycol)-dimethacrylate (PEGDMA), poly(propylene fumarate) (PPF) and oligo(poly(ethylene glyco) fumarate) (OPF) [481–485], and also natural hydrogels such as dextran, alginate, chitosan and hyaluronic acid synthesised from PEGDA/PEGDMA [486–489] and Michael-type addition reaction [490–492] and Schiff base-cross-linked hydrogels [465, 493–495]. In the case of enzyme-mediated cross-linking [458], transglutaminases (including Factor XIlla) and horseradish peroxidases (HRP) [459] are used for the catalysis of star-shaped PEG hydrogels [496] and tissue transglutaminase catalysed PEG hydrogels [497]. This also applies tyrosinase, phosphopantetheinyl transferase, lysyl oxidase, plasma amine oxidase, and phosphatases [498]. It made it possible in particular to develop new gels by engrafting tyramine groups into natural and synthetic polymers such as dextran, hyaluronic acid, alginate, cellulose, gelatin, heparin and PEG-PPO [499–505, 509].

Ionic cross-linking hydrogels include calcium-cross-linked alginate [459] and chitosan-polylysine, chitosan-glycerol phosphate salt and chitosan-alginate hydrogels [506–508]. Different synthetic and natural polymers were used for this purpose, including polyethylene glycol (PEG), and copolymers containing PEG [486, 510], hyaluronic acid (HA) [511] after an oxidation reaction through HA-tyramine conjugates [505] and as a result of the formation between HA–SH [492, 512] and Michael addition [491, 513], collagen and gelatin hydrogels mostly cross-linked using glutaraldehyde, genipin or water- soluble carbodiimides [513–515], chitosan [516–519], dextran 192 [520, 521] and alginate [522]. Hydrogels were used for reconstruction of the retina [523], ligament [524], fatty tissue [465], kidneys [525], muscles [526], blood vessels [527, 528], and also heart, neural cells, invertebral discs, bones and gristle [459]. Hydrogels were used to prevent adhesions [529, 530, 531], to promote cellular adhesion [490, 532, 533]. So-called strong hydrogels were developed to improve mechanical properties [534]. The three-dimensional representation is possible of placement of cells with energy in the hydrogel to vascular structures using a laser [535, 536], notably for recording directly the endothelial cell [535].

The general criteria of materials selection for tissue scaffolds relate to the material type and its structure, osteoconductivity ability, mechanical strength, ease of production and manipulation in clinical applications. **Table 1** presents numerous examples of the application of various bioactive and engineering materials, and their respective materials processing and tissue engineering technologies for manufacturing of the hybrid personalised implants and scaffolds [2].

Many layers of different types of cells at present can be three-dimensionally printed to directly create an organ, ensuring the highest currently possible degree of control over the structure of the regenerated tissues [537–542]. The first production system for three-dimensional printing of tissues was delivered only in 2009 based on the NovoGen bioprinting technology [543]. China has invested nearly 0.5 billion USD to establish 10 national institutes for development of organ

Fabrication stage	Investigated materials	Technologies applied	Areas of application
Fabrication of implant bearing structure	Ti, Ti alloys with V or Nb, Mg (possibly with additives of Ca, Zn and Mn), ceramic materials Al_2O_3 and ZrO_2, TiO_2, resorbable bioglass, e.g., Hench bioglass, from the CaO-SiO_2-P_2O_5 and SiO_2-Al_2O_3 system, hydroxyapatites, polymers, composites	Selective laser sintering, sintering, the use of organic matrix, lyophilisation of ceramic slip, rapid pressure assisted densification methods, such as rapid hot pressing/spark plasma sintering (SPS), skeletal casting, plastic working, cutting micro-treatment, computer-aided manufacturing, sol-gel methods, 3D printing, electrospinning, atomic layer deposition, and physical vapour deposition	Filling the losses of long bones, hip and knee joints, facial-skull bone, losses of joint cartilage cells, oesophagus losses and/or blood vessel losses, dental restorations, and skin restorations
Fabrication of porous implant part	Ti, Ti alloys with V or Nb, Mg (possibly with additives of Ca, Zn and Mn), ceramic materials Al_2O_3 and ZrO_2, TiO_2, resorbable bioglass, e.g., Hench bioglass, from the CaO-SiO_2-P_2O_5 I SiO_2-Al_2O_3 system, hydroxyapatites, polymers, composites		
Fabrication of coatings inside pores of porous implant part	A natural protein, synthetic and polysaccharide polymers, including thermosetting, including collagen, fibrin, alginate, silk, hyaluronic acid, chitosan, poly (lactic acid) (PLA), poly(glycolic acid) (PGA), polycaprolactone (PCL) and poly(propylene fumarate) (PPF), polyethylene glycol, Al_2O_3, resorbable bioglass, hydroxy carbonate apatite (HCA), calcium phosphate (CaPs), hydroxyapatite (HA), B-tricalcium phosphate (B-TCP), biphasic calcium phosphate, composites: collagen + hydroxy - apatite CaP-polymer tricalcium phosphate (TCP)-polycaprolactone (PCL), hydroxyapatite HA/poly(ester-urethane)(PU), collagen-bioglass	Infiltration, 3D printing, selective laser sintering, electrospinning, atomic layer deposition, physical vapour deposition, pressing, and sol-gel methods	
Fabrication and application of tissue cultures	Adipocytes, chondrocytes, osteoblasts, fibroblasts and skeletal myocytes	Cell transplantation, matrix implantation, cell implantation with matrix, breeding of xenogeneic and autologous cells and the stage of clinical activities	

Table 1. Examples of the application of various bioactive and engineering materials, and their respective material processing and tissue engineering technologies for manufacturing of the hybrid personalised implants and scaffolds [2].

printing [544], in which the printing of ears, liver and kidneys from living tissues was started in 2013. It is expected that fully functional printed organs can be achieved over the next dozens of years or so [545, 546]. In the meanwhile, there were reports that an Australian team obtained a kidney tissue print with this method for the first time [547]. An American team confirmed in 2014 it is ready to print a heart [548]. A three-dimensional structure is obtained by subsequent formation of layers of living tissues on the gel or sugar matrix substrate [549]. To use the three-dimensional printing technique, a polymer-cell mixture can be dosed, leading to the formation of cell hydrogel [550]. Microfluidics allows for the creation of three-dimensional systems of cells [551]. It is possible to obtain such cell systems from hydrogels by photopolymerisation of polymer solutions [552]. Using SFF techniques, including stereolithography techniques can create scaffolds made of PEG hydrogels also [440]. Vascularisation for organ printing remains a significant challenge in tissue engineering. The development of a vascular network in metabolically functional tissues enables the transport of nutrients and removal of wastes, ensuring maintenance of cells' viability for a long time [553]. Micro-formation techniques, by the three-dimensional printing of templates made of agarose fibres, are used for the creation of a microchannel network inside hydrogel products, including, in particular, inside star poly(ethylene glycol-co-lactide)acrylate (SPELA), methacrylated gelatin (GelMA), poly(ethylene glycol) dimethacrylate (PEGDMA) and poly(ethylene glycol) diacrylate (PEGDA) with different concentrations. In the last several years, the efficient formation of endothelial monolayers within the fabricated channels has also been confirmed [554]. Unfortunately, the progress in vascularisation control is limited, despite immense progress in the production of complicated tissue structures.

6. Contents of the book on regenerative medicine and dentistry

The book on Biomaterials in Regenerative Medicine contains a total of 18 chapters. Selected issues discussed in the previous five sub-chapters were presented in it. The literature review, in this first chapter, deals with technical, biological and medical aspects concerning materials and technologies for medical and dental implantable devices. In the second chapter, the metallic biomaterials and their application in regenerative medicine are presented in detail by the literature review. In addition to these two chapters, all the rest contain the results of their scientific research done by the authors of these chapters. Indeed, the originality of the presented research results constitutes the real value of the book at this moment passed to the readers' hands. The next chapters contain issues about additive technologies and laser manufacturing of materials used in regenerative medicine and mainly in regenerative dentistry. The third chapter is on microporous titanium-based materials located inside the pores by biocompatible thin films to facilitate the implantation and proliferation of living cells in the scaffolds thus produced. In turn, the fourth chapter includes mechanical properties comparison of engineering materials produced by additive and subtractive technologies for dental prosthetic restoration application. It discusses both solid and milled sintered titanium and its alloy, Co-Cr alloy, and sintered ZrO_2. In the fifth and sixth chapters, properties of Co-Cr dental alloys fabricated using an additive, technologies were presented. The progress of the application of 3D printing for tissue regeneration in oral and maxillofacial surgery is presented in chapter seven. In the eighth chapter, the issues of the tissue engineering and use of growth factors in bone regeneration are discussed. Laser processing was analysed in chapter nine concerning silicon for its synthesis as better biomaterials. Prospective of characterisation of the skin models and associated with it measurements and simulation of permeation and diffusion in 3D tissues are presented in the tenth chapter. The eleventh chapter also contains the skin regeneration problems explanation and also a description of the biomaterials for tendon and ligament regeneration. The next few chapters deal with natural and artificial gels of polymeric materials. In the twelfth chapter, authors described the hydrogels for regenerative medicine. Natural rubber latex biomaterials in bone regenerative medicine are presented in the thirteenth chapter. A systematic study of ethylene-vinyl acetate (EVA) in the manufacturing of protector devices for the orofacial system was described in the fourteenth chapter. The next few chapters outline other issues regarding regenerative medicine and tissue engineering. The fifteenth chapter includes tailoring bioengineered scaffolds for regenerative medicine. Biomaterials and stem cells as promising tools in tissue engineering and biomedical applications are the content of the sixteenth chapter. Identification of Fe_3O_4 nanoparticles biomedical purpose by magnetometric methods is the title of the seventeenth chapter. The eighteenth chapter applies to biomaterials for tissue engineering applications in diabetes mellitus.

The editor, publisher and the whole team of authors, by making this book available to the readers, deeply believe that the detailed information collected in the book, largely deriving from own and original research and R&D works pursued by the authors, will be beneficial for the readers to develop their knowledge and harmonise specific information concerning these topics, and will convince the engineers and medicals about the advantages of using the manufacturing and tissue engineering and advanced biomaterials in regenerative medicine and dentistry. On one hand, it makes possible gaining positive effects in the economic manufacturing of biomaterials and implantable devices; on the other hand, it will ameliorate the fate of many people affected by severe diseases. This awareness justifies the involvement in the execution of research and the effort put in describing their results in this book.

Notice

Fundusze Europejskie Inteligentny Rozwój ASKLEPIOS Unia Europejska

The study presented in the chapter was realised in the Medical and Dental Engineering Centre for Research, Design and Production ASKLEPIOS in Gliwice, Poland in the framework of the Project POIR.01.01.01-00-0485/16-00 on 'IMSKA-MAT Innovative dental and maxillofacial implants manufactured using the innovative additive technology supported by computer-aided materials design ADDMAT' co-financed by the National Centre for Research and Development in Warsaw, Poland.

Author details

Leszek A. Dobrzański

Address all correspondence to: leszek.adam@gmail.com

Science Centre of the Medical and Dental Engineering Centre for Research, Design and Production Asklepios, Gliwice, Poland

References

[1] Dobrzański LA. The concept of biologically active microporous engineering materials and composite biological-engineering materials for regenerative medicine and dentistry. Archives of Materials Science and Engineering. 2016;**80**(2):64-85. DOI: 10.5604/18972764.1229638

[2] Dobrzański LA. Overview and general ideas of the development of constructions, materials, technologies and clinical applications of scaffolds engineering for regenerative medicine. Archives of Materials Science and Engineering. 2014;**69**(2):53-80

[3] Dobrzański LA. Applications of newly developed nanostructural and microporous materials in biomedical, tissue and mechanical engineering. Archives of Materials Science and Engineering. 2015;**76**(2):53-114

[4] Dobrzański LA. Korszerű mérnöki anyagok kutatásának nanotechnológiai aspektusai. Miskolci Egyetem, Multidiszciplináris tudományok. 2016;**6**(1):21-44

[5] Dobrzański LA, Dobrzańska-Danikiewicz AD. Metalowe materiały mikroporowate i lite do zastosowań medycznych i stomatologicznych. Open Access Library. 2017;Annal **VII** (1):1-580 (ISBN 978-83-63553- 44-9; International OCSCO World Press, Gliwice)

[6] Dobrzański LA, editor. Powder Metallurgy – Fundamentals and Case Studies. Rijeka, Croatia: InTech; 2017. pp. 1-392 (ISBN 978-953-51- 3053-6)

[7] Dobrzański LA, editor. Polymer nanofibers produced by electrospinning applied in regenerative medicine. Open Access Library. 2015;**V**(3):1-168 (ISBN 978-83-63553-88-3)

[8] Dobrzański LA. Metale i ich stopy. Open Access Library. 2017;**VII**(2):1-982 (ISBN 978-83-63553-47-0; International OCSCO World Press, Gliwice)

[9] Nowacki J, Dobrzański LA, Gustavo F. Intramedullary implants for osteosynthesis of long bones. Open Access Library. 2012;**11**(17):1-150 (in Polish) (ISBN 978-83-63553-07-4)

[10] Dobrzański LA, Dobrzańska-Danikiewicz AD, Achtelik-Franczak A, Szindler M. Structure and properties of the skeleton microporous materials with coatings inside the pores for medical and dental applications. In: Muruganant M et al., editors. Frontiers in Materials Processing, Applications, Research and Technology. Springer Nature Singapore Pte Ltd; 2018. pp. 297-320. DOI: 10.1007/978-981-10-4819-7_26

[11] Dobrzański LA, Dobrzańska-Danikiewicz AD, Achtelik-Franczak A, Dobrzański LB, Szindler M, Gaweł TG. Porous selective laser melted Ti and Ti6Al4V materials for medical applications. In: Dobrzański LA, editor. Powder Metallurgy – Fundamentals and Case Studies. Rijeka, Croatia: InTech; 2017. pp. 161-181 (ISBN 978-953-51-3053-6, DOI: 10.5772/65375)

[12] Dobrzański LA, Matula G, Dobrzańska-Danikiewicz AD, Malara P, Kremzer M, Tomiczek B, Kujawa M, Hajduczek E, Achtelik-Franczak A, Dobrzański LB, Krzysteczko J. Composite materials infiltrated by aluminium alloys based on porous skeletons from alumina, Mullite and titanium produced by powder metallurgy techniques. In: Dobrzański LA,

editor. Powder Metallurgy – Fundamentals and Case Studies. Rijeka, Croatia: InTech; 2017. pp. 95-137 (ISBN 978-953-51-3053-6, DOI: 10.5772/65377)

[13] Dobrzański LA, Dobrzańska-Danikiewicz AD, Achtelik-Franczak A, Dobrzański LB, Hajduczek E, Matula G. Fabrication Technologies of the Sintered Materials Including Materials for medical and dental application. In: Dobrzański LA, editor. Powder Metallurgy – Fundamentals and Case Studies. Rijeka, Croatia: InTech; 2017. pp. 17-52 (ISBN 978-953-51-3053-6, DOI: 10.5772/65376)

[14] Dobrzański LA. Stan wiedzy o materialach stosowanych w implantologii oraz medycynie regeneracyjnej. In: Dobrzański LA, Dobrzańska-Danikiewicz AD, editors. Metalowe materiały mikroporowate i lite do zastosowań medycznych i stomatologicznych. Gliwice: International OCSCO World Press. Open Access Library, 2017;**VII**(1):117-120 (ISBN 978-83-63553-44-9)

[15] Dobrzański LA, Achtelik-Franczak A. Struktura i wlasnosci tytanowych szkieletowych materialów mikroporowatych wytworzonych metoda selektywnego spiekania laserowego do zastosowan w implantologii oraz medycynie regeneracyjnej. In: Dobrzański LA, Dobrzańska-Danikiewicz AD, editors. Metalowe materiały mikroporowate i lite do zastosowań medycznych i stomatologicznych. Gliwice: International OCSCO World Press, Open Access Library, 2017;**VII**(1):186-244 (ISBN 978-83-63553-44-9)

[16] Dobrzanska-Danikiewicz AD, Gawel TG, Kroll L, Dobrzanski LA. Nowe porowate materialy kompozytowe metalowo-polimerowe wytwarzane z udzialem selektywnego stapiania laserowego. In: Dobrzański LA, Dobrzańska-Danikiewicz AD, editors. Metalowe materiały mikroporowate i lite do zastosowań medycznych i stomatologicznych. Gliwice: International OCSCO World Press, Open Access Library. 2017;**VII**(1):245-288 (ISBN 978-83-63553-44-9)

[17] Dobrzanska-Danikiewicz AD, Dobrzanski LA, Szindler M, Achtelik-Franczak A, Dobrzanski LB. Obróbka powierzchni materialów mikroporowatych wytworzonych metoda selektywnego spiekania laserowego w celu uefektywnienia proliferacji zywych komórek. In: Dobrzański LA, Dobrzańska-Danikiewicz AD, editors. Metalowe materiały mikroporowate i lite do zastosowań medycznych i stomatologicznych. Gliwice: International OCSCO World Press, Open Access Library. 2017;**VII**(1):289-375 (ISBN 978-83-63553-44-9)

[18] Dobrzański LA, Achtelik-Franczak A. Struktura i wlasnosci materialów kompozytowych do zastosowan medycznych o osnowie odlewniczych stopów aluminium wzmacnianych tytanowymi szkieletami wytworzonymi metoda selektywnego spiekania laserowego. In: Dobrzański LA, Dobrzańska-Danikiewicz AD, editors. Metalowe materiały mikroporowate i lite do zastosowań medycznych i stomatologicznych. Gliwice: International OCSCO World Press, Open Access Library, 2017;**VII**(1):376-433 (ISBN 978-83-63553-44-9)

[19] Dobrzański LA. Autorska koncepcja rozwoju implanto-skafoldów oraz materialów biologiczno-inzynierskich do aplikacji w medycynie i stomatologii. In: Dobrzański LA,

Dobrzańska-Danikiewicz AD, editors. Metalowe materiały mikroporowate i lite do zastosowań medycznych i stomatologicznych. Gliwice: International OCSCO World Press, Open Access Library, 2017;**VII**(1):535-580 (ISBN 978-83-63553-44-9)

[20] Dobrzański LA, Pakuła D, Staszuk M. Chemical vapor deposition in manufacturing. In: Nee AYC, editor. Handbook of Manufacturing Engineering and Technology. London: Springer-Verlag; 2015. pp. 2755-2803 (ISBN 978-1-4471-4669-8)

[21] Dobrzański LA, Gołombek K, Lukaszkowicz K. Physical vapor deposition in manufacturing. In: Nee AYC, editor. Handbook of Manufacturing Engineering and Technology. London: Springer-Verlag; 2015. pp. 2719-2754 (ISBN 978-1-4471-4669-8)

[22] Dobrzański LA, Dobrzańska-Danikiewicz AD, Tański T, Jonda E, Drygała A, Bonek M. Laser surface treatment in manufacturing. In: Nee AYC, editor. Handbook of Manufacturing Engineering and Technology. London: Springer-Verlag; 2015. pp. 2677-2717 (ISBN 978-1- 4471-4669-8)

[23] Dobrzański LA, Dobrzańska-Danikiewicz AD. Foresight of the surface Technology in Manufacturing. In: Nee AYC, editor. Handbook of Manufacturing Engineering and Technology. London: Springer-Verlag; 2015. pp. 2587-2637 (ISBN 978-1-4471-4671-1)

[24] Dobrzański LA, Trzaska J, Dobrzańska-Danikiewicz AD. Use of neural networks and artificial intelligence tools for Modeling, characterization, and forecasting in material engineering. In: Hashmi S, editor. Comprehensive Materials Processing, G.F. Batalha (Ed.): Vol. 2: Materials Modeling and Characterization. Elsevier Ltd; 2014. pp. 161-198 (ISBN 978-0-08-096532-1)

[25] Dobrzański LA, Dobrzańska-Danikiewicz AD, Gaweł TG, Achtelik-Franczak A. Selective laser sintering and melting of pristine titanium and titanium Ti6Al4V alloy powders and selection of chemical environment for etching of such materials. Archives of Metallurgy and Materials. 2015;**60**(3):2039-2045

[26] Dobrzański LA, Dobrzańska-Danikiewicz AD, Malara P, Gaweł T, Dobrzański LB, Achtelik-Franczak A. Fabrication of scaffolds from Ti6Al4V powders using the computer aided laser method. Archives of Metallurgy and Materials. 2015;**60**(2):1065-1070

[27] Reimann Ł, Żmudzki J, Dobrzański LA. Strength analysis of a three-unit dental bridge framework with the finite element method. Acta of Bioengineering and Biomechanics. 2015;**17**(1):51-59

[28] Żmudzki J, Chladek G, Kasperski J, Dobrzański LA. One versus two implant-retained dentures: Comparing biomechanics under oblique mastication forces. Journal of Biomechanical Engineering. 2013;**135**(5):054503-1-054503-4

[29] Nowak AJ, Dobrzański LA. Geometrical structure investigation of the surface of internal oesophagus prosthesis. Archives of Materials Science and Engineering. 2017;**83**(2):79-85

[30] Dobrzański LA, Dobrzańska-Danikiewicz AD, Achtelik-Franczak A. The structure and properties of aluminium alloys matrix composite materials with reinforcement made of

titanium skeletons. Archives of Materials Science and Engineering. 2016;**80**(1):16-30. DOI: 10.5604/18972764.1229614

[31] Dobrzański LA, Hudecki A, Chladek G, Król W, Mertas A. Biodegradable and antimicrobial polycaprolactone nanofibers with and without silver precipitates. Archives of Materials Science and Engineering. 2015;**76**(1):5-26

[32] Kremzer M, Dobrzański LA, Dziekońska M, Macek M. Atomic layer deposition of TiO2 onto porous biomaterials. Archives of Materials Science and Engineering. 2015;**75**(2):63-69

[33] Dobrzański LA, Dobrzańska-Danikiewicz AD, Szindler M, Achtelik-Franczak A, Pakieła W. Atomic layer deposition of TiO_2 onto porous biomaterials. Archives of Materials Science and Engineering. 2015;**75**(1):5-11

[34] Dobrzański LA, Dobrzańska-Danikiewicz AD, Achtelik-Franczak A, Dobrzański LB. Comparative analysis of mechanical properties of scaffolds sintered from Ti and Ti6Al4V powders. Archives of Materials Science and Engineering. 2015;**73**(2):69-81

[35] Dobrzański LA, Hudecki A, Chladek G, Król W, Mertas W. Surface properties and antimicrobial activity of composite nanofibers of polycaprolactone with silver precipitations. Archives of Materials Science and Engineering. 2014;**70**(2):53-60

[36] Dobrzański LA, Hudecki A. Structure, geometrical characteristics and properties of biodegradable micro- and polycaprolactone nanofibers. Archives of Materials Science and Engineering. 2014;**70**(1):5-13

[37] Nowak AJ, Dobrzański LA, Rybczyński R, Mech R. Finite element method application for modelling of internal oesophageal prosthesis. Archives of Materials Science and Engineering. 2013;**64**(2):198-204

[38] Stefański T, Malara P, Kloc-Ptaszna A, Janoszka B, Postek-Stefańska L, Tyrpień-Golder K, Dobrzański LA. Erosive potential of calcium-supplemented citric acid on bovine enamel. Archives of Materials Science and Engineering. 2013;**64**(2):175-181

[39] Żmudzki J, Chladek G, Malara P, Dobrzański LA, Zorychta M, Basa K. The simulation of mastication efficiency of the mucous- borne complete dentures. Archives of Materials Science and Engineering. 2013;**63**(2):75-86

[40] Malara P, Kloc-Ptaszna A, Dobrzański LA, Górniak M, Stefański T. Water dilution and alcohol mixing on erosive potential of orange juice on bovine enamel. Archives of Materials Science and Engineering. 2013;**62**(1):15-21

[41] Chladek G, Żmudzki J, Malara P, Dobrzański LA, Krawczyk C. Effect of influence of introducing silver nanoparticles on tribological characteristics of soft liner. Archives of Materials Science and Engineering. 2013;**62**(1):5-14

[42] Reimann Ł, Dobrzański LA. Influence of the casting temperature on dental co-base alloys properties. Archives of Materials Science and Engineering. 2013;**60**(1):5-12

[43] Dobrzański LA, Reimann Ł, Krawczyk C. Effect of age hardening on corrosion resistance and hardness of CoCrMo alloys used in dental engineering. Archives of Materials Science and Engineering. 2012;**57**(1):5-12

[44] Szindler M, Dobrzański LA, Szindler MM, Pawlyta M, Jung T. Comparison of surface morphology and structure of Al2O3 thin films deposited by sol-gel and ALD methods. Journal of Achievements in Materials and Manufacturing Engineering. 2017;**2**(82):49-57. DOI: 10.5604/01.3001.0010.2354

[45] Dobrzański LA, Dobrzańska-Danikiewicz AD, Gaweł TG. Individual implants of a loss of palate fragments fabricated using SLM equipment. Journal of Achievements in Materials and Manufacturing Engineering. 2016;**77**(1):24-30. DOI: 10.5604/17348412.1229665

[46] Dobrzański LA, Dobrzańska-Danikiewicz AD, Gaweł TG. Ti6Al4V porous elements coated by polymeric surface layer for biomedical applications. Journal of Achievements in Materials and Manufacturing Engineering. 2015;**71**(2):53-59

[47] Nowak AJ, Dobrzański LA. Wear abrasive resistance of intracorporeal prosthesis of oesophagus. Journal of Achievements in Materials and Manufacturing Engineering. 2015; **69**(1):26-32

[48] Dobrzański LA, Adamiak M, Karoń M, Nieradka B. Influence of inorganic additives on morphology of electrospun fibres. Journal of Achievements in Materials and Manufacturing Engineering. 2015;**68**(2):64-71

[49] Hudecki A, Pawlyta M, Dobrzański LA, Chladek G. Micro and ceramic nanoparticles surface properties examination with gas adsorption method and microscopic transmission. Journal of Achievements in Materials and Manufacturing Engineering. 2013;**61**(2):257-262

[50] Dobrzański LA, Achtelik-Franczak A, Król M. Computer aided design in selective laser sintering (SLS) – Application in medicine. Journal of Achievements in Materials and Manufacturing Engineering. 2013;**60**(2):66-75

[51] Reimann Ł, Dobrzański LA, Nieradka B, Kusy M, Riedlmajer R. Influence the heat treatment of two base metal alloys used on dental prosthesis on corrosion resistance. Journal of Achievements in Materials and Manufacturing Engineering. 2013;**57**(2):83-90

[52] Dobrzański LA, Reimann Ł. Digitization procedure of creating 3D model of dental bridgework reconstruction. Journal of Achievements in Materials and Manufacturing Engineering. 2012;**55**(2):469-476

[53] Dobrzański LA, Dobrzańska-Danikiewicz AD, Gaweł TG. Computer-aided design and selective laser melting of porous biomimetic materials. Advances in Materials and Processing Technologies. 2017;**3**(1):70-82. DOI: 10.1080/2374068X.2016. 1247339

[54] Nowak AJ, Dobrzański LA. Badania struktury geometrycznej powierzchni wewnątrzustrojowej protezy przełyku. Przetwórstwo Tworzyw. 2015;**21**(4(166)):337-342

[55] Dobrzański LA, Wolany W. Biomaterials manufacturing using SLS method, 20th Jubilee international seminar of Ph.D. students, SEMDOK 2015, Terchova, Slovakia; 2015. pp. 109–112

[56] Dobrzański LA, Dobrzańska-Danikiewicz AD, Łukowiec D. Research methodology laser sintered biomaterials, 20th Jubilee international seminar of Ph.D. students, SEMDOK 2015, Terchova, Slovakia; 2015. pp. 57-66

[57] Czaja I, Hudecki A, Dobrzański LA. Examination of selected surface properties of solids by gas adsorption, 19th International seminar of Ph.D. students, SEMDOK 2014, Terchova, Slovakia; 2014. pp. 22-25

[58] Reimann Ł, Dobrzański LA. Forming the CoCrMo alloys properties by casting conditions, 18th International of PhD. students' seminar, SEMDOK 2013, Terchova, Slovakia; 2013. pp. 56-61

[59] Dobrzański LA, Reimann Ł, Krawczyk C. Heat Treatment Remanium 2000+ Cobalt Alloy Used as Substrate on Dentures, 14th International Materials Symposium, IMSP'2012, Proceedings, Denizli, Turkey; 2012. pp. 598-605

[60] Dobrzański LA, Reimann Ł. Influence of cobalt and chromium on properties of cobalt base alloys, Proceedings of the International Forum of Young Researchers, (Problemy Nedropol'zovaniâ, Meždunarodnyj Forum-Konkurs Molodyh Učenyh, Sbornik Naučnyh Trudov, Sankt-Peterburg), St Petersburg, Russia; 2012, Part II. pp. 7-10

[61] Dobrzański LA, Pusz A, Nowak AJ. Implanty wewnątrzustrojowe. Wady powierzchniowe w materiałach kompozytowych do implantów wewnątrzustrojowych, Materiały kompozytowe. Ludzie, Innowacje, Technologie. 2012;1:48-52

[62] Reimann Ł, Dobrzański LA. Zmiany odporności korozyjnej stopów na osnowie kobaltu po obróbce cieplnej. In: Pacyna J, XLI Szkoła Inżynierii Materiałowej, Kraków; 2013. pp. 294-299

[63] Dobrzańska-Danikiewicz A., Dobrzański LA, Gaweł TG. Scaffolds applicable as implants of a loss palate fragments, International Conference on Processing & Manufacturing of Advanced Materials. Processing, fabrication, properties, applications. THERMEC'2016, Graz, Austria; 2016. p. 193

[64] Hudecki A, Dobrzański LA. Biodegradable PVA and PEO polimeric nanofibers received in electrostatic field, Programme and Proceedings of the Twenty First International Scientific Conference "Achievements in Mechanical and Materials Engineering", AMME'2013, Gliwice – Kraków; 2013. p. 129

[65] Król M, Dobrzański LA, Kujawa M. Influence of technological parameters of the SLS method on AM steel parts, Programme and Proceedings of the Twenty First International Scientific Conference „Achievements in Mechanical and Materials Engineering", AMME'2013, Gliwice – Kraków; 2013. p. 87

[66] Reimann Ł, Żmudzki J, Dobrzański LA. Dobór materiałów na protezy stałe z wykorzystaniem system komputerowego, XVI Konferencja, Biomateriały i mechanika w stomatologii, Ustroń; 2016. p. 62

[67] Postek-Stefańska L, Stefański T, Malara P, Kloc-Ptaszna A, Dobrzański LA. Ocena biopolimerów polisacharydowych jako inhibitorów procesu roztwarzania szkliwa w kwasie cytrynowym, Biomateriały i mechanika w stomatologii, XV Konferencja, Program i streszczenia referatów, Ustroń; 2015. p. 66

[68] Reimann Ł, Dobrzański LA, Żmudzki J. Struktura i własności utwardzanego wydzieleniowo stopu CoCrMoW, Biomateriały i mechanika w stomatologii, XIII Konferencja, Program i streszczenia referatów, Ustroń; 2013. p. 66

[69] Reimann Ł, Dobrzański LA, Żmudzki J. Wpływ własności materiału metalowego na ugięcie trzypunktowego mostu protetycznego, Biomateriały i mechanika w stomatologii, XIII Konfrencja, Program i streszczenia referatów, Ustroń; 2013. p. 65

[70] Malara P, Kloc-Ptaszna A, Dobrzański LA, Stefański T, Górniak M, Chladek G. Wpływ wybranych dodatków na potencjał erozyjny soku pomarańczowego w stosunku do szkliwa zębów, Biomateriały i mechanika w stomatologii, XIII Konfrencja, Program i streszczenia referatów, Ustroń; 2013. p. 45

[71] Hudecki A, Dobrzański LA, Chladek. Własności nanowłókien polimerowych PLA i PCL z dodatkiem wypełniacza o własnościach przeciwdrobnoustrojowych otrzymywanych w polu elektrostatycznym, Biomateriały i mechanika w stomatologii, XIII Konfrencja, Program i streszczenia referatów, Ustroń; 2013. p. 34

[72] Hudecki A, Dobrzański LA, Chladek G. Własności mikro i nanowypełniaczy ceramicznych przewidywanych do poprawy własności materiałów polimerowych przeznaczonych dla stomatologii, Biomateriały i mechanika w stomatologii, XIII Konfrencja, Program i streszczenia referatów, Ustroń; 2013. p. 33

[73] Dobrzański LA, Achtelik-Franczak A. Zastosowanie narzędzi komputerowego projektowania w technologii selektywnego spiekania laserowego, Biomateriały i mechanika w stomatologii, XIII Konfrencja, Program i streszczenia referatów, Ustroń; 2013. p. 27

[74] Dobrzański LA, Achtelik-Franczak A. Zastosowanie technologii selektywnego laserowego spiekania proszków metali w medycynie, Biomateriały i mechanika w stomatologii, XIII Konfrencja, Program i streszczenia referatów, Ustroń; 2013. p. 26

[75] Reimann Ł, Dobrzański LA, Krawczyk C. Wpływ temperatury odlewania na własności stopów kobaltu, Biomateriały i mechanika w stomatologii. Program i materiały XII Konfrencji, Ustroń; 2012. p. 69

[76] Reimann Ł, Dobrzański LA. Odporność korozyjna stopów kobaltu stosowanych w inżynierii stomatologicznej, Proceedings of the Eighteenth International Scientific Conference on Contemporary Achievements in Mechanics, Manufacturing and Materials Science, CAM3S'2012, Gliwice – Ustroń; 2012, p. 82

[77] Dobrzański LA, Dobrzańska-Danikiewicz AD, Malara P, Achtelik-Franczak A, Dobrzański LB, Kremzer M. Sposób wytwarzania materiałów kompozytowych o mikroporowatej szkieletowej strukturze wzmocnienia, Zgłoszenie Patentowe P 417552 z dn. 13.06.2016

[78] Dobrzański LA, Dobrzańska-Danikiewicz AD, Malara P, Dobrzański LB, Achtelik-Franczak A. Kompozyty biologiczno-inżynierskie dla medycyny regeneracyjnej, Zgłoszenie Patentowe P 414723 z dn. 9.11.2015

[79] Dobrzański LA, Dobrzańska-Danikiewicz AD, Malara P, Gaweł TG, Dobrzański LB, Achtelik-Franczak A. Implanto-skafold kostny, Zgłoszenie Patentowe P 414424 z dn. 19.10.2015

[80] Dobrzański LA, Dobrzańska-Danikiewicz AD, Malara P, Gaweł TG, Dobrzański LB, Achtelik-Franczak A. Implanto-skafold lub proteza elementów anatomicznych układu stomatognatycznego oraz twarzoczaszki, Zgłoszenie Patentowe P 414423 z dn. 19.10.2015 1848

[81] Dobrzański LA, Dobrzańska-Danikiewicz AD, Malara P, Gaweł TG, Dobrzański LB, Achtelik A. Kompozyt wykonany z użyciem komputerowo wspomaganych metod laserowych na implanty twarzoczaszki oraz sposób jego wytwarzania (Composite produced using the computer-aided laser methods, intended for the facial skeleton implants and method for producing it), Zgłoszenie Patentowe P 411689 z dn. 23.03.2015, Biuletyn Urzędu Patentowego, Rok. 2016;**44**(20):6

[82] Dobrzański LA, Hudecki A. Sposób wytwarzania materiału kompozytowego o własnościach bioaktywnych i bakteriobójczych (Method for producing composite material with bioactive and bactericidal properties), Zgłoszenie Patentowe P 410452 z dn. 08.12.2014, Biuletyn Urzędu Patentowego, Rok. 2016;**44**(13):25

[83] Dobrzański LA, Hudecki A. Sposób wytwarzania materiału kompozytowego o własnościach bioaktywnych i bakteriobójczych (Method for producing composite material with bioactive and bactericidal properties), Zgłoszenie Patentowe P 410427 z dn. 08.12.2014, Biuletyn Urzędu Patentowego, Rok. 2016;**44**(13):25

[84] Dobrzański LA, Nowak A, Lampe P, Pusz A. Wewnątrzustrojowa proteza przełyku i sposób jej wykonania (Internal esophageal prosthesis and method of implementation thereof), Biuletyn Urzędu Patentowego, Rok. 2012;**40**(21):4-5, Opis patentowy PL 218012; 2014

[85] Dobrzańska-Danikiewicz AD, Achtelik-Franczak A, Wolany W, Dobrzański LA. Dental implants and bridges with scaffold structure – Gold Medal; 10th International Warsaw Invention Show IWIS 2016, Warsaw, Poland, 10–12.10.2016

[86] Dobrzańska-Danikiewicz AD, Achtelik-Franczak A, Wolany W, Dobrzański LA. Dental implants and bridges with scaffold structure –Gold Award, 9th Korea International Women's Invention Exposition, KIWIE 2016, Seoul, Korea, 16–19.06.2016

[87] Dobrzańska-Danikiewicz AD, Achtelik-Franczak A, Wolany W, Dobrzański LA. Dental implants and bridges with scaffold structure – The First Institute Inventors and Researches in I. R. Iran (FIRI) Award for the Best Invention, 9th Korea International Women's Invention Exposition, KIWIE 2016, Seoul, Korea, 16–19.06.2016

[88] Dobrzański LA, Dobrzańska-Danikiewicz AD, Malara P, Dobrzański LB, Achtelik-Franczak A, Gaweł TG. Implant-scaffold or Prosthesis Anatomical Structures of the Stomatognathic System and the Craniofacial – Gold Medal, International Exhibition of Technical Innovations, Patents and Inventions, INVENT ARENA 2016, Třinec, Czech Republic, 16–18.06.2016

[89] Dobrzański LA, Dobrzańska-Danikiewicz AD, Malara P, Dobrzański LB, Achtelik-Franczak A, Gaweł TG. Implant-scaffold or Prosthesis Anatomical Structures of the Stomatognathic System and the Craniofacial – Gold Medal, International Intellectual Property, Invention, Innovation and Technology Exposition, IPITEX 2016, Bangkok, Thailand, 2–6.02.2016

[90] Dobrzański LA, Dobrzańska-Danikiewicz AD, Malara P, Dobrzański LB, Achtelik-Franczak A, Gaweł TG. Implant-scaffold or Prosthesis Anatomical Structures of the Stomatognathic System and the Craniofacial – International Intellectual Property Network Forum (IIPNF) Leading Innovation Award, International Intellectual Property, Invention, Innovation and Technology Exposition, IPITEX 2016, Bangkok, Thailand, 2–6.02.2016

[91] Dobrzański LA, Dobrzańska-Danikiewicz AD, Malara P, Gaweł TG, Dobrzański LB, Achtelik-Franczak A. Global Inventions and Innovations Exhibitions Innova Cities Latino-America, ICLA 2015, Foz do Iguaçu, Brazil, 10–12.12.2015: The novel composite consisting of a metallic scaffold, manufactured using a computer aided laser method, coated with thin polymeric surface layer for medical applications – Semi Grand Prize;

[92] Dobrzański LA, Hudecki A. Composite material with bioactive and bacteriocidal properties and the way of its manufacturing – Silver Medal, Global Inventions and Innovations Exhibitions Innova Cities Latino-America, ICLA 2015, Foz do Iguaçu, Brazil, 10–12.12.2015

[93] Dobrzański LA, Dobrzańska-Danikiewicz AD, Malara P, Gaweł TG, Dobrzański LB, Achtelik-Franczak A. The novel composite consisting of a metallic scaffold, manufactured using a computer aided laser method, coated with thin polymeric surface layer for medical applications – Gold Medal, 9th International Warsaw Invention Show IWIS 2015, Warsaw, Poland, 12–14.10.2015

[94] Dobrzański LA, Hudecki A. Composite material with bioactive and bacteriocidal properties and the way of its manufacturing – Silver Medal, 9th International Warsaw Invention Show IWIS 2015, Warsaw, Poland, 12–14.10.2015

[95] Dobrzański LA, Hudecki A. Composite material with bioactive and bacteriocidal and the way of its manufacturing – Bronze Medal, 26th International Invention, Innovation & Technology Exhibition "ITEX 2015", Kuala Lumpur, Malaysia, 21–23.05.2015

[96] Nowak AJ, Dobrzański LA, Lampe P, Pusz A. Biocompatible composite material based on polymer matrix for the internal oesophageal prosthesis – Silver Medal, 64th International Trade Fair Ideas - Inventions - New Products, iENA 2012, Nuremberg, Germany, 1–4.11.2012

[97] Nowak AJ, Dobrzański LA, Lampe P, Pusz A. Biocompatible composite material based on polymer matrix for the internal oesophageal prosthesis with a specific functional properties – Gold Medal, 6th International Warsaw Invention Show IWIS 2012, Warsaw, Poland, 16–19.10.2012

[98] Nowak AJ, Dobrzański LA, Lampe P, Pusz A. Biocompatible composite material based on polymer matrix for the internal oesophageal prosthesis with a specific functional properties – Special Prize awarded by Chinese Innovation & Invention Society, Taiwan, 6th International Warsaw Invention Show IWIS 2012, Warsaw, Poland, 16–19.10.2012

[99] Nowak AJ, Dobrzański LA, Lampe P, Pusz A. Biocompatible composite material based on polymer matrix for the internal oesophageal prosthesis with a specific functional properties – Special Prize awarded by King Saud University, Saudi Arabia, 6th International Warsaw Invention Show IWIS 2012, Warsaw, Poland, 16–19.10.2012

[100] Nowak AJ, Dobrzański LA, Lampe P, Pusz A. Biocompatible composite material based on polymer matrix for the internal oesophageal prosthesis – Gold Medal, 8th Taipei International Invention Show & Technomart for Intellectual Property, Patents, Trademarks, Inventions, INST 2012, Taipei, Taiwan, 20–23.09.2012

[101] Dobrzański LA, Lampe P, Nowak AJ. Internal prosthesis of oesophagus from polymer based composite material reinforced with continuous aramid fibre – Medal "INVENTICA", 16th International Salon of Research, Innovation and Technological Transfer "INVENTICA 2012", Iasi, Romania, 13–15.06.2012

[102] Dobrzański LA, Nowak AJ, Lampe P, Pusz A. Biocompatible composite material based on polymer matrix for the internal oesophageal prosthesis – Bronze Medal, 23rd International Invention, Innovation & Technology Exhibition "ITEX 2012", Kuala Lumpur, Malaysia, 17–19.05.2012

[103] Dobrzański LA, et al. Badanie struktury i własności nowoopracowanych porowatych materiałów biomimetycznych wytwarzanych metodą selektywnego spiekania laserowego, BIOLASIN, Projekt UMO-2013/08/M/ST8/00818 Gliwice, 2013–2016

[104] Dobrzański LA, et al. Ustalenie istoty wpływu jednowymiarowych materiałów nanostrukturalnych na strukturę i własności nowo opracowanych funkcjonalnych materiałów nanokompozytowych i nanoporowatych, NANOCOPOR, Projekt UMO-2012/07/B/ST8/04070, Gliwice, 2013–2016

[105] Dobrzański LA, et al. Opracowanie nowego gradientowego materiału kompozytowego o osnowie polimerowej wzmacnianego włóknami aramidowymi oraz cząstkami proszku tytanu, przeznaczonego do produkcji protezy wewnątrzustrojowej przełyku, Projekt N507 422136, Gliwice, 2009–2011

[106] Dobrzański LA, et al. Opracowanie metodologii komputerowego wspomagania projektowania materiałowego, technologicznego oraz konstrukcyjnego stałych stomatologicznych protez wieloczłonowych w celu predykcji ich własności użytkowych, Projekt N507 438539, Gliwice, 2010–2013

[107] Dobrzański LA, et al. Foresight wiodących technologii kształtowania własności powierzchni materiałów inżynierskich i biomedycznych, FORSURF, Projekt UDA-POIG.01.01.01–00.23/08–00, Gliwice 2009–2012

[108] Centrum Projektowo-Badawczo-Produkcyjne Inżynierii Medycznej i Stomatologicznej ASKLEPIOS w Gliwicach, Projekt POIR.01.01.01–00-0485/16–00 pt. "IMSKA-MAT Innowacyjne implanto-skafoldy stomatologiczne i szczękowo-twarzowe wytwarzane z wykorzystaniem innowacyjnej technologii addytywnej wspomaganej komputerowym projektowaniem materiałowym ADD-MAT", Gliwice; 2017–2021, w toku

[109] Dobrzański LA. ONCO-ATIS Interaction of osteoblast natural cells ordered in the selectively laser sintered metallic materials as a basis for the creation of orthopedic and dental advanced types of implant-scaffalds, Project Proposal NCN 2017/26/A/ST8/00232, Gliwice-Cracow; 2017

[110] Yu C, Abbott PV. Australian Dental Journal. 2007;**52**(1):4-16

[111] Miyagi SPH et al. The Journal of Endocrinology. 2010;**36**(5):826-831

[112] Seltzer S, Farber PA. Oral Surgery, Oral Medicine, Oral Pathology and Oral Radiology. 1994;**78**:634-645

[113] Wójcicka A et al. Przegląd Epidemiologiczny. 2012;**66**(4):705-711

[114] Szymańska J, Szalewski L. Zdrowie Publiczne. 2011;**121**(1):86-89

[115] Bromblik A et al. Czasopismo Stomatologiczne. 2010;**63**(5):301-309

[116] Pawka B et al. Prob Hig Epid (in Polish). 2010;**91**(1):5-7

[117] Aas JA et al. Journal of Clinical Microbiology. 2008;**46**(4):1407-1417

[118] Petersen PE. Changing Oral Health Profiles of Children in Central and Estern Europe – Challenges for the 21st Century. Geneva, Switzerland: WHO Global Oral Health; 2003

[119] Jańczuk Z. Próchnica zębów, zapobieganie, klinika i leczenie. Warszawa: PZWL; 1994

[120] Scannapieco FA. Role of oral bacteria in respiratory infection. Journal of Periodontology. 1999;**70**:793-802

[121] Scannapieco FA et al. An Period. 2003;**8**:54-69

[122] Kustra P, Zarzecka J. Por Stom (in Polish). 2009;**9**(12):453-456

[123] Aleksander M et al. Oral and Maxillofacial Surgery. 2008;**12**(3):129-130

[124] Li X et al. Endodontics & Dental Traumatology. 1999;**15**:95-101

[125] Calayatud-Perez V et al. Journal of Cranio-Maxillofacial Surgery. 1993;**21**(3):127-129

[126] Kaiser LR. Topics in Health Care Financing. 1992;**18**:32-45

[127] Arsiwala A et al. Journal of Controlled Release. 2014;**189**:25-45

[128] Yang F et al. Biomaterials. 2005;**26**:5991-5998

[129] Atala A. Principles of Regenerative Medicine. Elsevier Academic Press; 2007

[130] USNIH. Regenerative Medicine, Report. US National Institutes of Health; 2006

[131] Cogle CR et al. Mayo Clinic Proceedings. 2003;**78**:993-1003

[132] Metallo CM et al. Journal of Cellular and Molecular Medicine. 2008;**12**:709-729

[133] Placzek MR et al. Journal of the Royal Society Interface. 2009;**6**:209-232

[134] Tavassoli M, Crosby WH. Science. 1968;**161**:54-56

[135] Caplan AI. Journal of Orthopaedic Research. 1991;**9**:641-650

[136] Pittenger MF et al. Science. 1999;**284**:143-147

[137] Culme-Seymour EJ et al. Regenerative Medicine. 2012;**7**:455-462

[138] Langer R, Vacanti JP. Science. 1993;**260**:920-926

[139] Viola J et al. The Emergence of Tissue Engineering as a Research Field. USA: The National Science Foundation; 2003

[140] Fung YC. UCSD #865023; 2001

[141] Lanza RP, Langer R, Vacanti J, editors. Principles of Tissue Engineering. San Diego: Academic Press; 2000

[142] Atala A, Lanza RP, editors. Methods of Tissue Engineering. San Diego: Academic Press; 2002

[143] Macarthur BD, Oreffo RO. Nature. 2005;**433**:19

[144] Trounson A et al. BMC Medicine. 2011;**9**(52):3

[145] Mason C et al. Cell Stem Cell. 2012;**11**:735-739

[146] Thomson JA et al. Science. 1998;**282**:1145-1147

[147] Shamblott MJ et al. Proceedings of the National Academy of Sciences of the United States of America. 1998;**95**:13726-13731

[148] Cetrulo CL Jr. Stem Cell Reviews. 2006;**2**:163-168

[149] Branch MJ et al. Investigative Ophthalmology & Visual Science. 2012;**53**:5109-5116

[150] Wang S et al. Journal of Hematology & Oncology. 2012;**5**:19

[151] AMERICORD. What is Cord Tissue? 2015. http://americordblood.com/cord-tissue-banking

[152] Le Blanc K et al. Scandinavian Journal of Immunology. 2003;**57**:11-20

[153] Maitra B et al. Bone Marrow Transplantation. 2004;**33**:597-604

[154] Aggarwal S, Pittenger MF. Blood. 2005;**105**:1815-1822

[155] Lee EH, Hui JH. Journal of Bone and Joint Surgery. British Volume (London). 2006;**88**:841-851

[156] Raff M. Annual Review of Cell and Developmental Biology. 2003;**19**:1-22

[157] Morrison SJ, Weissman IL. Immunity. 1994;**1**:661-673

[158] Bajada S et al. Journal of Tissue Engineering and Regenerative Medicine. 2008;**2**:169-183

[159] Johal KS et al. Regenerative Medicine. 2015;**10**:79-96

[160] Wilson CE et al. Journal of Materials Science. Materials in Medicine. 2002;**13**:1265-1269

[161] Kruyt MC et al. Tissue Engineering. 2003;**9**:327-336

[162] Helmlinger G et al. Nature Medicine. 1997;**3**:177-182

[163] Folkman J. The New England Journal of Medicine. 1971;**285**:1182-1186

[164] Mooney DJ et al. Cell Transplantation. 1994;**3**:203-210

[165] Li G et al. Cell Biology International. 2000;**24**:25-33

[166] Sanders JE et al. Microvascular Research. 2002;**64**:174-178

[167] Muschler GF et al. Journal of Bone and Joint Surgery. 2004;**86-A**:1541-1558

[168] Baksh D et al. Experimental Hematology. 2003;**31**:723-732. Stem Cells. 2007;**25**:1384-1392

[169] Caplan AI, Bruder SP. Trends in Molecular Medicine. 2001;**7**:259-264

[170] Zuk PA et al. Tissue Engineering. 2001;**7**:211-228

[171] Izadpanah R et al. Journal of Cellular Biochemistry. 2006;**99**:1285-1297

[172] Moll G et al. PLoS One. 2014;**9**:e85040

[173] Rayment EA, Williams DJ. Stem Cells. 2010;**28**:996-1004

[174] Thomas RJ et al. Cytotechnology. 2007;**55**:31-39

[175] Thomas RJ et al. Biotechnology and Bioengineering. 2009;**102**:1636-1644

[176] Thomas RJ et al. Biotechnology Letters. 2009;**31**:1167-1172

[177] Liu Y et al. Journal of Tissue Engineering and Regenerative Medicine. 2010;**4**:45-54

[178] Yang F, et al. 2008. In: Laurencin & Nair; 2008

[179] Peppas NA, Langer R. Science. 1994;**263**:1715-1720

[180] Hubbell JA. Biotechnology (N. Y). 1995;**13**:565-576

[181] Hubbell JA. Current Opinion in Biotechnology. 1999;**10**:123-129

[182] Healy KE et al. Annals of the New York Academy of Sciences. 1999;**875**:24-35

[183] Langer R, Tirrell DA. Nature. 2004;**428**:487-492

[184] Lütolf MP, Hubbell JA. Nature Biotechnology. 2005;**23**:47-55

[185] Elisseeff J et al. Stem Cells and Development. 2006;**15**:295-303

[186] Hwang NS et al. Stem Cells. 2006;**24**:284-291

[187] Yang F, et al. 2008. In: Laurencin & Nair; 2008

[188] Bettinger CJ, et al. 2008. In: Laurencin & Nair; 2008

[189] Noga M, et al. E-MRS Fall Meeting, Warsaw; 2013

[190] Rouwkema J et al. Trends in Biotechnology. 2008;**26**:434-441

[191] Bramfeldt H et al. Current Medicinal Chemistry. 2010;**17**:3944-3967

[192] Jain RK et al. Nature Biotechnology. 2005;**23**:821-823

[193] Bose S et al. Trends in Biotechnology. 2012;**30**:546-554

[194] Griss P, Heimke G. In: Williams D, editor. Biocompatibility of Implant Materials. London: Sector Publishing; 1976. pp. 52-68

[195] Taniguchi N et al. Materials Science and Engineering: C. 2016;**59**:690-701

[196] Robertson DM et al. Journal of Biomedical Materials Research. 1976;**10**:335-344

[197] Martin R. Critical Reviews in Biomedical Engineering. 1983;**10**:179-222

[198] Ramirez JM, Hurt WC. Journal of Periodontology. 1977;**48**:74-77

[199] Bentolila V et al. Bone. 1998;**23**:275-281

[200] Prendergast P, Huiskes R. Journal of Biomechanical Engineering. 1996;**118**:240-246

[201] Wang X, Ni Q. Journal of Orthopaedic Research. 2003;**21**:312-319

[202] Kufahl RH, Saha SA. Journal of Biomechanics. 1990;**23**:171-180

[203] Sietsema W. Bone. 1995;**17**:297-305

[204] Zardiackas LD et al. Journal of Biomedical Materials Research. 2001;**58**:180-187

[205] Wang X et al. Biomaterials. 2016;**83**:127-141

[206] Bernard S et al. Journal of the Mechanical Behavior of Biomedical Materials. 2013;**18**:12-19

[207] Hollister SJ. Advanced Materials. 2009;**21**:3330-3342

[208] Langer RS, Vacanti JP. Scientific American. 1999;**280**:86-89

[209] Langer R, Vacanti JP. Science. 1993;**260**:920-926

[210] Cima LG et al. Journal of Biomechanical Engineering. 1991;**113**:143-151

[211] Langer R et al. Biomaterials. 1990;**11**:738-745

[212] Karageorgiou V, Kaplan D. Biomaterials. 2005;**26**:5474-5491

[213] Story BJ et al. The International Journal of Oral & Maxillofacial Implants. 1998;**13**:749-757

[214] Lewandrowski KU et al. Journal of Biomaterials Science, Polymer Edition. 2000;**11**:879-889

[215] Bai F. Tissue Engineering. Part A. 2010;**16**:3791-3803

[216] Karp JM et al. Journal of Biomedical Materials Research. Part A. 2004;**71**:162-171

[217] Street J et al. Clinical Orthopaedics and Related Research. 2000;**378**:224-237

[218] Yan C et al. Materials and Design. 2014;**55**:533-541

[219] Lu L et al. Materials Research Bulletin. 2000;**35**(9):1555-1561

[220] Kumar S. Mod Char. 2003;**55**(10):43-47

[221] Mellor S et al. International Journal of Production Economics. 2014;**149**:194-201

[222] Guo N, Leu MC. Frontiers of Mechanical Engineering. 2013;**8**(3):215-243

[223] Melechow R et al. Tytan i jego stopy. Częstochowa: Wyd Pol Częst; 2004

[224] Kevorkijan VM. Journal of Computer Science and Technology. 1999;**59**:683-686

[225] Balla VK et al. Acta Biomaterialia. 2010;**6**(8):3349-3359

[226] Das K et al. Scra Mat. 2008;**59**(8):822-825

[227] Witte F et al. Journal of Biomedical Materials Research Part A. 2007;**81**(3):757-765

[228] Bandyopadhyay A et al. Acta Biomaterialia. 2010;**6**:1640-1648

[229] Nouri A et al. In: Mukherjee A, editor. Biomimetics Learning from Nature. Rijeka: InTech; 2010. pp. 415-450

[230] Yoshida K et al. Hut – Wiad Hut. 2011;**78**(1):153-156

[231] Lia JP et al. Biomacromolecules. 2007;**28**:2810-2820

[232] Choi JW et al. Biomedical Microdevices. 2010;**12**(5):875-886

[233] Wu G et al. The Journal of Prosthetic Dentistry. 2008;**100**(1):56-60

[234] Ciocca L et al. Medical & Biological Engineering & Computing. 2011;**49**(11):1347-1352

[235] Nałęcz M. Bioch i inż biom, t 6. In: Duch W, et al., EXIT, Warszawa; 2000

[236] Leiggener C et al. International Journal of Maxillofacial and Oral Surgery. 2009;**38**(2):187-194

[237] Jacobs PF. Rapid Prototyping & Manufacturing: Fundamentals of Stereolithography. Dearborn: SME publication; 1992

[238] Lee M et al. Biomaterials. 2005;**26**(20):4281-4289

[239] Cormier B et al. Rapid Prototyping Journal. 2004;**10**(1):35-41

[240] Rännar LE et al. Rapid Prototyping Journal. 2007;**13**(3):128-135

[241] Murr LE et al. Materials Science and Engineering: A. 2010;**527**(7–8):1861-1868

[242] Han GW et al. Materials Science and Engineering: A. 1997;**225**(1–2):204-207

[243] Murr LE et al. Acta Materialia. 2010;**58**(5):1887-1894

[244] Xue W et al. Acta Biomaterialia. 2007;**3**:1007-1018

[245] Osakada K, Shiomi M. International Journal of Machine Tools and Manufacture. 2006;**46**:1188-1193

[246] Kruth Mercelis JP et al. Rapid Prototyping Journal. 2005;**11**(1):26-36

[247] Bertol LS et al. Materials and Design. 2010;**31**:3982-3988

[248] Pattanayak DK et al. Acta Biomaterialia. 2011;**7**:1398-1406

[249] Abe F et al. Journal of Materials Processing Technology. 2001;**111**(1–3):210-213

[250] Darłak P, Dudek P. Odl: Nauka i praktyka. 2004;**1**:3-17

[251] Kruth JP et al. Journal of Materials Processing Technology. 2004;**149**(1–3):616-622

[252] Lethaus B et al. Journal of Maxillofacial and Oral Surgery. 2012;**40**(1):43-46

[253] Liao YS et al. International Journal of Advanced Manufacturing Technology. 2006;**27**(7–8):703-707

[254] Manakari V et al. Meta. 2017;**7**:2-35

[255] Jaegermann Z, Ślósarczyk A. Gęsta i porowata bioceramika korundowa w zastosowaniach medycznych. Kraków: Ucz. Wyd. Nauk.-Techn; 2007

[256] Hench LL, Andersson Ö. In: Hench LL, Wilson J, editors. An Introduction to Bioceramics. 1993. pp. 41-62

[257] Ciołek L et al. Engineering Biomaterials. 2009;**12**:91-93

[258] Ciołek L. Engineering Biomaterials. 2010;**13**:21-23

[259] Łączka M et al. Szkło Ceram. 1996;**47**:14-19

[260] Miguel BS et al. Journal of Biomedical Materials Research. Part A. 2010;**94**:1023-1033

[261] Wu C et al. Biomaterials. 2012;**33**(7):2076-2085

[262] Rezwan K et al. Biomaterials. 2006;**27**:3413-3431

[263] Karageorgiou V, Kaplan D. Biomaterials. 2005;**26**:5474-5491

[264] Wilson CE et al. Journal of Materials Science. Materials in Medicine. 2011;**22**:97-105

[265] Lichte P et al. Injury. 2011;**42**:569-573

[266] Deepthia S et al. International Journal of Biological Macromolecules. 2016;**93B**:1338-1353

[267] Lee S-H, Shin H. Advanced Drug Delivery Reviews. 2007;**59**:339-359

[268] Cheung H-Y et al. Composites: Part B. 2007;**38**:291-300

[269] Xue W et al. Journal of Biomedical Materials Research Part B. 2009;**91**:831-838

[270] Laschke MW et al. Acta Biomaterialia. 2010;**6**:2020-2027

[271] Banerjee SS et al. Acta Biomaterialia. 2010;**6**:4167-4174

[272] Tara MA et al. The International Journal of Prosthodontics. 2011;**24**:46-48

[273] Quante K et al. Dental Materials. 2008;**24**:1311-1315

[274] Śpiewak R, Piętowska J. ALERG-IM. 2006;**3**(34):5862

[275] Uter W et al. The European standard series in 9 European countries, 2002/2003 first results of the European surveillance system on contact allergies. Contact Dermatitis. 2005;**53**:136145

[276] Śpiewak R. Allergy. 2002;**25**:374381

[277] Obtułowicz K et al. Przegląd Lekarski. 2001;**58**(5):2427

[278] Antoszczyk G, Obtułowicz K. Post Derm Allergy. 2005;**22**:2936

[279] Śpiewak R, Brewczyński PZ. Polski Tygodnik Lekarski. 1993;**48**(28–30):651-652

[280] Rahilly G, Price N. Journal of Orthodontics. 2003;**30**:171174

[281] EU Parliament & Council (1994) Directive 94/27/EC of 30 June 1994 amending for the 12th time Directive 76/769/EEC... Of J L 188, 22.07.1994, 12

[282] Koutsoukis T et al. Journal of Prosthodontics. 2015;**24**:303-312

[283] Oyague RC et al. Journal of Dentistry. 2012;**40**:123-130

[284] Ucar Y et al. The Journal of Prosthetic Dentistry. 2009;**102**:253-259

[285] Castillo-De-Oyague R et al. Medicina Oral, Patología Oral y Cirugía Bucal. 2012;**17**: 610-617

[286] Castillo-Oyague R et al. Journal of Dentistry. 2013;**41**:90-96

[287] Oyague RC et al. Odontología. 2012;**100**:249-253

[288] Ortorp A et al. Dental Materials. 2011;**27**:356-363

[289] Kim KB et al. Journal of Advanced Prosthodontics. 2013;**5**:179-186

[290] Marciniak J. Biomateriały. Gliwice: Wyd Pol Śl; 2002

[291] Górny Z. Odlewnicze stopy kobaltu. Kraków: Inst Odl; 2008

[292] Klimek L, Zimmer D. The Dental Technician. 2015;**6**:26-30

[293] Surowska B et al. Post Nauki i Tech (in Polish). 2011;**11**:81-88

[294] Lucchetti MC et al. The Journal of Prosthetic Dentistry. 2015;**114**(4):602-608

[295] Kacprzyk B et al. Archives of Foundry Engineering. 2013;**13**(3):47-50

[296] Sokołowski J et al. J Stom. 2014;**67**(1):28

[297] Balla VK et al. Acta Biomaterialia. 2010;**6**:3349-3359

[298] Das K et al. Scripta Materialia. 2008;**59**:822-825

[299] Liu Y et al. Acta Materialia. 2016;**113**:56-67

[300] Challis VJ et al. Materials and Design. 2014;**63**:783-788

[301] Liu Y et al. Materials Science and Engineering: A. 2015;**642**:268-278

[302] Attar H et al. Materials Science and Engineering: A. 2015;**625**:350-356

[303] Gehrmann R et al. Materials Science and Engineering: A. 2005;**395**:338-349

[304] Kulekci MK. International Journal of Advanced Manufacturing Technology. 2008;**39**: 851-865

[305] Bettles CJ. Journal of Materials Engineering and Performance. 2008;**17**:297-301

[306] Yun Y et al. Materials Today. 2009;**12**:22-32

[307] Chai YC et al. Biomaterials. 2012;**33**:4044-4058

[308] Markwardt J et al. Journal of Biomedical Materials Research. Part A. 2014;**102**:1422-1430

[309] Pattanayak DK et al. Acta Biomaterialia. 2011;**7**:1398-1406

[310] Van Bael S et al. Acta Biomaterialia. 2012;**8**:2824-2834

[311] Bandyopadhyay A et al. Acta Biomaterialia. 2010;**6**:1640-1648

[312] Klimas J et al. Archives of Materials Science and Engineering. 2016;**78**(1):10-16

[313] Kirmanidou Y, et al. BioMed Res In, art.no.2908570. 2016

[314] Elias C et al. Journal of Metals. 2008;**60**(3):46-49

[315] Parchalskakowalik M, Klimek L. Inżynieria Materiałowa. 2013;**34**(5):526-529

[316] Dobrzański LA. Archives of Materials Science and Engineering. 2015;**76**(2):53-114

[317] Dobrzańska-Danikiewicz AD et al. Archives of Materials Science and Engineering. 2015;**76**(2):150-156

[318] Burnat B et al. Archives of Foundry Engineering. 2014;**14**(3):11-16

[319] Majkowska B et al. Archives of Metallurgy and Materials. 2015;**60**(2):755-758

[320] Klimas J et al. Archives of Materials Science and Engineering. 2016;**77**(2):65-71

[321] Jedynak B, Mierzwińska-Nastalska E. Dental Forum 2013;**41**(1):75-78

[322] Dobrzański LA et al. J Ach Mat Man Eng. 2015;**71**(2):53-59

[323] Dobrzański LB. PhD in progress, AGH; Kraków; 2017

[324] Emsley J. Nature's Building Blocks: An A-Z Guide to the Elements. Oxford: Oxford University Press; 2001

[325] Hong J et al. Thrombosis and Haemostasis. 1999;**82**(1):58-64

[326] Zięty A, Lachowicz M. Akt/PKB signaling pathway. 2014;**8**:187-192

[327] Okazaki Y et al. Materials Science and Engineering: A. 1966;**213**:138-147

[328] Semlitsch MF et al. Biomaterials. 1992;**13**(11):781-788

[329] Michalik R et al. Inżynieria Materiałowa. 1998;**19**(2):53-56

[330] Oliveira V et al. Brazilian Journal of Chemical Engineering. 1998;**15**(4)

[331] Al-Mobarak NA et al. Int J El-chem Sci 2011;**6**:2031-2042

[332] Dąbrowski R et al. Acta Bio-Optica et Informatica Medica, Biomedical Engineering. 2016;**22**(3):130-137

[333] Boehlert CJ et al. Materials Science and Engineering: C. 2008;**28**(3):323-330

[334] Walkowiak-Przybyło W et al. Journal of Biomedical Materials Research Part A. 2012; **100**(3):768-775

[335] Venkatarman BV, Sudha S. Asian Journal of Experimental Sciences. 2005;**19**(2):127-134

[336] Ngwa HA et al. Toxicology and Applied Pharmacology. 2009;**240**(2):273-285

[337] Williams DF. Biomaterials. 2008;**29**(20):2941-2953

[338] Steinemann SG. Injury. 1996;**27**(Suppl 3):SC16-SC22

[339] Braem A et al. Acta Biomaterialia. 2014;**10**:986-995

[340] Abdel-Hady Gepreel M, Niinomi M. Journal of the Mechanical Behavior of Biomedical Materials. 2013;**20**:407-415

[341] Khan MA et al. Biomaterials. 1999;**20**(7):631-637

[342] Paszenda Z et al. Journal of Achievements in Materials and Manufacturing Engineering. 2010;**38**(1):24-32

[343] Hong J et al. Thrombosis and Haemostasis. 1999;**82**(1):58-64

[344] Chlebus E et al. Materials Characterization. 2011;**62**:488-495

[345] Marcu T et al. Powder Metallurgy. 2012;**55**:309-314

[346] Szymczyk P et al. Acta of Bioengineering and Biomechanics. 2013;**15**:69-76

[347] Okrój W et al. Inż Biomaterials. 2005;**8**(43–44):13-16

[348] Tanaka Y et al. Journal of Artificial Organs. 2009;**12**(3):182-186

[349] Harris LG et al. Journal of Materials Science: Materials in Medicine. 2007;**18**(6):1151-1156

[350] Zhang LC et al. Scripta Materialia. 2011;**65**:21-24

[351] Hsu HC et al. Materials and Design. 2013;**47**:21-26

[352] Yao Y et al. Journal of Alloys and Compounds. 2014;**583**:43-47

[353] Zhuravleva K et al. Maturitas. 2013;**6**:5700-5712

[354] Leong KF et al. Biomaterials. 2003;**24**:2363-2378

[355] Nair LS, Laurencin CT. Advances in Biochemical Engineering/Biotechnology. 2006;**102**: 47-90

[356] Velema J, Kaplan D. Advances in Biochemical Engineering/Biotechnology. 2006;**102**:187-238

[357] Nowosielski R et al. Journal of Achievements in Materials and Manufacturing Engineering. 2013;**61**:367-374

[358] Witte F et al. Journal of Biomedical Materials Research. Part A. 2007;**81**(3):757-765

[359] Zhang X et al. Materials Science and Engineering: C. 2014;**42**:362-367

[360] Farraro KF et al. Journal of Biomechanics. 2014;**47**:1979-1986

[361] Jaworska L, De Benedetti B. priv. inf; 2014

[362] Li N, Zheng Y. Journal of Materials Science and Technology. 2013;**29**:489-502

[363] Kirkland NT et al. Acta Biomaterialia. 2012;**8**:925-936

[364] Wu G et al. Surface and Coating Technology. 2013;**233**:2-12

[365] Hornberger H et al. Acta Biomaterialia. 2012;**8**:2442-2455

[366] Yazdimamaghani M et al. Materials Letters. 2014;**132**:106-110

[367] Chen Z et al. Biomaterials. 2014;**35**:8553-8565

[368] Gu XN et al. Materials Science and Engineering: C. 2010;**30**:827-832

[369] Khanra AK et al. Materials Science and Engineering: A. 2010;**527**:6283-6288

[370] Khanra AK et al. Bulletin of Materials Science. 2010a;**33**:43-47

[371] Liu DB et al. Rare Metal Materials and Engineering. 2008;**37**:2201-2205

[372] Witte F et al. Biomaterials. 2007;**28**:2163-2174

[373] Ye XY et al. Journal of Materials Science. Materials in Medicine. 2010;**21**:1321-1328

[374] Razavi M et al. Materials Science and Engineering A. 2010;**527**:6938-6944

[375] Feng A, Han Y. Journal of Alloys and Compounds. 2010;**504**:585-593

[376] Brar HS et al. Journal of Metals. 2009;**61**:31-34

[377] Kirkland N. Corrosion Engineering Science and Technology. 2012;**47**:322-328

[378] Walker J et al. Journal of Biomedical Materials Research. Part B, Applied Biomaterials. 2014;**102B**(6):1316-1331

[379] Alvarez-Lopez M et al. Acta Biomaterialia. 2010;**6**:1763-1771

[380] Wong HM et al. Biomaterials. 2010;**31**:2084-2096

[381] Liu C et al. Materials Science and Engineering: A. 2007;**456**:350-357

[382] Vojtech D et al. Acta Biomaterialia. 2011;**7**:3515-3522

[383] Witte F et al. Current Opinion in Solid State & Materials Science. 2008;**12**:63-72

[384] Wan YZ et al. Materials and Design. 2008;**29**:2034-2037

[385] Gu X et al. Biomaterials. 2009;**30**:484-498

[386] Flaten T. Environmental Geochemistry and Health. 1990;**12**:152-167

[387] Wills M, Savory J. Lancet. 2001;**2**:29-34

[388] Flaten T. Brain Research Bulletin. 2001;**55**:187-196

[389] Pinto R et al. Electrochimica Acta. 2011;**56**:1535-1545

[390] Johnson I et al. Journal of Biomedical Materials Research. Part A. 2011:477-485

[391] Haley T. Journal of Pharmaceutical Sciences. 1965;**54**:663-670

[392] Zartner P et al. Catheterization and Cardiovascular Interventions. 2005;**66**:590-594

[393] Drynda A et al. Journal of Biomedical Materials Research. Part A. 2008;**91**:360-369

[394] Bruce D et al. Toxicology and Applied Pharmacology. 1963;**5**:750-759

[395] Gupta M et al. Magnesium, Magnesium Alloys, and Magnesium Composites. John Wiley & Sons; 2011. pp. 1-257, ISBN 9781118102701

[396] Savalani M et al. Journal of Engineering Manufacture. 2006;**220**:171-182

[397] Ng C et al. Applied Surface Science. 2011;**257**:7447-7454

[398] Ponader S et al. Journal of Biomedical Materials Research. Part A. 2010;**92**:56-62

[399] Zhang E et al. Materials Science and Engineering: C. 2009;**29**:987-993

[400] Xu L et al. Journal of Materials Science: Materials in Medicine. 2008;**19**:1017-1025

[401] Song GL, Atrens A. Advanced Engineering Materials. 1999;**1**:11-33

[402] Koh J, Choi D. Neuroscience. 1994;**60**:1049-1057

[403] Walker J et al. Chemico-Biological Interactions. 1989;**69**:279-291

[404] Alvarez K, Nakajima H. Maturitas. 2009;**2**:790-832

[405] Xin Y et al. Acta Biomaterialia. 2011;**7**:1452-1459

[406] Rettig R, Virtanen S. Journal of Biomedical Materials Research. Part A. 2008;**85**:167-175

[407] http://www.magnesium-elektron-powders.com/index.php/contacts

[408] Papadimitropoulos A et al. Biotechnology and Bioengineering. 2007;**98**:271-281

[409] Naito H et al. Tissue Engineering Part A. 2011;**17**:2321-2329

[410] Bose S, Tarafder S. Acta Biomaterialia. 2012;**8**:1401-1421

[411] Verron E et al. Drug Discovery Today. 2010;**15**:547-552

[412] Kimelman-Bleich N et al. Molecular Therapy. 2010;**19**:53-59

[413] Keeney M et al. Biomaterials. 2010;**31**:2893-2902

[414] Dagalakis N et al. Journal of Biomedical Materials Research. 1980;**14**:511-528

[415] Mooney DJ et al. Cell Transplantation. 1994;**3**:203-210

[416] Wald HL et al. Biomaterials. 1993;**14**:270-278

[417] Tay CY et al. Small. 2011;**7**:1361-1378

[418] Wang GJ et al. Microsystem Technologies. 2005;**12**:120-127

[419] Lima MJ et al. Materials Science and Engineering: C. 2014;**42**:615-621

[420] Zhang S et al. Biomaterials. 1999;**20**:1213-1220

[421] Schvartzman M, Wind SJ. Nano Letters. 2009;**9**:3629-3634

[422] Cao H et al. Applied Physics Letters. 2002;**81**:174-176

[423] Chou SY, et al. 41st Int Conf Electron, Ion, and Photon Beam; 1997

[424] Kim J et al. Tissue Engineering Part A. 2014;**20**:2127-2130

[425] Delamarche E et al. Science. 1997;**276**:779-781

[426] Mazzoli A. Medical & Biological Engineering & Computing. 2013;**51**:245-256

[427] Darsell J et al. Journal of the American Ceramic Society. 2003;**86**:1076-1080

[428] Bose S et al. Scripta Materialia. 1999;**41**:1009-1014

[429] Hutmacher DW et al. Journal of Biomedical Materials Research. 2001;**55**:203-216

[430] Bettinger CJ, et al. 2008. In: Laurencin & Nair; 2008

[431] Yeong W-Y et al. Rapid Prototyping Journal. 2006;**12**:229-237

[432] Dutta Roy T et al. Journal of Biomedical Materials Research. Part A. 2003;**67**:1228-1237

[433] Sachs EM, et al. 1993. U.S. Patent 5,204,055

[434] Khalyfa A et al. Journal of Materials Science. Materials in Medicine. 2007;**18**:909-916

[435] Manjubala I et al. Journal of Materials Science. Materials in Medicine. 2005;**16**:1111-1119

[436] Sachlos E et al. Biomaterials. 2003;**24**:1487-1497

[437] Sawkins MJ et al. Recent Patents on Biomedical Engineering. 2013;**6**:3-21

[438] Lee KW et al. Biomacromolecules. 2007;**8**:1077-1084

[439] Cooke MN et al. Journal of Biomedical Materials Research Part B. 2003;**64**:65-69

[440] Arcaute K et al. Annals of Biomedical Engineering. 2006;**34**:1429-1441

[441] Dhariwala B et al. Tissue Engineering. 2004;**10**:1316-1322

[442] Liao H-T, et al. BioMed Res Inter. 2014; Article ID 321549

[443] Hoque ME et al. Journal of Biomaterials Science. Polymer Edition. 2005;**16**:1595-1610

[444] Sun JJ et al. Journal of Materials Science. Materials in Medicine. 2007;**18**:1017-1023

[445] Crump SS. U.S. Patent 5,121,329A; 1992

[446] Karoluk M, Pawlak A, Chlebus E. XI Konf. Nauk. im. Prof. D. Tejszerskiej, Ustroń; 2014. pp. 53-54

[447] Williams JM et al. Biomaterials. 2005;**26**:4817-4827

[448] Tan KH et al. Bio-medical Materials and Engineering. 2005;**15**:113-124

[449] Shuai C et al. Nanotechnology. 2011;**22**:285703

[450] Wiria FE et al. Acta Biomaterialia. 2007;**3**:1-12

[451] Chua CK et al. Journal of Materials Science. Materials in Medicine. 2004;**15**:1113-1121

[452] Sill TJ, von Recum HA. Biomaterials. 2008;**29**:1989-2006

[453] Murugan R, Ramakrishna S. Tissue Engineering. 2007;**13**:1845-1866

[454] Matthews JA et al. Biomacromolecules. 2002;**3**:232-238

[455] Hartgerink JD et al. Proceedings of the National Academy of Sciences of the United States of America. 2002;**99**:5133-5138

[456] Zhang S et al. Seminars in Cancer Biology. 2005;**15**:413-420

[457] Gelain F et al. Macromolecular Bioscience. 2007;**7**:544-551

[458] Um SH et al. Nature Materials. 2006;**5**:797-801

[459] Hunt JA et al. Journal of Materials Chemistry B. 2014;**2**:5319-5338

[460] Qin S et al. Advanced Materials. 2006;**18**:2905-2909

[461] Guan J et al. Biomacromolecules. 2008;**9**:1283-1292

[462] Lee SB et al. Journal of Applied Polymer Science. 2004;**92**:2612-2620

[463] Liu R et al. Colloid & Polymer Science. 2009;**287**:627-643

[464] Wang LQ et al. Reactive and Functional Polymers. 2002;**53**:19-27

[465] Tan H et al. Biomaterials. 2009;**30**:2499-2506

[466] Ohya S et al. Journal of Artificial Organs. 2004;**7**:181-186

[467] Gong CY et al. International Journal of Pharmaceutics. 2009;**365**:89-99

[468] Shim WS et al. Biomaterials. 2006;**27**:5178-5185

[469] Hiemstra C et al. Biomacromolecules. 2006;**7**:2790-2795

[470] Jeong B et al. Macromolecules. 1999;**32**:7064-7069

[471] Lau BK et al. Journal of Polymer Science. Part B. 2004;**42**:2014-2025

[472] Park K et al. Key Engineering Materials. 2007;**342-343**:301-304

[473] Nagase K et al. Biomacromolecules. 2008;**9**:1340-1347

[474] Kurata K, Dobashi A. Journal of Macromolecular Science, Part A. 2004;**41**:143-164

[475] Xu C, Kopecek J. Polymer Bulletin. 2007;**58**:53-63

[476] Loo Y et al. Biotechnology Advances. 2012;**30**:593-603

[477] Zhang S. Nature Biotechnology. 2003;**21**:1171-1178

[478] Gabriel SA et al. Science. 2004;**303**:1352-1355

[479] Chen R, Hunt JA. Journal of Materials Chemistry. 2007;**17**:3974-3979

[480] Bokhari MA et al. Biomaterials. 2005;**26**:5198-5208

[481] Hui JH et al. Clinical Orthopaedics. 2013;**471**:1174-1185

[482] Killion JA et al. Journal of Materials Science. 2012;**47**:6577-6585

[483] Ye C et al. Acta Polymerica Sinica. 2012;**10**:1143-1150

[484] Lihong H et al. Macromolecules. 2007;**40**:4429-4438

[485] Yu Q et al. Macromolecules. 2001;**34**:1612-1618

[486] Rouillard AD et al. Tissue Engineering Part C. 2011;**17**:173-179

[487] Ibrahim S et al. Acta Biomaterialia. 2011;**7**:653-665

[488] Kamoun EA, Menzel H. Journal of Applied Polymer Science. 2010;**117**:3128-3138

[489] Choi JS, Yoo HS. Journal of Biomedical Materials Research. Part A. 2010;**95**:564-573

[490] Shu XZ et al. Journal of Biomedical Materials Research. Part A. 2004;**68**:365-375

[491] Lütolf MP et al. Advanced Materials. 2003;**15**:888-892

[492] Overstreet DJ et al. Journal of Polymer Science. Part B. 2012;**50**:881-903

[493] Wang DA et al. Nature Materials. 2007;**6**:385-392

[494] Nishi KK, Jayakrishnan A. Biomacromolecules. 2004;**5**:1489-1495

[495] Maia J et al. Polymer. 2005;**46**:9604-9614

[496] Sperinde JJ, Griffith LG. Macromolecules. 2000;**33**:5476-5480

[497] Ryan BJ et al. Trends in Biotechnology. 2006;**24**:355-363

[498] Moreira Teixeira LS et al. Biomaterials. 2012;**33**:1281-1290

[499] Park KM et al. Biomacromolecules. 2010;**11**:706-712

[500] Fernandez C et al. Carbohydrate Research. 2006;**341**:1253-1265

[501] Amini AA, Nair LS. Journal of Bioactive and Compatible Polymers. 2012;**27**:342-355

[502] Sakai S et al. Acta Biomaterialia. 2009;**5**(2):554-559

[503] Sakai S, Kawakami K. Journal of Biomedical Materials Research. Part A. 2008;**85**:345-351

[504] Wennink JWH et al. Macromolecular Symposia. 2011;**309-310**:213-221

[505] Jin R et al. Tissue Engineering Part A. 2010;**16**:2429-2440

[506] Rowley JA et al. Biomaterials. 1999;**20**:45-53

[507] Berger J et al. European Journal of Pharmaceutics and Biopharmaceutics. 2004a;**57**:19-34

[508] Crompton KE et al. Biomaterials. 2007;**28**:441-449

[509] Patil JS et al. Digest Journal of Nanomaterials and Biostructures. 2010;**5**:241-248

[510] Tessmar JK, Gopferich AM. Macromolecular Bioscience. 2007;**7**:23-39

[511] Baier Leach J et al. Biotechnology and Bioengineering. 2003;**82**:578-589

[512] Young JL et al. Acta Biomaterialia. 2013;**9**:7151-7157

[513] Shu XZ et al. Biomaterials. 2004;**25**:1339-1348

[514] Hwang YJ, Lyubovitsky JG. Biopolymers. 2013;**99**:349-356

[515] Macaya D et al. Advanced Functional Materials. 2011;**21**:4788-4797

[516] Fathima NN et al. International Journal of Biological Macromolecules. 2004;**34**:241-247

[517] Lahiji A et al. Journal of Biomedical Materials Research. 2000;**51**:586-595

[518] Hong Y et al. Acta Biomaterialia. 2007;**3**:23-31

[519] Berger J et al. European Journal of Pharmaceutics and Biopharmaceutics. 2004;**57**:35-52

[520] Leach JB et al. Journal of Biomedical Materials Research. Part A. 2004;**70**:74-82

[521] Jukes JM et al. Tissue Engineering Part A. 2010;**16**:565-573

[522] Jin R et al. Biomaterials. 2007;**28**:2791-2800

[523] Mosahebi A et al. Tissue Engineering. 2001;**7**:525-534

[524] Ballios BG et al. Biomaterials. 2010;**31**:2555-2564

[525] Noth U et al. Cytotherapy. 2005;**7**:447-455

[526] Gao J et al. Biomaterials. 2012;**33**:3673-3681

[527] Hume SL et al. Acta Biomaterialia. 2012;**8**:2193-2202

[528] Liu Y, Chan-Park MB. Biomaterials. 2009;**30**:196-207

[529] Shinohara S et al. Journal of Bioscience and Bioengineering. 2013;**116**:231-234

[530] Yeo Y et al. Biomaterials. 2006;**27**:4698-4705

[531] Sawada T et al. Human Reproduction. 2001;**16**(2):353-356

[532] Rowley JA et al. Biomaterials. 1999;**20**:45-53

[533] Hern DL, Hubbell JA. Journal of Biomedical Materials Research. 1998;**39**:266-276

[534] Kaneko T et al. Macromolecules. 2005;**38**:4861-4867

[535] Nahmias Y et al. Biotechnology and Bioengineering. 2005;**92**:129-136

[536] Odde DJ, Renn MJ. Biotechnology and Bioengineering. 2000;**67**:312-318

[537] Boland T et al. The Anatomical Record. Part A, Discoveries in Molecular, Cellular, and Evolutionary Biology. 2003;**272**:497-502

[538] Mironov V et al. Trends in Biotechnology. 2003;**21**:157-161

[539] Mironov V et al. Biofabrication. 2009;**1**:1-16

[540] Mironov V et al. Current Opinion in Biotechnology. 2011;**22**(5):667-673

[541] Ringeisen BR et al. MRS Bulletin. 2013;**38**(10):834-843

[542] Ozbolat IT, Yin Y. IEEE Transactions on Biomedical Engineering. 2013;**60**(3):691-699

[543] Writers S. http://lifescientist.-com.au/content/biotechnology/news/invetech-helps-bring-bio-printers-to-life-413047968. 2009

[544] CFR. http://www.cfr.org/technology-and-science/3d-printing-challenges-opportunities-international-relations/-p31709. 2013

[545] Quigley JT http://thediplomat.com/2013/08/-chinese-scientists-are-3d-printing-ears-and-livers-with-living-tissue/. 2013

[546] 3ders.org. http://www.3ders.org/articles/20130815-how-do-they-3d-print-kidney-in-china.html. 2013

[547] Littre M. http://www.dailymail.co.uk/sciencetech/article-2637158/Humans-fitted-kidneys-3D-printers.html. 2014

[548] Hoying J, et al. http://www.techrepublic.com/article/new-3d-bioprinter-to-reproduce-human-organs/. 2014

[549] BBC. http://www.bbc.com/news/technology-18677627; 2012

[550] Landers R, Mülhaupt R. Macromolecular Materials and Engineering. 2000;**282**:17-21

[551] Tan W, Desai TA. Biomaterials. 2004;**25**:1355-1364

[552] Albrecht DR et al. Nature Methods. 2006;**3**:369-375

[553] Rzeszutko L et al. Kardiologia Polska. 2014. DOI: 10.5603/KP.a2014.0147

[554] Bertassoni LE et al. Lab on a Chip. 2014;**14**:2202-2211

Chapter 2

Up-to-Date Knowledge and Outlooks for the Use of Metallic Biomaterials: Review Paper

Ildiko Peter, Ladislau Matekovits and Mario Rosso

Additional information is available at the end of the chapter

http://dx.doi.org/10.5772/intechopen.69970

Abstract

In all cases, when a material has to be used in medical applications, the knowledge of its physical, chemical and biological properties is of fundamental significance, since the direct contact between the biological system and the considered device could generate reactions whose long-term effects must be clearly quantified. The class of materials that exhibits characteristics that allow their use for the considered applications are commonly called *biomaterials*. Patients suffering from different diseases generate a great demand for real therapies, where the use of biomaterials are mandatory. Commonly, metallic biomaterials are used because their structural functions; the high strength and resistance to fracture they can offer, provide reliable performance primarily in the fields of orthopedics and dentistry. In metals, because of their particular structure, plastic deformation takes place easier, inducing good formability in manufacturing. The present paper is not encyclopaedic, but reports in the first part some current literature data and perspectives about the possibility of use different class of metallic materials for medical applications, while the second part recalls some results of the current research in this field carried out by the authors.

Keywords: biomaterials, metallic biomaterials, morphology, biocompatibility, implanted antenna

1. Introduction

1.1. Outline

Generally, materials are playing an important role in our society. Apart other materials, biomaterials became key elements for the human well-being. The beginning of biomaterials goes back to thousands of years (the use of metals in dental implants dates back to 200 A.D.), while after World War II, a real expansion of biomaterials can be considered [1].

Actually, biomaterials can be employed for both short-term or permanently, amplifying or partially or completely replacing some organs, tissues, or other elements of the body with the aim to save or to improve the people's well-being. The numbers of the aged people are quickly increasing that in turn requires a greater call to substitute failed body parts with artificial devices made of biomaterials. Additionally, the ongoing tendency to move in the direction of using short-term implants and/or degradable ones constitutes one of the key progresses in biomaterials research. Further important issues are correlated to the possibility to reduce as much as possible implant related infections by means of infection-resistant biomaterials, which usually involves the application of suitable coatings. The use of any biomaterial is influenced by different features, i.e., composition, mechanical strength, rate of degradation (in case of temporary application), growth rate of the human cells, osteoinduction, osteoconduction, etc. [2].

The present review is not encyclopedic and does not completely cover any features about biomaterials, regenerative medicine (RM), and tissue engineering (TE) aspects, but it shows some essential features belonging to such a wide-ranging subject. Some experimental results here presented deal with the outcome of some of the investigations performed by the authors on the development of metallic biomaterials and their possible use for medical implant development. It also includes the description of a multidisciplinary application where a data transmission system is referred to that has been possible to be developed because of the existence of large dimension implanted device. Nowadays, other similar applications are under focus: controlled drug delivery from swallowed capsules by means of external (to the body) control through a communication system that requires the use of an antenna, explicitly a metallic device, is just one of the many applications that could be nominated.

1.2. Metallic biomaterials

Metallic biomaterials show excellent structural functions, superior than ceramic and polymeric biomaterials and for a long time they have been usually employed to substitute unhealthy natural parts or to repair different organs of the human body. Extension of metal made implants was firstly determined by the request to repair bone and later on their application for orthopedic purpose has been increased together with the short-term applications of pins and screws, followed by the use of permanent implants for total joint substitution. Further evolution involves the use of metals and their alloys for dental applications and recently there is a higher tendency for their use in non-conventional reconstructive surgery of organs and hard tissue. There are only some areas where the use of metallic biomaterials is exploited such as joint prostheses, bone fixation plates, pins and screws, dental implants and materials, ocular, contact and intra-ocular lenses, vascular grafts, urinary and intra-vascular catheters, surgical meshes, and voice prostheses.

Within the different class of biomaterials, firstly, *stainless steel* has been used for medical purpose. This is principally due to their low cost, good corrosion resistance conferred by the presence of Cr, allowing the development of the passive and protective oxide layer. **Table 1** reports the composition of the most important/recommended stainless steels used for the manufacturing of orthopedic implants. ASTM F138 and F139 reveal higher biomedical interest because of a better fatigue strength, higher ductility, and better machinability [3], but the high Ni content determines possible toxicity and allergy [4].

Steels (ASTM)	C	Mn	P	S	Si	Cr	Ni	Mo	N	Cu
F138	0.03	2.0	0.025	0.01	0.75	17.0–19.0	13.0–15.0	2.25–3.0	0.10	0.50
F1314	0.03	4.0–6.0	0.025	0.01	0.75	20.5–23.5	11.5–13.5	2.0–3.0	0.20–0.40	0.50
F1586	0.08	2.0–4.25	0.025	0.01	0.75	19.5–22.0	9.0–11.0	2.0–3.0	0.25–0.50	0.25
F2229	0.08	21–24	0.03	0.01	0.75	19.0–23.0	0.10	0.50–1.50	0.90	0.25

Table 1. Composition (wt%) of some austenitic stainless steels [5].

However, according to some research, e.g., [6], stainless steels can suffer pitting, crevice, corrosion fatigue, stress corrosion cracking, and galvanic corrosion within the human body accelerated by their reduced wear resistance. Additionally, appearance of allergic reaction limits their use as orthopedic joint prosthesis [7, 8]. Release of metallic ions, as corrosion product, in the human body can reduce the lifetime of the implant leading to an additional surgical procedure [9–12]. If high amount of Ni ions is released in the tissues, development of the most common contact allergy [13, 14] and cancer [15, 16] can be favored.

The corrosion resistance of stainless steels can be modified by the (i) different oxygen concentration (generally, low oxygen content contributes to a higher corrosion rate, because the growth of the protective oxide layer is delayed on the surfaces of the metallic implant), (ii) pH value which can be altered from a neutral condition to lower values due to the inflammatory cell secretions, and (iii) presence of aqueous ions level in the surrounding body fluid. If possible, the concentration of the released metal ions has to be as reduced as possible, and it has to be innocuous for the human body during a long service period [17]. The high elastic modulus of stainless steel is considered as one of the highest difficulties leading to the damage of the bone by stress shielding which is directly connected to the failure of the implant.

During the time, *Cobalt-Chromium alloys* have been developed, which reveal some superior properties and some drawbacks relating to their production costs than stainless steels [18, 19].

The alloys belonging to this family show high corrosion resistance attributable to the natural growth of the passive oxide film on the top of the metallic device which is resistant also in the chloride-rich background [20–23].

Generally, two basic types of such alloys are commonly employed:

- Cobalt-Chromium-Molybdenum (CoCrMo) alloy and
- Cobalt-Nickel-Chromium-Molybdenum (CoNiCrMo) alloy.

Table 2 reports the composition of the most important alloys belonging to these classes used for the production of orthopedic implants.

CoCrMo has been used for many years in dental applications and for the manufacturing of artificial joints, while for prosthetic stem productions, for hip and knee joints the alloy containing Ni has been employed [21]. CoCrMo alloy, because of its high corrosion and wear resistance, is used for joint prostheses for the femoral head in combination with an ultra-high molecular weight polyethylene cup.

Alloys (ASTM)	Cr	Mo	Ni	Fe	C	Si	Mn	W	P	S	Other
F15	27–30	5–7	1.0	0.01	0.35 max	1.0	1.0	0.2	0.02	0.01	0.25N 0.3Al 0.01B
F799 (low C)	26–30	5–7	1.0	0.01	0.05	1.0	1.0	-	-	-	0.25N
F799 (high C)	26–30	5–7	1.0	0.01	0.25	1.0	1.0	-	-	-	0.25N
F563	18–22	3–4	15–25	0.01	0.05	0.5	1.0	3.0–4.0	-	0.01	0.50–3.5 Ti
F562	19–21	9–10.5	33–37	0.01	0.25 max	0.15	0.15	-	0.015	0.01	1.0 Ti
F90	19–21	-	9–11	0.01	0.05–0.15	0.40	1.0–2.0	14–16	0. 04	0.03	-
F1058	19–21	6-8	14–16	0.01	0.15	1.2	1.0–2.0	-	0.015	0.015	0.10 Be 39.0-41.0 Co

Table 2. Composition (wt%) of some CoCrMo and Co-Ni alloys [5].

However, Ni and Co ions, when released, are susceptible to induce allergic responses: Ni is carcinogenic and causes a toxicity problem, which can be avoided using an as much as possible low level of Ni.

Actually, *Ti and its alloys* have been received much more attention, even if they have been used since the late 1960s. Due to the fact that Ti is completely inert and resistant to corrosion in any fluids in the human body and tissues, these alloys exhibit the best biocompatibility among metallic biomaterials combined to good fatigue resistance, excellent in vivo corrosion resistance which is acquired by the growth on their surface of the stable passive oxide layer. In addition, showing some properties which are close to the human bones, i.e., strength, density, and relatively low elastic modulus, these alloys began to be extensively employed for joint replacements. Lower elastic moduli than other metallic biomaterial allow obtaining less stress-shielding phenomenon. Generally, Ti and its alloys are used for bone fixation, joint substitution, pacemakers, implants for dental, artificial heart valves, components and stents in rapid blood centrifuges due to their chemical stability and particular high strength [24]. Central position has been made for a commercially *pure Ti* (cp-Ti, ASTM F67), *Ti6Al4V alloy* (ASTM F136) strengthened by alloys obtained by the modification of them and *Ti-based shape memory alloys* (SMA) [25, 26]. Cp-Ti is typically used in dental and spinal surgery, Ti6Al4V for the manufacturing of artificial knee, hip and shoulder joints and for bone fixators, while SMAs are employed as orthodontic equipment and temporary spine and long bones fixations.

According to ISO 5832-2, four grades of cp-Ti are recognized for medical applications. Cp-Ti can contain a minor level of interstitial elements (O, N, and H) affecting the mechanical properties by interstitial solid solution strengthening. Cp-Ti is available in four grades, where (i) grade I corresponds to the lowest O content and yield strength and at the same time to the highest ductility and (ii) grade IV reveals the highest O content and the highest strength combined with the lowest ductility. **Table 3** reports the grades of different cp-Ti with the maximum limits of the interstitial elements and some mechanical properties. Due to the capacity for stimulating fast osseointegration, coming from the incorporation of OH^- ions inside the passive TiO_2 layer and due to the reaction of the hydroxylated surface area with the inorganic phase constituents (prevalently $Ca^{2+}$and $(PO_4)^{3-}$) present in the bone, cp-Ti is principally used for endosseous dental implants manufacturing.

Ti6Al4V alloy reveals outstanding mechanical strength, high biocompatibility and consents good implant-bone integration. However, over the years, some weaknesses during the use of such alloy have been noticed: (i) it has higher elastic modulus than bone, producing stress-shielding effect, (ii) it shows lower wear resistance too, and (iii) the release of Al and/or V can determine permanent diseases, i.e., osteomalacia and Alzheimer's disease [27].

The research community has dedicated high attention to these problems, and the trend is to develop new Ti alloys, by alloying the basic alloy and reducing as much as possible or even totally removing the release of any toxic elements. In this scenario, alternatives alloy with ($\alpha + \beta$) structure, without V have been obtained, i.e., Ti6Al7Nb, Ti5Al2.5Fe, Ti6Al6Nb1Ta, and Ti5Al3Mo4Zr, which at the beginning have been developed for airspace application. These alloys reveal higher corrosion resistance, fatigue properties than other biomaterials, principally governed by the size and the distribution of the α and β phases [28]. Such structure can be obtained during mechanically working till to obtain the preferred form followed by fast cooling at room temperature and annealing for recrystallization within the ($\alpha + \beta$) two-phase field and determine a fatigue strength more than 650 MPa.

β-Ti and near β-Ti alloys, i.e., Ti12Mo6Zr-2Fe, 35.5Nb7.3Zr5.7Ta, and Ti13Nb13Zr, introduced in the 1990s and their continuous improvement have been one of the leading topics of orthopedic materials research, because they contain superior amount of β-stabilizing elements and they are characterized by a considerably lower elastic modulus than other biometals (44–51 GPa for water-quenched and cold-worked Ti-13Nb-13Zr alloy compared to 110 GPa for Ti6Al4V). Additionally, good formability, high hardenability, excellent corrosion resistance, and better notch sensitivity (measure of how sensitive a material to notches or geometric discontinuities is) than the ($\alpha + \beta$) Ti alloys characterize such alloys. In order to amplify the wear resistance, surface hardening can be carried out by aging in an oxygen-rich environment

Ti grade	O	N	H	σ_{yield} (MPa)	$\sigma_{ultimate}$ (MPa)	%E
1	0.18	0.03	0.015	170	240	24
2	0.25	0.03	0.015	275	345	20
3	0.35	0.05	0.015	380	450	18
4	0.40	0.05	0.015	483	550	15

Table 3. Maximum limits of the interstitial elements and mechanical properties of different cp-Ti grades.

determining the growth of a hard oxide film on the surface with an interstitial solid solution strengthening of the subsurface section atmosphere [29].

A class of special Ti alloys, developed in Japan, with exclusive physical-mechanical properties and large range of possibilities in medical applications are Gum alloys, belonging to the β-type of Ti alloys, fundamentally expressed as Ti (Ta, Nb, and V) + (Zr, Hf, and O). These alloys have excellent mechanical behavior at room temperature: an ultra-low Young's modulus (60–70 GPa) and a non-linear elastic behavior, an extended elastic limit, ultra-high strength (>1 GPa), superplastic-like deformability, Invar-like thermal expansion, and Elinvar-like thermal dependence of the elastic modulus making available several prospects for their application [30]. Such alloy, with no hazardous elements, combines extremely low elastic modulus with extremely high strength and can be considered a perfect candidate for many medical implants.

The capacity to return to the initial memorized shape varying the temperature is the most important features of *shape memory alloys*. Such a particular characteristic can be exploited in critical medical application, when recovering the original shape after large deformations induced by mechanical load and for conserving the deformed shape up to the heat-induced recovery of the original shape has primary importance [31]. Generally, in orthopedic, dental, and cardiovascular uses NiTi alloy (Nitinol) is employed because its good corrosion resistance and pseudoelastic property (reversible elastic response to an applied stress). However, the study [32] reports that at temperatures equal to the human body the formation of a passive layer on NiTi is not as much protective than the passive layer on Ti6Al4V. Additionally, there are some limitations too in the widespread use of such alloy because even if some studies have revealed that NiTi alloy is biocompatible [31, 33, 34], the possible release of Ni ions represents a serious apprehension [13, 15] because Ni is highly allergenic element, and it can induce intense inflammatory reactions.

Zr, Ta, and Nb have been employed as constituent elements of Ti alloys for medical applications because of their good biocompatibility and high corrosion resistance. Zr belongs to the same group as Ti and exhibits similar properties [35]. Recently, Zr-Nb [36, 37] and Zr-Mo [38] have been developed. Ta and Nb are non-toxic elements, and they exhibit very similar physical and chemical properties. Ta exhibits excellent chemical stability and good biocompatibility similar to that of Ti:Ta has been employed in dental and orthopedic applications such as radiographic bone markers, vascular clips, and as a material for cranial defect repair and nerve repair since the 1940s [39, 40]. In particular, porous Ta has been developed for bone in-growth applications, i.e., hip and knee arthroplasty, spinal surgery, and bone graft substitutes [41, 42]. Introduction of Ta in TiNi shape memory alloy determines a low incidence of artefacts, according to [43, 44], the blood compatibility of Ta is higher than that of other metals.

Recently, among other research, *biodegradable metals and their alloys* have attracted high interest, because of their possibility of degradation in biological surroundings. The main benefit of this class of materials is related to the fact that after the tissue has appropriately restored and they are no longer useful, the removal of follow-up surgery is avoided allowing higher protection since physical irritation and chronic inflammatory local reactions are totally excluded [45–47].

Fe, Mg, and Zn are all vital nutritional elements for a healthy body and generally reveal excellent biocompatibility in the human body, with no any sign of local or systemic toxicity biodegradable materials than polymers for load-bearing applications. Both pure Fe and pure Mg have been reported to possess excellent biocompatibility in the human body and show no signs of local or systemic toxicity [45, 48, 49].

Mg and its alloys (i.e., (AZ91, WE43, AM50, and LAE442) can be considered as favorable alloys for medical applications, and they have already been positively verified *in vivo* and in clinical studies and it has been reported that development of new bone is facilitated [50–52], when they are implanted as bone fixtures. The elastic modulus of Mg (41–45 GPa) is closer to that of natural bone (3–20 GPa). However, the mechanical properties of such alloys, i.e., strength and the elongation to fracture are not always acceptable which combined to the development of significant quantities of hydrogen limits their widespread use. Some studies [53–55] have reported as alternative solution the use of Fe-based alloys. However, the implants realized using such alloys reveal some reactions like those developed in long-term uses which is predominantly due to the low-degradation rate of pure Fe in biological atmosphere [56].

In Refs. [53, 54] the authors have developed Fe35Mn alloy with a higher degradation rate compared to the pure Fe; however, compared to Mg-based alloys, the interested value is still at least one order of magnitude inferior.

Even with the significant evolution carried out over the time on the implants made of by biodegradable alloys, many tasks are still unexplained.

1.3. Biomaterials in tissue engineering and regenerative medicine

Current evolutions of biomaterials and the constantly increasing demand for new technologies have transformed the field of TE and RM too. In such areas, biomaterials have received more and more attention and can be placed in pole position for repairing tissue functions by joining materials design and engineering with cell therapy. RM is considered new frontline of medical research; however, the idea of generating synthetic tissues is not so recent, since it goes back to 1938 [57]. Biomaterials can offer physical sustain for engineered tissues and can control and guide the cells. RM involves the replacement or restoration of human cells, tissue or organs, to repair or create typical function [58, 59].

Biomaterials engineering involves production, processing, characterization, and knowledge of innovative materials, including metallic ones. The most important limitations of RM is related to the generally health-care programs: the inclination to substitute conservative methodologies with new therapies is not so immediate. A large use of RM would be possible when a deep knowledge related to all connecting aspects (positive and negative outcomes) will be absolutely solved [60]. For the use on large scale RM, the ideas developed at laboratory level have to be inserted and totally transformed into widespread commercial products. In the past century, achievements in medicine have included different approaches and fundamentally, three different lines can be really able to follow the objective of RM: (i) cell-based therapy (meaning of introducing new and healthy cells in pathologic tissues), (ii) use of both biological or synthetic materials (use of cells and extra cellular matrix providing structural and functional support), and (iii) implantation of scaffolds seeded with cells (mixture of the earlier mentioned two possibilities) [61].

Generally, metallic ions have a significant role for some day-to-day living biological processes, being essential for many biochemical reactions: i.e. K, Na, Ca, Mg. Ca^{2+} ions level governs the intra- and inter-cellular communications, blood coagulation and muscle contraction, Mg^{2+} ions support photosynthesis, the electron transfer routes are generally based on Fe proteins, transport of oxygen requires Fe and Cu containing proteins, Zn has the main role in regulation of DNA transcription are only some cases which highlight the importance of them [62–66].

In addition, metallic ions have interesting capacity and can be exploited as therapeutic agents in tissue engineering. There are a large number of metallic ions which are vital cofactors of enzymes (Co, Cu, Fe, Mn, Zn, Ag, Sr, V, and Ga), and their specific properties can be correlated to their use for therapeutic purpose. The geometry and valence of metal ions, their hydrolytic and redox activity, Lewis acidity, radiochemical properties directly involve fast kinetics, higher affinity to interact with other ions with high flexibility to be inserted in engineered biomaterials allowing the modification/revision of cellular functions and their metabolism. Actually, there is a continuous growth in medical field of the use of metal ions in RM and tissue engineering, with a special attention concerned to their therapeutic properties. However, the possible toxicity of metal ions in case of their eventually local release has to be carefully considered, when exploiting them [67–70].

2. Real case study

Some fundamental aspects related to metallic biomaterials and their use in RM and TE are integrated with some experimental results obtained by the authors [71–75] during the years spent studying the development and characterization of metallic materials for medical purpose. In particular, some results, (1) prevalently related to the corrosion resistance, which can be directly associated with their biocompatibility of some (i) Ti-enriched Co-based traditional alloys to be submitted for dental application and (ii) TiNb alloys for load-bearing implant production will be reminded here; (2) the third line (iii) involves the possibility of using an implanted two-element antenna array placed on a metallic cylinder (metal made implant) embedded in a polymer dielectric that acts as the substrate between the ground plane and radiators with the purpose to use it for in-body communications.

2.1. Some details and the corrosion resistance of the experimentally prepared biometals

Cold crucible levitation melting technique with induction heating system has been used for the metallic alloys manufacturing, which guarantee high purity melts, important features for materials used for surgical implants. Enhancement with Ti has been performed in the case of the Co-based basic alloy. **Table 4** reports the chemical compositions of the alloys for dentistry, while **Table 5** reports the composition of the alloy used for load-bearing implant production.

Alloys	Chemical composition (wt%)					
	Cr	Mo	Si	Ti	Other	Co
CoCrMoTi4	26.5	6.0	1.0	4.0	1.5	bal.
CoCrMoTi6	28.0	4.5	1.0	6.0	0.5	bal.

Table 4. Chemical composition of the experimentally developed alloys.

Alloy	Ti	Nb	Ta	Zr	O
TiNb	71.4	24	1	3	0.6

Table 5. Chemical composition (at %) of the Ti-Nb alloy.

Wirobond®280 (Co60.9, Cr25, Mo4.8, W6.2, Ga2.9) and Ti6Al4V have been considered as reference material. The microstructures of the alloys investigated are reported in **Figure 1**, where a relatively homogeneous structure has been observed. Major details are reported in Refs. [71–73].

Following some structural and microstructural characterization (**Figure 1** reports the general morphology of the investigated alloys) combined with some mechanical properties evaluation, which can be found in [71–73], some words are spent about the corrosion behavior, verified by static immersion test, of the metallic alloys prepared, because such line is directly associated to the biocompatibility behavior of the metal alloys. The test has been carried out according to the route specified in the Standard ISO 10271/2011 at 37°C (±1°C). According to the standard, immersion of the samples has been realized in acid solution (7.5 ml lactic acid, 5.85 g NaCl, 300 ml H_2O di grade 2 purity, and 700 ml H_2O) simulating the biological environment. The pH value has been corrected to arrive to the neutral situation. The samples have been monitored after the permanence in the acid solution for 28 days. As a reference solution, an acid solution maintained in the same condition without any metallic alloy inside has been used. The sample surfaces, following corrosion test, have been investigated by scanning electron microscopy

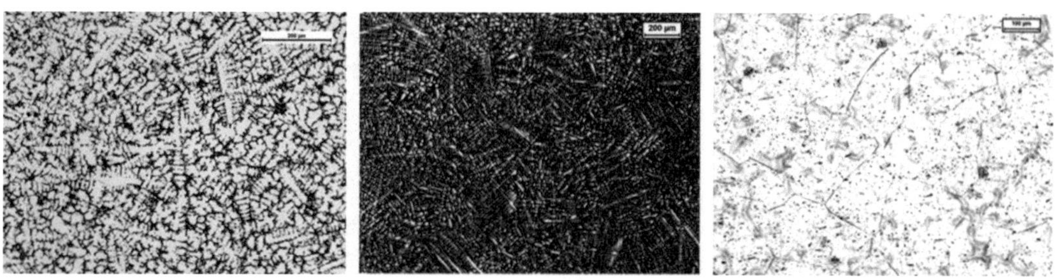

Figure 1. Optical micrographs of the investigated alloys.

analysis. The hydrogen ions measurements in the solution with the metallic alloys inside and the weight loss measurements are directly correlated to damage arisen and indicate the progress of the corrosion which involves the metallic alloys. After a regular time, the weights of the metallic samples have been measured, and the pH values have been monitored. Comparison of the results obtained for the solution with and with no metallic alloys inside has been carried out. The investigated alloys do not present any significant weight alteration after 28 days and no significant release of metal ions was observed (**Figure 2**). The same results have been confirmed by the pH value measurements (**Figure 3**). When the dissolution of the passive film developed on the surface of the alloy has a very low rate, the layer is able to protect the alloy. The release of metal ions is lower than 0.1% within the week, and this is below the threshold level indicated in the Standard (according to the Standard ISO 10271/2011 the allowed deviation from the original state could be only 1%).

Microstructural observation and compositional analysis after corrosion test confirm the analytical results, since on the surface of the alloys extracted from the physiological solution, no evidence of corrosion product has been found. The experimental alloys are used for some crown (**Figure 4** left) and load bearing implant (**Figure 4** right) manufacturing. The research in this direction is ongoing to further investigate additional performance of the alloys as much as possible in functional environment.

2.2. Some details about the use of the implant itself as ground plane for implanted antenna

Biotelemetry, increasingly exploited in the recent healthcare systems, involves the application of telemetry in medicine and health care, allowing remote checking of various vital functions

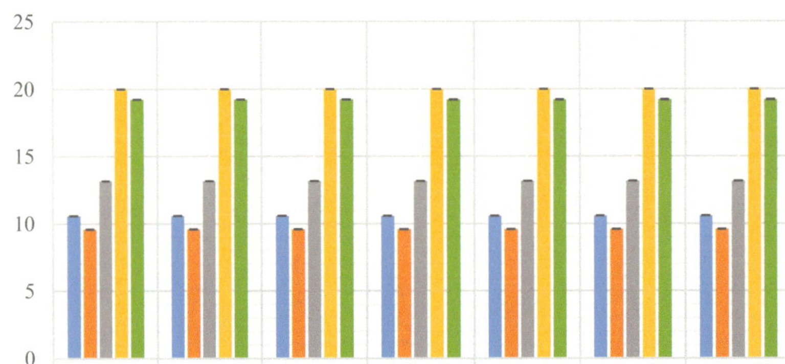

	Initial	After 24 h	After 48 h	After 7 days	After 15 days	After 21 days	After 28 days
CoCrMo6Ti6	10.588	10.588	10.588	10.588	10.588	10.588	10.588
CoCrMo6Ti4	9.5857	9.5857	9.5856	9.5855	9.585	9.585	9.5846
TiNb	13.1669	13.1638	13.16	13.16	13.16	13.16	13.1581
Ti6Al4V	20.00	20.00	20.00	20.00	20.00	20.00	20.00
Wirobond	19.2316	19.2316	19.2316	19.2205	19.2187	19.2185	19.2181

CoCrMo6Ti6 CoCrMo6Ti4 TiNb Ti6Al4V Wirobond

Figure 2. Weight loss measurement evaluation results during 28 days of investigation.

http://dx.doi.org/10.5772/intechopen.69970

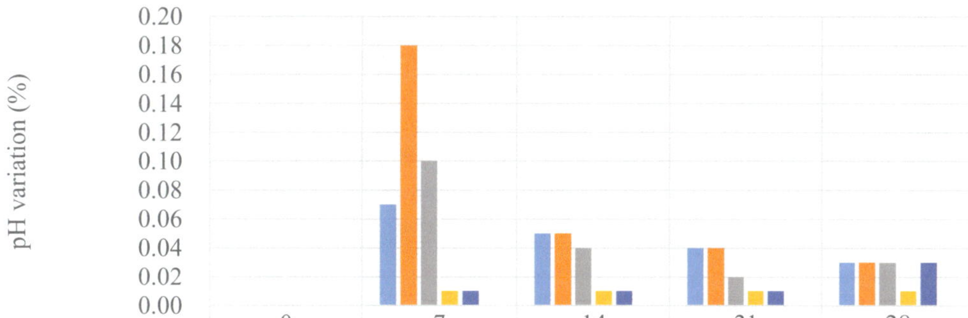

	0	7	14	21	28
CoCrMoTi4	0.00	0.07	0.05	0.04	0.03
CoCrMoTi6	0.00	0.18	0.05	0.04	0.03
Wirobond®280	0.00	0.10	0.04	0.02	0.03
TiNb	0.00	0.01	0.01	0.01	0.01
Ti6Al4V	0.00	0.01	0.01	0.01	0.03
Ref.sol.	0.00	0.00	0.00	0.00	0.00

Figure 3. Variation of the pH value during 28 days of investigation.

(i.e., temperature, blood pressure, and cardiac beat) in patients. The use of such technologies, firstly, can considerably decrease hospitalization time and then allows the constant collection of the patient's biological data. When the information has to be obtained from inside the body, it should be firstly revealed by a dedicated sensor and in a second step to be transmitted to an external (to the body) receiver. This operation requires the use of antennas that are made of conductive materials, i.e., metal. Such devices are in a direct contact with the different tissues, so special attention to their biocompatibility has to be paid. On the other hand, to increase the efficiency of the transmission, directive antennas are to be used, which implicitly means a large dimension. Such space is not always available inside the body. An alternative solution consists of using low-profile printed antennas, but they require the use of a large ground

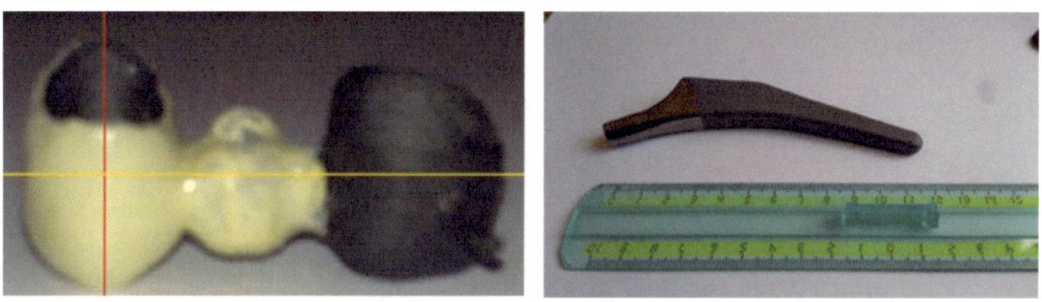

Figure 4. Photographs of the produced elements for medial application.

plane. A recently proposed solution by the author consists of the use of the implant itself as ground plane for the antenna. The rectangular radiator is made of a similar metal as the implant, i.e., biocompatible, and it is conformal to the cylindrical bone structure. The reported results in [74] refer to the case when a piece of bone is fully substituted by a biometallic cylinder. To maintain the distance between the ground plane and radiator constant, a biocompatible polymer, namely polydimethylsiloxane (PDMS), has been considered both as dielectric and superstrate. The geometry is presented in **Figure 5**.

The thickness of each of the two PDMS layers is of $h_{PDMS} = 2$ mm with respect to the radius of the overall geometry of 106 mm (similar to an adult arm). Details on the dielectric properties of the tissues and materials involved in the numerical experiment can be found in [74].

As a result, a wide beam, approximately 100° half power beam width (HPBW) radiation pattern in the orthogonal to the bone axis plane has been obtained. Such characteristics allow a significant freedom to the patient to move without losing the communication link between the implanted antenna and fixed external base station.

However, considering the low efficiency, basically due to the lossy medium, and of the deep positioning of the antenna (with respect to subcutaneous solutions), and exploiting the low profile configuration, as a second step two radiators have been inserted, aiming to realize a two-element array with scanning possibility. It should be mentioned that this solution does not require additional space, since the second radiator is embedded in the already present

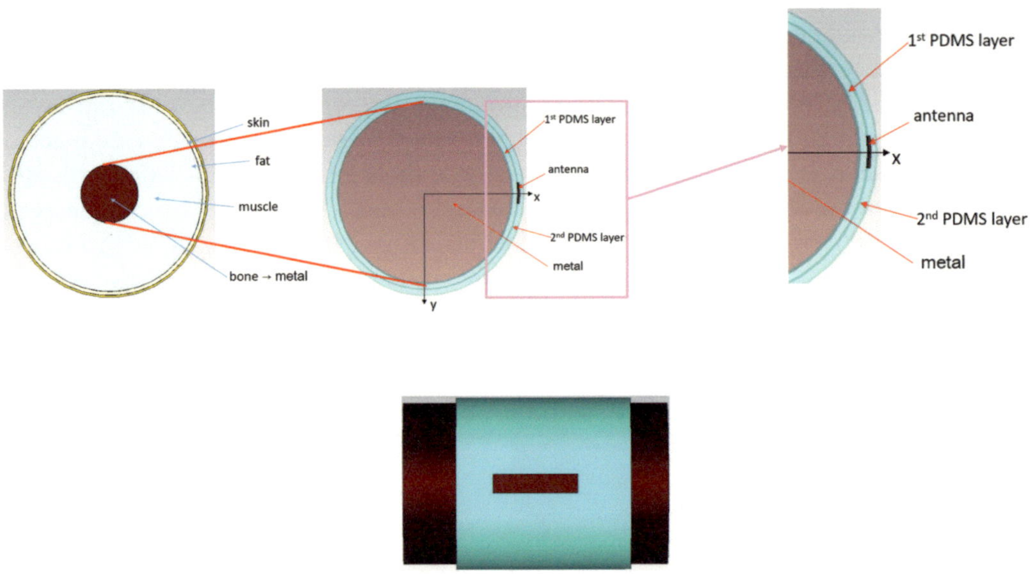

Figure 5. Numerical adult leg model: the bone is substituted by an equal diameter cylinder (left); the conformal antenna is positioned on a constant thickness PDMS dielectric layer of ring shape around the biometallic implant (central), side view of the structure (with removed tissues and outer PDMS layer for a better rendering (right).

PDMS layer. Because of the available space, different configurations have been tested to reduce the mutual coupling between the two radiators. The parametric study allows quantifying the coupling for different angular distances between the two radiators in the presence and absence of the coupling reduction elements. In all cases, metallic parts made of the same biometal have been considered. Due to the low coupling a more directive, i.e., higher gain, system can be built. The higher gain associated to the possibility to scan the beam further increases the freedom of movement of the patient the system is implanted in. All these features could be obtained at a very low extra space demand, basically due to the exploitation of the metallic implant as ground plane of the planar antennas.

From the manufacturing point of view, the proposed solution, i.e., antenna and coupling reduction structure, can be implemented on the biometallic cylinder before it is implanted. Moreover, the metallic cylinder could also incorporate the necessary electronics, sensors, beam forming network, etc. and power supply (batteries).

Similar solution using two dipole antennas has also been discussed in Ref. [75] where conical implant has been considered. The presence of the bone surrounding the implant eliminates the use of any additional materials, i.e., the PDMS layers, in the previous case. In particular, two orthogonal dipoles in turn-style like configuration have been considered. This solution allows to generate a circularly polarized radiated field, further increasing the freedom of movement of the patients.

3. Conclusions

The aim of this chapter was twofold. The first section was dedicated to offering a short outlook and to demonstrating on how metals and their alloys can be used for medical purpose focusing on their role and their influence on the surrounding body environment. The chapter has illustrated some significant features which are appropriate to such a widespread topic. Secondly, some experimental results achieved by the authors over time and illustrated here were proposed to partially support the theoretical line by providing a real case study.

In particular, some results about the biocompatibility of Ti enriched Co-based traditional alloys and TiNb alloy for medical implant production and the possibility of using an implanted two-element antenna array placed on a metal made implant inserted in a polymer dielectric acting as substrate among the ground plane, and radiators with the purpose to use it for in-body communications were highlighted.

There is a continuous need to obtain a deep understanding about the roles and the effect of the metallic ions in body environment. The interdisciplinary character of such research was pointed out with the aim to encourage the teamwork between material scientists, electromagnetic engineers, biologist, tissue engineers, and biomedical and medical researchers to develop new biomaterials and to acquire sufficient data about biological processes as a function of foreign biomaterial introduced in the human body.

Author details

Ildiko Peter[1]*, Ladislau Matekovits[2] and Mario Rosso[1]

*Address all correspondence to: ildiko.peter@polito.it

1 Department of Applied Science and Technology, Politecnico di Torino, Torino, Italy

2 Department of Electronics and Telecommunications, Politecnico di Torino, Torino, Italy

References

[1] Temenoff JS, Mikos AG. Biomaterials: The Intersection of Biology and Materials Science. Upper Saddle River, NJ [u.a.] Pearson Prentice Hall, 2008

[2] Yarlagadda P, Chandrasekharan M, Shyan JYM. Recent advances and current developments in tissue scaffolding. Bio-Medical Materials and Engineering. 2005;**15**:159-177

[3] Özbek I, et al. Characterization of borided AISI 316L stainless steel implant. Vacuum. 2002;**65**(3-4):521-525

[4] Singh R, Dahotre N. Corrosion degradation and prevention by surface modification of biometallic materials. Journal of Materials Science: Materials in Medicine. 2007;**18**(5):725-751

[5] Davis JR, editor. Metallic materials. In: Handbook of Materials for Medical Devices. Materials Park, OH: ASM International; 2003

[6] Kilner T, Pilliar RM, Weatherly GC, Allibert C. Phase identification and incipient melting in a cast co-Cr surgical implant alloy. Journal of Biomedical Materials Research. 1982;**16**:63-79

[7] Mishra AK, Hamby MA, and Kaiser WB. Metallurgy, microstructure, chemistry and mechanical properties of a new grade of cobalt-chromium alloy before and after porous-coating. In: Disegi JA, Kennedy RL, Pilliar R, editors. Cobalt-Base Alloys for Biomedical Applications, ASTM STP 1365. West Conshohocken, PA: American Society for Testing and Materials; 1999. pp. 71-88

[8] Jones DA. Principles and Prevention of Corrosion. Vol. 1. Upper Saddle River, NJ: Prentice Hall; 1996; Yu J, Zhao Z, Li L. Corrosion fatigue resistances of surgical implant stainless steels and titanium alloy. Corrosion Science. 1993;**35**(1):587-597

[9] Okazaki Y, Gotoh E. Comparison of metal release from various biocompatible metals in vitro. Biomaterials. 2005;**26**(1):11-21

[10] Jacobs JJ, et al. Metal release in patients who have had a primary total hip arthroplasty. A prospective, controlled, longitudinal study. The Journal of Bone and Joint Surgery. 1998;**80**(10):1447-1458

[11] Jacobs JJ, et al. Metal release and excretion from cementless titanium alloy total knee replacements, Clinical Orthopaedics and Related Research. 1999;**358**:173-180

[12] Brayda-Bruno M, et al. Evaluation of systemic metal diffusion after spinal pedicular fixation with titanium alloy and stainless steel system: A 36-month experimental study in sheep. The International Journal of Artificial Organs. 2001;**24**(1):41-49

[13] Hanawa T. Metal ion release from metal implants. Materials Science and Engineering: C. 2004;**24**:745-752

[14] Vahter M, Berglund M, Akesson A, Liden C. Metals and women's health. Environmental Research. 2002;**88**:145-155

[15] IARC Monographs on the Evaluation of Carcinogenic Risks to Humans, Surgical Implants and Other Foreign Bodies. Vol. 74. Lyon: World Health Organization (WHO) International Agency For Research On Cancer; 1999. p. 65

[16] Bal W, Kozoowski H, Kasprzak K. Promising in-vitro performances of nickel-free nitrogen containing stainless steels for orthopaedic applications. Journal of Inorganic Biochemistry. 2000;**79**:213-221

[17] Sumita M, et al. Failure Processes in biometallic materials. Comprehensive Structural Integrity. 2003;**9**:131-167

[18] Ong K, Kurtz SM, Lau E, Parvizi J, Mont MA, Manley MT. Migratory patterns of primary total hip and knee patients influence hospital utilization. Poster No. 951, 60th Annual Meeting of the Orthopaedic Research Society, New Orleans, LA, March 15-18, 2014

[19] Thomann J. Uggowitzer, Wear- corrosion behaviour of biocompatible austhenitic stainless steel. Wear. 2000;**239**:48-58

[20] Navarro M, et al. Biomaterials in orthopaedics. Journal of the Royal Society Interface. 2008;**5**(27):1137-1158

[21] Alvarado J, et al. Biomechanics of hip and knee prostheses1. 2003. Applications of Engineering Mechanics in Medicine, GED – University of Puerto Rico Mayaguez. December 2003

[22] Vidal CV, Muñoz AI. Effect of thermal treatment and applied potential on the electrochemical behaviour of CoCrMo biomedical alloy. Electrochimica Acta. 2009;**54**(6):1798-1809

[23] Oztürk O, Türkan U, Eroglu AE. Metal ion release from nitrogen ion implanted CoCrMo orthopedic implant material. Surface and Coatings Technology. 2006;**200**(20):5687-5697

[24] Park JB, Lakes RS. Hard tissue replacement-II: Joints and teeth. Biomaterials. 2007:**2**(3) 395-458

[25] Kocich R, Szurman I, Kursa M, Fiala J. Investigation of influence of preparation and heat treatment on deformation behaviour of the alloy NiTi after ECAE. Materials Science and Engineering: A. 2009;**512**:100-104

[26] Thomasová M, Seiner H, Sedlák P, Frost M, Ševcík M, Szurman I, et al. Evolution of macroscopic elastic moduli of martensitic polycrystalline NiTi and NiTiCu shape memory alloys with pseudoplastic straining. Acta Materialia. 2017;**123**:146-156

[27] Walker PR, LeBlanc J, Sikorska M. Effects of aluminum and other cations on the structure of brain and liver chromatin. Biochemistry. 1989;**28**(9):3911-3915

[28] Pilliar RM, Weatherly GC. Developments in implant alloys. CRC Critical Reviews in Biocompatibility. 1986;**1**:371-403

[29] Mishra AK, Davidson JA, Poggie RA, Kovacs P, Fitzgerald TJ. Mechanical and tribological properties and biocompatibility of diffusion hardened Ti-13Nb-13Zr – A new titanium alloy for surgical implants. In: Medical Applications of Titanium and its Alloys: The Material and Biological Issues ASTM STP1272. 1996. pp. 96-113

[30] Furuta T, Kuramoto S, Hwang J, Nishino K, Saito T. Elastic deformation behavior of multi-functional Ti–Nb–Ta–Zr–O alloys. Materials Transactions. 2005;**46**(12):3001-3007. DOI: 10.2320/matertrans.46.3001

[31] Petrini L, Migliavacca F. Biomedical applications of shape memory alloys. Stress. Hindawi Publishing Corporation Journal of Metallurgy. 2011;**2011**(2011): pp 1-15. DOI:10.1155/2011/50148

[32] Rondelli G. Corrosion resistance tests on NiTi shape memory alloy. Biomaterials. 1996;**17**(20):2003-2008

[33] Golish SR, Mihalko WM. Principles of biomechanics and biomaterials in orthopaedic surgery. Instructional Course Lectures. 2011;**60**:575

[34] Henderson E, Buis A. Nitinol for prosthetic and orthotic applications. Journal of Materials Engineering and Performance. 2011;**20**(4):663-665

[35] Kobayashi E, Doi H, Yoneyama T, Hamanaka H, Matsumoto S, Kudaka K. Evaluation of mechanical properties of dental-cast Ti–Zr based alloys. Journal of Dental Materials. 1995;**14**:321-328

[36] Nomura N, Tanaka Y, Suyalatu, Kondo R, Doi H, Tsutsumi Y, et al. Effects of phase constitution of Zr–Nb alloys on their magnetic susceptibilities. Materials Transactions. 2009;**50**:2466-2472

[37] Kondo R, Nomura N, Suyalatu, Tsutsumi Y, Doi H, Hanawa T. Microstructure and mechanical properties of as-cast Zr–Nb alloys. Acta Biomaterialia. 2011;**7**:4278-4284

[38] Suyalatu, Kondo R, Tsutsumi Y, Doi H, Nomura N, Hanawa T. Effects of phase constitution on magnetic susceptibility and mechanical properties of Zr-rich Zr–Mo alloys. Acta Biomaterialia. 2011;**7**:4259-4266

[39] Findlay DM, Welldon K, Atkin GJ, Howie DW, Zannettino ACW, Bobyn D. The proliferation and phenotypic expression of human osteoblasts on tantalum metal. Biomaterials. 2004;**25**:2215-2227

[40] Levine BR, Sporer S, Poggie RA, Valle CJD, Jacobs JJ. Experimental and clinical performance of porous tantalum in orthopedic surgery. Biomaterials. 2006;**27**:4671-4681

[41] Bobyn JD, Stackpool GJ, Hacking SA, Tanzer M, Krygier JJ. Characteristics of bone in growth and interface mechanics of a new porous tantalum biomaterial. The Journal of Bone and Joint Surgery. 1999;**81B**:907-914

[42] Bobyn JD, Poggie RA, Krygier JJ, Lewallen DG, Hanssen AD, Lewis RJ, et al. Clinical validation of a structural porous tantalum biomaterial for adult reconstruction. The Journal of Bone and Joint Surgery. 2004;**86**:123-129

[43] DePalma VA, Baier RE, Ford JW, Gott VL, Furuse A. Investigation of the surface properties of several metals and their relation to blood compatibility. Journal of Biomedical Materials Research. 1972;**6**:37-75

[44] Mani G, Feldman MD, Patel D, Agrawal CM. Coronary stents: A materials perspective. Biomaterials. 2007;**28**:1689-1710

[45] Staiger MP, Pietak AM, Huadmai J, Dias G. Magnesium and its alloys as orthopedic biomaterials: A review. Biomaterials. 2006;**27**:1728-1734

[46] Fare S, Ge QA, Vedani M, Vimercati G, Gastaldi D, Migliavacca F, et al. Evaluation of material properties and design requirements for biodegradable magnesium stents. Materia-Brazil. 2010;**15**:103-112

[47] Moravej M, Mantovani D. Biodegradable metals for cardiovascular stent application: Interests and new opportunities. International Journal of Molecular Sciences. 2011;**12**:4250-4270

[48] Schinhammer M, Hanzi AC, Loffler JF, Uggowitzer PJ. Design strategy for biodegradable Fe-based alloys for medical applications. Acta Biomaterialia. 2010;**6**:1705-1713

[49] Vojtech D, Kubasek J, Serak J, Novak P. Mechanical and corrosion properties of newly developed biodegradable Zn-based alloys for bone fixation. Acta Biomaterialia. 2011;**7**:3515-3522

[50] Witte F, Kaese V, Haferkamp H, Switzer E, Meyer-Lindenberg A, Wirth CJ, et al. In vivo corrosion of four magnesium alloys and the associated bone response. Biomaterials. 2005;**26**:3557-3563

[51] Witte F, et al. In vitro and in vivo corrosion measurements of magnesium alloys. Biomaterials. 2006;**27**:1013-1018

[52] Zartner P, Buettner M, Singer H, Sigler M. First biodegradable metal stent in a child with congenital heart disease: Evaluation of macro and histopathology. Catheterization and Cardiovascular Interventions. 2007;**69**:443-446

[53] Hermawan H, Dubé D, Mantovani D. Development of degradable Fe-35Mn alloy for biomedical application. Advanced Materials Research. 2007;**15-17**:107-112 [THERMEC 2006 Supplement]

[54] Hermawan H, Alamdari H, Mantovani D, Dube D. Iron-manganese: New class of metallic degradable biomaterials prepared by powder metallurgy. Powder Metallurgy. 2008;**51**:38-45

[55] Zhu S, Huang N, Xu L, Zhang Y, Liu H, Sun H, et al. Biocompatibility of pure iron: In vitro assessment of degradation kinetics and cytotoxicity on endothelial cells.Materials Science and Engineering C; **29**(2009):1589-1592

[56] Peuster M, Hesse C, Schloo T, Fink C, Beerbaum P, von Schnakenburg C. Long-term biocompatibility of a corrodible peripheral iron stent in the porcine descending aorta. Biomaterials. 2006;**27**:4955-4962

[57] Aida L. Alexis Carrel (1873-1944): Visionary vascular surgeon and pioneer in organ transplantation. Journal of Medical Biography. 2014;**22**(3):172-175

[58] Mason C, Dunnill P. A brief definition of regenerative medicine. Regenerative Medicine. 2008;**3**(1):1-5

[59] Langer R, Vacanti JP. Tissue engineering. Science. 1993;**260**:920-926

[60] Gardner J, Webster A. The social management of biomedical novelty: Facilitating translation in regenerative medicine. Social Science & Medicine. 2016;**156:**90-97

[61] Atala A, Bauer SB, Soker S, Yoo JJ, Retik AB. Tissue-engineered autologous bladders for patients needing cystoplasty. Lancet. 2006;**367**(9518):1241-1246

[62] Silva JJRFD, Williams RJP. The Biological Chemistry of the Elements. Oxford: Clarendon Press; 1991

[63] Glusker JR. Structural aspects of metal liganding to functional groups in proteins. Advances in Protein Chemistry. 1991;**42**:1-73

[64] Atala A. Regenerative medicine strategies. Journal of Pediatric Surgery. 2012;**47**(1):17-28

[65] Mouriño V, Boccaccini A. Bone tissue engineering therapeutics: Controlled drug delivery in three-dimensional scaffolds. Journal of the Royal Society Interface. 2010;**7**:209-222

[66] Mouriño V, Newby P, Phisbin F, Cattalini P, Lucangioli S, Boccaccini AR. Preparation, characterization and in vitro studies of gallium crosslinked alginate/bioactive glass composite films. Soft Matter. 2011;**7**:6705-6712

[67] Rutherford JC, Bird AJ. Metal-responsive transcription factors that regulate iron, zinc, and copper homeostasis in eukaryotic cells. Eukaryotic Cell. 2004;**3**:1-13

[68] Barralet J, Gbureck U, Habibovic P, Vorndran E, Gerard C, Doillon CJ. Angiogenesis in calcium phosphate scaffolds by metallic copper ion release. Tissue Engineering Part A. 2009;**15**:1601-1609

[69] Kawashita M, Tsuneyama S, Miyaji F, Kokubo T, Kozuka H, Yamamoto K. Antibacterial. silver-containing silica glass prepared by sol-gel method. Biomaterials. 2000;**21**:393-398

[70] Volker A, Thorsten B, Peter S, Michael W, Peter S, Elvira D, Eugen D, Reinhard S. An in vitro assessment of the antibacterial properties and cytotoxicity of nanoparticulate silver bone cement. Biomaterials. 2004;**25**:4383-4391

[71] Peter I, Rosso M, Toppi A, Dan I, Ghiban B. Investigation on cobalt based alloy modified by titanium for dental application. Archives of Materials Science and Engineering. 2013;**61**(2):62-68. ISSN:1897-2764

[72] Peter I, Rosso M. Study of Ti-enriched CoCrMo alloy for dental application. IEEE Access, The Journal for Rapid Open Access Publishing. 2015;**3**:73-80

[73] Ghiban A, Ghiban B, Peter I, Rosso M, Dan I, Tiganescu TV. Structural behaviour of CoCrMoTi(Zr) alloys for dental applications. Revista de Chimie. 2016;**67**(6):1131-1136

[74] Matekovits L, Huang J, Peter I, Esselle KP. Mutual coupling reduction between implanted microstrip antennas on a cylindrical Bio-metallic ground plane. IEEE Access. 2017; **5**. DOI: 10.1109/ACCESS.2017.2703872. (early access: http://ieeexplore.ieee.org/stamp/stamp.jsp?arnumber=7927379)

[75] Li J, Peter I, Matekovits L. Circularly polarized implanted antenna with conical Bio-Metallic ground plane. In: Proceedings of IASTEDs 13th International Conference on Biomedical Engineering (BioMed 2017); Feb. 20-22, 2017; Innsbruck, Austria. IEEE; 2017. pp. 265-269

Chapter 3

Microporous Titanium-Based Materials Coated by Biocompatible Thin Films

Anna D. Dobrzańska-Danikiewicz,
Leszek A. Dobrzański, Marek Szindler,
Lech B. Dobrzański, Anna Achtelik-Franczak and
Eugeniusz Hajduczek

Additional information is available at the end of the chapter

http://dx.doi.org/10.5772/intechopen.70491

Abstract

This chapter presents the outcomes of numerous own works concerning constructional solutions and fabrication technologies of a new generation of custom, original, hybrid, microporous high-strength engineering and biological materials with microporous rigid titanium and Ti6Al4V alloy skeletons manufactured by Selective Laser Sintering (SLS), whose pores are filled with living cells. The so constructed and fabricated implants, in the connection zone with bone stumps, contain a porous zone, with surface treatment inside pores, enabling the living tissues to grow into. As the adhesion and growth of living cells are dependent on the type and characteristic of the substrate it is necessary to create the most advantageous proliferation conditions of living cells inside the pores of a microporous skeleton made of titanium and Ti6Al4V alloy. In order to improve the proliferation conditions of cells ensured by a fully compatible substrate, internal coatings with TiO_2, Al_2O_3 oxides and $Ca_{10}(PO_4)_6(OH)_2$ hydroxyapatite of the surface of pores of a microporous skeleton made of titanium and Ti6Al4V alloy with SLS was used. Two technologies have been chosen for the deposition of thin coatings onto the internal surfaces of pores: Atomic Layer Deposition (ALD) and the sol-gel of deep coating from the liquid phase.

Keywords: implant-scaffold, scaffold, engineering biological composite, titanium, SLS, thin film, ALD, dip coating

1. The basis of biological interaction of living cells with a substrate made of metallic micro-skeletons manufactured by selective laser sintering and coated inside pores using atomic layer deposition or the sol-gel methods

The continuity of tissues requiring tissue regeneration in order to restore their normal condition is disrupted as a result of bodily injuries related to numerous accidents, most often work accidents, traffic accidents and sports accidents, as well as due to surgical interventions resulting from the employed therapeutic methods of treating numerous disorders, most often cancerous diseases or removal of inflammatory conditions [1]. One of the most significant and costly problems of modern medicine is the necessity to replace or supplement organs or tissues to prevent the biological and social degradation of patients and to restore their living functions, either normal functions or such acceptably similar to normal, resulting from a growing number of cases of organ or tissue loss or damage in the human population due to post-injury or post-resection losses as well as those originating from the operative treatment of cancerous tumours or inflammation processes and as a result of other disorders, and also work, traffic and sports accidents. The most widespread cases are caused by a strong development of civilisational diseases, including cancer, and the incidence rate of malignant cancers has been regularly rising. The number of traffic accident victims has been growing substantially, as well [2]. One of the fundamental tasks globally is to improve the society's condition of health, medical care and health safety. The idea is to overcome problems which occur more and more often and to protect against serious health risks, especially such as civilisational diseases, pandemics and bioterrorism and to support research, particularly research into the development of modern technologies, serving to advance medicine and healthcare, most of all to replace the lost tissues and organs by technical means, to ensure more complete prophylactics of diseases and safe treatment of patients, which are one of the important indicators of economic welfare. Bone reconstruction, for example, of legs and hands and in the craniofacial area, as well as skin and other soft tissue reconstruction, and also the reconstruction of oesophagus and blood vessels, are often required due to the consequences of civilisational diseases and accidents, including malignant cancers. Patients' healthcare expectations are also growing, and economic aspects at the global scale call for the efficient elimination of disabilities, in particular motoric disabilities, and the restoration of the previously handicapped persons to physical fitness and usually most often to full, or at least partial, professional activity, which considerably lessens pressure on the diminishing resources of social insurance funds. It is vital to reduce waiting times for treatment, to lower prices and to improve availability of medical products or services and therapy, to reduce the risk of treatment failure, in particular by tailor-made personalised medical products according to a patient's individual anatomical features, and ultimately to reduce therapy discomfort for patients and their family.

The development of regenerative medicine started nearly a quarter of century ago along with the works [3], consisting of treatment by replacing old and sick cells with young cells using tissue engineering methods and cell-based therapies or gene therapy and creating numerous new opportunities in counteracting diseases and their consequences [4–8]. On the other hand, tissue engineering – introduced somewhat earlier [9] – consists of the development of biological substitutes for restoration, maintenance or improvement of functions of tissues or entire organs [10, 11], involving the construction and fabrication of scaffolds maintaining the developing tissues and involving the production of a replacement tissue for clinical use [12] as a substitute of damaged tissues or entire organs [13, 14], capable of restoring, maintaining or improving the functions of particular tissues or organs [10]. It is obviously a continued endeavour to develop an engineering material with its properties corresponding to the tool being replaced, supplemented or aided, which does not cause an immunological response of the immunity system and expediting wound healing and not causing implant rejection. The advancement of tissue engineering methods poses further challenges for biomaterials which should not only be fully compatible, but when used for three-dimensional scaffolds, should ensure conditions for cellular cultures, taking into account the possibility of controlled growth, division and differentiation of different types of cells and the impact of various environmental factors on their living functions, and also for transporting medicines in a patient's organism.

Among the biocompatible materials currently and widely used in many branches of industry are titanium and its alloys. These materials are characterised by excellent corrosion resistance in the majority of aggressive environments. Titanium has two allotrope types: Tiα and Tiβ. The α allotrope type endures to the temperature of 882°C and crystallises in a hexagonal structure with a compact lattice (A3). The β allotrope type endures to the temperature of 882–1068°C being a melting point and it crystallises in a regular structure with a centred spatial lattice (A2) [15]. Titanium alloys are classified as single-phase α and β, double-phase $\alpha + \beta$ and pseudo-α and pseudo-β. Double-phase $\alpha + \beta$ alloys are the most popular, widely used group of titanium alloys with good strength and plastic properties. This group includes the most popular Ti6Al4V titanium alloy. Titanium and titanium alloys, conventionally manufactured by casting and plastic treatment [16], are used in the arms, chemical, automotive, aviation, power and transport sector, as well as in architecture and sports [17–20]. Pristine titanium and its alloys, independent of the wide engineering application, are often used in medicine in order to replace damaged tissues. From many years, these materials have been used for endoprosthesis of hip-joints and knee-joints, bone plates, screws for fracture repair and heart valve prostheses [21–24]. Moreover, artificial heart is currently developed using these materials [25]. Recently, for medical applications, materials sintered from powders of pristine titanium, Ti6Al4V and Ti6Al7Nb are also applied [21–25]. **Table 1** shows comparative analysis of these materials.

A structure of the extracellular matrix (ECM) occurring in living organisms, and acting as a natural scaffold of cells [26], can be imitated by biomaterials which, for this reason, can find their application as a substrate for the controlled breeding of living cells. Capable of growth, division and differentiation are such living cells only, which are attached to a substrate and have undergone adsorption on a substrate surface. In natural conditions, this is possible owing to the presence of the extracellular matrix. The coincidence of a metallic substrate with living cells is a very vital aspect. It is also vital to determine the potential favourable effect of indirect layers – in the form of thin coatings deposited onto the inner surfaces of pores of a metallic

No.	Evaluation criteria	Ti	Ti6Al4V	Ti6Al7Nb
1	The current use of this material in medicine	High	Very high	Minor
2	Powder granulation, μm	10–45	15–45	0–100
3	Density, g/cm³	4.51	4.43	4.52
4	Melting point, °C	1600–1660	1660	1520–1580
5	Price, €/kg (state for 2014 year)	400	500	250
6	Availability	High	High	Low
7	Powder application in SLS/SLM process	Yes	Yes	No

Table 1. Comparative heuristic analysis of pristine titanium, Ti6Al4V and Ti6Al7N powders.

skeleton made of titanium and its alloys produced by SLS – on the potential adhesion and proliferation of living cells. Not all the factors are known, which are decisive for the influence of a substrate made of engineering materials on the attachment, division and growth of living cells, despite the fact that intensive research has been conducted in this field. A material's specificity is one of the most essential factors crucial for the breeding or growth of cells on the surface of engineering materials [27, 28]. The domains ensuring the adhesion of cells to a substrate are very important in interactions between living organisms and engineering materials in cultures of living stem cells and in implantology. The one which is best recognised is arginyl glycyl aspartic acid (RGD), which is a tripeptide composed of L-arginine, glycine, and L-aspartic acid, as well as other proteins, including integrins, cadherins, selectins and immunoglobulin-like proteins [29–37]. After placing an implant in an organism, its surface is covered with a thin layer of water within a few seconds, and then, within a few seconds to several hours, with a layer of specific proteins [38] from those contained in physiological fluids [39]. A weakly vascularised, fibrous layer of cells [29, 40] is growing on such cells within several minutes to several days, and finally a cytoskeleton of cells is reorganised, leading notably to the flattening of cells [41]. Surface wettability influences the ability of adhesion proteins to attach to the substrate, and their affinity to the material surface has influence on the structure and composition of a layer of proteins [38]. The free enthalpy of surface and its wettability influence the growth of cells, but do not have an effect on their shape and orientation [42–44]. The adhesion domains placed on the surface of engineering biomaterials influence the behaviour of cells due to an effect on the functioning of integrins [29]. Cells' material attachability depends on proteins' substrate adhesion [29–35, 37–39, 42, 45]. An excessively hydrophilic and hydrophobic character of the surface with the wetting angle of over 80 or below 15° [46] is sometimes unfavourable for the growth of cells, although surface hydrophobicity may, in the initial phase, support cells' adhesion, when hydrophilicity may support their division and multiplication [47]. In general, cells are preferentially attaching, dividing and growing on hydrophilic surfaces of a material, whilst their ability of adhesion to a substrate material is diminishing on hydrophobic surfaces [38, 39, 47–53].

As flexibility of different types of cells varies [54–66], their morphology and living functions depend on the stiffness of the substrate on which they are grown [54–69]. Usually cells' differentiation ability increases due to increased material stiffness, although exceptions to this rule exist [55, 57, 58, 60–66]. Because cells are interacting with the substrate surface [55, 57, 58, 60–66, 70], the cells growing on a stiff substrate exhibit greater rigidity [64, 67, 71], a more organised cellular cytoskeleton [55, 59, 63–67] and greater flattening [55, 57, 60, 63–66, 70, 72] than those growing on a more elastic substrate. Cells, especially fibroblasts and cells of smooth muscles, exhibit mechanotaxis (durotaxis), moving towards a substrate with greater rigidity [57, 62, 64–66, 72, 73]. Increased substrate rigidity is decisive for the reduced ability of cells to migrate [65], and different types of cells react differently to differentiated rigidity of a substrate made of engineering materials [54]. A material's surface topography influences cells' adhesion ability [74, 75], and they show an ability of adaptation to a material's surface specificity [53, 76–79]. Surface topography is dictating proteins' adhesion and creation of bonds, which is essential for material biocompatibility, influencing also morphology, spatial orientation, division and differentiation of cells [38, 80]. Cells' behaviour is also influenced by surface texture and smoothness [2, 38, 74, 76, 77, 81–84]. Cell adhesion is more difficult on less developed surfaces with higher smoothness [85], whereas the division of osteoblasts is faster on surfaces with smaller smoothness [86–88], opposite to fibroblasts which are proliferating fastest on smoother surfaces [56, 89–92]. The flattening of cells is decreasing as the smoothness of the substrate surface is deteriorating [86, 88, 91, 92].

Additive technologies [93–95] can be employed, in particular, for producing different implants, including dental implants and bridges, individualised implants of the upper jaw bone, hip joint and skull fragments using suitable biomaterials. An exceptional usefulness of additive technologies of producing solid and microporous materials in medicine and dentistry has been confirmed by comparing powder metallurgy technologies, casting technologies, metallic foam manufacturing technologies and additive manufacturing technologies with procedural benchmarking techniques [96, 97] using a universal scale of relative states for comparative evaluation [96–99]. Considering the additive technologies applied most widely in industry, only few have found their application in prosthetics, especially in prosthodontics, that is, electron beam melting (EBM) [100–105], and also 3D printing for production of indirect models, although selective laser sintering/selective laser melting offers broadest opportunities (SLS/SLM) [100, 106–125]. The sintering/melting of grains takes place by remelting the particles of a new powder on the surface with the existing piece of an item being constituted by a laser beam moving in a programmed fashion in line with the pre-defined geometrical characteristics of a metal element being produced. SLS/SLM technologies are so attractive due to the opportunities offered by 3D design with the use of CAD methods and due to the related overall control over the materials fabricated, both, in terms of the structure, sizes and repeatability of geometric features. Microskeletons with controlled sizes and shapes of pores can also be manufactured.

A microporous element manufactured by SLS can be further worked and combined with other materials, for example, by infiltration or internal treatment of pores' surface in an appropriately chosen technological process [117–121, 126–128]. Titanium and titanium alloys from Al, Nb and Ta [1, 2, 15, 26, 127–138], well tolerated by a human organism, have been long used in medicine and dentistry for prosthetic and implantological purposes, also fabricated by SLS. Titanium matrix materials do not cause allergic reactions and are stainless, feature high strength and hardness and also thermal conductivity several times lower than traditional prosthetic materials [139]. Titanium is a very thrombogenic material [140], and biocompatibility, especially thrombocompatibility [141] of this material can be enhanced by introducing alloy elements. Pores are dimensionally adapted to be filled by the reconstructed cells and their migration and also neovascularisation [142] for preventing blood clots [143]. Their section cannot be too small to prevent their sealing [144]. A porous structure of scaffolds should secure the diffusion of nutrients and metabolism products. Permanently fixated scaffolds do not ensure scaffold removal as is the case with scaffolds removable during regeneration in the natural condition [143–145]. Other metallic materials can also be utilised for this purpose, for example, polymer materials [40, 81, 146–153] or polymer matrix composite materials with a fraction of metals or ceramic particles [81, 150]. Unlike metals and their alloys, polymer materials can be not only biocompatible, but also biodegradable and bioresorbable [29, 46, 154, 155], and also in some cases osteoinductive [156] and can reduce thrombogenic properties of the material [40]. Another concept relates, however, to covering the surface of metallic materials with other materials, also as a result of working an internal surface of pores [125, 157]. Porous metallic materials, chiefly Ti and Ta [158] and Mg [159], are used for non-biodegradable scaffolds primarily due to relatively high compressive strength and fatigue strength [160, 161]. Coatings of internal surfaces of pores are used, however, because they meet the imposed requirements better than metals and their alloys [40, 162, 163]. The following is necessary, especially, not to cause disease-related changes: the lack of a toxic effect of the implant material, the lack of its mutagenic properties, the lack of its effect on the composition of body fluids, both by the material being implanted as well as in the case of polymer materials, also by its decomposition products [40, 81, 147].

Original own works [117–120, 126, 164–180] have been undertaken concerning constructional solutions and fabrication technologies of a new generation of custom, original, hybrid, microporous high-strength engineering and biological materials with microporous rigid titanium and titanium alloy skeletons manufactured by selective laser sintering, whose pores are filled with living cells (**Figure 1**). This will ensure the natural ingrowth of a living tissue, at least in the connection zone of prosthetic/implant elements, with bone or organ stumps, and will eliminate the need to apply for patients the mechanical elements which are positioning and fixating the implants. The so constructed and fabricated implants, in the

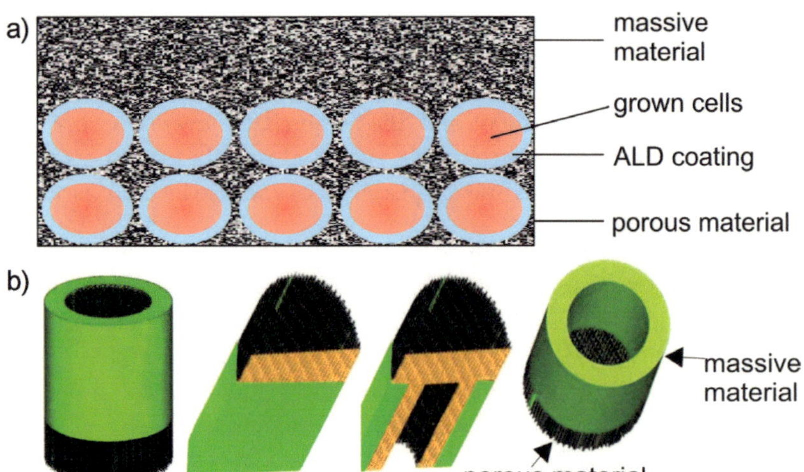

Figure 1. Chart of constructional assumptions (a) hybrid and multilayer biologically active microporous composite engineering materials consisting of biologically active cellular structures and of implant-scaffolds; (b) implant-scaffolds with microporous zones acting as scaffolds.

connection zone with bone stumps, contain a porous zone, with surface treatment inside pores, enabling the living tissues to grow and they remain in an organism permanently and do not require re-operation.

As the adhesion and growth of living cells are dependent on the type and characteristics of the substrate, in order to implement the planned concept of manufacturing engineering and biological materials and implant-scaffolds, it is required to seek the most advantageous proliferation conditions of living cells inside the pores of a microporous skeleton made of titanium and titanium alloys, which is the underlying scope of the research presented in this chapter. It appears that the improvement of proliferation conditions of cells is ensured by a substrate made of fully compatible materials, including TiO_2, Al_2O_3 oxides and a hydroxyapatite, $Ca_{10}(PO_4)_6(OH)_2$. For this reason, it was decided to employ these materials as coatings of internal surfaces of pores of a microporous skeleton fabricated by SLS. Two technologies have been chosen, respectively, for the deposition of thin coatings onto the internal surfaces of pores, ensuring the uniform thickness of coatings on all the walls and openings of a substrate being coated, even with highly complicated shapes, that is, the so-called atomic layer deposition (ALD) [96, 124, 126, 128, 181–184] (**Figure 2**) and the sol-gel technology of coating deposition from the liquid phase by the immersion method [96, 181, 185–194] (**Figure 3**).

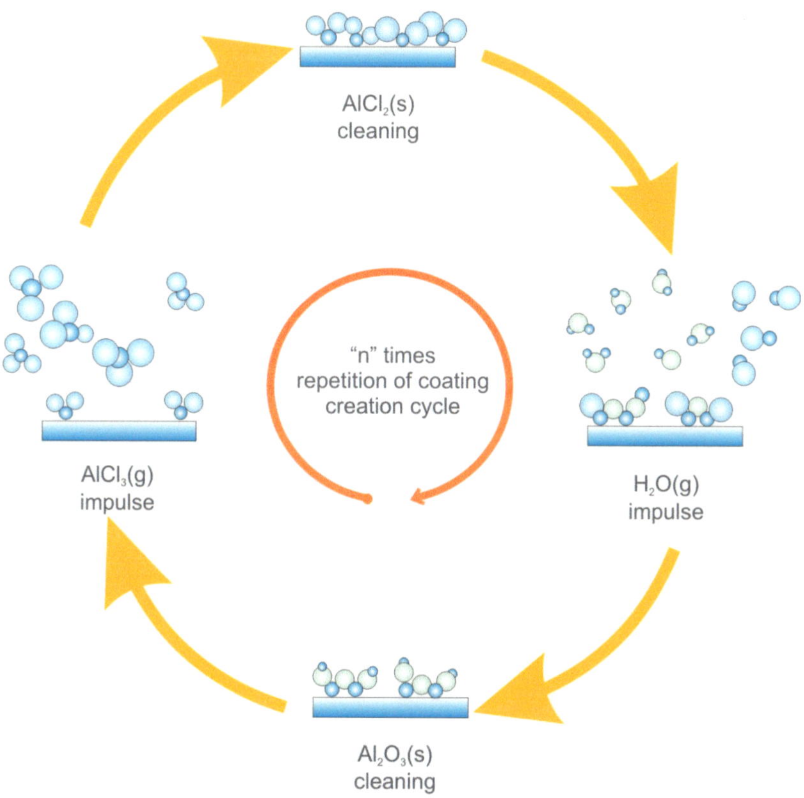

Figure 2. Sequence of the phenomena associated with ALD technology [96].

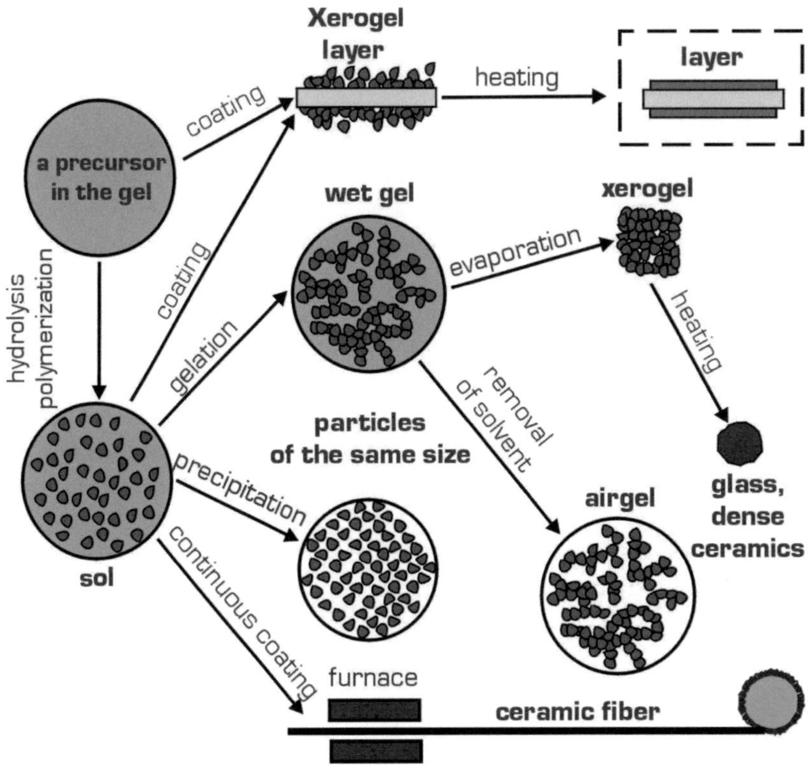

Figure 3. Sequence of the phenomena associated with the sol-gel technology of coating deposition [96].

2. Structure and properties of selectively laser sintered microskeletons made of titanium and Ti6Al4V alloy

The elements used for the research were produced by an additive method by powder sintering in an SLS process. Selective laser sintering can be grouped into a stage of designing a given element, the outcome of which is a 3D CAD model in the *.stl* format, then transferred into a machine's software, and another stage, at which an item designed virtually in advance is produced for real, a layer by layer, until a final product is obtained (**Figure 4**).

The following was used for model designing: Solid Works 2015 Pro and 3D Marcarm Engineering AutoFab (Software for Manufacturing Applications) software, version Autofab MCS 2.0 and Autofab MTT 64 and the CAD/CAM tools available in, respectively, SLM 250H system by MTT Technologies Group and in AM 125 system by Renishaw for selective laser sintering. The software enables to select model dimensions, constructional features, the type of the model volume filling either as solid or porous and to choose the size of a cell unit making up the entire model. A set of "hexagon cross" unit cells was used, selected in a geometrical analysis, and as a result of preliminary studies, from the set available in the software. By duplicating them, the entire elements were designed, consisting of nodes and single lattice fibres, linking the particular skeleton node (**Figure 5**). Fabrication conditions were also selected, at the stage of virtual design of elements, achievable in, respectively, SLM 250H

http://dx.doi.org/10.5772/intechopen.70491

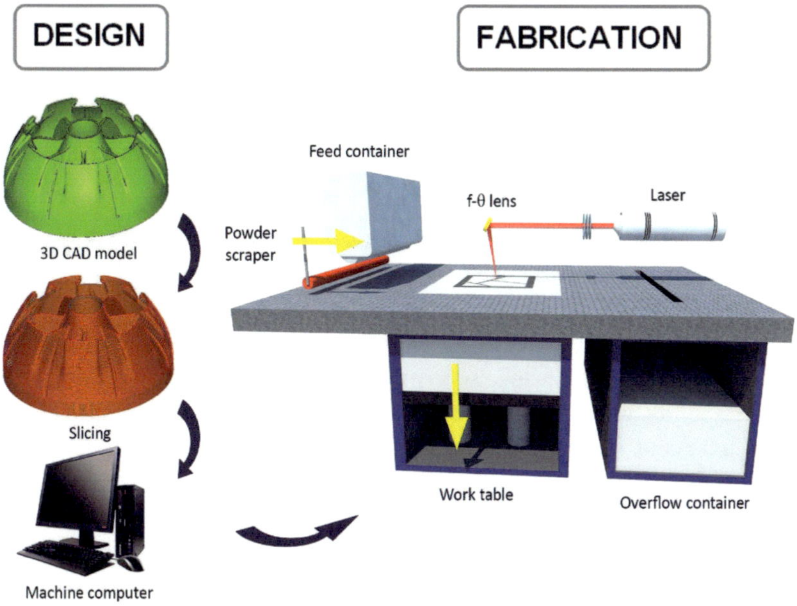

Figure 4. Selective laser sintering technology process diagram [195].

system by MTT Technologies Group and AM 125 by Renishaw for selective laser sintering, including layer thickness, laser power, laser beam diameter, scanning rate, distance between particular remelting paths. A spatial orientation of unit cells of 45° relative to the *x*-axis of the system of coordinates was also chosen experimentally, because other orientations analysed initially turned out to be less advantageous due to the mechanical properties of the elements produced (**Figure 5**). The structures of microporous skeletons with appropriately differentiated average sizes of pores of ~450, ~350 and ~250 μm, with suitable sizes of a unit cell of 700, 600 and 500 μm, were defined by reproducing and assuming suitable values characterising the spatial lattice, such as height, depth and width. Solid specimens were also made for comparative purposes.

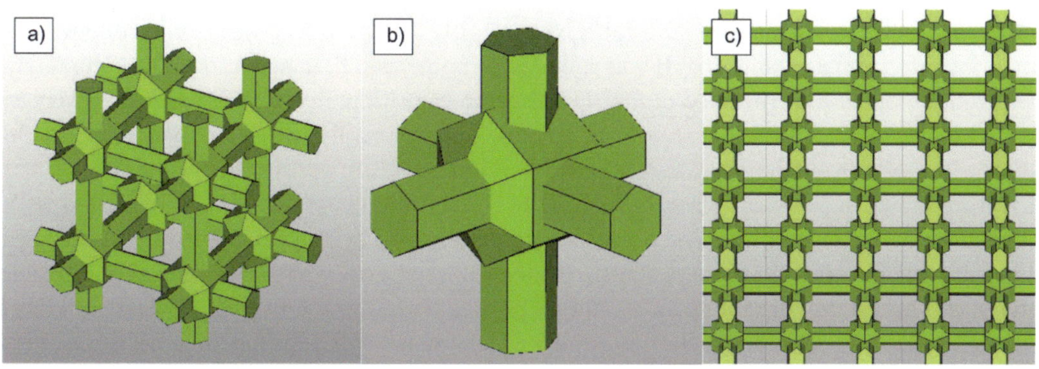

Figure 5. (a, b) Hexagon cross unit cell; (c) image of structure of computer models presenting the arrangement of unit cells in the space of the system of coordinates at the angle of 45° relative to the *x*-axis.

It was very important to establish the correct laser power value, ranging between 50 and 200 W, and the laser beam diameter, of 30–150 μm, and also the distance between laser beams and distance between laser remelting paths equal to or smaller than the laser beam diameter in case of solid elements, as opposite to the variant when it is larger than the laser beam diameter, which is decisive for the porosity of the produced element. Such conditions were selected as a result of preliminary studies.

When a model is designed and the selected fabrication conditions are taken into account, it is transferred to the software of the SLM 250H system machine by MTT Technologies Group or AM 125 by Renishaw, where selective laser sintering takes place. The YFL fibre laser, with an active material doped with Ytterbium and the maximum power of, respectively, 400 and 200 W, was used in the devices. The skeleton microporous materials, fabricated by SLS, exhibit porosity which depends on the manufacturing conditions, including mainly laser power and laser beam diameter and the distance between laser beams and the distance between laser remelting paths. Solid titanium with the density of 4.51 g/cm^3, corresponding to the density of solid titanium given in literature, can be achieved if the laser power of 110 W is applied. The smallest porosity of 61–67% corresponds to the selectively laser sintered elements with the highest mass and the average pore size of ~250 μm, whereas the highest porosity of 75–80% corresponds to elements with the smallest mass and the average pore size of ~450 μm. The porosity of 70–75% was obtained for the average pore size of ~350 μm.

Two types of powders with a spherical shape were used, respectively, for selective laser sintering to produce solid specimens and microporous skeletons (**Figure 6a,b**) and with the composition shown in **Table 2**, also confirmed with spectral examinations with the energy dispersive spectrometry (EDS) method (**Figure 6c,d**):

- titanium powder with Grade 4 and grain size of up to 45 μm, oxygen concentration reduced to 0.14%, the aim of which is to ensure process safety,
- Ti6Al4V alloy powder with the grain diameter of 15–45 μm, for medical applications.

Figure 7a and **b** and shows a surface structure of pristine titanium and Ti6Al4V alloy manufactured by selective laser sintering with laser beam size of 50 μm and laser power of 110 W. Chemical composition, examined with an EDS spectrometer, shows that Ti only exists in the first case in the sample (**Figure 7c**), and Ti, Al and V (**Figure 7d**) were identified in the other case, which corresponds to the data given in **Table 2** for chemical composition of powders used for selective laser sintering. It was revealed in both cases that, apart from the completely sintered materials with revealed paths of laser beam transition, fine powder particles exist, as well, which – after sintering – should be removed mechanically or by chemical etching. On the surface of porous titanium microskeletons (**Figures 8–10**), apart from sintered titanium, there are grains of powder loosely bonded with a skeleton, which were not fully melted with a titanium microskeleton; the results of local remelting also exist, as indicated by the surface topography of porous titanium skeletons for the different arrangement of cell units. In order to remove such unfavourable surface effects, porous titanium, after selective laser sintering, was subjected to preliminary cleaning in an isopropyl alcohol solution with an ultrasound

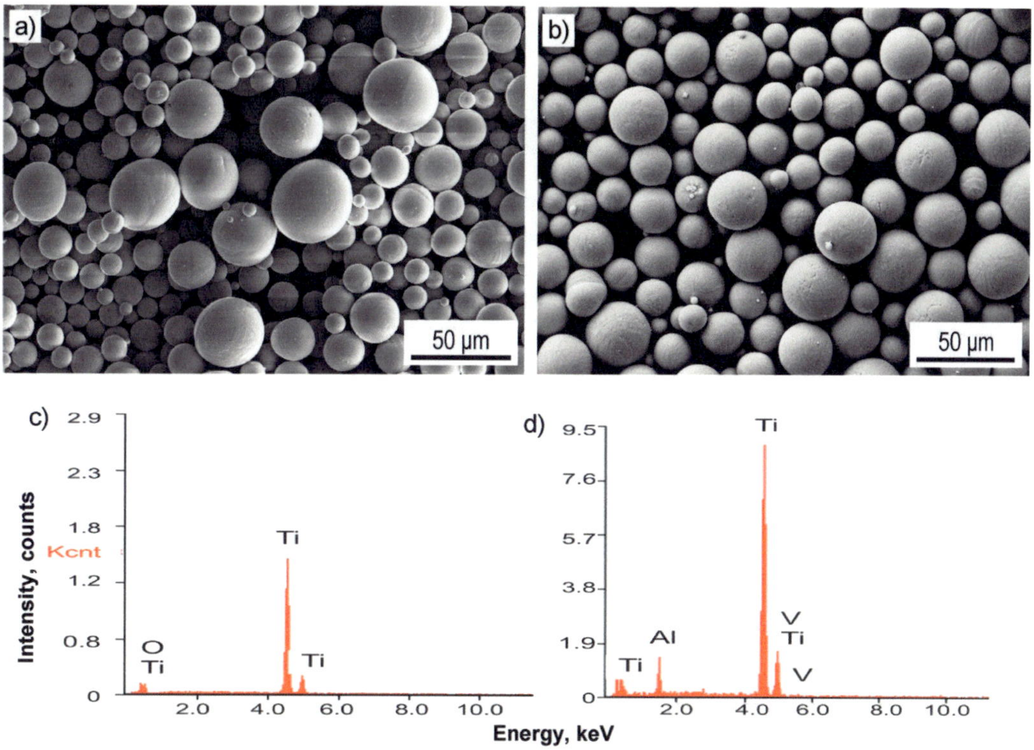

Figure 6. Powder of (a) Ti, (b) Ti6Al4V alloy (SEM), (c, d) relevant diagrams of EDS [119].

Powder	Mass concentration of elements, %									
	Al	V	C	Fe	O	N	H	Others total	Others each	Ti
Ti	–	–	0.01	0.03	0.14	0.01	0.004	<0.4	<0.01	Remainder
Ti6Al4V	6.35	4.0	0.01	0.2	0.15	0.02	0.003	≤0.4	≤0.1	

Table 2. Chemical composition of the powders used for selective laser sintering.

washer. After getting rid of excessive powder from the pores of the titanium skeleton, it was subjected to etching in an aqua regia solution with the fraction volume of 3:1 HCl:HNO_3, for 1 hour with an ultrasound washer, to etch the surface remelting not removed in preliminary cleaning and the fine powder particles not attached permanently to the previously constituted titanium microskeleton. The samples were subjected to etching in an aqua regia solution, as a result of which about 3% of the sintered material mass was removed (**Figure 11**). The roughness of skeletons before etching is much higher than after etching **(Figure 11).** A 14% aqueous hydrofluoric acid solution can also be used alternatively for etching.

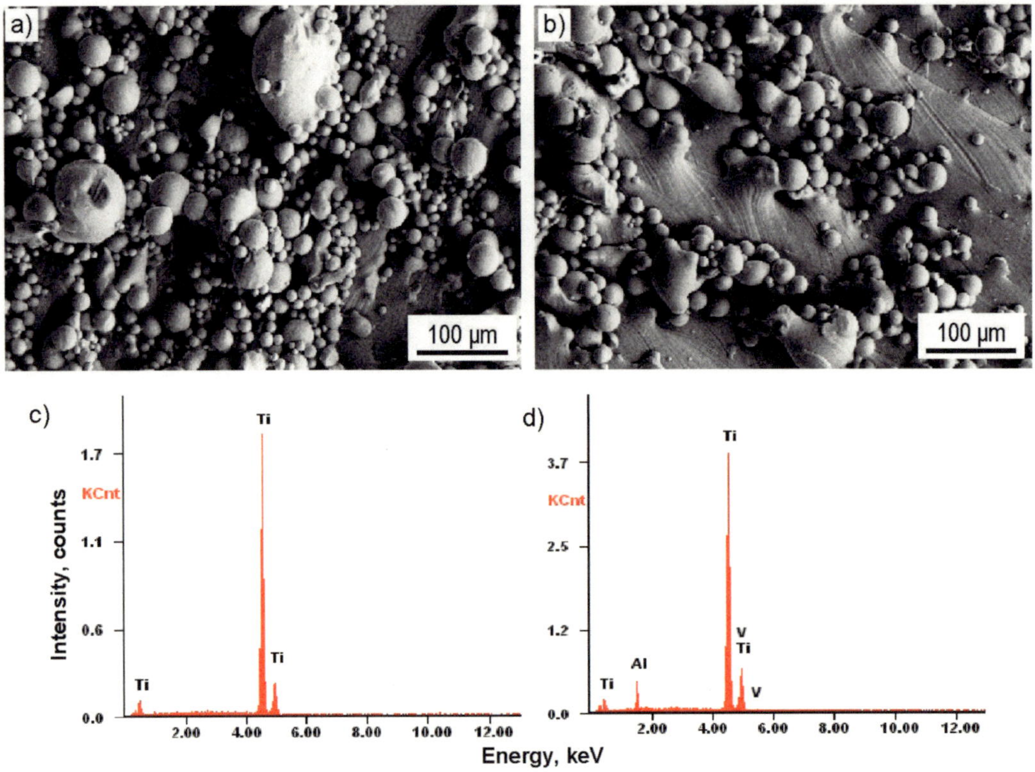

Figure 7. Surface structure of solid samples manufactured by selective laser sintering with the use of laser beam of 50 μm with laser power of 110 W; SEM of (a) solid titanium, (b) Ti6Al4V alloy, (c, d) respectively, EDS charts [119].

The structural examinations of selectively laser sintered titanium and Ti6Al4V alloy were carried out in a transmission electron microscope, TITAN 80–300, by FEI. Titanium and Ti6Al4V alloy have a crystalline structure, hence for the appropriately high microscope resolution, one can observe the rows of atoms arranged parallel to each other, both, when the microscope is in the TEM transmission mode and in the STEM scanning-transmission mode (**Figure 12**).

It is not possible to use standard normalised conditions of mechanical tests due to a porous structure of the analysed materials, which are not covered by the system of EN standards, and also due to the manufacturing costs and time of the samples for tests of mechanical properties and due to a limited size of a working chamber and manufacturing device efficiency. A custom, own method was therefore established for performing such examinations with a universal tensile testing machine, Zwick 020, in the conditions principally corresponding to static tensile tests, three-point bending and compression tests. Moreover, specially adapted miniaturised samples were also produced for examinations of strength properties (**Figure 13**), with the dimensions of measuring parts of, respectively, 3 × 3 × 15 mm, 3 × 10 × 35 mm (for support spacing of 30 mm) and 10 × 10 × 10 mm.

The mechanical properties of solid and porous materials fabricated by SLS, that is, sintered titanium and sintered Ti6Al4V alloy, were compared each time. The results of examinations of, respectively, tensile strength, bending strength and comprehensive strength were presented in

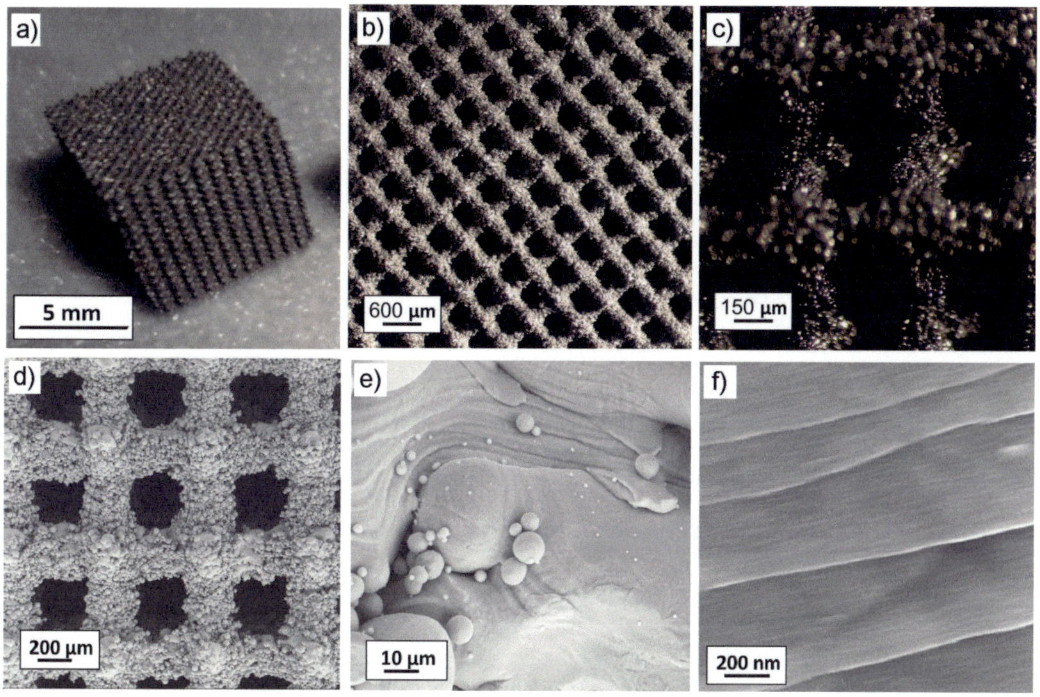

Figure 8. Titanium scaffold (a) visible with bare eye, (b, c) stereoscope microscope, (d–f) SEM.

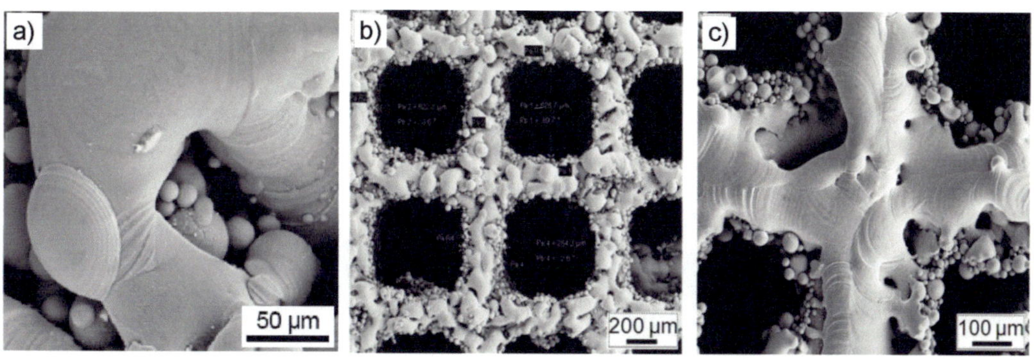

Figure 9. Surface typography of porous titanium skeletons with different pore size: (a) 350 µm, (b) 450 µm, (c) 630 µm; SEM.

such order. The impact of laser power ranging 70–110 W was investigated in the first place on tensile strength values of sintered titanium and sintered Ti6Al4V alloy. The results of examinations for five samples, for each treatment option of each of such materials, are presented in **Figure 14**.

Figure 15 presents the comparison of diagrams of dependency between tensile stress and elongation for solid and porous laser sintered Ti and Ti6Al4V alloy samples with the pore size of approx. 250 µm, subjected to static tensile tests. The results were obtained to compare the conditions of

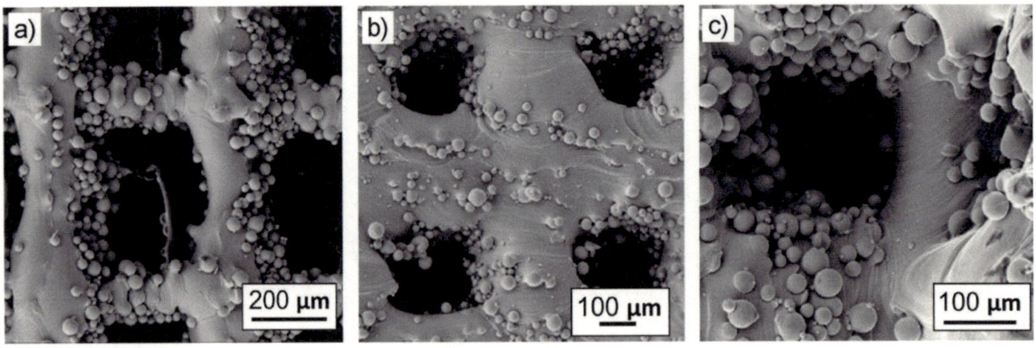

Figure 10. Surface topography of microskeletons made of Ti6Al4V alloy manufactured as multiplication of different unit cells presented with different magnificence (a, b, c); SEM images.

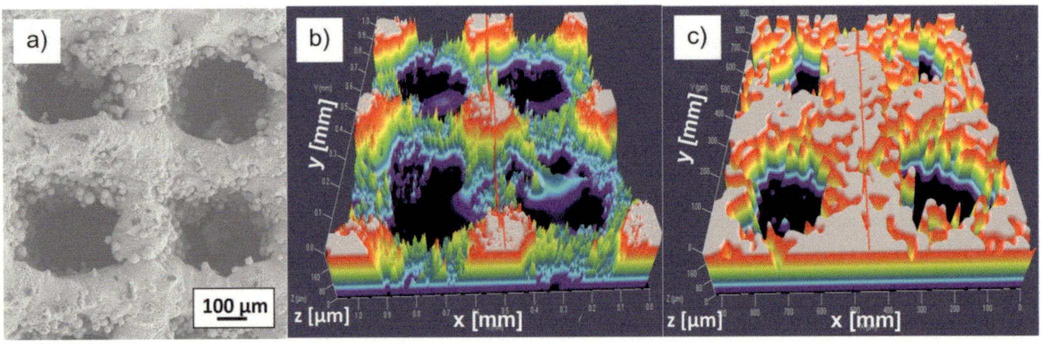

Figure 11. Surface topography of porous titanium skeletons with the pore size of ~450 µm (a) after etching in aqua regia solution (SEM), (b) after ultrasound cleaning, (c) after etching in aqua regia solution; (b, c) laser confocal microscope.

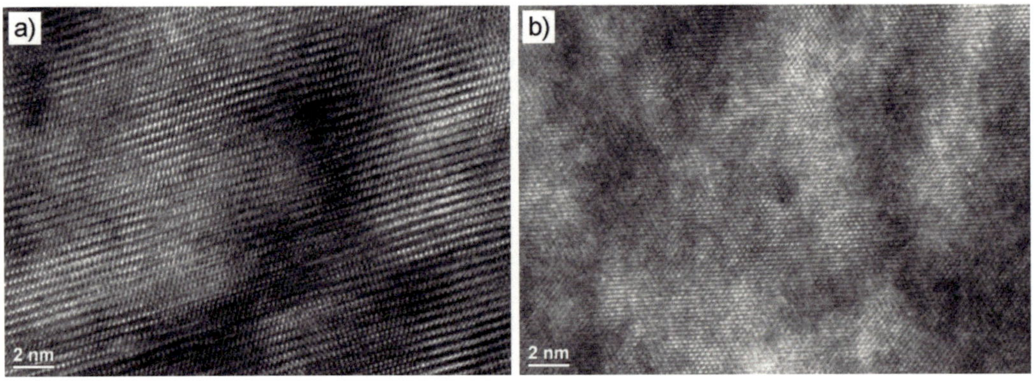

Figure 12. Crystalline structure of: (a) pristine titanium, (b) Ti6Al4V alloy; HRTEM image.

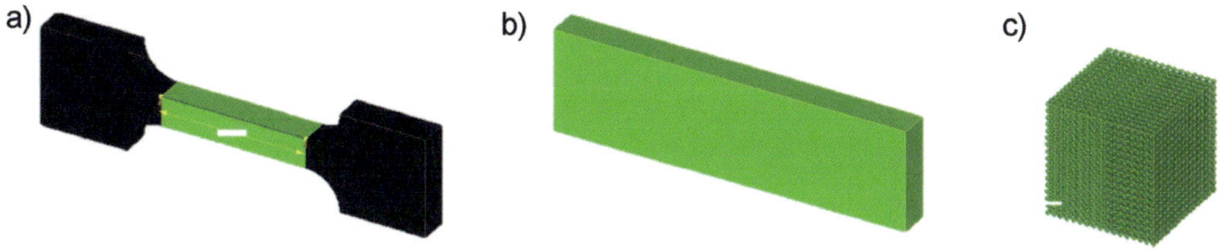

Figure 13. Image of computer model of solid samples for static test: (a) tensile test, (b) three-point bending test, (c) compression test.

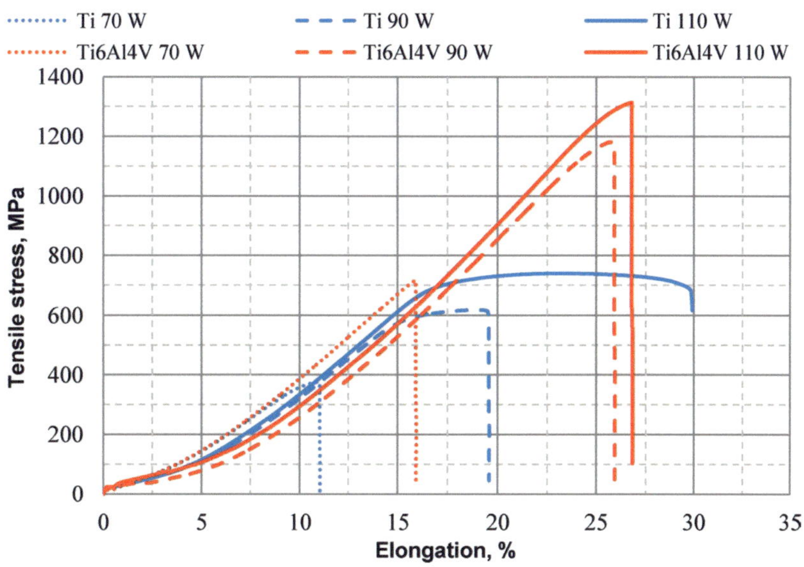

Figure 14. Comparison of diagrams of dependency between tensile stress and elongation for solid samples made of Ti6Al4V alloy and pristine titanium sintered at different laser powers [119].

selective laser sintering, that is, with the laser power of 60 W, which is favourable for porous materials. The examinations pinpoint, however, they are completely unacceptable for solid materials.

Bending strength tests were performed the same as tensile strength tests. The geometrical characteristics of the samples given in **Figure 13** were selected. The tests were carried out in relation to the investigated solid materials, that is, sintered titanium, sintered Ti6Al4V alloy as well as in relation to porous materials manufactured by SLS. The results concerning such tests were obtained as previously, for comparable conditions of selective laser sintering, that is, for the laser power of 60 W. The impact of laser power ranging 70–110 W on bending strength values of sintered titanium and sintered Ti6Al4V alloy for five samples for each treatment variant of each of the materials is shown in **Figure 16**. **Figure 17** compares the diagrams of dependency between bending stress and deflection for solid and porous Ti and Ti6Al4V alloy samples with the pore size of approx. 250 μm, selectively laser sintered in the given conditions.

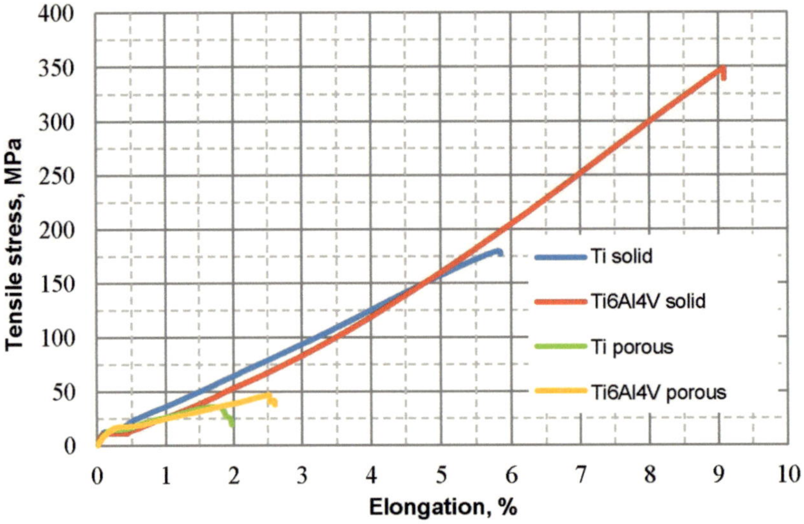

Figure 15. Comparison of diagrams of dependency between tensile stress and elongation for solid and porous laser sintered Ti and Ti6Al4V alloy samples with the pore size of approx. 250 μm [119].

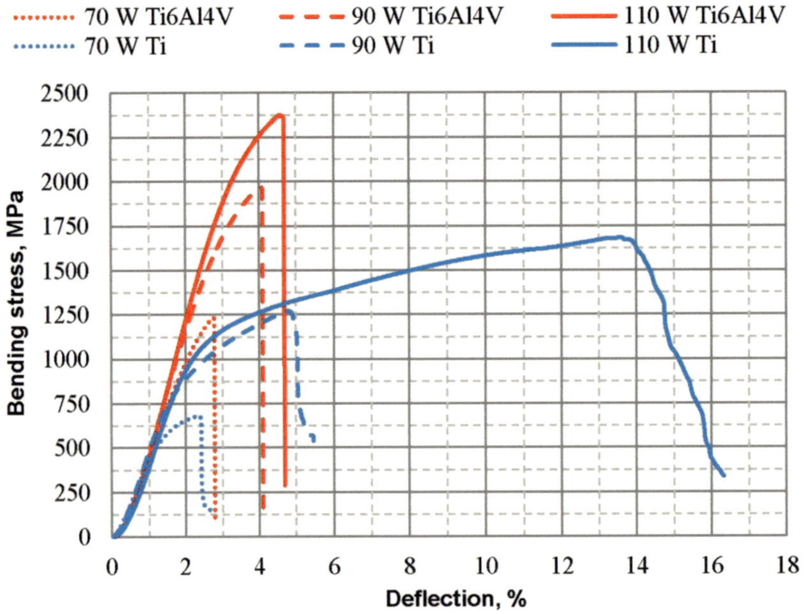

Figure 16. Comparison of diagrams of dependency between bending stress and deflection for solid samples made of Ti6Al4V alloy and pristine titanium sintered at different laser powers [119].

The results of compressive strength tests are also presented for solid and porous Ti and Ti6Al4V alloy samples with the pore size of approx. 250 μm, selectively laser sintered with the laser power of 60 W (**Figure 18**).

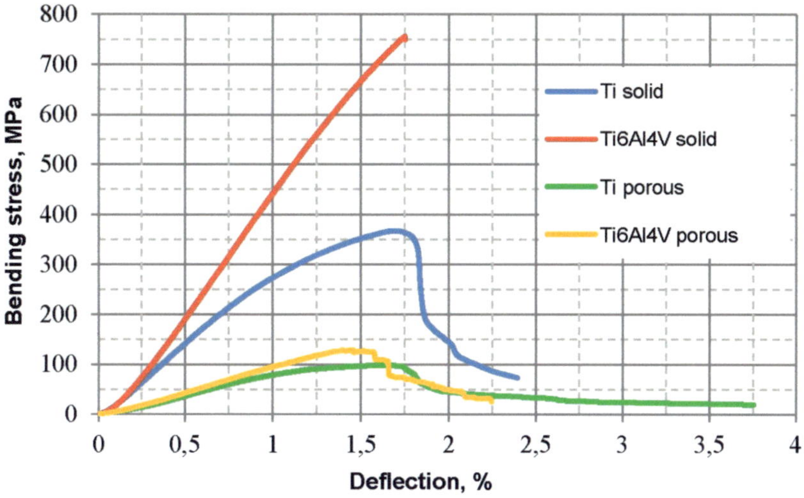

Figure 17. Comparison of diagrams of dependency between bending stress and deflection for solid and porous selectively laser sintered Ti and Ti6Al4V alloy samples with the pore size of approx. 250 μm [119].

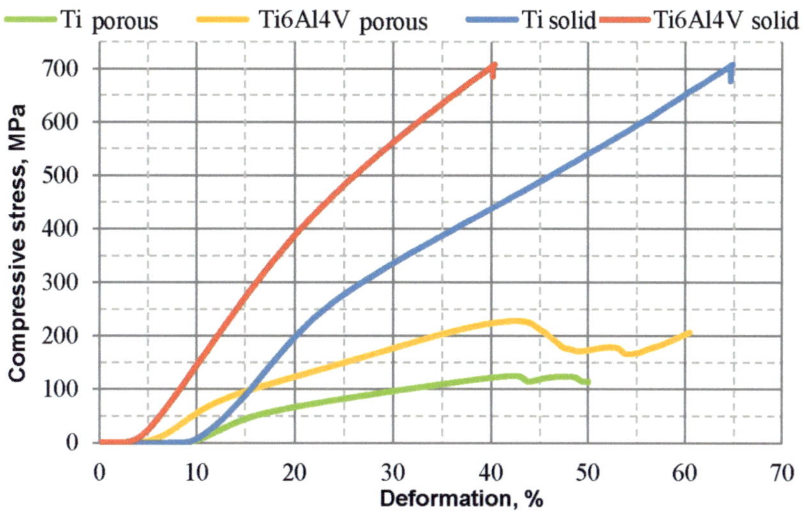

Figure 18. Comparison of diagrams of dependency between compressive stress and deformation for solid and porous selectively laser sintered Ti and Ti6Al4V alloy samples [119].

3. Structure and properties of ALD coatings on the substrate from selectively laser sintered microskeletons made of titanium and Ti6Al4V alloy

The ALD technique enables to deposit a chosen TiO_2 coating very uniformly across the entire surface of a part being treated, also if this part has a porous structure, as is the case with scaffolds. Changes in the sample colour, depending on the number of the executed ALD cycles, hence depending on the thickness of the deposited TiO_2 layer, is an interesting phenomenon observed with a bare eye (**Figure 19**) and in a light stereoscopic microscope, Discovery V12 Zeiss, allowing to view colourful magnified images. An uncoated element has silver-metallic colour, and when subjected to surface treatment by the ALD method, it becomes, respectively: brown-gold (500 cycles), navy blue (1000 cycles) and light blue with silver shade (1500 cycles) (**Figure 19**).

Scaffolds manufactured by the selective laser sintering method from powders of titanium and biocompatible Ti6Al4V titanium alloy, then coated with a thin layer of Al_2O_3 in the process of deposition of single atomic layers, ALD, were examined – analogously as TiO_2 layers – by means of a stereomicroscope, Discovery V12 Zeiss, allowing to identify that – along with the changing deposition thickness of an ALD layer – the colour of scaffolds is changing. The scaffolds, onto which Al_2O_3 layers were deposited in 500 cycles, are dark brown; such onto which Al_2O_3 layers were deposited in 1000 cycles are navy blue, and such onto which Al_2O_3 layers were deposited in 1500 cycles are dark blue (**Figure 20**). The differences in colours are also visible with a bare eye.

The measurements of layers' thickness with a spectroscope ellipsometer for each sample with deposited TiO_2 layers were performed in 25 places and statistical calculations were carried

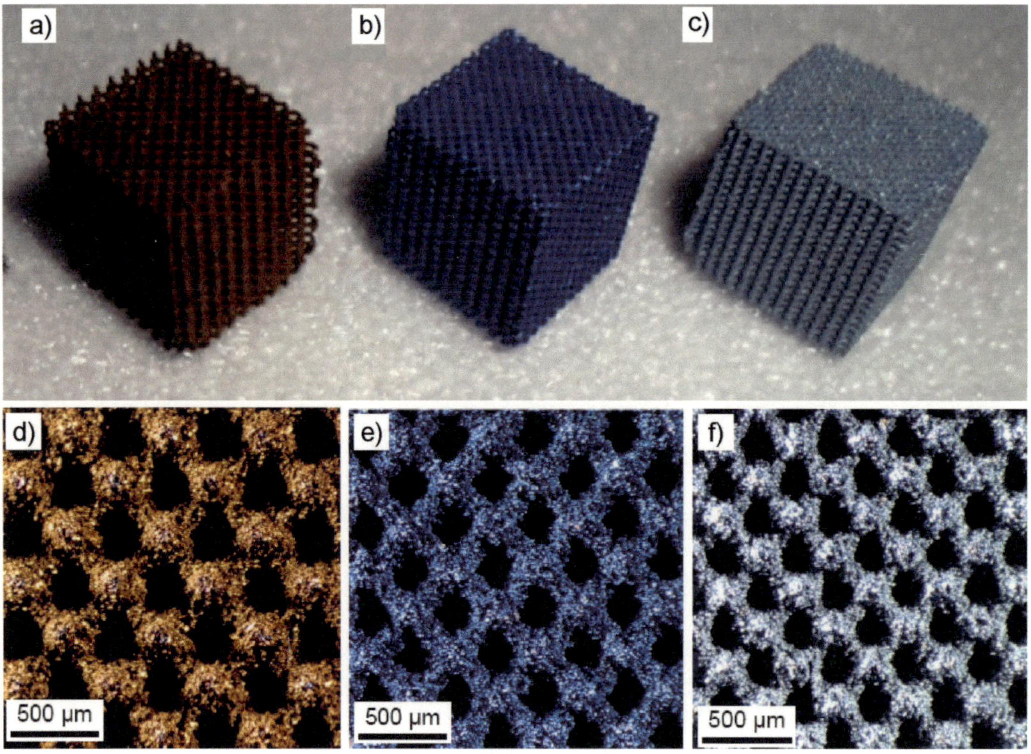

Figure 19. Titanium scaffolds coated with TiO_2 layers deposited by ALD; (a–c) cubic scaffolds viewed with bare eye, (d–f) stereoscopic scaffold images made with the magnification of 32× after: (a, d) 500 cycles, (b, e) 1000 cycles, (c, f) 1500 cycles.

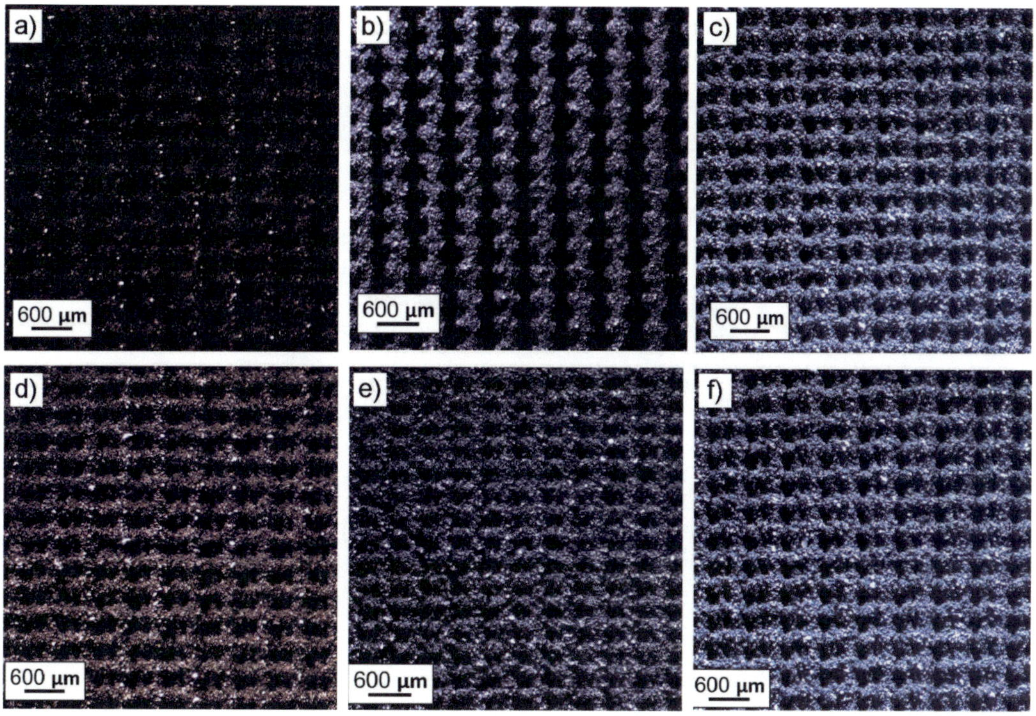

Figure 20. Surface topography of scaffolds manufactured from: (a–c) pristine titanium; (d–f) Ti6Al4V alloy; and coated with a layer of Al_2O_3 in: (a, d) 500; (b, e) 1000; (c, f) 1500 cycles; stereoscope microscope.

out, which has permitted to create a series of 2D maps of thickness distribution of the deposited atomic layers (**Figure 21**). The average thickness of TiO_2 layers deposited by ALD technique for the analysed cases of 500, 1000 and 1500 cycles is, respectively, 56, 99 and 149 nm. The difference in the thickness of the deposited TiO_2 layers on the studied area does not exceed 2 nm, which can be analysed in detail by studying layer thickness distribution maps. The best results were obtained for a layer deposited in 1000 cycles. A difference in the thickness of the deposited layer in this case does not exceed 1.1 nm across the entire area of the surface-treated item.

The topography of the scaffolds' surface coated with TiO_2 layers, deposited by the ALD method, was examined by means of an atomic force microscope, AFM XE-100 Park System, in two and three dimensions (**Figure 22**). There are irregularities with a nanometric scale on the scaffold surface, the number of which is rising proportionally to the number of the deposited layers. In particular, a layer deposited in 500 cycles has a rather uniform granular structure and the larger clusters of atoms are occurring on it only occasionally. In the case of a layer deposited in 1000 cycles, clusters of atoms with a diameter of about 1 µm occur every several microns. The biggest clusters of atoms, forming "islands" with the length of up to several micrometres, occur in the case of a layer deposited in 1500 cycles.

The detailed surface morphology examinations of the produced TiO_2 layers were performed with an electron scanning microscope, Supra 35, by Zeiss, with the accelerating voltage of

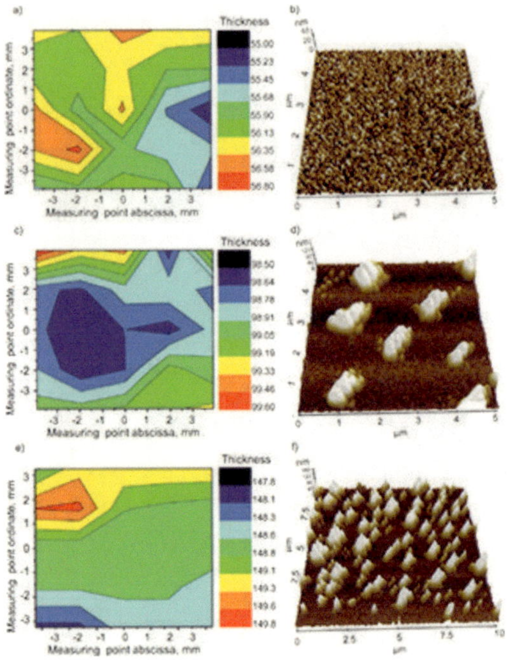

Figure 21. Titanium scaffolds coated with TiO_2 layers deposited by ALD after: (a, b) 500 cycles, (c, d) 1000 cycles, (e, f) 1500 cycles; (a, c, e) thickness deposition maps; (b, d, f) AFM image of 3D surface topography.

Figure 22. Scaffold surface with TiO_2 layer deposited in 1500 cycles presented with different magnificence (a, b, c); SEM images.

10–20 kV (**Figure 22**). Secondary Electrons detection with SE detectors by In Lens was used to obtain surface topography images. A nanometric thickness of TiO_2 layers deposited by ALD results in the fact that the layers can be observed in a scanning electron microscope only for very high magnifications of 150kx (**Figure 22**). A clear difference between a scaffold surface without surface treatment and scaffold surface covered with a TiO_2 layer in an ALD process can be observed only when such high magnifications are used. The scaffold surface, immediately following fabrication, is smooth with clear longitudinal bands arranged every several dozen/several hundreds of nanometres, corresponding to the laser activity direction.

The deposited atomic TiO_2 layer, when magnified by approx. 150kx, is visible as a "sheep", that is, a set of numerous adjacent oval granules of which only a few have a larger diameter.

The detailed surface morphology examinations of thin Al_2O_3 layers produced in a process of deposition of single atomic ALD layers, the same as TiO_2 layers, was examined by means of a scanning electron microscope, Supra 35, by Zeiss, with the accelerating voltage of 2–10 kV (**Figures 23** and **24**) with different magnification of up to 100kx inclusive. The examinations performed with the highest magnification allow to spot a clear difference between the scaffold surface without surface treatment and the scaffold surface with a deposited layer of aluminium oxide, which is coated with convexities with the size of up to approx. 10 nm if 500 cycles are used, to approx. 200 nm if layers are deposited in 1500 cycles (**Figures 23** and **24**).

A TiO_2 layer deposited by ALD onto a surface of a scaffold made of pristine titanium is of an amorphous structure, opposite to a crystalline titanium structure clearly shown in TEM images (**Figure 25**), as confirmed by TiO_2 layer examinations with a transmission electron microscope, TITAN 80–300 by FEI. Depending on the number of cycles, the thickness of a TiO_2 layer deposited by ALD varies between several dozens to hundred and a few dozens of nanometres. The examinations of TiO_2 layers with a transmission electron microscope on samples in a form of thin foils prepared with a focused ion beam microscope (FIB) with the use of gallium arsenide with deposition of a thin layer of platinum show that the both chemical

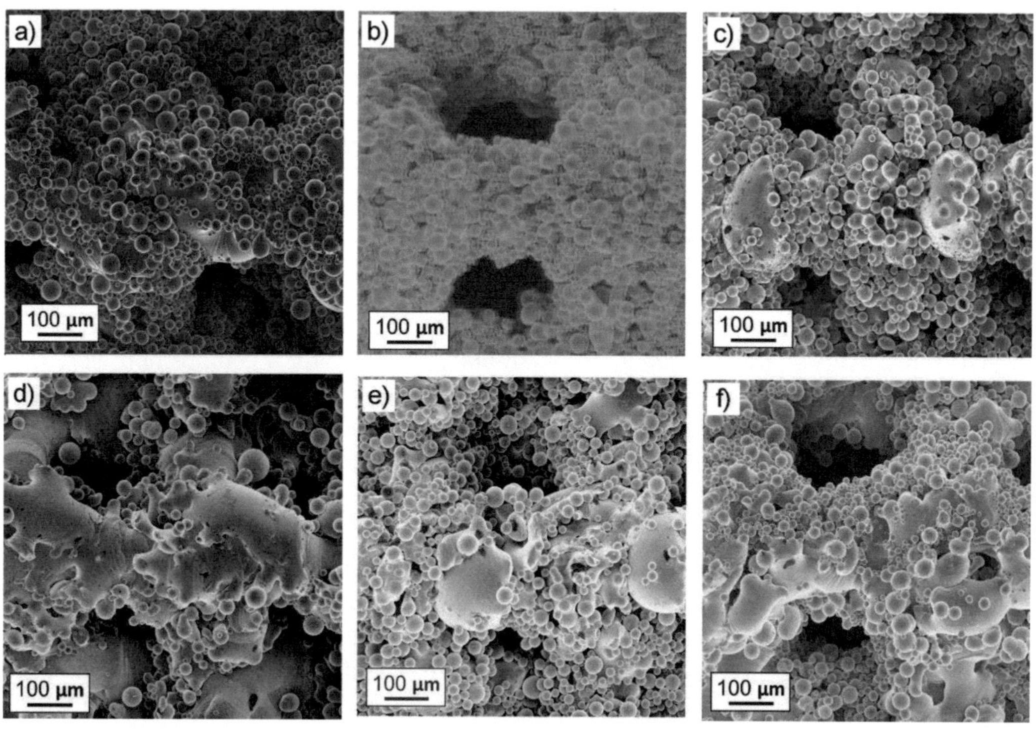

Figure 23. Surface topography of scaffolds manufactured from: (a–c) pristine titanium; (d–f) Ti6Al4V alloy and coated with Al_2O_3 layer in: (a, d) 500; (b, e) 1000; (c, f) 1500 cycles; SEM.

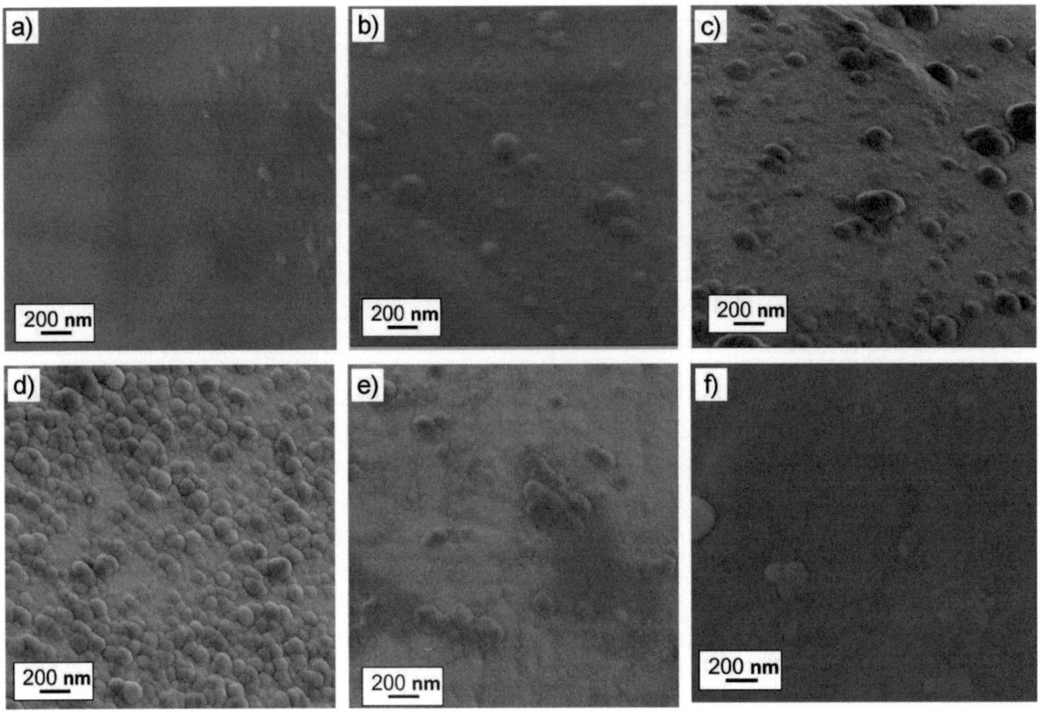

Figure 24. High-resolution surface topography of scaffolds manufactured from: (a–c) pristine titanium; (d–f) Ti6Al4V alloy and coated with Al_2O_3 layer in: (a, d) 500; (b, e) 1000; (c, f) 1500 cycles; SEM.

elements, apart from titanium and oxygen, are found on the EDS chart from an area situated on the periphery of the deposited layer, located in the direct neighbourhood of the protected layer of platinum (**Figure 25**). The EDS examinations of the deposited layers carried out in the region closer to the substrate reveal the presence of titanium and oxygen only (**Figure 25**), as is also confirmed by a prepared distribution map of chemical elements (**Figure 26**).

In order to confirm the presence of layers consisting of TiO_2 on the surface of a scaffold fabricated from TiAl6V4 powder, qualitative examinations of chemical composition were performed with the EDS scattered X-ray radiation spectroscopy method using an EDS (energy dispersive spectrometer), and a scaffold not containing a TiO_2 layer was considered a reference material. An analysis of the diagrams shown in **Figure 27** indicates that a spectrum was recorded in both cases with reflexes distinctive for titanium, aluminium and vanadium, whereas a reflex coming from titanium and oxygen exists additionally in a material covered with a layer of ALD, which corresponds to the occurrence of a TiO_2 layer on the surface of this material. Examinations were also undertaken using the inVia Reflex device by Renishaw, being an automated Raman system. A Raman spectrum obtained with the wave dispersion method, after base line correction, within the spectral range of 150–3200 cm^{-1}, is coming from a material being a scaffold, while a Raman spectrum coming from a material coated with a thin layer of titanium dioxide is shown in **Figure 27**, which – just like an EDS analysis – confirms the presence of titanium, aluminium and vanadium in the reference material; moreover

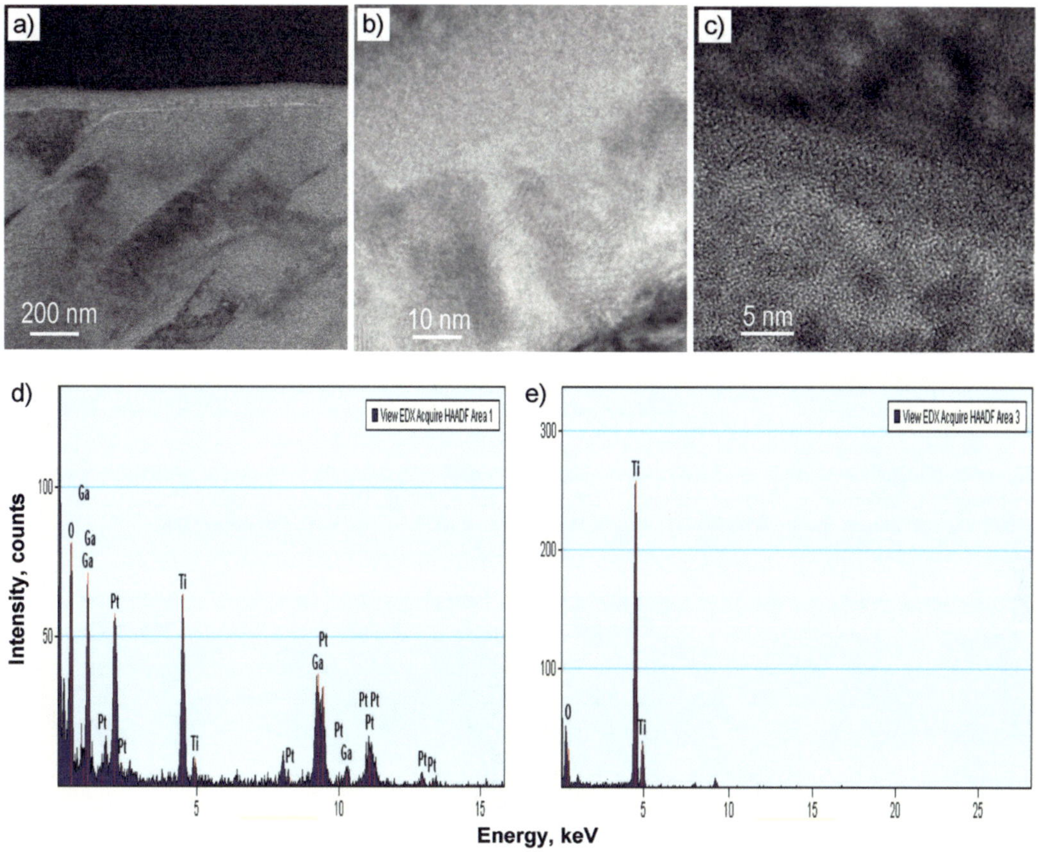

Figure 25. Amorphous TiO_2 layer deposited onto pristine titanium with a crystalline structure in a technological process lasting 1500 cycles; (a–c) HRTEM; (d, e) qualitative analysis of EDS chemical composition of TiO_2 layer: (d) from the peripheral region contiguous to the protective platinum layer; (c) near a titanium substrate.

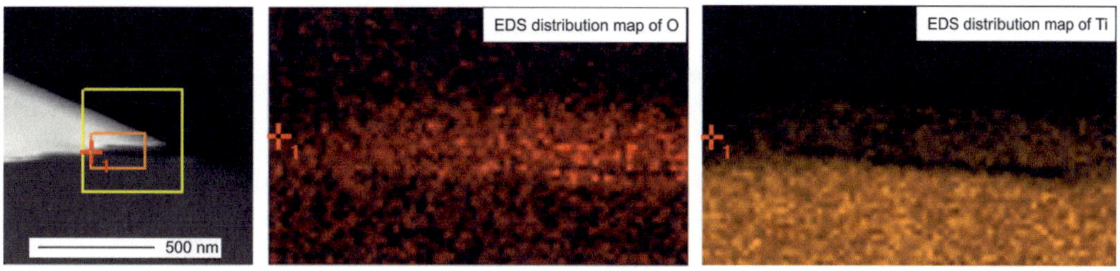

Figure 26. Distribution map of chemical elements present in the TiO_2 layer situated near titanium substrate.

oxygen, present in the surface layer, is found – apart from such chemical elements – in a material coated with a TiO_2 layer.

It was found, with special WiRETM 3.1 software, that a layer deposited by the ALD method is anatase, being a polymorphous type of titanium dioxide. **Figure 28** shows the images of

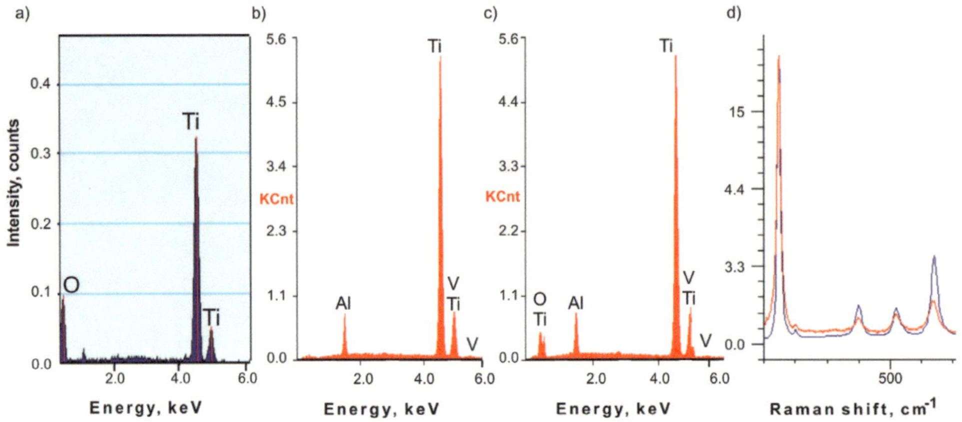

Figure 27. (a) Qualitative analysis of EDS chemical composition of TiO_2 layer; (b, c) results of qualitative analysis of chemical composition of: (b) TiAl6V4 without ALD layer, (c) TiAl6V4 with TiO_2 layer deposited by ALD method after 1500 cycles; (d) Raman spectrum of Ti6Al4V scaffold deposited by ALD method with TiO_2 layer [124].

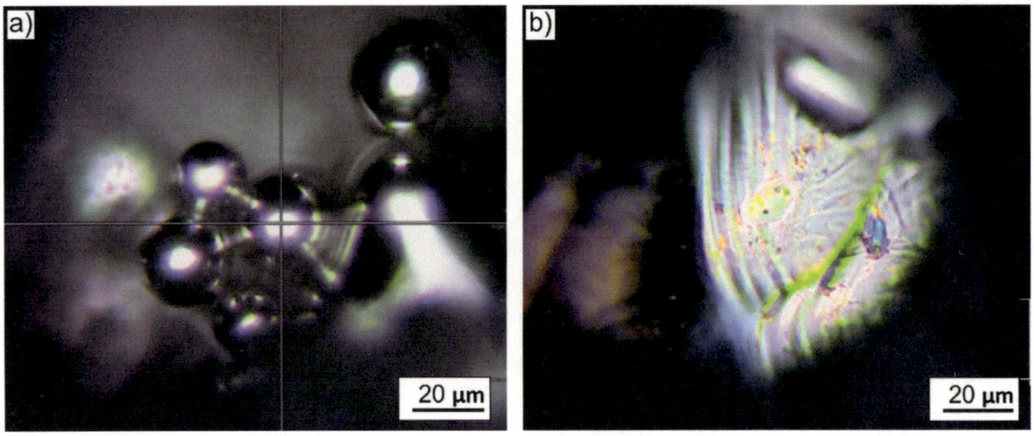

Figure 28. Points where Raman spectra were recorded, originating from: (a) reference Ti6Al4V scaffold, (b) Ti6Al4V scaffold covered with a layer of TiO_2; images coming from a confocal microscope.

points where spectra were recorded, originating from a reference Ti6Al4V scaffold and a scaffold covered with a layer of TiO_2; the images were made with a confocal microscope being a constituent part of the inVia Reflex device.

Structural examinations were performed with the X-ray diffraction (XRD) method (**Figure 29**) with an X′Pert Pro X-ray diffractometer by Panalytical (CuKα radiation, $\lambda = 1.54050 \cdot 10^{-10}$ m) using filtered radiation of a copper lamp with the voltage of 45 kV and a filament current of 35 mA, and the deposited TiO_2 layers were examined with the razing-incidence method due to the small thickness of not more than 150 nm, thus extinguishing

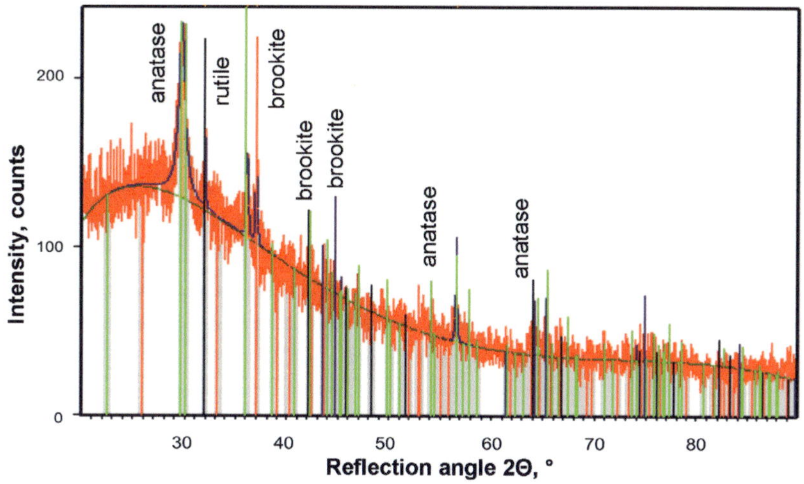

Figure 29. X-ray diffraction pattern of TiO_2 layer applied by atomic layer deposition (ALD) performed with the razing-incidence X-ray diffraction method.

the peaks coming from the substrate. Reflexes coming from three polymorphous variants of titanium dioxide, that is, anatase, rutile and brookite, were identified in the examinations.

In order to confirm that the observed ALD layers are fabricated from Al_2O_3 aluminium oxide, qualitative examinations were performed of chemical composition with the EDS scattered X-ray radiation spectroscopy method and were displayed as diagrams in **Figure 30**. For thin Al_2O_3 layers, deposited on a scaffold made of titanium, spectra were recorded with reflexes characteristic for aluminium and oxygen, coming from a layer and for titanium coming from the substrate (**Figure 30a**). A spectrum was recorded for a Ti6Al4V scaffold, with

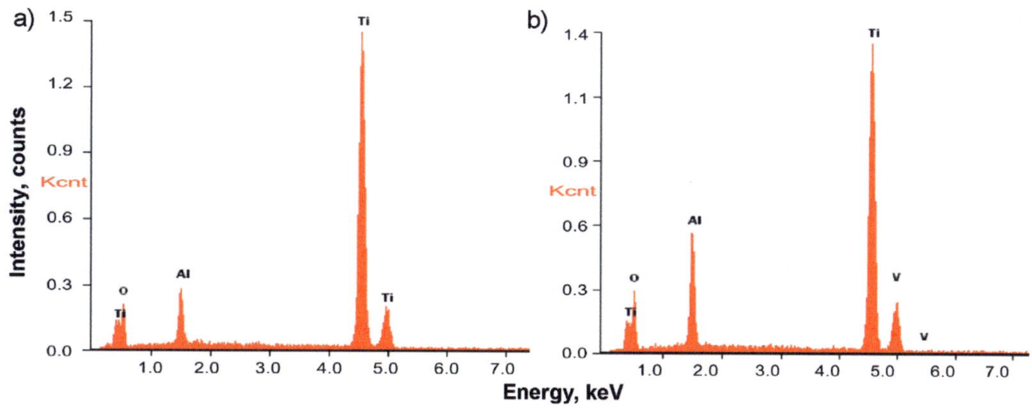

Figure 30. EDS spectrum made for scaffolds manufactured from: (a) pristine titanium; (b) Ti6Al4V titanium alloy; and coated with Al_2O_3 layers in 500 cycles.

reflexes characteristic for titanium and vanadium, coming from the substrate and oxygen from the layer. The reflex recorded from aluminium comes from the coating and substrate (**Figure 30b**). Distribution maps of chemical elements were additionally performed in the investigated materials. The presence of titanium, aluminium and oxygen was found in a titanium scaffold coated with aluminium oxide, and additionally vanadium is also found in a scaffold produced from Ti6Al4V. Supplementary X-ray examinations were also performed as a result of which no crystalline phase of aluminium oxide was found for ALD layers, which indicates their amorphous form. Such a result is expected due to the analogy with TiO_2 layers – deposited with the same method – subjected to examinations at a nanometric scale. An amorphous structure of the deposited layers is clearly seen in TEM images as opposed to crystalline titanium being the substrate.

4. Structure and properties of layers of hydroxyapatite deposited with the sol-gel immersion method on microporous selectively laser sintered skeletons made of titanium and Ti6Al4V alloy

Thin layers of hydroxyapatite were deposited with the immersion sol-gel technique (dip coating). The solution was prepared using hydroxyapatite nanopowder (HA), polyethylene glycol (PEG), glycerine and ethyl alcohol. The scaffolds manufactured from pristine titanium and Ti6Al4V alloy, coated with a layer of hydroxyapatite, manufactured by the sol-gel technique, were examined by means of a stereomicroscope microscope, Discovery V12 ZEISS. The detailed surface morphology examinations of the produced layers were undertaken with an electron scanning microscope Supra 35 by Zeiss. SEM images of thin layers were presented, on which hydroxyapatite particles are visible (**Figures 31** and **32**). The shape of the majority of hydroxyapatite particles is oval, however, a large content of particles with an elongated shape, with rounded, sometimes sharpened edges, is observed. The size of hydroxyapatite particles can be estimated at 20–80 nm. In case of sol-gel layers deposited on a scaffold made of titanium, spectra were recorded with reflexes characteristic for calcium, phosphorus and

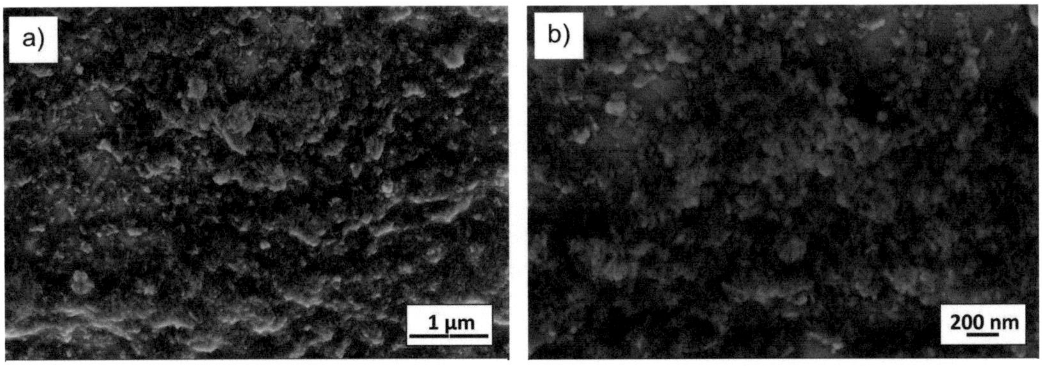

Figure 31. Scaffold made of Ti with the sol-gel layer of hydroxyapatite deposited after 10 immersions presented with different magnificence (a, b); SEM images.

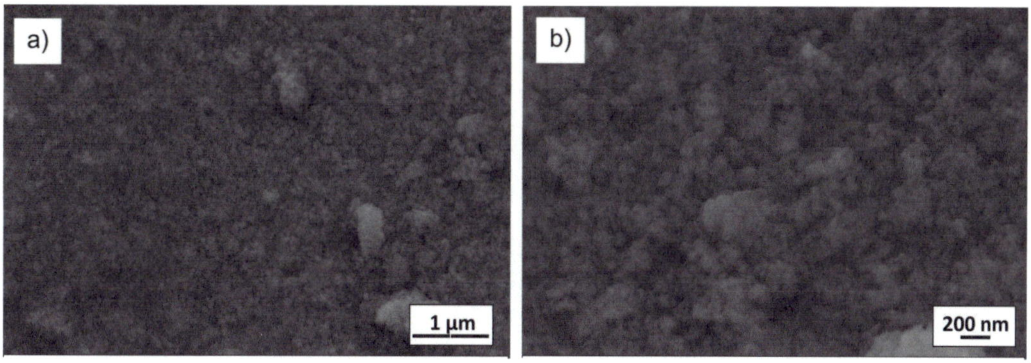

Figure 32. Scaffold made of Ti6Al4V alloy with the sol-gel layer of hydroxyapatite deposited after 10 immersions presented with different magnificence (a, b); SEM images.

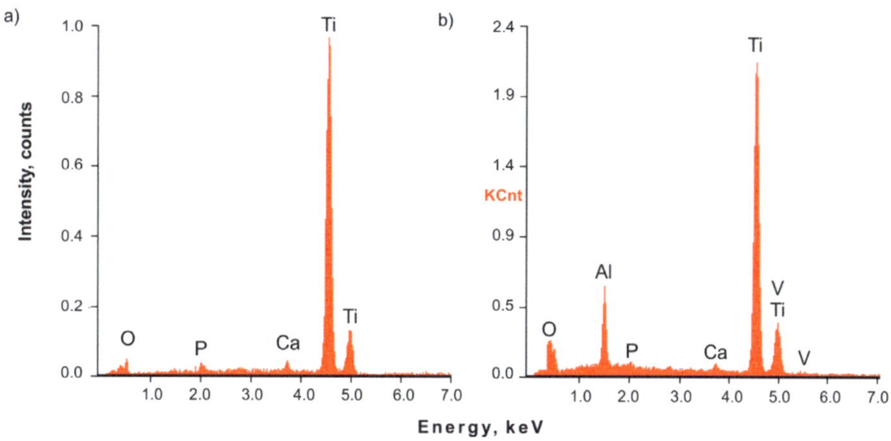

Figure 33. EDS spectrum of scaffold made of: (a) Ti, (b) Ti6Al4V: With a sol-gel layer of hydroxyapatite deposited after 10 immersions.

oxygen coming from a layer, and being the main hydroxyapatite components, and a reflex for titanium coming from the substrate (**Figure 33**). Similarly, for a scaffold made of Ti6Al4V alloy, spectra were recorded with reflexes characteristic for calcium, phosphorus and oxygen coming from the layer, and titanium, aluminium and vanadium, coming from the substrate. In addition, distribution maps of chemical elements were additionally performed in the investigated samples. Titanium was identified in a pure Ti scaffold, and also vanadium if Ti6Al4V alloy was also used. In the samples covered with a sol-gel layer, apart from the chemical elements coming from the substrate (e.g. titanium or titanium, aluminium and vanadium), calcium, phosphorus and oxygen were additionally identified, coming from a layer of hydroxyapatite (**Figure 33**). Structural examinations of sol-gel layers were performed by means of an X-ray structure analysis. Reflexes were identified coming from hydroxyapatite (**Figure 34**).

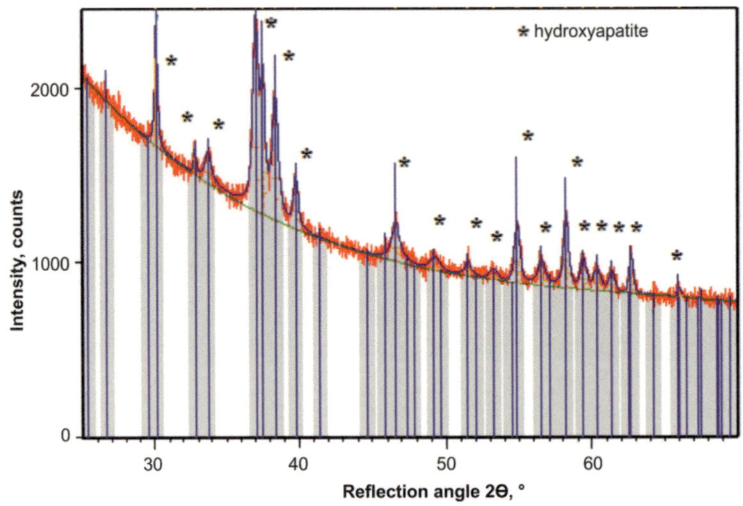

Figure 34. X-ray diffraction pattern of hydroxyapatite layer made with the sol-gel method.

5. Final remarks

One of the many elements largely influencing the quality of the society's life is the standard of healthcare. For this reason, all the activities in this sphere are considered as priorities in social policy in many countries and by international organisations. The life quality of patients after mechanical injuries, with tumorous and genetic diseases and with lost teeth greatly depends on the technical level, innovativeness and avant-garde solutions for medical and dental implants. It is obvious that the successfulness of the medical care provided depends on competencies of the doctors performing diagnostics, taking relevant decisions, applying appropriate therapies and related surgical procedures. The issue is a synergy of the actions jointly undertaken by medical doctors and engineers. The aspects of engineering-assisted medicine relate, both, to the constructional, material and technological design of, in particular, medical and dental implants. The development of medical implants is largely dependent upon advancements in biomaterials engineering, characterised by the required biological compatibility and harmonious interaction with living matter, without acute or chronic reactions, nor an inflammatory condition of the surrounding tissues after introduction into an organism. The advanced research efforts described in this chapter concern the engineering aspects of development of innovative implant-scaffolds, which enable the adhesion and proliferation of a patient's living cells into an intentionally designed porous structure of an implant which, on one hand, acts as an element transferring high stresses existing in an organism, and on the other hand acts as a scaffold into which a patient's cells are growing into. Metallic materials, despite their disadvantages, such as insufficient corrosive resistance and insufficient biotolerance in some areas of applications, are characterised by a pool of very advantageous mechanical properties. High fatigue corrosion resistance, brittle cracking resistance and tensile and bending strength should be considered especially significant. However, weak susceptibility of metallic materials as a substrate for cells' development is undoubtedly an issue. An effect of a substrate on the proliferation of living cells has been described in detail based on comprehensive literature studies. This aspect is very complex. Porous metallic materials are an attractive implantation material due to better, as compared to traditional alloys, accommodation of the elasticity modulus to the bone and because bone tissue can grow on pores and by ensuring appropriate implant fixation to the bone. Dedicated technologies ensuring surface treatment inside pores with the size of 200–400 μm are fully innovative and require very advanced nanotechnological instrumentation. Titanium and titanium alloys belong to a group of metallic materials applied for many years in the fabrication of implants for bone surgery, maxillo-facial surgery and prosthodontics. The principal reasons include relatively low density, a beneficial strength-to-yield stress limit ratio, good corrosive resistance and best biocompatibility in this group of biomaterials. Titanium and titanium alloys are considered as such allowing to eliminate the risk associated with a harmful effect of chemical elements occurring as alloy additives in metallic materials, and concerns and controversies around some of them have not been scientifically proven until now.

A porous scaffold, with its dimensions and shape perfectly suited to a patient's tissue loss, made of a biocompatible material (Ti or Ti6AlV4), additionally coated with a nanometric layer of osteoconductive titanium oxide, aluminium oxide or hydroxyapatite inside pores, seems to be a breakthrough solution. It is a highly hybridised and composite engineering material, fabricated by a hybrid technology combining avant-garde and experimental additive technologies of selective laser sintering SLS, in conjunction with ALD and sol-gel technologies appropriate for nanotechnology. Modern CAD/CAMD software allows to convert the data acquired at a clinical stage into a 3D solid model of a patient's lost tissues. The model is then converted into a porous model through the multiplication of a unit cell whose dimensions and shape may be designed according to a patient's individual preferences. The pores existing in the material structure have the diameter of up to 500 μm and should be open, because a scaffold, in its intended conditions of use, is to grow through a patient's living tissue.

This chapter presents the outcomes of numerous author's complementary technological, structural and strength investigations which are a basis for optimised selection of engineering materials, adequate technologies and constructional assumptions for completely new and innovative products. Biomimetic, light, porous, rough and biocompatible materials with unique mechanical and functional properties finding their application for completely innovative scaffolds and implant-scaffolds can be manufactured owing to a custom combination of advanced methods of computer aided materials design and selective laser sintering (SLS) of titanium and Ti6AlV4 alloy powders with avant-garde deposition methods of single atomic ALD or sol-gel methods. A series of the investigations performed to date, in which technological conditions were established for the fabrication of this type of coatings inside pores, was completed successfully, which is presented thoroughly in this chapter. The role of the manufactured layers, with their thickness which can be programmed in advance, and not only controlled *post factum*, is to enhance osteoconduction of the materials which are to be ultimately placed in a human organism. The application of the technologies and materials described is of fundamental significance to ensure the synergy of clinical effects obtained by medical doctors and dentists by classical prosthetics and implantation and the natural nesting and proliferation of living tissues in a microporous bonding zone with scaffolds or implant-scaffolds created from the newly developed engineering materials. It is the authors' intention to launch such new products soon for broad application in medicine and regenerative and intervention dentistry. Scaffolds will be fabricated this way in the form of microporous skeletons, and, optionally, implant-scaffolds consisting of a solid core and a microporous, strongly developed surface layer, connected in a hybrid way into the solid whole. The results of the investigations conducted allow to perform the currently pursued author's biological research pertaining to the nesting and proliferation of living tissues in the micropores of the created porous microskeletons and to assess the deposition of internal surfaces of micropores with layers supporting the growth of living tissues. Another report will be drafted following the completion of the research, concerning the details of the biological and clinical investigations performed, supplementary to the research and engineering works presented in this chapter.

Acknowledgements

Fundusze Europejskie Inteligentny Rozwój ASKLEPIOS Unia Europejska

The research presented in the chapter was realised in connection with the Project POIR. 01.01.01-00-0485/16-00 on "IMSKA-MAT Innovative dental and maxillofacial implants manufactured using the innovative additive technology supported by computer-aided materials design ADDMAT" realised by the Medical and Dental Engineering Centre for Research, Design and Production ASKLEPIOS in Gliwice, Poland and co-financed by the National Centre for Research and Development in Warsaw, Poland.

Author details

Anna D. Dobrzańska-Danikiewicz[2*], Leszek A. Dobrzański[1], Marek Szindler[3], Lech B. Dobrzański[1], Anna Achtelik-Franczak[1] and Eugeniusz Hajduczek[3#]

*Address all correspondence to: anna.dobrzanska.danikiewicz@gmail.com

1 Medical and Dental Engineering Centre for Research, Design and Production Asklepios, Gliwice, Poland

2 University of Zielona Góra, Faculty of Mechanical Engineering, Zielona Góra, Poland

3 Faculty of Mechanical Engineering, Silesian University of Technology, Gliwice, Poland

Deceased author

References

[1] Anselme K, Bigerelle H. Topography effects of pure titanium substrates on human osteoblast long-term adhesion. Acta Biomaterialia. 2005;**1**:211-222. DOI: 10.1016/j.actbio.2004.11.009

[2] Huang H-H, Ho C-T, Lee TH, Lee T-L, Liao KK, Chen FL. Effect of surface roughness of ground titanium on initial cell adhesion. Biomolecular Engineering. 2004;**21**:93-97. DOI: 10.1016/j.bioeng.2004.05.001

[3] Kaiser LR. The future of multihospital systems. Topics in Health Care Financing. 1992;**18**:32-45

[4] Cogle CR, Guthrie SM, Sanders RC, Allen WL, Scott EW, Petersen BE. An overview of stem cell research and regulatory issues. Mayo Clinic Proceedings. 2003;**78**:993-1003

[5] Metallo CM, Azarin SM, Ji L, De Pablo JJ, Palecek SP. Engineering tissue from human embryonic stem cells. Journal of Cellular and Molecular Medicine. 2008;**12**:709-729

[6] Atala A, Lanza R, Thomson JA, Nerem R, editors. Principles of Regenerative Medicine. Second ed. San Diego: Academic Press; 2011. p. 1202

[7] Regenerative Medicine 2006, Report. US National Institutes of Health. 2006. Available from: http://stemcells.nih.gov/staticresources/info/scireport/PDFs/Regenerative_Medicine_2006.pdf [Accessed: 2017-03-14]

[8] Placzek MR, Chung I-M, Macedo HM, Ismail S, Mortera Blanco T, Lim M, Cha JM, Fauzi I, Kang Y, Yeo DCL, Ma CYJ, Polak JM, Panoskaltsis N, Mantalaris A. Stem cell bioprocessing: Fundamentals and principles. Journal of The Royal Society Interface. 2009;**6**:209-232

[9] Fung YC. A Proposal to the National Science Foundation for an Engineering Research Center at UCSD. Center for the Engineering of Living Tissues. UCSD #865023; 2001

[10] Langer R, Vacanti JP. Tissue engineering. Science. 1993;**260**:920-926

[11] Viola J, Lal B, Grad O. The Emergence of Tissue Engineering as a Research Field. USA: The National Science Foundation; 2003. Available from: https://www.nsf.gov/pubs/2004/nsf0450/start.htm Accessed: 2017-03-17

[12] Mac Arthur BD, Oreffo ROC. Bridging the gap. Nature. 2005;**433**:19

[13] Atala A, Lanza RP, editors. Methods of Tissue Engineering. San Diego: Academic Press; 2002. p. 1285

[14] Lanza RP, Langer R, Vacanti J, editors. Principles of Tissue Engineering. San Diego: Academic Press; 2000. p. 995

[15] Dobrzańska-Danikiewicz AD, Gaweł TG, Wolany W. Ti6Al4V titanium alloy used as a modern biomimetic material. Archives of Materials Science and Engineering. 2015;**76**:150-156

[16] Dobrzański LA, Basis of Material and Metal Science: Engineering materials with basis of materials design. WNT: Warszawa; 2002. 1500 p. ISBN 83-204-2793-2 (in Polish)

[17] Caiazzo F, Cardaropoli F, Alfieri V, et al. Disk-laser welding of Ti-6Al-4V titanium alloy plates in T-joint configuration. Procedia Engineering. 2017;**183**:219-226

[18] Donachie MJ, Donachie SJ. Superalloys: A Technical Guide. Materials Park, Ohio: ASM International; 2002

[19] Shin H, Jo S, Mikos AG. Biomimetic materials for tissue engineering. Biomaterials. 2003;**24**:4353-4364

[20] Farrell T. Superalloy materials now cost competitive in vacuum furnace hot-zone construction. Industrial Heating. 2005;**72**:129-133

[21] Ramakrishnaiah R, al Kheraif AA, Mohammad A, et al. Preliminary fabrication and characterization of electron beam melted Ti–6Al–4V customized dental implant. Saudi Journal of Biological Sciences. 2017;**24**:787-796

[22] Parthasarathy J, Starly B, Raman S, Christensen A. Mechanical evaluation of porous titanium (Ti6Al4V) structures with electron beam melting (EBM). Journal of the Mechanical Behavior of Biomedical Materials. 2010;**3**:249-259

[23] Toh WQ, Sun Z, Tan X, et al., Comparative study on tribological behavior of Ti-6Al-4V and Co-Cr-Mo samples additively manufactured with electron beam melting. Proceedings of the 2nd International Conference on Progress in Additive Manufacturing (Pro-AM 2016); 2016. pp. 342-348

[24] Ceretti E, Gay JDC, Rodriguez CA, JVL d S. Biomedical Devices: Design, Prototyping, and Manufacturing. John Wiley & Sons, Inc., Hoboken, New Jersey, USA; 2017

[25] Elias C, Lima J, Valiev R, Meyers M. Biomedical applications of titanium and its alloys. Journal of Metals. 2008;**60**:46-49

[26] Mrksich M. A surface chemistry approach to studying cell adhesion. Chemical Society Reviews. 2000;**29**:267-273. DOI: 10.1039/a705397e

[27] Ross AM, Jiang Z, Bastmeyer M, Lahann J. Physical aspects of cell culture substrates: Topography, roughness, and elasticity. Small. 2012;**8**:336-355. DOI: 10.1002/smll.201100934

[28] Xu Y, Shi Y, Ding S. A chemical approach to stem-cell biology and regenerative medicine. Nature. 2008;**453**:338-344. DOI: 10.1038/nature07042

[29] Griffith LG. Polymeric biomaterials. Acta Materialia. 2000;**48**:263-277. DOI: 10.1016/S1359-6454(99)00299-2

[30] Dee KC, Andersen TT, Bizios R. Osteoblast population migration characteristics on substrates modified with immobilized adhesive peptides. Biomaterials. 1999;**20**:221-227

[31] Hynes RO. Cell adhesion: Old and new questions. Trends in Cell Biology. 1999;**9**:M33-M37

[32] Benoit DSW, Anseth KS. The effect on osteoblast function of colocalized RGD and PHSRN epitopes on PEG surfaces. Biomaterials. 2005;**26**:5209-5220. DOI: 10.1016/j.biomaterials.2005.01.045

[33] Garcıa AJ, Vega MD, Boettiger D. Modulation of cell proliferation and differentiation through substrate-dependent changes in fibronectin conformation. Molecular Biology of the Cell. 1999;**10**:785-793

[34] Spargo BJ, Testoff MA, Nielsen TB, Stenger DA, Hickman JJ, Rudolph AS. Spatially controlled adhesion, spreading, and differentiation of endothelial cells on self-assembled molecular monolayers. Proceedings of the National Academy of Sciences of the USA. 1994;**91**:11070-11074

[35] Fuller G, Shields D. Molecular Basis of Cell Biology. Chapter 8. Warszawa: Wydawnictwo Lekarskie PZWL; 2005. 300 p. ISBN: 83-200-3259-8 (in Polish)

[36] Petersen S, Gattermayer M, Biesalski M. Hold on at the right spot: Bioactive surfaces for the design of live-cell micropatterns. Advances in Polymer Science. 2011;**240**:35-73. DOI: 10.1007/12_2010_77

[37] Lee MH, Ducheyne P, Lynch L, Boettiger D, Composto RJ. Effect of biomaterial surface properties on fibronectin–$\alpha5\beta1$ integrin interaction and cellular attachment. Biomaterials. 2006;**27**:1907-1916. DOI: 10.1016/j.biomaterials.2005.11.003

[38] Roach P, Eglin D, Ronde K, Perry CC. Modern biomaterials: A review – Bulk properties and implications of surface modifications. Journal of Materials Science, Materials in Medicine. 2007;**18**:1263-1277. DOI: 10.1007/s10856-006-0064-3

[39] McClary KB, Ugarova T, Grainger DW. Modulating fibroblast adhesion, spreading, and proliferation using self-assembled monolayer films of alkylthiolates on gold. Journal of Biomedical Materials Research. 2000;**50**:428-439

[40] Florjańczyk Z, Penczka S, editors. Chemistry of Polymers. Volume 3. Natural Polymers and Polymers with Special Properties. Warszawa: Oficyna Wydawnicza Politechniki Warszawskiej; 1999. 25 p. ISBN: 83-7207-029-6 (in Polish)

[41] Scotchford CA, Gilmore CP, Cooper E, Leggett GJ, Downes S. Protein adsorption and human osteoblast-like cell attachment and growth on alkylthiol on gold self-assembled monolayers. Journal of Biomedical Materials Research. 2002;**59**:84-99

[42] Comelles J, Estévez M, Martínez E, Samitier J. The role of surface energy of technical polymers in serum protein adsorption and MG-63 cells adhesion. Nanomedicine: Nanotechnology, Biology, and Medicine. 2010;**6**:44-51. DOI: 10.1016/j.nano.2009.05.006

[43] Yang L, Li Y, Sheldon BW, Webster TJ. Altering surface energy of nanocrystalline diamond to control osteoblast responses. Journal of Materials Chemistry. 2012;**22**:205-214. DOI: 10.1039/C1JM13593G

[44] Lai H-C, Zhuang L-F, Liu X, Wieland M, Zhang Z-Y, Zhang Z-Y. The influence of surface energy on early adherent events of osteoblast on titanium substrates. Journal of Biomedical Materials Research, Part A. 2009;**93**:289-296. DOI: 10.1002/jbm.a.32542

[45] Yap FL, Zhang Y. Protein and cell micropatterning and its integration with micro/nanoparticles assembly. Biosensors and Bioelectronics. 2007;**22**:775-788. DOI: 10.1016/j.bios.2006.03.016

[46] Fabianowski W, Polak B, Lewandowska-Szumieł M. Polymers used for bone reconstruction - evaluation of selected polymeric substrates in osteoblast culture in vitro. Polimery. 2004;**49**:522-529 (in Polish)

[47] Papenburg BJ, Rodrigues ED, Wessling M, Stamatialis D. Insights into the role of material surface topography and wettability on cell-material interactions. Soft Matter. 2010;**6**:4377-4388. DOI: 10.1039/B927207K

[48] Curran JM, Chen R, Hurt JA. Controlling the phenotype and function of mesenchymal stem cells in vitro by adhesion to silane-modified clean glass surfaces. Biomaterials. 2005;**26**:7057-7067. DOI: 10.1016/j.biomaterials.2005.05.008

[49] Keselowsky BG, Collard DM, Garcia AJ. Integrin binding specificity regulates biomaterial surface chemistry effects on cell differentiation. Proceedings of the National Academy of Sciences of the USA. 2005;**102**:5953-5957. DOI: 10.1073/pnas.0407356102

[50] Lan MA, Gersbach CA, Michael KE, Keselowsky BG, Garcia AJ. Myoblast proliferation and differentiation on fibronectin-coated self assembled monolayers presenting different surface chemistries. Biomaterials. 2005;**26**:4523-4531. DOI: 10.1016/j.biomaterials. 2004.11.028

[51] Liu ZZ, Wong ML, Griffiths LG. Effect of bovine pericardial extracellular matrix scaffold niche on seeded human mesenchymal stem cell function. Scientific Reports. 2016;**6**(37089):1-12. DOI: 10.1038/srep37089

[52] Sakakibara K, Hill JP, Ariga K. Thin-film-based nanoarchitectures for soft matter: Controlled assemblies into two-dimensional worlds. Small. 2011;**7**:1288-1308. DOI: 10.1002/smll.201002350

[53] Faid K, Voicu R, Bani-Yaghoub M, Tremblay R, Mealing G, Py C, Barjovanu R. Rapid fabrication and chemical patterning of polymer microstructures and their applications as a platform for cell cultures. Biomedical Microdevices. 2005;**7**:179-184. DOI: 10.1007/s10544-005-3023-8

[54] Wells RG. The role of matrix stiffness in regulating cell behavior. Hepatology. 2008;**47**:1394-1400. DOI: 10.1002/hep.22193

[55] Rehfeldt F, Engler AJ, Eckhardt A, Ahmed F, Discher DE. Cell responses to the mechanochemical microenvironment – Implications for regenerative medicine and drug delivery. Advanced Drug Delivery Reviews. 2007;**59**:1329-1339. DOI: 10.1016/j. addr.2007. 08.007

[56] Marklein RA, Burdick JA. Controlling stem cell fate with material design. Advanced Materials. 2010;**22**:175-189. DOI: 10.1002/adma.200901055

[57] Kim HD, Peyton SR. Bio-inspired materials for parsing matrix physicochemical control of cell migration: A review. Integrative Biology. 2012;**4**:37-52. DOI: 10.1039/c1ib00069a

[58] Ehrbar M, Sala A, Lienemann P, Ranga A, Mosiewicz K, Bittermann A, Rizzi CS, Weber FE, Lutolf MP. Elucidating the role of matrix stiffness in 3D cell migration and remodelling. Biophysical Journal. 2011;**100**:284-293. DOI: 10.1016/j.bpj.2010.11.082

[59] Engler AJ, Sen S, Sweeney HL, Discher DE. Matrix elasticity directs stem cell lineage specification. Cell. 2006;**126**:677-689. DOI: 10.1016/j.cell.2006.06.044

[60] Moore SW, Sheetz MP. Biophysics of substrate interaction: Influence on neural motility, differentiation, and repair. Developmental Neurobiology. 2011;**71**:1090-1101. DOI: 10.1002/dneu.20947

[61] Lutolf MP, Gilbert PM, Blau HM. Designing materials to direct stem-cell fate. Nature. 2009;**462**:433-441. DOI: 10.1038/nature08602

[62] Lo CM, Wang HB, Dembo M, Wang Y-L. Cell movement is guided by the rigidity of the substrate. Biophysical Journal. 2000;**79**:144-152. DOI: 10.1016/S0006-3495(00)76279-5

[63] Kollmannsberger P, Bidan CM, Dunlop JWC, Fratzl P. The physics of tissue patterning and extracellular matrix organisation: How cells join forces. Soft Matter. 2011;**7**:9549-9560. DOI: 10.1039/C1SM05588G

[64] Breuls RGM, Jiya TU, Smit TH. Scaffold stiffness influences cell behavior: Opportunities for skeletal tissue engineering. The Open Orthopaedics Journal. 2008;**2**:103-109. DOI: 10.2174/1874325000802010103

[65] Nemir S, West JL. Synthetic materials in the study of cell response to substrate rigidity. Annals of Biomedical Engineering. 2010;**38**:2-20. DOI: 10.1007/s10439-009-9811-1

[66] Bhatia SK. Engineering Biomaterials for Regenerative Medicine: Novel Technologies for Clinical Applications. New York, Dordrecht, Heidelberg, London: Springer Science+Business Media; 2012. p. 346. DOI: 10.1007/978-1-4614-1080-5

[67] Engler AJ, Griffin MA, Sen S, Bönnemann CG, Sweeney HL, Discher DE. Myotubes differentiate optimally on substrates with tissue-like stiffness: Pathological implications for soft or stiff microenvironments. The Journal of Cell Biology. 2004;**166**:877-887. DOI: 10.1083/jcb.200405004

[68] Miroshnikova YA, Jorgens DM, Spirio L, Auer M, Sarang-Sieminski AL, Weaver VM. Engineering strategies to recapitulate epithelial morphogenesis within synthetic 3 dimensional extracellular matrix with tunable mechanical properties. Physical Biology. 2011;**8**(026013). DOI: 10.1088/1478-3975/8/2/026013

[69] Zaman MH, Trapani LM, Sieminski AL, MacKellar D, Gong H, Kamm RD, Wells A, Lauffenburger DA, Matsudaira P. Migration of tumor cells in 3D matrices is governed by matrix stiffness along with cell-matrix adhesion and proteolysis. Proceedings of the National Academy of Sciences of the USA. 2006;**103**:10889-10894. DOI: 10.1073/pnas.0604460103

[70] Guo W-H, Frey MT, Burnham NA, Wang Y-L. Substrate rigidity regulates the formation and maintenance of tissues. Biophysical Journal. 2006;**90**:2113-2130. DOI: 10.1529/biophysj.105.070144

[71] D'Andrea Markert L, Lovmand J, Duch M, Pedersen FS. Topographically and chemically modified surfaces for expansion or differentiation of embryonic stem cells. In: Atwood C, editor. Methodological Advances in the Culture, Manipulation and Utilization of Embryonic Stem Cells for Basic and Practical Applications. Rijeka: InTech; 2011. p. 139-158. DOI: 10.5772/15346

[72] Georges PC, Janmey PA. Cell type-specific response to growth on soft materials. Journal of Applied Physiology. 2005;**98**:1547-1553. DOI: 10.1152/japplphysiol.01121.2004

[73] Sochol RD, Higa AT, Janairo RRR, Lib S, Lin L. Unidirectional mechanical cellular stimuli via micropost array gradients. Soft Matter. 2011;**7**:4606-4609. DOI: 10.1039/c1sm05163f

[74] Yang Y, Leong KW. Nanoscale surfacing for regenerative medicine. WIREs Nanomedicine and Nanobiotechnology. 2010;**2**:478-495. DOI: 10.1002/wnan.74

[75] Folch A, Toner M. Microengineering of cellular interactions. Annual Review of Biomedical Engineering. 2000;**2**:227-256. DOI: 10.1146/annurev.bioeng.2.1.227

[76] Zahor D, Radko A, Vago R, Gheber LA. Organization of mesenchymal stem cells is controlled by micropatterned silicon substrates. Materials Science and Engineering: C. 2007;**27**:117-121. DOI: 10.1016/j.msec.2006.03.005

[77] Poellmann MJ, Harrell PA, King WP, Wagoner Johnson AJ. Geometric microenvironment directs cell morphology on topographically patterned hydrogel substrates. Acta Biomaterialia. 2010;**6**:3514-3523. DOI: 10.1016/j.actbio.2010.03.041

[78] Owen GR, Jackson J, Chehroudi B, Burt H, Brunette DM. A PLGA membrane controlling cell behaviour for promoting tissue regeneration. Biomaterials. 2005;**26**:7447-7756. DOI: 10.1016/j.biomaterials.2005.05.055

[79] Eklblad T, Liedberg B. Protein adsorption and surface patterning. Current Opinion in Colloid & Interface Science. 2010;**15**:499-509. DOI: 10.1016/j.cocis.2010.07.008

[80] Roach P, Parker T, Gadegaard N, Alexander MR. Surface strategies for control of neuronal cell adhesion: A review. Surface Science Reports. 2010;**65**:145-173. DOI: 10.1016/j.surfrep.2010.07.001

[81] Olędzka E, Sobczak M, Kołodziejski WL. Polymers in medicine – Review of past achievements. Polimery. 2007;**52**:793-803 (in Polish)

[82] Tanaka M. Design of novel 2D and 3D biointerfaces using self-organization to control cell behaviour. Biochimica et Biophysica Acta. 1810;**2011**:251-258. DOI: 10.1016/j.bbagen.2010.10.002

[83] Diener A, Nebe B, Becker P, Bech U, Lüthen F, Neumann HG, Rychly J. Control of focal adhesion dynamics by material surface characteristics. Biomaterials. 2005;**26**:383-392. DOI: 10.1016/j.biomaterials.2004.02.038

[84] Fan YW, Cui FZ, Hou SP, Xu QY, Chen LN, Lee IS. Culture of neural cells on silicon wafers with nano-scale surface topography. Journal of Neuroscience Methods. 2002;**120**:17-23. DOI: 10.1016/S0165-0270(02)00181-4

[85] Ricci JL, Grew JC, Alexander H. Connective-tissue responses to defined biomaterial surfaces. I. Growth of rat fibroblast and bone marrow cell colonies on microgrooved substrates. Journal of biomedical materials research. Part A. 2007;**85**:313-325. DOI: 10.1002/jbm.a.31379

[86] Kieswetter K, Schwartz Z, Humrnert TW, Cochran DL, Simpson J, Dean SDD, Boyan BD. Surface roughness modulates the local production of growth factors and cytokines by osteoblast-like MG-63 cells. Journal of Biomedical Materials Research 1996;**32**:55-63. DOI: 10.1002/(SICI)1097-4636(199609)32:1<55::AID-JBM7>3.0.CO;2-O

[87] Washburn NR, Yamada KM, Simon CG Jr, Kennedy SB, Amis EJ. High-throughput investigation of osteoblast response to polymer crystallinity: Influence of nanometer-scale roughness on proliferation. Biomaterials. 2004;**25**:1215-1224. DOI: 10.1016/j.biomaterials.2003. 08.043

[88] Kunzler TP, Drobek T, Schuler M, Spencer ND. Systematic study of osteoblast and fibroblast response to roughness by means of surface-morphology gradients. Biomaterials. 2007;**28**:2175-2182. DOI: 10.1016/j.biomaterials.2007.01.019

[89] Chung T-W, Wang S-S, Wang Y-Z, Hsieh C-H, Fu E. Enhancing growth and proliferation of human gingival fibroblasts on chitosan grafted poly (ε-caprolactone) films is influenced by nano-roughness chitosan surfaces. Journal of Materials Science: Materials in Medicine. 2009;**20**:397-404. DOI: 10.1007/s10856-008-3586-z

[90] Xu C, Yang F, Wang S, Ramakrishna S. In vitro study of human vascular endothelial cell function on materials with various surface roughness. Journal of Biomedical Materials Research, Part A. 2004;**71**:154-161. DOI: 10.1002/jbm.a.30143

[91] Takamori ER, Cruz R, Gonçalvez F, Zanetti RV, Zanetti A, Granjeiro JM. Effect of roughness of zirconia and titanium on fibroblast adhesion. Artificial Organs. 2008;**32**:305-309. DOI: 10.1111/j.1525-1594.2008.00547.x

[92] Łopacińska JM, Gradinaru C, Wierzbicki R, Købler C, Schmidt MS, Madsen MT, Skolimowski M, Dufva M, Flyvbjerg H, Mølhvae K. Cell motility, morphology, viability and proliferation in response to nanotopography on silicon black. Nanoscale. 2012;**4**:3739-3745. DOI: 10.1039/c2nr11455k

[93] Bandyopadhyay A, Espana F, Balla VK, Bose S, Ohgami Y, Davies NM. Influence of porosity on mechanical properties and in vivo response of Ti6Al4 implants. Acta Biomaterialia. 2010;**6**:1640-1648

[94] Nouri A, Hodgson PD, Wen C. Biomimetic porous titanium scaffolds for orthopedic and dental applications. In: Mukherjee A, editor. Biomimetics Learning from Nature. Rijeka: InTech; 2010. p. 415-450

[95] Yoshida K, Saiki Y, Ohkubo S. Improvement of drawability and fabrication possibility of dental implant screw made of pure titanium. Hutnik – Wiadomości Hutnicze. 2011;**78**:153-156

[96] Dobrzańska-Danikiewicz AD. The book of critical technologies of surface and properties formation of engineering materials. Open access. Library. 2012;**6**:1-823 (in Polish)

[97] Dobrzańska-Danikiewicz AD. Computer integrated development prediction methodology in materials surface engineering. Open access. Library. 2012;**1**:1-289 (in Polish)

[98] Dobrzańska-Danikiewicz AD. Foresight of material surface engineering as a tool building a knowledge based economy. Materials Science Forum. 2012;**706-709**:2511-2516. DOI: 10.4028/www.scientific.net/MSF.706-709.2511

[99] Dobrzański LA, Dobrzańska-Danikiewicz AD. Foresight of the surface Technology in Manufacturing. In: Nee AYC, editor. Handbook of Manufacturing Engineering and Technology. London: Springer-Verlag; 2015. p. 2587-2637 978-1-4471-4671-1

[100] Guo N, Leu MC. Additive manufacturing: Technology, applications and research needs. Frontiers of Mechanical Engineering. 2013;**8**:215-243. DOI: 10.1007/s11465-013-0248-8

[101] Cormier D, Harrysson O, West H. Characterization of H13 steel produced via electron beam melting. Rapid Prototyping Journal. 2004;**10**:35-41

[102] Rännar LE, Glad A, Gustafson CG. Efficient cooling with tool inserts manufactured by electron beam melting. Rapid Prototyping Journal. 2007;**13**:128-135

[103] Murr LE, Gaytan SM, Medina F, Martinez E, Martinez JL, Hernandez DH, Machado BI, Ramirez DA, Wicker RB. Characterization of Ti6Al4V open cellular foams fabricated by additive manufacturing using electron beam melting. Materials Science and Engineering A. 2010;**527**:1861-1868

[104] Han GW, Feng D, Yin M, Ye WJ. Ceramic/aluminum co-continuous composite synthesized by reaction accelerated melt infiltration. Materials Science and Engineering A. 1997;**225**:204-207

[105] Murr LE, Gaytan SM, Ceylan A, Martinez E, Martinez JL, Hernandez DH, Machado BI, Ramirez DA, Medina F, Collins S. Characterization of titanium aluminide alloy components fabricated by additive manufacturing using electron beam melting. Acta Materialia. 2010;**58**:1887-1894

[106] Kumar S. Selective laser sintering: A qualitative and objective approach. Modeling and Characterization. 2003;**55**:43-47

[107] Xue W, Vamsi KB, Bandyopadhyay A, Bose S. Processing and biocompatibility evaluation of laser processed porous titanium. Acta Biomateralia. 2007;**3**:1007-1018

[108] Osakada K, Shiomi M. Flexible manufacturing of metallic products by selective laser melting of powder. International Journal of Machine Tools & Manufacture. 2006;**46**:1188-1193

[109] Kruth J-P, Mercelis P, Van Vaerenbergh J, Froyen L, Rombouts M. Binding mechanisms in selective laser sintering and selective laser melting. Rapid Prototyping Journal. 2005;**11**:26-36. DOI: 10.1108/13552540510573365

[110] Bertol LS, Júnior WK, da Silva FP, Kopp CA. Medical design: Direct metal laser sintering of Ti-6Al-4V. Materials and Design. 2010;**31**:3982-3988

[111] Pattanayak DK, Fukuda A, Matsushita T, Takemoto M, Fujibayashi S, Sasaki K, Nishida N, Nakamura T, Kokubo T. Bioactive Ti metal analogous to human cancellous bone: Fabrication by selective laser melting and chemical treatments. Acta Biomaterialia. 2011;**7**:1398-1406

[112] Abe F, Osakada K, Shiomi M, Uematsu K, Matsumoto M. The manufacturing of hard tools from metallic powders by selective laser melting. Journal of Materials Processing Technology. 2001;**111**:210-213. DOI: 10.1016/S0924-0136(01)00522-2

[113] Darłak P, Dudek P. High porous materials - manufacturing methods and applications. Molding: Science and. Practice. 2004;**1**:3-17 (in Polish)

[114] Kruth JP, Froyen L, Van Vaerenbergh J, Mercelis P, Rombouts M, Lauwers B. Selective laser melting of iron-based powder jet. Journal of Materials Processing Technology. 2004;**149**:616-622. DOI: 10.1016/j.jmatprotec.2003.11.051

[115] Lethaus B, Poort L, Böckmann R, Smeets R, Tolba R, Kessler P. Additive manufacturing for microvascular reconstruction of the mandiblein 20 patients. Journal of Cranio-Maxillo-Facial Surgery. 2012;**40**:43-46. DOI: 10.1016/j.jcms.2011.01.007

[116] Liao YS, Li HC, Chiu YY. Study of laminated object manufacturing with separately applied heating and pressing. International Journal of Advanced Manufacturing Technology. 2006;**27**:703-707. DOI: 10.1007/s00170-004-2201-9

[117] Dobrzański LA, Dobrzańska-Danikiewicz AD, Malara P, Gaweł TG, Dobrzański LB, Achtelik-Franczak A. Fabrication of scaffolds from Ti6Al4V powders using the computer aided laser method. Archives of Metallurgy and Materials. 2015;**60**:1065-1070. DOI: 10.1515/amm-2015-0260

[118] Dobrzański LA, Dobrzańska-Danikiewicz AD, Achtelik-Franczak A, Dobrzański LB, Szindler M, Gaweł TG. Porous selective laser melted Ti and Ti6Al4V materials for medical applications. In: Dobrzański LA, editor. Powder Metallurgy – Fundamentals and Case Studies. Rijeka: INTECH; 2017. pp. 161-181. DOI: 10.5772/65375

[119] Dobrzański LB. Structure and Properties of Engineering Materials for Prosthetic Restorations of Stomatognitive System Manufactured by Additive and Defective Methods. [Ph.D. Thesis in Progress]. Kraków: AGH University of Science and Technology; 2017

[120] Dobrzański LA, Dobrzańska-Danikiewicz AD, Achtelik-Franczak A, Dobrzański LB. Comparative analysis of mechanical properties of scaffolds sintered from Ti and Ti6Al4V powders. Archives of Materials Science and Engineering. 2015;**73**:69-81

[121] Dobrzański LA. Overview and general ideas of the development of constructions, materials, technologies and clinical applications of scaffolds engineering for regenerative medicine. Archives of Materials Science and Engineering. 2014;**69**:53-80

[122] Dobrzański LA, Dobrzańska-Danikiewicz AD, Gaweł TG, Achtelik-Franczak A. Selective laser sintering and melting of pristine titanium and titanium Ti6Al4V alloy powders and selection of chemical environment for etching of such materials. Archives of Metallurgy and Materials. 2015;**60**:2039-2045. DOI: 10.1515/amm-2015-0346

[123] Karoluk M, Pawlak A, Chlebus E. Using of SLM additive technology in Ti-6Al-7Nb titanium alloy processing for biomedical applications. XI Scientific Conference named after Prof. Dagmara Tejszerska "May-Picnic of Young Biomechanics". Ustroń, Poland; 2014. pp. 53-54 (in Polish)

[124] Dobrzański LA, Dobrzańska-Danikiewicz AD, Achtelik-Franczak A, Szindler M. Structure and properties of the skeleton microporous materials with coatings inside the pores for medical and dental applications. International Conference on Frontiers in Materials Processing, Applications, Research, & Technology, FiMPART'2015; 12-15 Juni 2015; Hyderabad, India. Singapore: Springer; 2017 (in press)

[125] Dobrzański LA & Dobrzańska-Danikiewicz AD (eds.), Microporous and solid metallic materials for medical and dental application, Open Access Library, Annal **VII**(1) 2017. pp. 1-580 (in Polish)

[126] Achtelik-Franczak A. Composite engineering materials with matrix made from selective laser sintered microporous titanium [Ph.D. thesis]. Gliwice, Poland: Silesian university of Technology; 2016 (in Polish)

[127] Dobrzański LA, Dobrzańska-Danikiewicz AD, Gaweł TG. Ti6Al4V porous elements coated by polymeric surface layer for biomedical applications. Journal of Achievements in Materials and Manufacturing Engineering. 2015;**71**:53-59

[128] Dobrzański LA, Dobrzańska-Danikiewicz AD, Szindler M, Achtelik-Franczak A, Pakieła W. Atomic layer deposition of TiO_2 onto porous biomaterials. Archives of Materials Science and Engineering. 2015;**75**:5-11

[129] Klimas J, Łukaszewicz A, Szota M, Laskowski K. Work on the modification of the structure and properties of Ti6Al4V titanium alloy for biomedical applications. Archives of Materials Science and Engineering. 2016;**78**:10-16. DOI: 10.5604/18972764.1226308

[130] Kirmanidou Y, Sidira M, Drosou M-E, Bennani V, Bakopoulou A, Tsouknidas A, Michailidis N, Michalakis K. New Ti-alloys and surface modifications to improve the mechanical properties and the biological response to orthopedic and dental implants: A review. BioMed Research international. 2016. Art no. 2908570. DOI: 10.1155/2016/2908570

[131] Parchańska-Kowalik M, Klimek L. Wpływ obróbki chemicznej na stan powierzchni tytanu. Inżynieria Materiałowa. 2013;**34**:526-529

[132] Dobrzański LA. Applications of newly developed nanostructural and microporous materials in biomedical, tissue and mechanical engineering. Archives of Materials Science and Engineering. 2015;**76**:53-114

[133] Burnat B, Parchańska-Kowalik M, Klimek L. The influence of chemical surface treatment on the corrosion resistance of titanium castings used in dental prosthetics. Archives of Foundry Engineering. 2014;**14**:11-16. DOI: 10.2478/afe-2014-0052

[134] Majkowska B, Jażdżewska M, Wołowiec E, Piekoszewski W, Klimek L, Zieliński A. The possibility of use of laser-modified Ti6Al4V alloy in friction pairs in endoprostheses. Archives of Metallurgy and Materials. 2015;**60**:755-758. DOI: 10.1515/amm-2015-0202

[135] Klimas J, Łukaszewicz A, Szota M, Szota K. Characteristics of titanium grade 2 and evaluation of corrosion resistance. Archives of Materials Science and Engineering. 2016;**77**:65-71. DOI: 10.5604/18972764.1225596

[136] Jedynak B, Mierzwińska-Nastalska E. Titanium - properties and use in dental prosthetics. Dental. Forum. 2013;**41**:75-78 (in Polish)

[137] Emsley J. Nature's Building Blocks: An A-Z Guide to the Elements. Oxford: Oxford University Press; 2001. 538 p 978-0198503408

[138] Hong J, Andersson J, Ekdahl KN, Elgue G, Axen N, Larsson R, Nilsson B. Titanium is a highly thrombogenic biomaterial: Possible implications for osteogenesis. Thrombosis and Haemostasis. 1999;**82**:58-64

[139] PN-EN ISO 10993-4:2017. Biological Evaluation of Medical Devices. Part 4: Selection of Tests for Interaction with Blood, Technical Committee ISO/TC 194, ICS: 11.100.20, Available from: https://www.iso.org/standard/63448.html Accessed: 2017-08-23

[140] Street J, Winter D, Wang JH, Wakai A, McGuinness A, Redmond HP. Is human fracture hematoma inherently angiogenic? Clinical Orthopaedics and Related Research. 2000;**378**:224-237

[141] Karp JM, Sarraf F, Shoichet MS, Davies JE. Fibrin-filled scaffolds for bone-tissue engineering: An in vivo study. Journal of Biomedical Materials Research, Part A. 2004;**71**:162-171

[142] Yang F, Neeley WL, Moore MJ, Karp JM, Shukla A, Langer R. Tissue engineering: The Therapeutic Strategy of the Twenty-First Century. In: Laurencin CT, Nair LS, editors. Nanotechnology and Tissue Engineering: The Scaffold. Boca Raton, FL: CRC Press Taylor & Francis Group; 2008. p. 3-32

[143] Leong KF, Cheah CM, Chua CK. Solid freeform fabrication of three-dimensional scaffolds for engineering replacement tissues and organs. Biomaterials. 2003;**24**:2363-2378

[144] Nair LS, Laurencin CT. Polymers as biomaterials for tissue engineering and controlled drug delivery. Advances in Biochemical Engineering, Biotechnology. 2006;**102**:47-90

[145] Velema J, Kaplan D. Biopolymer-based biomaterials as scaffolds for tissue engineering. Advances in Biochemical Engineering, Biotechnology. 2006;**102**:187-238

[146] Kannan RY, Salacinski HJ, Sales KM, Butler PE, Seifalian AM. The endothelialization of polyhedral oligomeric silsesquioxane nanocomposites: An in vitro study. Cell Biochemistry and Biophysics. 2006;**45**:129-136

[147] Nowacka M. Biomaterials used in cellular engineering and regenerative medicine. Wiadomości Chemiczne. 2012;**66**:909-933 (in Polish)

[148] Ulery BD, Nair LS, Laurencin CT. Biomedical applications of biodegradable polymers. Journal of Polymer Science, Part B, Polymer Physics. 2011;**49**:832-864. DOI: 10.1002/polb.22259

[149] Ghanbari H, Kidane AG, Buriesci G, Ramesh B, Darbyshire A, Seifalian AM. The anti-calcification potential of a silsesquioxane nanocomposite polymer under in vitro conditions: Potential material for synthetic leaflet heart valve. Acta Biomaterialia. 2010;**6**:4249-4260. DOI: 10.1016/j.actbio.2010.06.015

[150] Motwani MS, Rafiei Y, Tzifa A, Seifalian AM. In situ endothelialization of intravascular stents from progenitor stem cells coated with nanocomposite and functionalized biomolecules. Biotechnology and Applied Biochemistry. 2011;**58**:2-13. DOI: 10.1002/bab.10

[151] Kidane AG, Burriesci G, Edirisinghe M, Ghanbari H, Bonhoeffer P, Seifalian AM. A novel nanocomposite polymer for development of synthetic heart valve leaflets. Acta Biomaterialia. 2009;**5**:2409-2417. DOI: 10.1016/j.actbio.2009.02.025

[152] Słomkowski S. Hybrid polymeric materials for medical applications. Polimery. 2006;**51**:87-94

[153] Kidane AG, Burriesci G, Cornejo P, Dooley A, Sarkar S, Bonhoeffer P, Edirisinghe M, Seifalian AM. Current developments and future prospects for heart valve replacement therapy. Journal of biomedical materials research, part B, applied. Biomaterials. 2008;**88**:290-303. DOI: 10.1002/jbm.b.31151

[154] Middleton JC, Tipton AJ. Synthetic biodegradable polymers as orthopedic devices. Biomaterials. 2000;**21**:2335-2346

[155] An YH, Woolf SK, Friedman RJ. Pre-clinical in vivo evaluation of orthopaedic bioabsorbable devices. Biomaterials. 2000;**21**:2635-2652

[156] Robinson BP, Hollinger JO, Szachowicz EH, Brekke J. Calvarial bone repair with porous D,L-polylactide. Otolaryngology - Head and Neck Surgery. 1995;**112**:707-713. DOI: 10.1016/S0194-59989570180-X

[157] Das K, Balla VK, Bandyopadhyay A, Bose S. Surface modification of laser-processed porous titanium for load-bearing implants. Scripta Materialia. 2008;**59**:822-825

[158] Balla VK, Bodhak S, Bose S, Bandyopadhyay A. Porous tantalum structures for bone implants: Fabrication, mechanical and in vitro biological properties. Acta Biomaterialia. 2010;**6**:3349-3359

[159] Witte F, Ulrich H, Palm C, Willbold E. Biodegradable magnesium scaffolds: Part II: Periimplant bone remodelling. Journal of Biomedical Materials Research, Part A. 2007;**81**:757-765

[160] Bose S, Roy M, Bandyopadhyay A. Recent advances in bone tissue engineering scaffolds. Trends in Biotechnology. 2012;**30**:546-554

[161] Yun Y, Dong Z, Lee N, Liu Y, Xue D, Guo X, Kuhlmann J, Doepke A, Halsall HB, Heineman W, Sundaramurthy S, Schulz MJ, Yin Z, Shanov V, Hurd D, Nagy P, Li W, Fox C. Revolutionizing biodegradable metals. Materials Today. 2009;**12**:22-32. DOI: 10.1016/S1369-7021(09)70273-1

[162] Vert M. Not any new functional polymer can be for medicine: What about artificial biopolymers? Macromolecular Bioscience. 2011;**11**:1653-1661. DOI: 10.1002/mabi.201100224

[163] Li B, Ma Y, Wang S, Moran PM. Influence of carboxyl group density on neuron cell attachment and differentiation behavior: Gradient-guided neurite outgrowth. Biomaterials. 2005;**26**:4956-4963. DOI: 10.1016/j.biomaterials.2005.01.018

[164] Gaweł TG. Manufacturing technology obtaining and reserach of innovative porous biomimetic materials. Ph.D. Thesis (in progress). Gliwice, Poland: Silesian University of Technology; 2017 in Polish

[165] Dobrzański LA, Dobrzańska-Danikiewicz AD, Achtelik-Franczak A, Dobrzański LB, Hajduczek E, Matula G. Fabrication technologies of the sintered materials including materials for medical and dental application In: Dobrzański LA, editor. Powder Metallurgy – Fundamentals and Case Studies. Rijeka: INTECH; 2017. p. 17-52. DOI: 10.5772/65376

[166] Dobrzański LA, Dobrzańska-Danikiewicz AD, Malara P, Dobrzański LB, Achtelik-Franczak A. Biological and engineering composites for regenerative medicine. Patent Application P. 2015;**414723**(9):11

[167] Dobrzański LA, Dobrzańska-Danikiewicz AD, Malara P, Gaweł TG, Dobrzański LB, Achtelik-Franczak A. Bone implant scaffold. Patent Application P. 2015;**414424**(19):10

[168] Dobrzański LA, Dobrzańska-Danikiewicz AD, Malara P, Gaweł TG, Dobrzański LB, Achtelik-Franczak A. Implant scaffold and a prosthesis of anatomical elements of a dental system and craniofacial bone. Patent Application P. 2015;**414423**(19):10

[169] Dobrzański LA, Dobrzańska-Danikiewicz AD, Malara P, Gaweł TG, Dobrzański LB, Achtelik A. Composite fabricated by computer-aided laser methods for craniofacial implants and its manufacturing method. Patent Application P. 2015;**411689**(23):03

[170] Dobrzański LA, Dobrzańska-Danikiewicz AD, Malara P, Achtelik-Franczak A, Dobrzański LB, Kremzer M. Sposób wytwarzania materiałów kompozytowych o mikro-porowatej szkieletowej strukturze wzmocnienia. Patent Application P. 2016;**417552**(13):06

[171] Dobrzański LA, Dobrzańska-Danikiewicz AD, Malara P, Dobrzański LB, Achtelik-Franczak A, Gaweł TG. Implant-scaffold or prosthesis anatomical structures of the Stomatognathic system and the craniofacial. Gold medal on international exhibition of technical innovations, patents and inventions, INVENT ARENA 2016; Třinec, Czech Republic. 2016;**6**:16-18

[172] Dobrzański LA, Dobrzańska-Danikiewicz AD, Malara P, Dobrzański LB, Achtelik-Franczak A, Gaweł TG. Implant-scaffold or prosthesis anatomical structures of the Stomatognathic system and the craniofacial. Gold Medal and International Intellectual Property Network Forum (IIPNF) Leading Innovation Award on International Intellectual Property, Invention, Innovation and Technology Exposition, IPITEX 2016. Bangkok, Thailand. 2016;**2**:2-6

[173] Dobrzański LA, Dobrzańska-Danikiewicz AD, Malara P, Gaweł TG, Dobrzański LB, Achtelik-Franczak A. The novel composite consisting of a metallic scaffold, manufactured using a computer aided laser method, coated with thin polymeric surface layer for medical applications. Gold Medal on 9th International Warsaw Invention Show IWIS 2015. Warsaw, Poland; 12-14.10.2015

[174] Dobrzański LA, Dobrzańska-Danikiewicz AD, Malara P, Gaweł TG, Dobrzański LB, Achtelik-Franczak A. The novel composite consisting of a metallic scaffold, manufactured using a computer aided laser method, coated with thin polymeric surface layer for medical applications. Semi grand prize on global inventions and innovations exhibitions Innova cities Latino-America, ICLA 2015. Foz do Iguaçu, Brazil; 2015;**12**:10-12

[175] Dobrzański LA. Fabrication, structure and mechanical properties of laser sintered materials for medical applications. Invited lecture on XXV International Materials Research Congress. Cancun, Mexico; 14-19 August 2016

[176] Dobrzański LA. Application of the additive manufacturing by selective laser sintering for constituting implantscaffolds and hybrid multilayer biological and engineering composite materials. Keynote lecture on international conference on Processing & Manufacturing of advanced materials THERMEC'2016, processing, fabrication, properties, applications. Graz, Austria; 29 May – 3 June 2016

[177] Dobrzański LA. Metallic implants-scaffolds for dental and orthopaedic application. Invited lecture on 9° COLAOB - Congresso Latino-Americano de Orgãos Artificiais e Biomateriais; Foz do Iguaçu, Brazil; 24-27 August 2016

[178] Dobrzański LA et al. Investigations of Structure and Properties of Newly Created Porous Biomimetic Materials Fabricated by Selective Laser Sintering, BIOLASIN. Project UMO-2013/08/M/ST8/00818. Gliwice: Silesian University of Technology; 2013-2016

[179] Kremzer M, Dobrzański LA, Dziekońska M, Macek M. Atomic layer deposition of TiO_2 onto porous biomaterials. Archives of Materials Science and Engineering. 2015;**75**:63-69

[180] Project POIR.01.01.01-00-0485/16-00 IMSKA-MAT Innovative dental and maxillo-facial implant-scaffolds manufactured using the innovative technology of additive computer-aided materials design ADD-MAT. Design, Research and production Centre of Medical and Dental Engineering ASKLEPIOS in Gliwice; 2017-2021. In progress

[181] Dobrzański LA, Dobrzańska-Danikiewicz AD. Forming of Structure and Properties of Engineering Materials Surface. Gliwice: Wydawnictwo Politechniki Śląskiej; 2013. 492 p (in Polish)978-83-7880-077-7

[182] Kim TH, Lee KM, Hwang J, Hong WS. Nanocrystalline silicon films deposited with a modulated hydrogen dilution ratio by catalytic CVD at 200°C. Current Applied Physics. 2009;**9**:e108-e110

[183] Publication of the IC Knowledge LLC. Technology Backgrounder: Atomic Layer Deposition. Georgetown, MA; 2004

[184] Dobrzański LA, Szindler M, Szindler MM. Surface morphology and optical properties of Al_2O_3 thin films deposited by ALD method. Archives of Materials Science and Engineering. 2015;**73**:18-24

[185] Bala H. Introduction into Materials Chemistry. Warszawa: WNT; 2003. 445 p (in Polish)8320427991

[186] Pajdowski L. General Chimistry. 11th ed. Wydawnictwo Naukowe PWN: Warszawa; 2002. 616 p (in Polish)8301123567

[187] Jasińska B. General Chimistry. Kraków: Wydawnictwa AGH; 1998. 315 p (in Polish)

[188] Głuszek J. Oxidative Protective Coatings Obtained by sol-gel Method. Oficyna Wydawnicza Politechniki Wrocławskiej: Wrocław; 1998. 154 p (in Polish)8370853617

[189] Wrigth JD, Sommerdijk NAJM. Sol-gel Materials, Chemistry and Applications. Amsterdam: Gordon and Breach Science Publishers; 2001. 125 p 90-5699-326-7

[190] Hwang K, Song J, Kang B, Park Y. Sol-gel derived hydroxyapatite films on alumina substrates. Surface and Coatings Technology. 2000;**123**:252-255

[191] Dong Y, Zhao Q, Wu S, Lu X. Ultraviolet-shielding and conductive double functional films coated on glass substrates by sol-gel process. Journal of Rare Earths. 2010;**28**:446-450

[192] Langlet M, Kim A, Audier M, Guillard C, Herrmann JM. Transparent photocatalytic films deposited on polymer substrates from sol-gel processed titania sols. Thin Solid Films. 2003;**429**:13-21

[193] Seo M, Akutsu Y, Kagemoto H. Preparation and properties of Sb-doped SnO_2/metal substrates by sol-gel and dip coating. Ceramics International. 2007;**33**:625-629

[194] Łączka M, Terczyńska A, Cholewa-Kowalska K. Gel coatings on the glass, part 1. Świat Szkła. 2008;**9**:52-55 (in Polish)

[195] Dobrzański LA, editor. Powder Metallurgy – Fundamentals and Case Studies. Rijeka: InTech; 2017. 392 p. DOI: 10.5772/61469

Mechanical Properties Comparison of Engineering Materials Produced by Additive and Subtractive Technologies for Dental Prosthetic Restoration Application

Lech B. Dobrzański

Additional information is available at the end of the chapter

http://dx.doi.org/10.5772/intechopen.70493

Abstract

Comparative investigations are presented into the structure and properties of selected engineering materials used for dental prosthetic restorations manufactured alternatively by the subtractive method by milling on computerised numerical control (CNC) milling machines and by the additive selective laser sintering (SLS) method of solid and porous elements using computer-aided design/computer-aided manufacturing (CAD/CAM) techniques; moreover, an original technology was presented of manufacturing the elements used in prosthodontics, produced with titanium and Ti6Al4V alloy powders with the SLS technique. Suitability was confirmed of applying the manufacturing technologies used in prosthodontics from powders by the SLS technique. The results were compared of the executed tests of strength micro-samples and of the selected prosthetic Bridges. The SLS technology for titanium, and even more for Ti6Al4V alloy, ensures the achievement of mechanical properties comparable or better than a reference Co-Cr alloy commonly used for prosthetic restorations, including prosthetic bridges fabricated by milling solid discs with CNC milling machines. For all the examined engineering materials, the milled and then sintered ZrO_2 material exhibits the lowest strength properties. The results presented in this chapter can be directly applied in dental practice.

Keywords: prosthodontics, machining, additive machining, selective laser sintering, titanium and its alloys , Co-Cr alloys, zirconium oxide

1. Justification and scope of comparative investigations of selected alloys and technologies of dental manufacturing of prosthetic restorations

Practical dental engineering encounters numerous problems with selecting the most suitable material and technology of dental manufacturing of a prosthetic restoration by a dental surgeon conducting treatment. One of the important reasons is inability to compare directly the properties and structure of analogous prosthetic restorations but produced using various technologies and from different alternative engineering materials accepted for medical, especially dental, uses as compared to the restorations used most often until now. It is beyond any doubt that dental surgeons are responsible for the strategy and quality of dental treatment. The literature studies carried out as well as the Author's long-term own engineering experience in running medical practice and a dental engineering workshop show that—parallel to clinical experience of a dentist conducting treatment of dental system diseases—the successfulness of such activities also depends on the results of a dental engineer's activities, consisting of the appropriate execution of prosthetic restorations and implants, depending on the selection of the most appropriate engineering material and appropriate manufacturing technology, ensuring, respectively, the required biocompatibility, mechanical properties and corrosion resistance, as well as high production accuracy and aesthetic requirements. Such activities require, therefore, close cooperation of a dentist and a dental engineer having extensive and interdisciplinary engineering knowledge of the principles of construction, technology and science of materials and computer-aided methods of engineering and dental works.

The dental system, especially teeth, is still subject to numerous disease processes, especially tooth decay, so widespread in many countries and practically across the globe [1–4]. This disease has a considerable social range and is a serious medical problem, not only due to local mouth cavity disorders but also due to a high risk of systemic complications stemming from the dental pulp disease, and is a direct cause of a majority of tooth losses among a wide group of people in many countries. The extraction of unhealthy teeth was a dominant practice in such cases for a long time. There may also be inborn causes of missing teeth, for example, due to missing tooth buds. The basic method of producing permanent prosthetic restorations, which is still often used, is to cast metal alloys requiring successive machining, mainly to remove elements of the gate system, and to ensure required surface roughness and dimensional accuracy. Apart from conventional technologies of manufacturing prosthetic restorations including especially powder metallurgy, casting technologies, metallic foam manufacturing, also additive technologies of fabricating solid and microporous materials in medicine and dentistry are exceptionally useful. The potential of additive technologies is much larger than of other technologies and is represented chiefly by opportunities ensured by three-dimensional designing and the related control of the structure, sizes and shape of the materials produced and the repeatability of geometric features of the elements produced, and because no wastes are produced and because the simplicity of the technology is comprised of two key stages only: designing and manufacturing.

Considering the additive technologies employed most extensively in industry, only few have found their application in prosthetics, with selective laser sintering/selective laser melting (SLS/SLM) offering the biggest opportunities [5–15]. This issue is even more important to clarify whether such additive technologies, applied in dental engineering, are competitive for conventional technologies, because original own works [5, 16–36] have been undertaken concerning constructional solutions and fabrication technologies of a new generation of custom, original, hybrid, microporous, high-strength engineering and biological materials with microporous rigid titanium and titanium alloy skeletons manufactured by selective laser sintering, whose pores are filled with living cells. The aim of such constructional and technological solutions is to ensure conditions for natural ingrowth of living tissues in the connection zone of prosthetic/implant elements, with bone or organ stumps, to eliminate the need to apply for patients the mechanical elements which are positioning and fixating the implants. In order to manufacture engineering and biological materials and implant scaffolds, it is required to seek the most advantageous proliferation conditions of living cells inside the pores of a microporous skeleton made of titanium and titanium alloys, because the adhesion and growth of living cells are dependent on the type and characteristics of the substrate. A porous zone requires surface treatment inside pores to improve proliferation conditions of living tissues, which are not advantageous for a substrate made of titanium or Ti6Al4V alloy. It appears that the improvement of proliferation conditions of cells is ensured by a substrate made of fully compatible materials, including TiO_2, Al_2O_3 oxides and a hydroxyapatite, $Ca_{10}(PO_4)_6(OH)_2$, used as coatings of internal surfaces of pores of a microporous skeleton fabricated by SLS. Thin layers inside pores were deposited by two technologies ensuring the uniform thickness of coatings on all the walls and openings of a substrate, that is, the so-called atomic layer deposition (ALD) [19, 37–43], and the sol-gel technology of coating deposition from the liquid phase by the immersion method [37, 40, 44–52]. In order to be able to accomplish the above-mentioned general aims of the investigations, it has to be confirmed that additive technologies, especially SLS, do not lead to unacceptable deterioration of useful properties of the so-produced prosthetic restorations.

Apart from permanent prosthetic restorations made traditionally on a metallic substructure, new technologies of introducing ceramic materials have been launched, including so-called fully porcelain systems made of zirconium oxide and aluminium oxide. For dimensional accuracy and aesthetics as well as biocompatibility and dental strength of prosthetic restorations, the constant development of manufacturing technologies of prosthetic restorations is required along with the evolvement of the engineering materials currently employed in prosthetics. Highly varied conditions of joining the artificially produced prosthetic restorations and implants with a body's natural tissues, mainly in the oral cavity, are driving especially high requirements for the engineering material applied for prosthetic restorations and implants. The reason for this is not only special conditions of using various prosthetic and implantological restorations but also different economic and technological conditions and differentiated availability and level of clinical methods and also important aesthetic reasons. Computer-aided design/computer-aided manufacturing (CAD/CAM) techniques [53–59], allowing to improve dimensional accuracy of the fabricated prosthetic restorations, have become increasingly vital over the last decade. CAD/CAM techniques find their application in dental engineering in machining, in plastic forming, in additive manufacturing technologies and in special technologies and both in metal and ceramics working. Computer-aided design enables, in particular, to produce a digital prosthetic model by scanning with an intra-cavity scanner and also to customise the designing and manufacturing of each prosthetic restoration under constant supervision of a doctor conducting the treatment. Cone beam computed tomography (CBCT), used often in dental surgeries these days as the basic diagnostic tool for comprehensive dental treatment, mainly implantological, prosthetic and endodontic treatment, creates excellent possibilities in this field. A widespread application of CBCT also permits to utilise the information acquired about the condition of a patient's tissues in planning and performing implant and prosthetic treatment [55, 60].

The assumptions made require special interest in the selected engineering materials finding application in dental engineering, including mainly casting alloys of cobalt, titanium and its alloys and ceramics on the zirconium oxide matrix [61–72]. The essential aspects underlying the investigations described here were the questions whether—in each case—it is feasible to apply prosthetic restorations made on a metallic substructure of zirconium oxide, guaranteeing currently best aesthetic results, for prosthetic restorations and to explain if it is possible to use a zirconium oxide substructure in the side sections and in circular restorations together with recognising the limitations for the use of this material. Another practical aspect is to develop a methodology for application of titanium and its alloys as compared to Co-Cr alloys of the Vitalium type for implantoprosthetic restorations so as to guarantee the lowest possible mass of the completely manufactured prosthetic restoration while ensuring aesthetic requirements and the maximum, possible strength properties. For production of metallic substructures, it is necessary to ensure space for veneering porcelain, among others, in intermolecular spaces on abutments between crowns and bridge spans but also on the vestibular space, so that the dark colour of the substructure does not influence the colour of the ready veneering ceramics. Considering that disadvantages and limitations are known of casting and machining technologies such as computerised numerical control (CNC) milling currently used in dental engineering laboratories, it was decided to compare the state-of-the-art manufacturing techniques currently applied in dental engineering, including completely innovative additive techniques which—due to high implementation price, but very low price of producing a single prosthetic point—are rarely used in practice. As Co-Cr alloys are most often used in dental practice and, therefore, most often described in the literature, such were used as a reference point for the works performed already described in literature. This will allow practitioners, especially dentists, to easily transfer the results presented in this article to everyday uses and to consider such results for designing the extensiveness of prosthetic restorations and for cost valuation. The matter is very complex and requires multi-criteria optimisation, which cannot be performed individually by any prosthetic laboratory. The matter requires, therefore, methodologically planned research which can demonstrate—to the practising therapists and prosthetic workshops cooperating with them—the advantages of some material and technological solutions over those commonly in use these days. This will surely have a major impact on the outcomes and durability of dental treatment and on satisfaction, comfort and patients' improved health condition. The aspects highlighted were the reasons for outlining the topic of the chapter presenting the results of the work [18], after making a thorough literature analysis of the status of research into the dental system, its diseases and treatment methods of tooth losses as well as materials and technologies applied in prosthetics. A concept presented in this article is covered by a broader range of works followed in the recent years [5, 7, 16, 17, 21–26, 34, 55, 56, 73–79], in which the author has participated actively. The works concern principally the possibility of producing solid and porous titanium parts by additive technology, for use in dentistry and reconstructive medicine.

This chapter depicts the outcomes of comparative investigations into the structure and properties of selected engineering materials used for dental prosthetic restorations manufactured alternatively by the subtractive method by milling on CNC milling machines and by the additive SLS method of solid and porous items using CAD/CAM techniques; moreover, an original technology was presented of manufacturing the elements used in prosthodontics, produced with titanium and Ti6Al4V alloy powders with the SLS technique. The assumptions made in this chapter are focussed on the selected engineering materials finding application in dental engineering, including mainly casting alloys of cobalt, titanium and its alloys and ceramics on the zirconium oxide matrix.

Materials for milling were delivered in the form of discs with the diameter of 98.3 mm and height of 10–16 mm, and the chemical composition, according to suppliers' attestations, is shown in **Table 1**.

Two types of powders with a spherical shape were used, respectively, for selective laser sintering and with the composition shown in **Table 2**, also confirmed with spectral examinations with the energy dispersive spectrometry (EDS) method:

- Titanium powder with Grade 4 and grain size of up to 45 μm, oxygen concentration reduced to 0.14%, while average oxygen content in titanium powders is about 0.5%, the aim of which is to ensure process safety,
- Ti6Al4V alloy powder with the grain diameter of 15–45 μm, for medical applications.

The results of comparative investigations are presented into the structure and properties of the mentioned engineering materials used for dental prosthetic restorations manufactured alternatively by the subtractive method by milling on CNC milling machines and by the additive SLS method of solid and porous items using CAD/CAM techniques; moreover, an original developed

Material	Mass concentration/fraction, %					
Ti grade 2	Ti	Others N, C, H, Fe, O altogether				
	> 99.6	< 0.4				
Ti6Al4V	Ti	Al	V	Others N, C, H, Fe, O altogether		
	89.4	6.2	4.0	≤ 1.0		
Co-Cr	Co	Cr	W	Mo	Si	Others C, Fe, Mn, N altogether
	59.0	25.0	9.5	3.5	1.0	< 1.0
ZrO_2	$ZrO_2 + HfO_2 + Y_2O_3$					Including Y_2O_3
	> 99.0					4.5-6.0

Table 1. Chemical composition of the examined materials manufactured by CAD/CAM by milling solid discs.

Powder	Mass concentration of elements, %									
	Al	V	C	Fe	O	N	H	Others total	Others each	Ti
Ti	–	–	0.01	0.03	0.14	0.01	0.004	< 0.4	< 0.01	remainder
Ti6Al4V	6.35	4.0	0.01	0.2	0.15	0.02	0.003	≤ 0.4	≤ 0.1	

Table 2. Chemical composition of the powders used for selective laser sintering.

technology was presented of manufacturing the elements used in prosthodontics, produced with titanium and Ti6Al4V alloy powders with the SLS technique. Experimental comparative investigations were carried out of selected engineering materials recommended for clinical use for permanent dental prostheses manufactured by CAD/CAM. The properties and structure were also compared directly of analogous prosthetic restorations but produced using the various technologies mentioned and from different alternative engineering materials already mentioned, as well, as compared to the restorations made of Co-Cr alloy by milling solid discs, used most often until now. In order to create a ranking of the materials and technologies applied, analogous elements were produced by milling from solid discs made of ZrO_2 ceramics and then by sintering. Such material and technology are commonly regarded by dentist practitioners as the most avant-garde solution. Elements were produced from titanium and Ti6Al4V alloy by machining by way of milling solid discs, and analogous elements were made from such materials by the additive method in an SLS process. An original technology was established of manufacturing elements used in prosthodontics, produced with titanium and Ti6Al4V alloy powders with the SLS technique.

2. Additive and machining technology of fabrication of materials and dental restorations and research methodology

The elements used for the investigations were fabricated by the machining method by milling solid discs on a CNC milling machine and by the additive SLS method. In the case of the engineering materials manufactured by the machining method by milling on CNC milling machines, technological conditions of manufacturing were chosen experimentally. The technological processes mentioned have no influence on any structural changes, except that ceramic ZrO_2 material of the Cercon type has to be sintered already after milling, so the selection of manufacturing conditions does not have to be verified experimentally and structural changes do not have to be confronted with the mechanical properties obtained of the elements produced. The situation is opposite if the SLS additive technology is applied. Investigations over the selection of technological conditions of selective laser sintering involve the selection of laser power, laser beam diameter, a dependency between the laser beam diameter and the distance between laser beams and the distance between laser remelting paths to desired geometrical features of solid and porous elements produced from titanium and Ti6Al4V alloy and the structure formed depending on such technological conditions and its influence on mechanical properties. Each time before producing the examined elements, design was performed using the CAD method, and relevant data in the stereolithography (STL) format is transferred for the execution of computer-aided manufacturing in a relevant technological device.

The solid elements used for the investigations, fabricated by the machining method by milling solid discs, were prepared with special CAD software, dental wings operation system (DWOS), by Dental Wings and 3D Marcam Engineering AutoFab software (software for manufacturing applications) for geometrically simpler shapes of test specimens.

The software features the width offset of the tools used for milling, to ensure the highest possible design reproduction in reality. The so-designed elements were verified by the software for surface cohesion and developed as ready designs in the stereolithography (STL) format. The design files are then implemented to a CAM module, Mayka Dental, by Picasoft. The CAM module enables to position the design prepared on a material disc according to the optimum three-axial milling axis and ensures the placement of the milled elements in a way least interfering with the design and to determine the optimum axis of precision five-axial milling to exploit optimally the capabilities of the milling machine. The so-prepared design is transformed into a machine language, and the file produced is transferred to the milling machine's software.

Another group of the examined elements was manufactured by an additive technology of selective laser sintering by designing a given element virtually, in order to develop a 3D CAD model in the STL format, transferred to the machine's software and then by producing in reality the so-designed element layer by layer until producing a ready product. 3D Marcam Engineering AutoFab (software for manufacturing applications) software was used for model designing as a CAD/CAM tool. The software enables to select model dimensions, constructional features and the type of the model volume filling either as solid or porous and to choose the size of a cell unit making up the entire model. A 3D model featuring a layered structure is produced as a result of designing. The surface of the elements modelled in the STL format, which is accommodated to SLS manufacturing conditions, is shown with a net of triangles, with the accuracy of the surface being represented being inversely proportional to their size.

The model size and structure have to be assumed each time and divided into layers with the expected thickness. The number of layers in a virtual model matches the actually designed number of powder layers when the given element is really produced. A set of "hexagon cross" unit cells was used, selected in a geometrical analysis and as a result of preliminary studies, from the set available in the software. By duplicating them, the entire elements were designed, consisting of nodes and single-lattice fibres, linking the particular skeleton nodes (**Figure 1**).

Fabrication conditions were also selected, at the stage of virtual design of elements, available in the selective laser sintering system, including layer thickness, laser power, laser beam diameter, scanning rate and distance between particular remelting paths. A spatial orientation of unit cells of 45° relative to the axis x of the system of coordinates was also selected experimentally, because other orientations analysed initially, including at the angle of 0°, at the angle of 45° relative to the axis y, at the angle of 45° relative to the axis x and 45° to the axis y, at the angle of 45° relative to the axis y and 45° relative to the axis x, turned out to be less advantageous due to the mechanical properties of the elements produced (**Figure 1**).

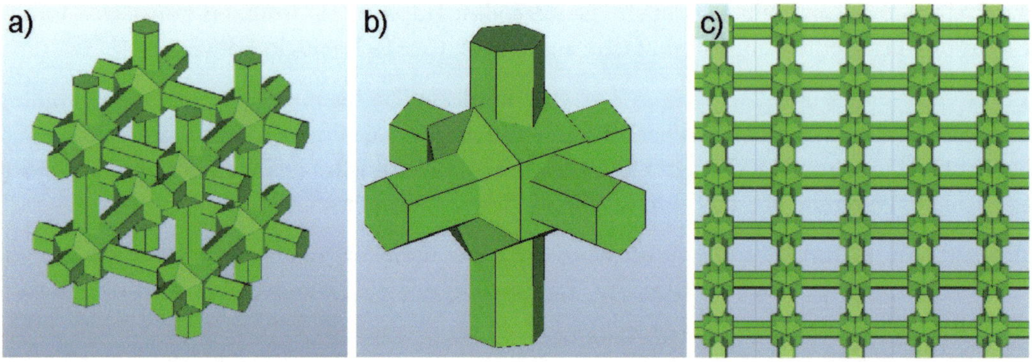

Figure 1. (a, b) "Hexagon cross unit cell"; (c) the arrangement of unit cells in the space of the system of coordinates at the angle of 45° relative to the axis x.

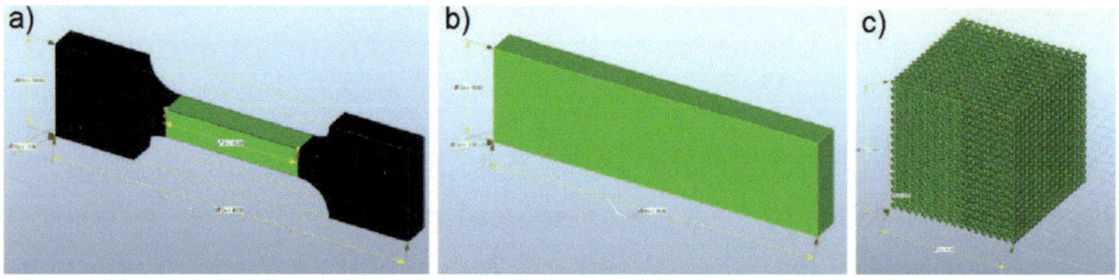

Figure 2. Computer model image of solid samples for static test: (a) tensile test, (b) three-point bending test and (c) compression test.

Micro-samples for static tensile, bending and compression tests were specially selected for the investigations (**Figure 2**), made of all the chosen materials manufactured using, respectively, the additive SLS method from powders with individual CAD/CAM and with the machining technology using CAD/CAM techniques by milling solid discs on CNC milling machines. Micro-samples for tensile strength, bending strength and compression strength tests were also designed for the conditions given above with the pore size of 200–250 μm.

In order to perform bending strength tests, models were created of three-point prosthetic bridges from three different sections of a dental arch (**Figure 3**), including the front, front-wing and wing section, taking into account first of all the highest possible stresses which may occur in such type of prosthetic restorations. They were made as solid from all the examined engineering materials for bending strength tests with the three-point bending method with the both analysed technologies, that is, machining method by milling solid discs on a CNC milling

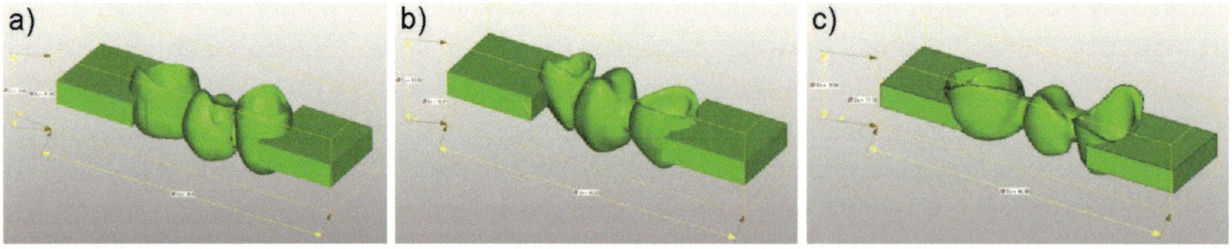

Figure 3. Computer model of bridges nos. 1, 2 and 3 for bending strength tests with the three-point bending method.

machine and by the additive SLS method. The bending strength was compared to dental bridges made of solid materials, that is, sintered titanium, sintered Ti6Al4V alloy, milled Ti6Al4V alloy, milled Co-Cr alloy and milled and sintered ZrO_2 sinter. The results of strength tests of dental bridges selectively sintered by laser in optimised conditions were compared with the results of tests for bridges milled with CAD/CAM methods from discs of other tested materials, additionally sintered after milling in the case of ZrO_2. The samples prepared correspond to the prosthetic restorations designed in the author's own clinical practice and in a dental engineering laboratory for particular patients and used by them. Geometrical features were taken into account, such as the fitting of a prosthetic restoration to a pillar, of 0.03 mm; the width of the threshold area; the space for prosthetic cement in the occlusal region, of 0.1 mm; the reduction of the crown shape considering the space for porcelain, of 1.5 mm; and the minimum thickness of the substructure according to the material manufacturer's recommendations, of 0.3 mm, for a substructure made of Co-Cr alloys, titanium and its alloys and 0.6 mm for a zirconium oxide substructure. The software features the width offset of the tools used for milling.

The so-prepared micro-samples and bridges were verified by software for surface cohesion and prepared as a ready design in the STL format (stereolithography), processed into a machine language of a CNC milling machine by Yenadent D43W or AM 125 machine by Renishaw for selective powder laser sintering. After designing the shape of the elements produced, short machining of such elements takes place with a CNC milling machine from a disc of the selected material and additional sintering in a resistance furnace for ceramic prostheses (**Table 3**). In the case of additive treatment, after designing the produced element the same as previously, the particular layers of metals and alloys are joined layer by layer in the form of powder in an SLS process (**Table 3**).

As reference results, the results were used of tests of micro-samples and prosthetic bridges fabricated by the machining method with CAD/CAM techniques by milling—with CNC milling machines—solid discs made of Co-Cr alloy of the Vitalium type, used most often in dental treatment, mainly because the alloys are commonly used in casting methods. In order to create a ranking of the materials and technologies used, analogous elements were produced by milling from solid discs made of ZrO_2 ceramics and then by sintering. Titanium elements were fabricated by the machining method by milling solid discs and by the additive SLS method. Elements from Ti6Al4V powder by SLS (**Table 3**) were also produced for comparative purposes. Apart from examinations of solid materials, examinations of porous elements with the pore size of 200–250 μm manufactured by SLS from titanium and Ti6Al4V alloy powder were also carried out for comparative purposes.

Milling from disc on CNC milling machine			Manufacturing technology
Co-Cr	ZrO_2	Ti6Al4V	Solid engineering material
Sintered solid engineering material		Ti6Al4V	Ti
Manufacturing technology		Selective laser sintering of powders	

Table 3. Diagram of selection of engineering materials and manufacturing technologies of the examined elements.

3. Comparison of strength properties of the examined solid engineering materials for dental prosthetic restorations manufactured by additive and machining methods

The investigations of mechanical properties of the examined engineering materials were performed in two series in each case. According to the assumed aim of the investigations, mechanical properties were compared in the first series, starting with tensile strength, each time mechanical properties of solid materials produced by the SLS technique for different laser power, that is, sintered titanium and sintered Ti6Al4V alloy. Mechanical properties were compared in the second series of all the examined solid materials, that is, sintered titanium, sintered Ti6Al4V alloy, milled Ti6Al4V alloy, milled Co-Cr alloy and milled and milled ZrO_2 sinter. The results of the examinations were presented in such order for, respectively, tensile strength, bending strength and comprehensive strength.

The impact of laser power, ranging 70–110 W, was investigated in the first place on tensile strength values of sintered titanium and sintered Ti6Al4V alloy (**Figure 4**). The results of tensile strength tests for sintered titanium, sintered Ti6Al4V alloy, obtained in optimum conditions of production, milled Ti6Al4V alloy, milled Co-Cr alloy and milled ZrO_2 sinter, are shown in **Figure 5**.

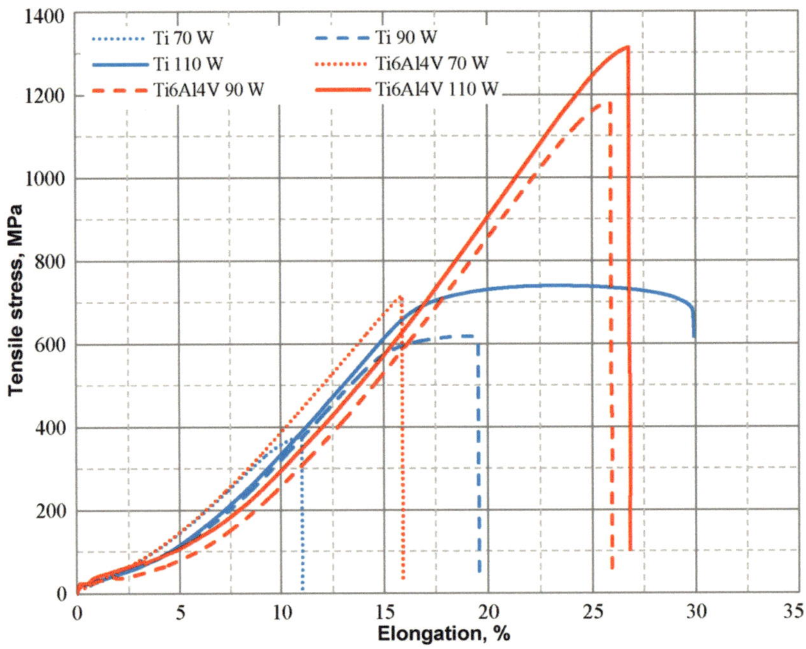

Figure 4. Comparison of diagrams of dependency between tensile stress and elongation for solid samples made of Ti6Al4V alloy and pristine titanium sintered at different laser powers.

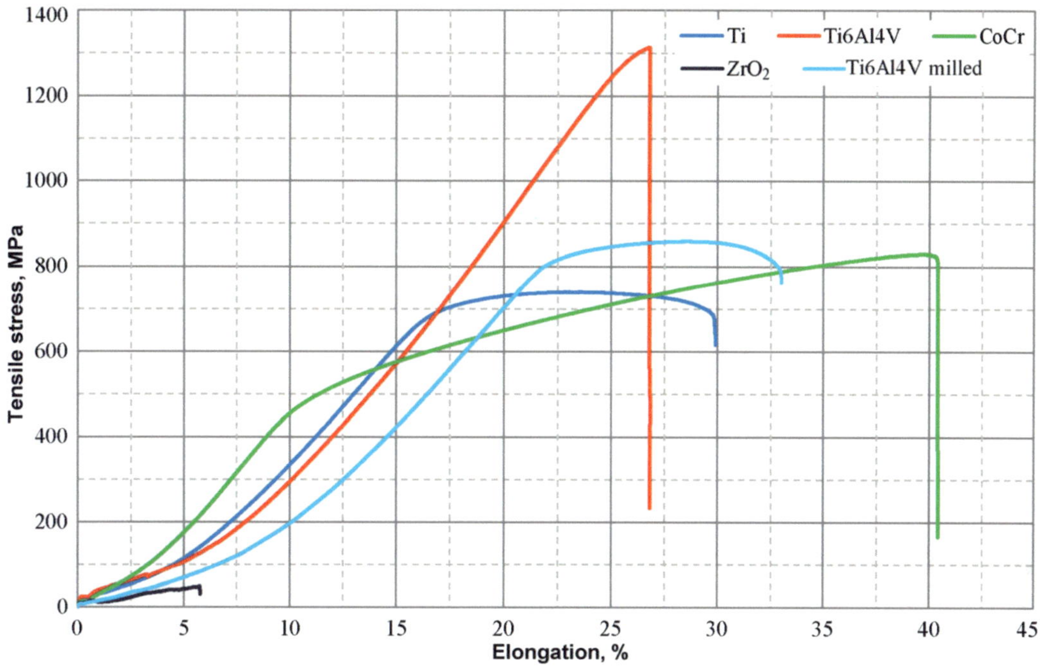

Figure 5. Comparison of diagrams of dependency between tensile stress and elongation for various examined materials, that is, for sintered Ti, sintered Ti6Al4V alloy, milled Co-Cr alloy, milled ZrO_2 sinter and milled Ti6Al4V alloy.

The test results given in **Figure 5** are an interesting reason for application of the tested materials in dental prosthetics. A result of approx. 47 MPa for tensile strength of milled and then sintered ZrO_2 alloys is certainly concerning. Selectively laser sintered titanium exhibits the tensile strength of approx. 740 MPa, slightly smaller than specific for milled Co-Cr alloy samples, approx. 825 MPa, and Ti6Al4V alloy, approx. 858 MPa, and much lower than the tensile strength of approx. 1312 MPa of the selectively laser sintered Ti6Al4V alloy. It is the highest tensile strength value for all the examined engineering materials.

Bending strength tests were performed the same as tensile strength tests. The investigations were performed in relation to solid materials manufactured by SLS for different laser power, that is, sintered titanium and sintered Ti6Al4V alloy (**Figure 6**), as well as in relation to the investigated solid materials, that is, sintered titanium, sintered Ti6Al4V alloy, milled Ti6Al4V alloy, milled Co-Cr alloy and milled ZrO_2 sinter (**Figure 7**). It is noted that the bending strength of the sintered titanium of approx. 1682 MPA is only about 3% smaller than the bending strength of about 1740 MPa of the Co-Cr alloy but strongly smaller than the bending strength of approx. 1959 MPa of the milled Ti6Al4V alloy and much higher than the bending strength of approx. 2374 MPa of this alloy after SLS. The smallest is the bending strength of approx. 605 MPa of sintered ZrO_2 alloys milled from discs and then sintered.

Figure 8 is the comparison of diagrams of dependency between compressive stress and deformation for selectively laser sintered Ti samples and Ti6Al4V alloy.

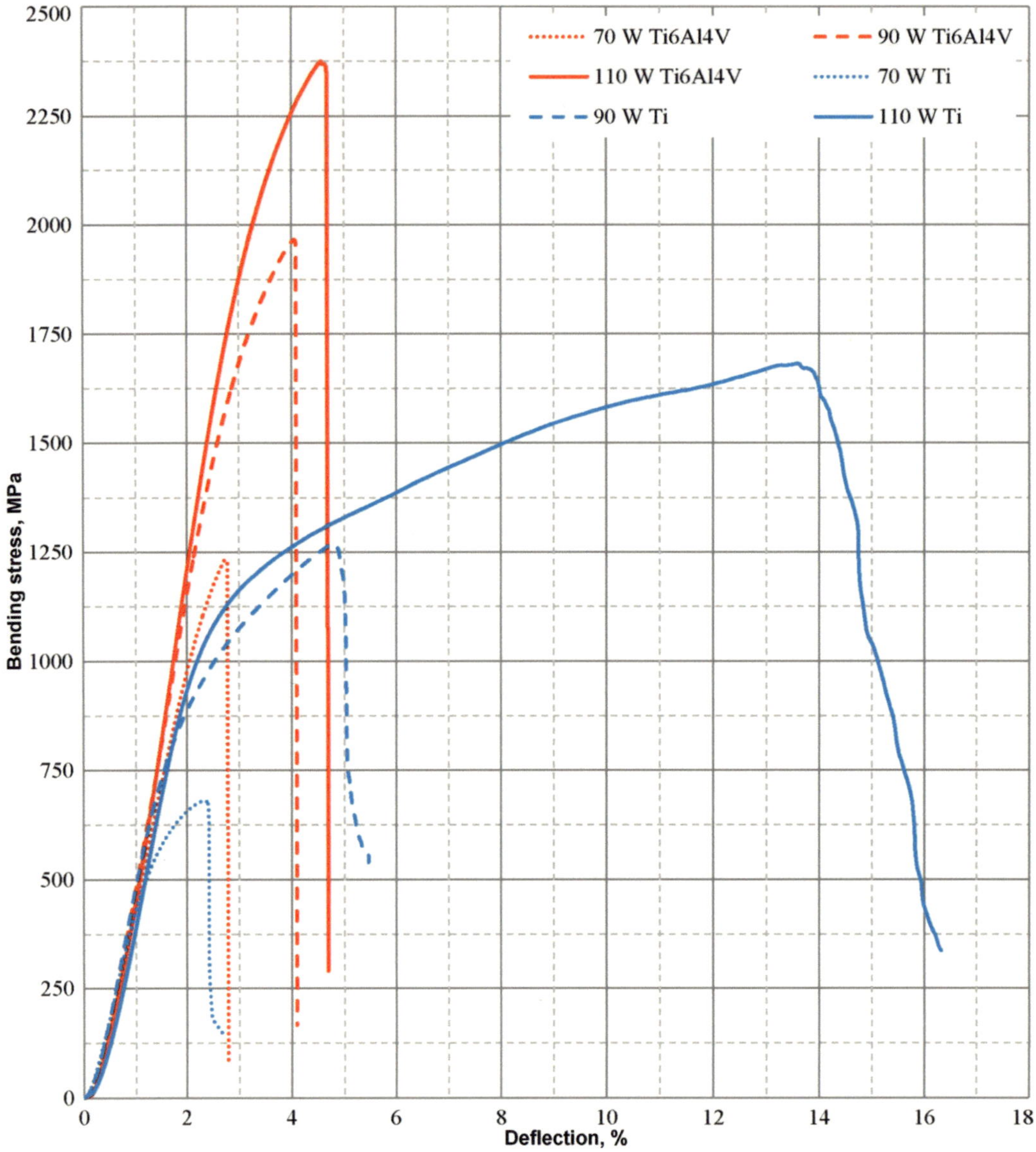

Figure 6. Comparison of diagrams of dependency between bending stress and bending for solid samples made of Ti6Al4V alloy and pristine titanium sintered at different laser powers.

4. Comparison of results of strength tests of dental bridges manufactured from the investigated materials with the additive and machining technology

The results of comparative investigations are presented of mechanical properties of dental bridges manufactured from the investigated engineering materials. The results of strength

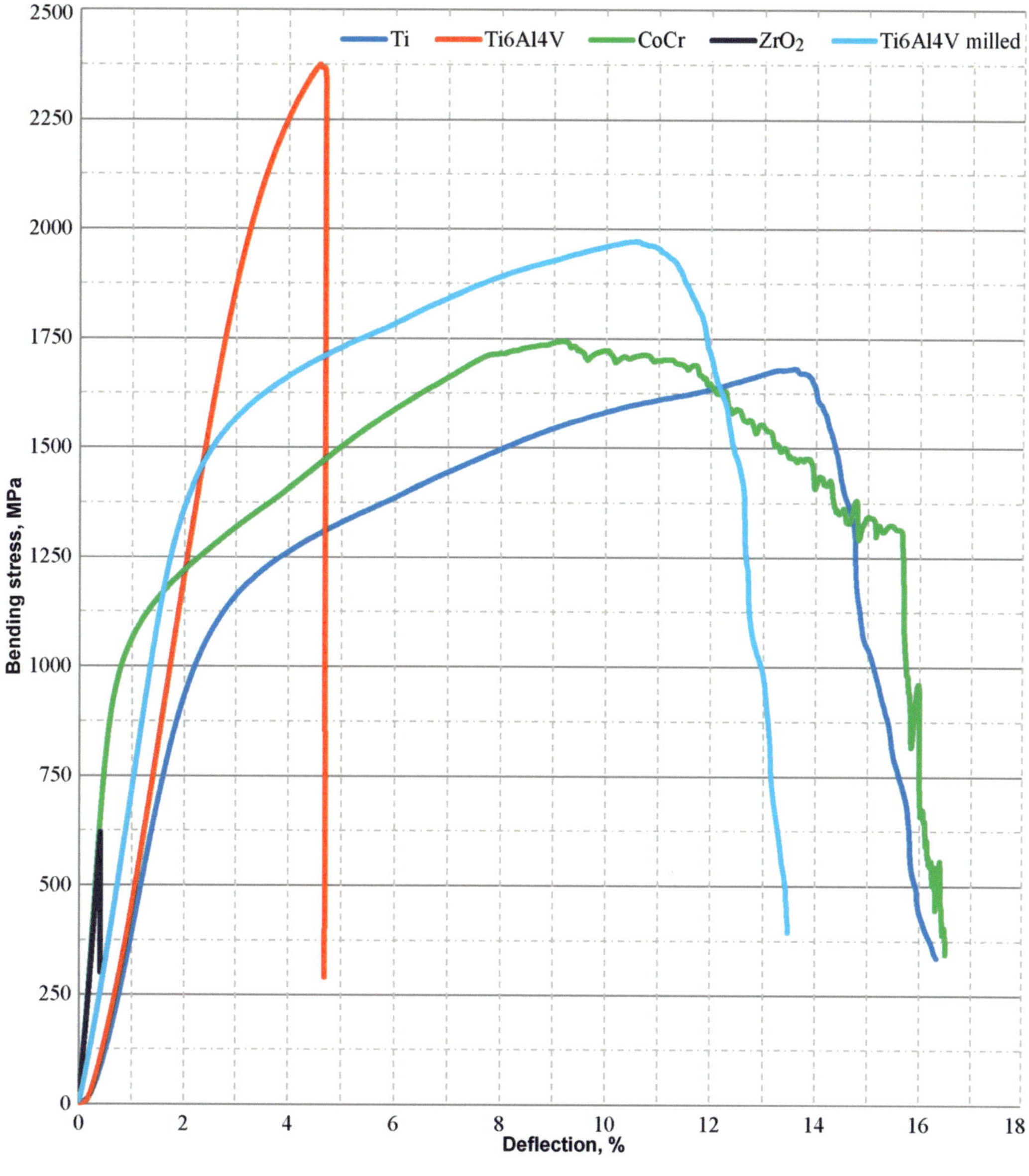

Figure 7. Comparison of diagrams of dependency between tensile stress and bending for various examined materials, that is, for sintered Ti, sintered Ti6Al4V alloy, milled Co-Cr alloy, milled ZrO_2 sinter and milled Ti6Al4V alloy.

tests of dental bridges selectively sintered by laser in optimised conditions were compared with the results of tests for bridges milled with CAD/CAM methods from discs of other tested materials, additionally sintered after milling in the case of ZrO_2. The impact of laser power, ranging 70–110 W, was investigated in the first place on bending strength values of three bridges manufactured by SLS from sintered titanium and sintered Ti6Al4V alloy. The results of bending strength tests of a dental bridge no. 1 produced from titanium and Ti6AlV4 alloy were compared directly in **Figure 9**. Regardless the material produced by selective laser

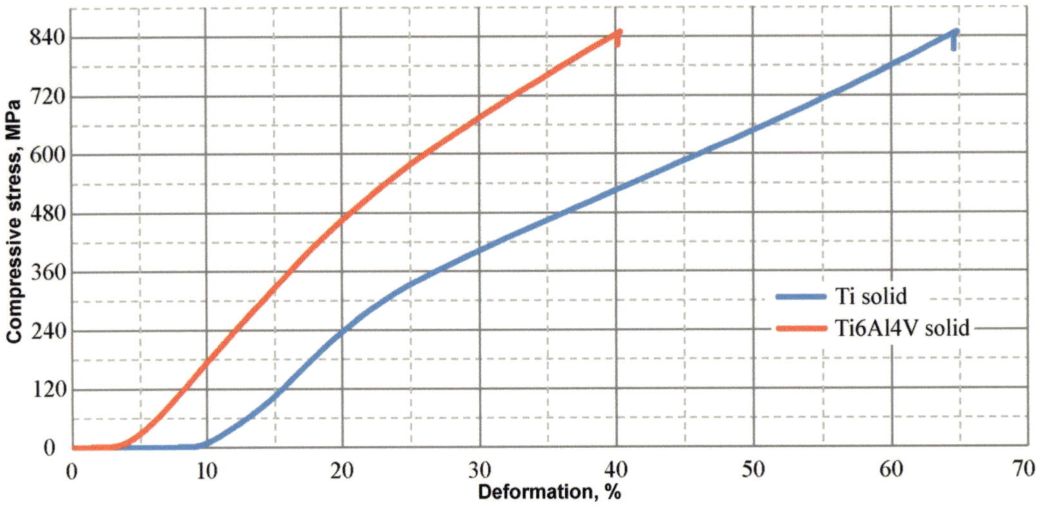

Figure 8. Comparison of diagrams of dependency between compressive stress and deformation for selectively laser sintered Ti and Ti6Al4V alloy samples.

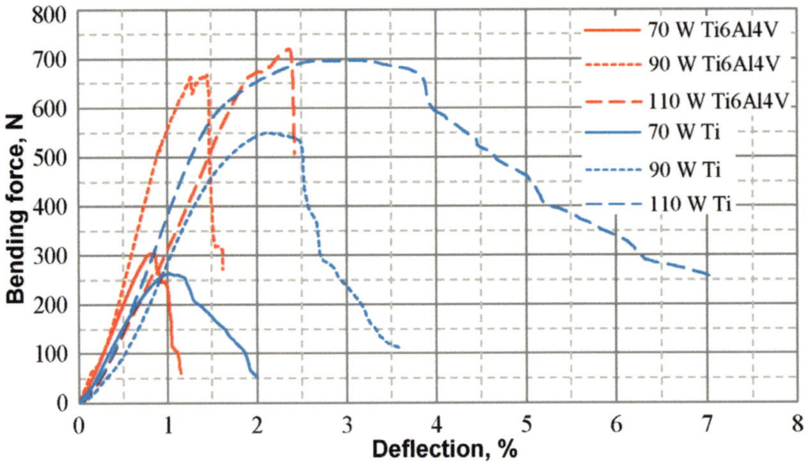

Figure 9. Comparison of diagrams of dependency between bending strength and bending of solid bridge no. 1 manufactured by SLS from titanium and Ti6Al4V alloy for different laser power.

sintering, the maximum bending strength strongly depends on laser power. Sintering with the laser power of 70 W ensures small maximum bending force of approx. 260 and approx. 310 N, respectively, for titanium and Ti6Al4V alloy and is rising almost proportionally to, respectively, approx. 700 and approx. 730 N with laser power increase to 110 W. The maximum bending of the produced dental bridges depends very strongly on the applied laser power and a selectively sintered material. For titanium, this bending—depending on the laser power—varies between approx. 0.9 and approx. 2.4%, when it is much higher for Ti6Al4V alloy and ranges within approx. 2 to approx. 7%, and bending strength is systematically falling as bending is rising (**Figure 9**).

The analogous strength properties were obtained for a dental bridge no. 2 produced from titanium and Ti6AlV4 (**Figure 10**). In the case of a titanium bridge, if laser power is increased from 70 to 110 W, this has a considerable effect on the value of the maximum bending strength, which increases on average from approx. 350 to approx. 1315 N. In the case of a bridge made of Ti6Al4V alloy produced by SLS, if laser power is raised in the same range, the maximum bending strength rises from approx. 525 to approx. 1560 N. In the case of Ti6Al4V alloy, each time, for each laser power, the maximum bending strength causing sample fracture is higher than for bridges produced from titanium, and this takes place in each case for relatively smaller bending. For example, selectively laser sintered solid bridges with laser power of 110 W from Ti6Al4V alloy are damaged at approx. 3% bending, when bending for titanium exceeds 4.5% (**Figure 10**).

As laser power is increased from 70 to 110 W, the value of the maximum bending strength for dental bridge no. 3, produced from titanium, increases from approx. 240 to approx. 580 N, in

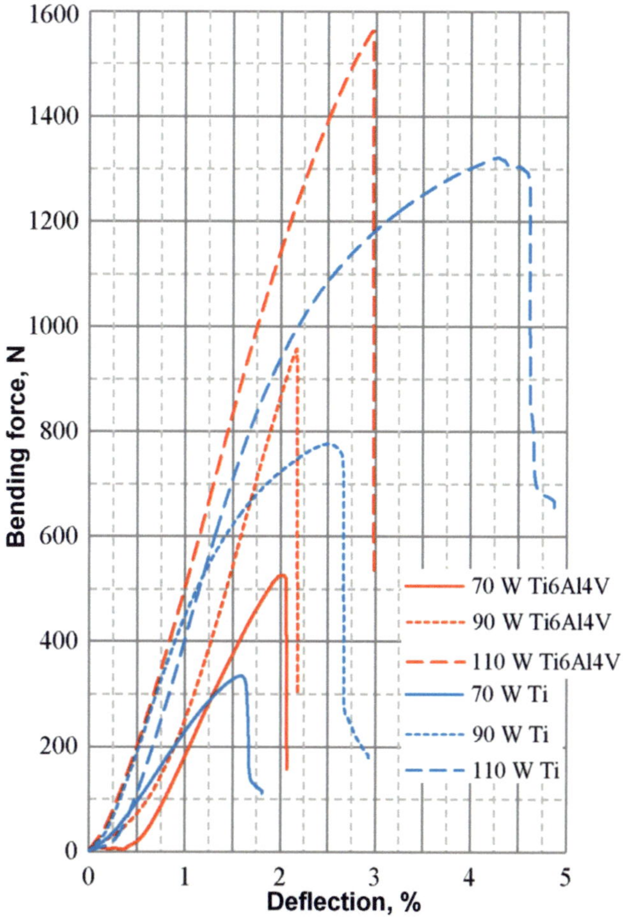

Figure 10. Comparison of diagrams of dependency between bending strength and bending of solid bridge no. 2 manufactured by SLS from titanium and Ti6Al4V alloy for different laser power.

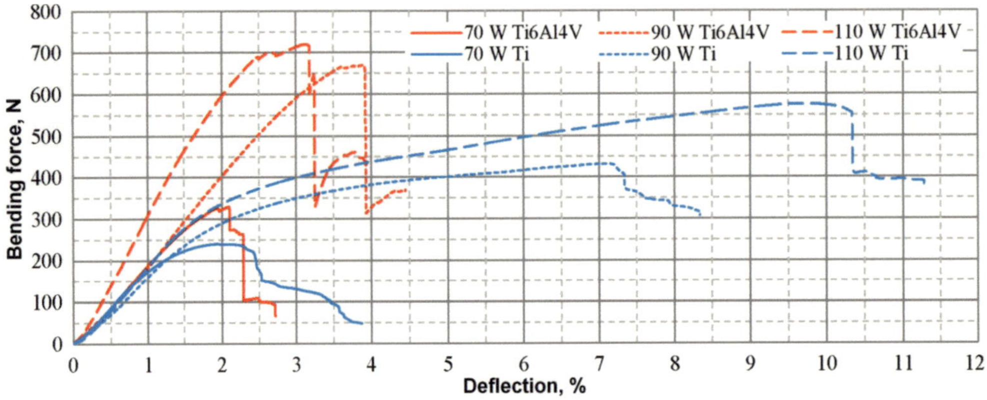

Figure 11. Comparison of diagrams of dependency between bending strength and bending of solid bridge no. 3 manufactured by SLS from titanium and Ti6Al4V alloy for different laser power.

the case of a bridge made of Ti6Al4V alloy, from approx. 340 to approx. 720 N (**Figure 11**). In the case of selectively laser sintered Ti6Al4V alloy, each time, for each laser power, bending strength causing sample fracture is higher than for bridges produced from titanium, and this takes place in each case for relatively smaller bending (**Figure 11**).

Diagrams were compared of dependency between bending strength and bending of solid bridges nos. 1, 2 and 3 manufactured from all the examined materials, that is, from selectively laser sintered titanium and Ti6Al4V alloy and milled from Ti6Al4V alloy discs, Co-Cr alloy and milled ZrO_2 sinter and then sintered (**Figure 12**). The maximum bending strength highly relies on the dental bridge type and varies from approx. 580 N for bridge no. 3 to approx. 1320 N for bridge no. 2. On the other hand, bridge no. 3 exhibits the maximum bending of approx. 10.4% while bridge no. 2 smallest bending of approx. 4.6%. Bridge no. 1 shows indirect properties. In the case of solid bridges nos 1, 2 and 3 manufactured by SLS from Ti6Al4V alloy with laser power of 110 W (**Figure 12**), the maximum bending strength of approx. 1570 N occurs for bridge no. 2, and the two other bridges show the strength of approx. 720 N. Maximum bending for bridges nos 3, 2 and 1 is, respectively, approx. 3.2, approx. 3 and approx. 2.4%. In the case of bridges nos. 2, 1 and 3 manufactured by CAD/CAM by milling discs from Ti6Al4V alloy, as presented in **Figure 12**, the maximum bending strength is much higher and is, respectively, approx. 2970, approx. 2030 and 1030 N, and maximum bending is, respectively, approx. 5, approx. 3.4 and 5.6%. Diagrams were also compared of dependency between bending strength and bending of solid bridges nos. 1, 2 and 3 manufactured by CAD/CAM by milling discs from Co-Cr alloy and ZrO_2 sinter (**Figure 12**). The comparison of diagrams of dependency between bending strength and bending of solid bridges nos. 1, 2 and 3 manufactured from all the examined materials, that is, from selectively laser sintered titanium and Ti6Al4V alloy and with the CAD/CAM method by milling discs from Co-Cr alloy and ZrO_2 sinter, shows that the highest maximum bending strengths are specific for bridge no. 2 regardless the material applied, and highest bending is seen for bridge no. 3.

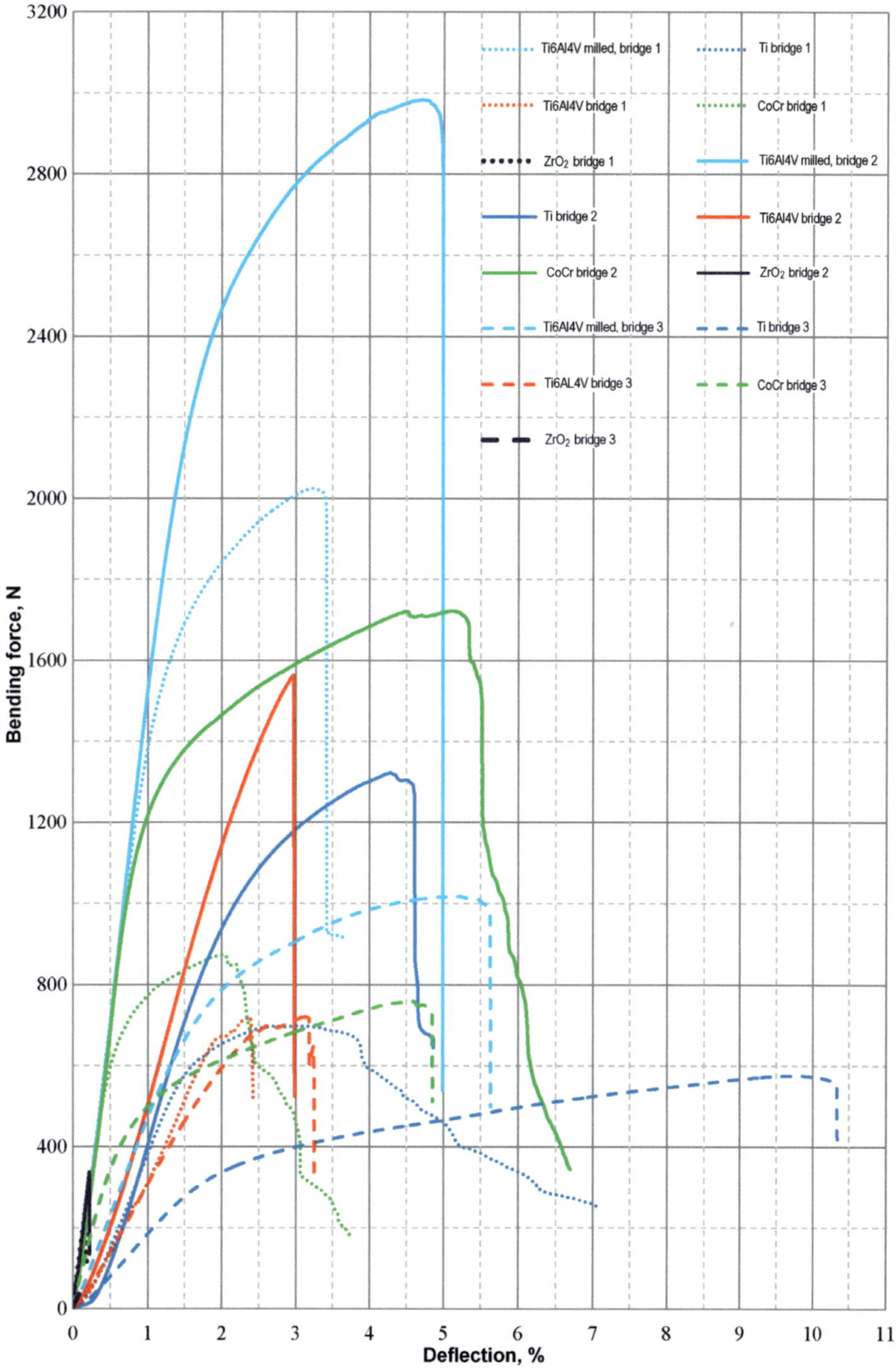

Figure 12. Comparison of diagrams of dependency between bending strength and bending of solid bridges nos. 1, 2 and 3 manufactured from all the examined materials, that is, from selectively laser sintered titanium and Ti6Al4V alloy and with the CAD/CAM method by milling discs from Ti6Al4V alloy, Co-Cr alloy and sintered ZrO_2 sinter.

5. Structure of the investigated engineering materials manufactured by additive and machining technologies

It was revealed in examinations of the structure of the investigated materials in a scanning electron microscope that in the case of Co-Cr and Ti6Al4V alloy, the surface of the samples produced by CAD/CAM, by milling solid discs, exhibits roughness distinctive for precision machining (**Figures 13** and **14**), the same as for ceramic ZrO_2 material which is sintered after performing machining (**Figure 13**). The examinations of chemical composition with the EDS method confirm the presence, in the first case, of Co, Cr and W (**Figure 13**) and in the second case of, respectively, Ti, Al and V (**Figure 14**) and in the third case of Zr and O (**Figure 13**), corresponding to the chemical composition of the materials applied. **Figure 14** shows a surface structure of pristine titanium and Ti6Al4V alloy manufactured by selective laser sintering with

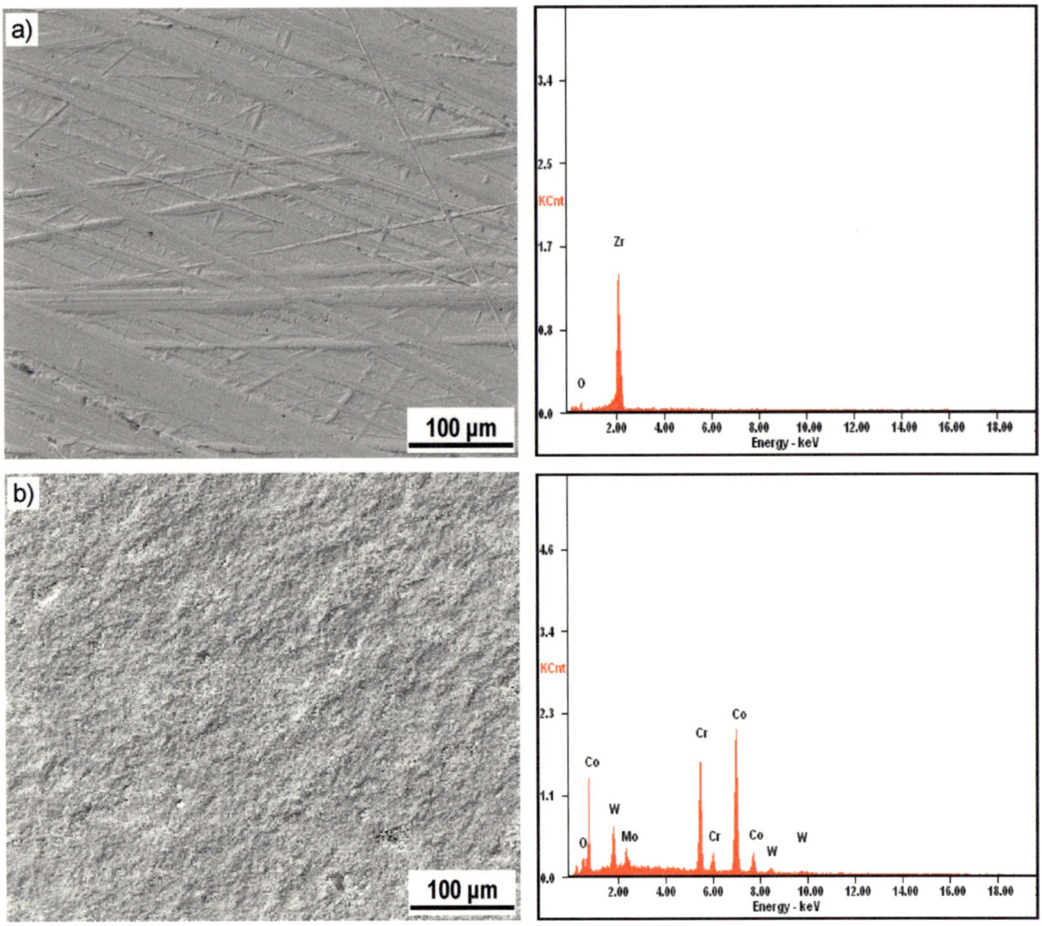

Figure 13. Surface structure of the samples produced by CAD/CAM by milling solid discs with the results of examinations of EDS chemical composition of (a) Co-Cr alloy, (b) sintered ZrO_2 sinter after machining (SEM).

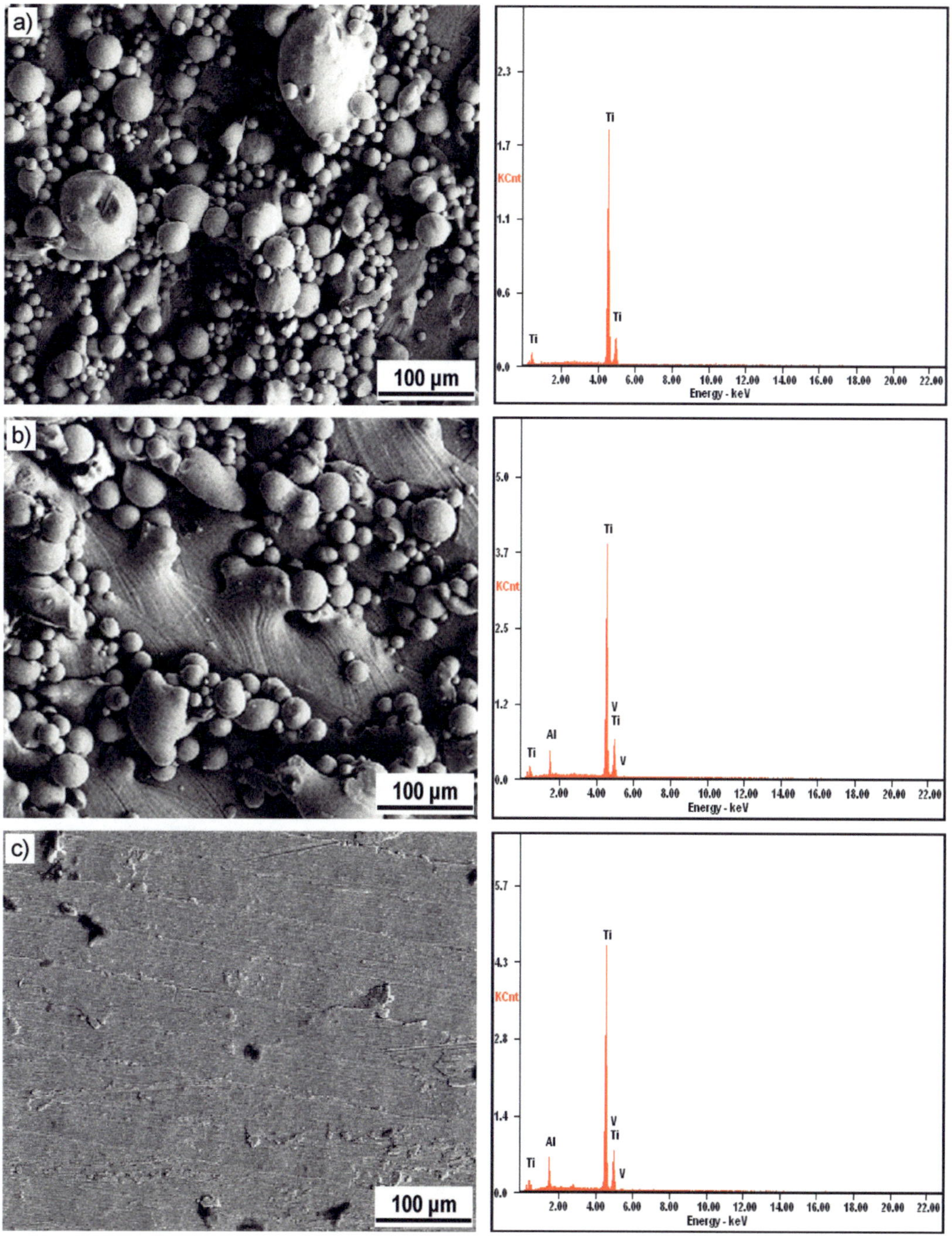

Figure 14. Surface structure of samples manufactured by selective laser sintering with the use of laser beam of 50 μm with laser power of 110 W of (a) solid titanium, (b) Ti6Al4V alloy and (c) Ti6Al4V alloy by CAD/CAM method by milling solid discs with results of examinations of EDS chemical composition (SEM).

the laser beam size of 50 μm and laser power of 110 W. It was revealed in the both cases that apart from the completely sintered materials, with the shown paths of laser beam transition, fine powder particles exist, as well, which—after sintering—should be removed mechanically or by chemical etching.

X-ray tests allow to establish the phase composition of the investigated materials based on the data derived from international files. The following was identified, respectively, in particular samples: Co and Cr in Co-Cr, ZrO_2 alloy, Ti in pristine titanium and $Al_{0.3}Ti_{0.7}$ phase and $Ti_{0.8}V_{0.2}$ in Ti6Al4V alloy irrespective of the manufacturing technology by milling from discs or selective laser sintering.

Investigations were carried out into the structure of the examined engineering materials subjected to the examinations of mechanical properties, and the structure was compared, especially of fractures of the examined materials produced by SLS, that is, sintered titanium and sintered Ti6Al4V alloy, and also manufactured by CAD/CAM by milling discs from solid Ti6Al4V, Co-Cr alloys and ZrO_2 sinter milled and then sintered. The selectively laser sintered materials have a continuous structure without pores, as for titanium (**Figure 15a**), in which

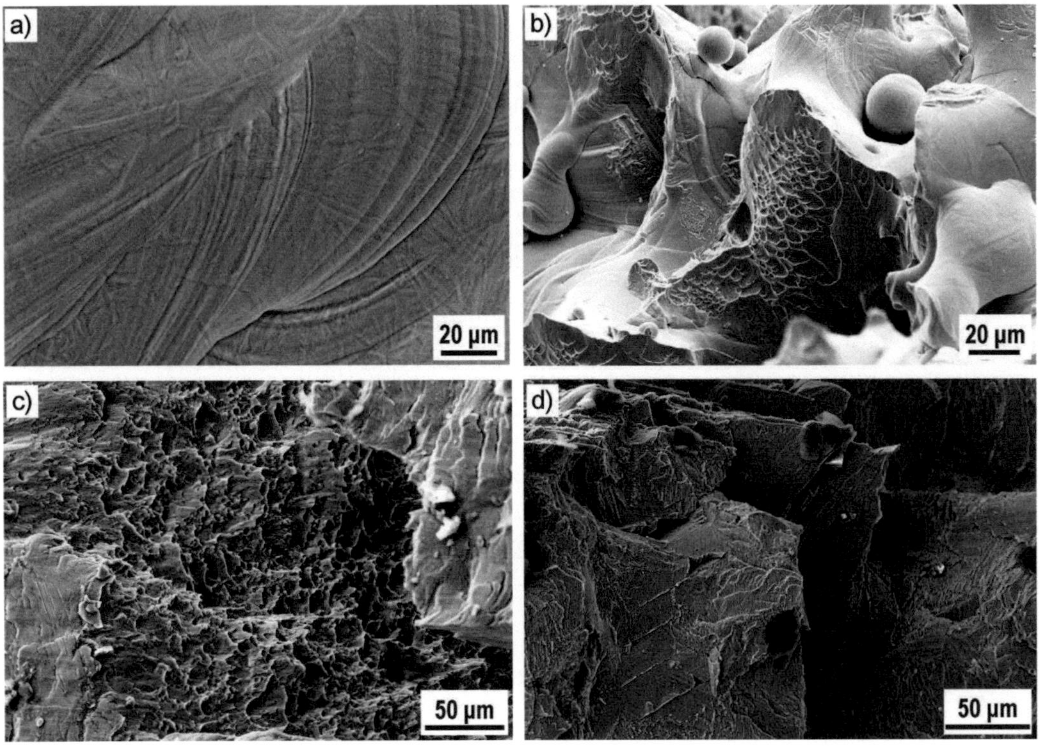

Figure 15. Surface structure of solid samples manufactured by selective laser sintering (a) with the use of laser beam of 50 μm with laser power of 110 W, (b) fracture structure after a static bending test of samples of solid Ti6Al4V alloy manufactured by selective laser sintering with laser beam size of 50 μm after sintering with laser power of 70 W and (c, d) fracture structure after a static bending test of the samples produced by CAD/CAM by milling discs from solid alloys of (c) Ti6Al4V and (d) Co-Cr; SEM.

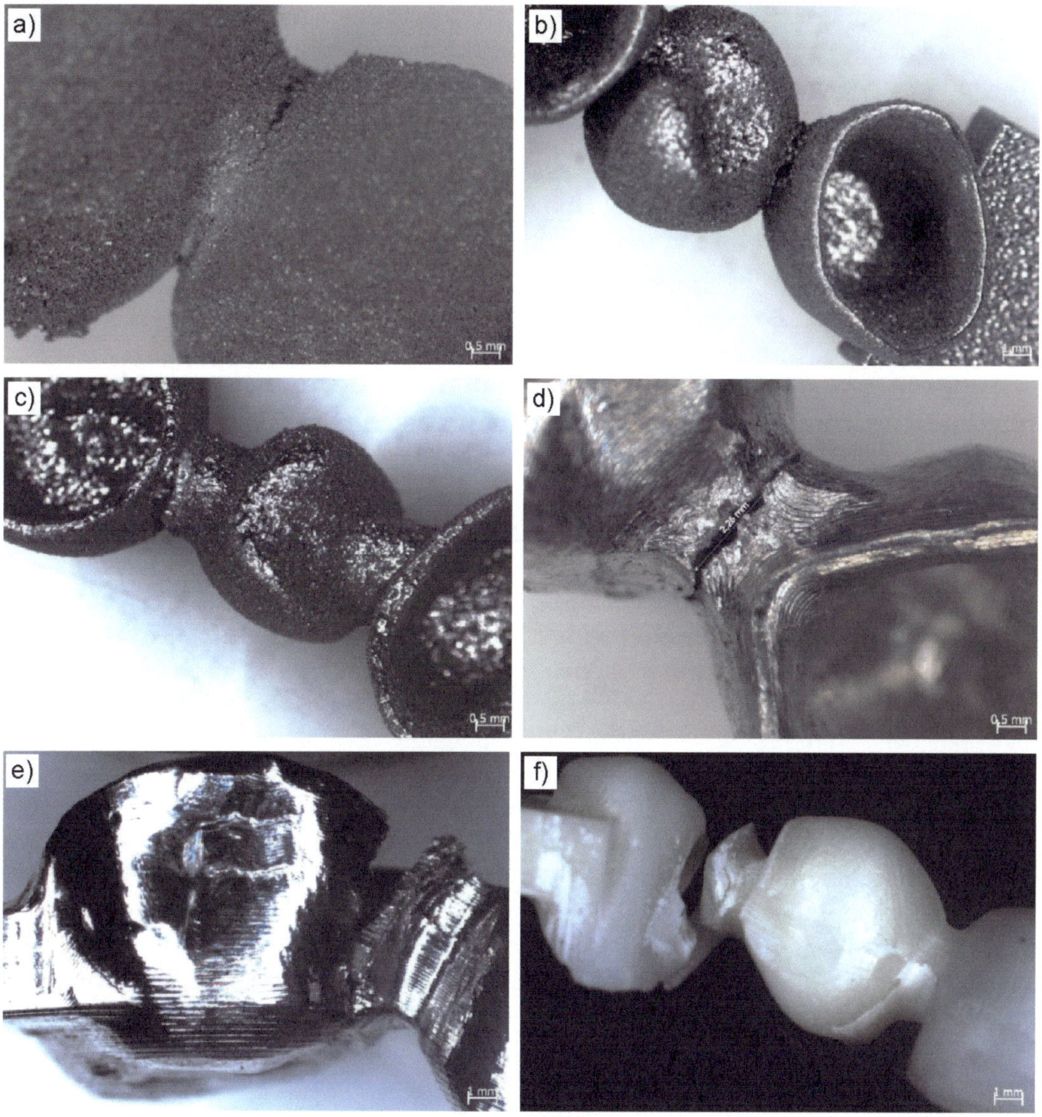

Figure 16. View of dental bridges after static bending test, manufactured from (a,b) titanium by SLS, (a) bridge no. 2 and (b) bridge no. 1; (c) from Ti6Al4V alloy by SLS bridge no. 3; and (d–f) fabricated by CAD/CAM by disc milling (d) from Ti6Al4V alloy, bridge no. 2; (e) from Co-Cr alloy, bridge no. 3; and (f) from ZrO_2 sinter manufactured by CAD/CAM by disc milling and then sintering, bridge no. 3; stereoscope microscope.

particular laser paths can be distinguished, provided that laser power for a laser beam with a diameter of 50 μm is 110 W. It is revealed by analysing a fracture of titanium, laser sintered with the laser power of 70 W, subjected to a static bending test, that—especially—finest powder particles are not sintered fully and also that a ductile fracture occurs in bridges of the sintered material. The samples sintered with the laser power of 110 W are solid and show a ductile fracture. A fracture structure of the samples from the laser sintered Ti6Al4V alloy after a static bending test is analogous to this obtained for titanium (**Figure 15b**) and—after sintering with the laser power of 110 W—shows a ductile fracture. A ductile fracture also exists in the case of fractographic tests of Ti6Al4V alloy but in the form of discs for sample milling (**Figure 15c**). The samples milled from Co-Cr alloy show, however, a brittle fracture, only with the local presence of a ductile fracture (**Figure 15d**).

In the case of all the examined metallic materials, and independently from the applied manufacturing technology, a fracture of samples occurs most often on a root of the abutments linking successive teeth, while in the case of bridges made of ZrO_2, such bridges undergo brittle cracking in different places (**Figure 16**).

6. Suitability of application of additive manufacturing methods of prosthetic restorations of the dental system as compared to machining methods

Comparative investigations are presented in this chapter of the structure and properties of selected engineering materials used for dental prosthetic restorations manufactured alternatively by the machining method by milling on computerised numerical control (CNC) milling machines and by the additive selective laser sintering (SLS) method of solid and porous elements using CAD/CAM techniques; moreover, an original technology was presented of manufacturing the elements used in prosthodontics, produced with titanium and Ti6Al4V alloy powders with the SLS technique. The suitability was confirmed of applying the manufacturing technologies used in prosthodontics from powders by the SLS technique. The results of the executed tests of strength micro-samples and of the selected prosthetic bridges were compared. The bending strength of the titanium sintered selectively by laser, of the milled Ti6Al4V alloy and of the milled Co-Cr alloy is within the range of 1739–1959 MPa and tensile strength is within 739 and 858 MPa. It points out that the materials can be employed as prosthetic restorations, similarly as a selectively laser sintered Ti6Al4V alloy with the bending strength of approx. 2374 MPa and tensile strength of approx. 1312 MPa and pristine Ti with the bending strength of approx. 1682 MPa and tensile strength of 739 MPa. The SLS technology for titanium, and even more for Ti6Al4V alloy, ensures the achievement of mechanical properties comparable or better than a reference Co-Cr alloy which is commonly used for prosthetic restorations, including prosthetic bridges fabricated by milling solid discs with CNC milling machines. For all the examined engineering materials, the milled and then sintered ZrO_2 material exhibits the lowest strength properties. The average tensile strength of this material is approx. 47 MPa, that is, it is nearly 20 times smaller than of Co-Cr alloy, and the average bending strength is smaller nearly 3 times than this of Co-Cr alloy and is approx. 605 MPa. The maximum elongation does not exceed, respectively, 6%, and maximum bending is approx. 0.3%, and the both values are much smaller than such appropriate for Co-Cr alloy. The microscopic examinations of the surface of prosthetic bridges fabricated from this material by the machining method of disc milling and then sintering show more advantageous roughness than bridges produced from Co-Cr alloy, and their fracture after bending is decisively brittle, and cracking takes place in different places, not as in the case of Co-Cr alloy. Prosthetic restorations on a substructure of sintered ZrO_2 are, however, used more and more often, especially because it is necessary to ensure the best aesthetic values of the solution employed.

As a zirconium oxide substructure can be completely white, but also after colouring with special paints before sintering, it can have a colour consistent with the VITA colour palette, a fully natural aesthetic effect can be achieved of a ready prosthetic restoration. It is also noteworthy that a translucence effect can be achieved, especially on secant edges of such a prosthetic restoration, responding to the most demanding patients' needs. Such effects can be attained on a substructure made of Co-Cr alloy. On the other hand, it is practically impossible to use titanium as a substructure for prosthetic crowns and bridges because the material tends to change the shade of porcelain to darker and colder; hence it is very difficult to ensure the manufacturing repeatability of prosthetic restorations and their quality. A solution ensuring the best possible aesthetic values is to use zirconium oxide as a substructure. Due to a substantial technological and therapeutic risk, a solution with the lowest strength properties is most expensive, though. Therefore, a dental engineer needs to be aware of the limitations of this material and has to design prosthetic restorations on a zirconium oxide substructure differently than when metallic materials are used. The author of this work prepares in his daily engineering practice all kinds of bridges with ¾ crowns, that is, such which are made on all the walls, except for the vestibular wall and the chewing surface, with a full structure, as so-called full contoured. Such a crown is coated from the vestibular part with veneering porcelain to ensure the best aesthetic values, and special paints and glaze are applied onto the other surfaces. The so-designed bridge, in the example shown in this work, would have walls with a thickness of approx. 2.5 mm instead of 0.6 mm, and a section area of abutments would be approx. 20–22 mm^2 instead of 10–12 mm^2, as for the designs presented in this work and essentially relating to metallic materials. A much greater active section will allow to transfer higher loads, accordingly. The designer also has to keep in mind to support bridge spans on the periodontal tissue, never using the saddleback-like gum shape, but using an elliptic support with subgingival correction in each case. The so-designed prosthetic restoration will surely feature much higher strength properties than a prosthetic restoration with a reduced structure, being a reference point used in this work. Unfortunately, prosthetic restorations made on a zirconium oxide substructure with a simplified structure, as so-called caps with simple abutments, are still commonly used by dentists and dental technicians. The test results obtained seem not to confirm the applicability of the bridges, made of this material, designed this way. This puts into question a potential application of ZrO_2 sinter bridges in dental practice without using a special design method of such dental restorations. This fact can be confirmed by cases, known from medical practice, when such bridges are damaged during normal use by patients. The results presented in this chapter can be directly applied in dental practice to remove consequences of teeth lost due to tooth decay.

Acknowledgements

Fundusze Europejskie Inteligentny Rozwój ASKLEPIOS Unia Europejska

The research presented in the chapter was realised in connection with the Project POIR.01.01.01-00-0485/16-00 on "IMSKA-MAT Innovative dental and maxillofacial implants manufactured using the innovative additive technology supported by computer-aided materials design ADDMAT" realised by the Medical and Dental Engineering Centre for Research, Design and Production ASKLEPIOS in Gliwice, Poland and co-financed by the National Centre for Research and Development in Warsaw, Poland.

Author details

Lech B. Dobrzański

Address all correspondence to: lbdobrzanski@gmail.com

Medical and Dental Engineering Centre for Research, Design and Production ASKLEPIOS, Gliwice, Poland

References

[1] Kanasi E, Ayilavarapu S, Jones J. The aging population: Demographics and the biology of aging. Periodontology 2000. 2016;**72**:13-18. DOI: 10.1111/prd.12126

[2] Petersen PE, Ogawa H. Prevention of dental caries through the use of fluoride – The WHO approach. Community Dental Health. 2016;**33**:66-68. DOI: 10.1922/CDH_ Petersen03

[3] Oral Health Country/Area Profile Project CAPP, Centre for Oral Health Sciences. Malmö, Sweden: Malmö University; 2016. Available from: http://www.mah.se/CAPP/ Accessed: 2017-03-10

[4] Petersen PE. Oral Health Surveys: Basic Methods. Fifth Edition. Geneva, Switzerland: World Health Organization; 2013. 125 p. ISBN: 978-92 4-154864-9

[5] Dobrzański LA, Dobrzańska-Danikiewicz AD, Achtelik-Franczak A, Dobrzański LB. Comparative analysis of mechanical properties of scaffolds sintered from Ti and Ti6Al4V powders. Archives of Materials Science and Engineering. 2015;**73**:69-81

[6] Pattanayak DK, Fukuda A, Matsushita T, Takemoto M, Fujibayashi S, Sasaki K, Nishida N, Nakamura T, Kokubo T. Bioactive Ti metal analogous to human cancellous bone: Fabrication by selective laser melting and chemical treatments. Acta Biomaterialia. 2011; **7**:1398-1406

[7] Dobrzański LA, Dobrzańska-Danikiewicz AD, Gaweł TG, Achtelik-Franczak A. Selective laser sintering and melting of pristine titanium and titanium Ti6Al4V alloy powders and selection of chemical environment for etching of such materials. Archives of Metallurgy and Materials. 2015;**60**:2039-2045. DOI: 10.1515/amm-2015-0346

[8] Lu L, Fuh J, Chen Z, Leong CC, Wong YS. In situ formation of TiC composite using selective laser melting. Materials Research Bulletin. 2000;**35**:1555-1561

[9] Kruth J-P, Mercelis P, Van Vaerenbergh J, Froyen L, Rombouts M. Binding mechanisms in selective laser sintering and selective laser melting. Rapid Prototyping Journal. 2005; **11**:26-36. DOI: 10.1108/13552540510573365

[10] Bertol LS, Júnior WK, da Silva FP, Kopp CA. Medical design: Direct metal laser sintering of Ti-6Al-4V. Materials and Design. 2010;**31**:3982-3988

[11] Kumar S. Selective laser sintering: A qualitative and objective approach. Modeling and Characterization. 2003;**55**:43-47

[12] Guo N, Leu MC. Additive manufacturing: Technology, applications and research needs. Frontiers of Mechanical Engineering. 2013;**8**:215-243. DOI: 10.1007/s11465-013-0248-8

[13] Xue W, Vamsi KB, Bandyopadhyay A, Bose S. Processing and biocompatibility evaluation of laser processed porous titanium. Acta Biomateralia. 2007;**3**:1007-1018

[14] Osakada K, Shiomi M. Flexible manufacturing of metallic products by selective laser melting of powder. International Journal of Machine Tools & Manufacture. 2006;**46**:1188-1193

[15] Mellor S, Hao L, Zhang D. Additive manufacturing: A framework for implementation. International Journal of Production Economics. 2014;**149**:194-201

[16] Dobrzański LA, Dobrzańska-Danikiewicz AD, Malara P, Gaweł TG, Dobrzański LB, Achtelik-Franczak A. Fabrication of scaffolds from Ti6Al4V powders using the computer aided laser method. Archives of Metallurgy and Materials. 2015;**60**:1065-1070. DOI: 10.1515/amm-2015-0260

[17] Dobrzański LA, Dobrzańska-Danikiewicz AD, Achtelik-Franczak A, Dobrzański LB, Szindler M, Gaweł TG. Porous selective laser melted Ti and Ti6Al4V materials for medical applications. In: Dobrzański LA, editor. Powder Metallurgy – Fundamentals and Case Studies. Rijeka: InTech; 2017. p. 161-181. DOI: 10.5772/65375

[18] Dobrzański LB. Struktura i własności materiałów inżynierskich na uzupełnienia protetyczne układu stomatognatycznego wytwarzane metodami przyrostowymi i ubytkowymi [Ph.D. Thesis in progress]. Kraków: AGH University of Science and Technology; 2017

[19] Achtelik-Franczak A. Inżynierskie materiały kompozytowe o wzmocnieniu z mikroporowatego tytanu selektywnie spiekanego laserowo [Ph.D. Thesis]. Gliwice, Poland: Silesian University of Technology; 2016

[20] Gaweł TG. Opracowanie technologii wytwarzania oraz badanie struktury i własności innowacyjnych porowatych materiałów biomimetycznych [Ph.D. Thesis in progress]. Gliwice, Poland: Silesian University of Technology; 2017

[21] Dobrzański LA, Dobrzańska-Danikiewicz AD, Achtelik-Franczak A, Dobrzański LB, Hajduczek E, Matula G. Fabrication technologies of the sintered materials including materials for medical and dental application. In: Dobrzański LA, editor. Powder Metallurgy – Fundamentals and Case Studies. Rijeka: InTech; 2017. p. 17-52. DOI: 10.5772/65376

[22] Dobrzański LA, Dobrzańska-Danikiewicz AD, Malara P, Dobrzański LB, Achtelik-Franczak A. Biological and engineering composites for regenerative medicine. Patent Application P 414723.9.11.2015

[23] Dobrzański LA, Dobrzańska-Danikiewicz AD, Malara P, Gaweł TG, Dobrzański LB, Achtelik-Franczak A. Bone Implant Scaffold. Patent Application P 414424.19.10.2015

[24] Dobrzański LA, Dobrzańska-Danikiewicz AD, Malara P, Gaweł TG, Dobrzański LB, Achtelik-Franczak A. Implant Scaffold and a Prosthesis of Anatomical Elements of a Dental System and Craniofacial Bone. Patent Application P 414423.19.10.2015

[25] Dobrzański LA, Dobrzańska-Danikiewicz AD, Malara P, Gaweł TG, Dobrzański LB, Achtelik A. Composite Fabricated by Computer-Aided Laser Methods for Craniofacial Implants and its Manufacturing Method. Patent Application P 411689.23.03.2015

[26] Dobrzański LA, Dobrzańska-Danikiewicz AD, Malara P, Achtelik-Franczak A, Dobrzański LB, Kremzer M. Sposób wytwarzania materiałów kompozytowych o mikro-porowatej szkieletowej strukturze wzmocnienia. Patent Application P 417552.13.06.2016

[27] Dobrzański LA, Dobrzańska-Danikiewicz AD, Malara P, Dobrzański LB, Achtelik-Franczak A, Gaweł TG. Implant-Scaffold or Prosthesis Anatomical Structures of the Stomatognathic System and the Craniofacial. Gold Medal on International Exhibition of Technical Innovations, Patents and Inventions. Třinec, Czech Republic: INVENT ARENA; 2016 16-18.06.2016

[28] Dobrzański LA, Dobrzańska-Danikiewicz AD, Malara P, Dobrzański LB, Achtelik-Franczak A, Gaweł TG. Implant-Scaffold or Prosthesis Anatomical Structures of the Stomatognathic System and the Craniofacial. Gold Medal and International Intellectual Property Network Forum (IIPNF) Leading Innovation Award on International Intellectual Property, Invention, Innovation and Technology Exposition. Bangkok, Thailand: IPITEX; 2016 2-6.02.2016

[29] Dobrzański LA, Dobrzańska-Danikiewicz AD, Malara P, Gaweł TG, Dobrzański LB, Achtelik-Franczak A. The novel composite consisting of a metallic scaffold, manufactured using a computer aided laser method, coated with thin polymeric surface layer for medical applications. Gold Medal on 9th International Warsaw Invention Show IWIS 2015. Warsaw, Poland; 12-14.10.2015

[30] Dobrzański LA, Dobrzańska-Danikiewicz AD, Malara P, Gaweł TG, Dobrzański LB, Achtelik-Franczak A. The novel composite consisting of a metallic scaffold, manufactured using a computer aided laser method, coated with thin polymeric surface layer for medical applications. In: . Semi Grand Prize on Global Inventions and Innovations Exhibitions Innova Cities Latino-America. Foz do Iguaçu, Brazil: ICLA; 2015 10-12.12.2015

[31] Dobrzański LA. Fabrication, structure and mechanical properties of laser sintered materials for medical applications. Invited lecture on XXV International Materials Research Congress. Cancun, Mexico; 14-19 August 2016

[32] Dobrzański LA. Application of the additive manufacturing by selective laser sintering for constituting implants caffolds and hybrid multilayer biological and engineering

composite materials. Keynote lecture on International Conference on Processing & Manufacturing of Advanced Materials THERMEC'2016, Processing, Fabrication, Properties, Applications. Graz, Austria; 29 May–3 June 2016

[33] Dobrzański LA. Metallic Implants-Scaffolds for Dental and Orthopaedic Application. Invited Lecture on 9° COLAOB - Congresso Latino-Americano de Orgãos Artificiais e Biomateriais. Brazil: Foz do Iguaçu; 24-27 August 2016

[34] Dobrzański LA et al. Investigations of Structure and Properties of Newly Created Porous Biomimetic Materials Fabricated by Selective Laser Sintering, BIOLASIN. Project UMO-2013/08/M/ST8/00818. Gliwice: Silesian University of Technology; 2013-2016

[35] Kremzer M, Dobrzański LA, Dziekońska M, Macek M. Atomic layer deposition of TiO_2 onto porous biomaterials. Archives of Materials Science and Engineering. 2015;**75**:63-69

[36] Project POIR.01.01.01-00-0485/16-00 IMSKA-MAT Innowacyjne implanto-skafoldy stomatologiczne i szczękowo-twarzowe wytwarzane z wykorzystaniem innowacyjnej technologii addytywnej wspomaganej komputerowym projektowaniem materiałowym ADD-MAT. Gliwicach: Centrum Projektowo-Badawczo-Produkcyjne Inżynierii Medycznej i Stomatologicznej ASKLEPIOS; 2017–2021 in progress

[37] Dobrzańska-Danikiewicz AD. The book of critical technologies of surface and properties formation of engineering materials. Open Access Library. 2012;**6**:1-823 (in Polish)

[38] Dobrzański LA, Dobrzańska-Danikiewicz AD, Achtelik-Franczak A, Szindler M. Structure and properties of the skeleton microporous materials with coatings inside the pores for medical and dental applications. International Conference on Frontiers in Materials Processing, Applications, Research, & Technology, FiMPART'2015; 12–15 June 2015; Hyderabad, India. Singapore: Springer; 2017 (in press)

[39] Kim TH, Lee KM, Hwang J, Hong WS. Nanocrystalline silicon films deposited with a modulated hydrogen dilution ratio by catalytic CVD at 200°C. Current Applied Physics. 2009;**9**:e108-e110. DOI: 10.1016/j.cap.2008.12.041

[40] Dobrzański LA, Dobrzańska-Danikiewicz AD. Kształtowanie struktury i własności powierzchni materiałów inżynierskich. Wydawnictwo Politechniki Śląskiej: Gliwice; 2013. 492 p 978-83-7880-077-7

[41] Technology Backgrounder: Atomic Layer Deposition. Georgetown, MA: Publication of the IC Knowledge LLC; 2004

[42] Dobrzański LA, Szindler M, Szindler MM. Surface morphology and optical properties of Al_2O_3 thin films deposited by ALD method. Archives of Materials Science and Engineering. 2015;**73**:18-24

[43] Dobrzański LA, Dobrzańska-Danikiewicz AD, Szindler M, Achtelik-Franczak A, Pakieła W. Atomic layer deposition of TiO_2 onto porous biomaterials. Archives of Materials Science and Engineering. 2015;**75**:5-11

[44] Wrigth JD, Sommerdijk NAJM. Sol-Gel Materials, Chemistry and Applications. Amsterdam: Gordon and Breach Science Publishers; 2001. 125 p 90-5699-326-7

[45] Bala H. Wstęp do chemii materiałów. Warszawa: WNT; 2003. 445 p 8320427991

[46] Langlet M, Kim A, Audier M, Guillard C, Herrmann JM. Transparent photocatalytic films deposited on polymer substrates from sol-gel processed titania sols. Thin Solid Films. 2003;**429**:13-21. DOI: 10.1016/S0040-6090(02)01290-7

[47] Pajdowski L. Chemia ogólna. Wyd. 11. Wydawnictwo Naukowe PWN: Warszawa; 2002. 616 p 8301123567

[48] Głuszek J. Tlenkowe powłoki ochronne otrzymywane metodą sol-gel. Oficyna Wydawnicza Politechniki Wrocławskiej: Wrocław; 1998. 154 p 8370853617

[49] Hwang K, Song J, Kang B, Park Y. Sol–gel derived hydroxyapatite films on alumina substrates. Surface and Coatings Technology. 2000;**123**:252-255

[50] Łączka M, Terczyńska A, Cholewa-Kowalska K. Powłoki żelowe na szkle, Część 1. Świat Szkła. 2008;**9**:52-55

[51] Dong Y, Zhao Q, Wu S, Lu X. Ultraviolet-shielding and conductive double functional films coated on glass substrates by sol-gel process. Journal of Rare Earths. 2010;**28**:446-450. DOI: 10.1016/S1002-0721(10)60276-1

[52] Seo M, Akutsu Y, Kagemoto H. Preparation and properties of Sb-doped SnO_2/metal substrates by sol–gel and dip coating. Ceramics International. 2007;**33**:625-629. DOI: 10.1016/j.ceramint.2005.11.013

[53] Majewski S. Nowe technologie wytwarzania stałych uzupełnień zębowych: galwanoforming, technologia CAD/CAM, obróbka tytanu i współczesne systemy ceramiczne. Protetyka Stomatologiczna. 2007;**57**:124-131

[54] Miyazaki T, Hotta Y. CAD/CAM systems available for the fabrication of crown and bridge restorations. Australian Dental Journal. 2011;**56**:97-106. DOI: 10.1111/j.1834-7819.2010.01300.x

[55] Malara P, Dobrzański LB. Computer-aided design and manufacturing of dental surgical guides based on cone beam computed tomography. Archives of Materials Science and Engineering. 2015;**76**:40-149

[56] Malara P, Dobrzański LB, Dobrzańska J. Computer-aided designing and manufacturing of partial removable dentures. Journal of Achievements in Materials and Manufacturing Engineering. 2015;**73**:157-164

[57] Galanis CC, Sfantsikopoulos MM, Koidis PT, Kafantaris NM, Mpikos PG. Computer methods for automating preoperative dental implant planning: Implant positioning and size assignment. Computer Methods and Programs in Biomedicine. 2007;**86**:30-38. DOI: 10.1016/ j.cmpb.2006.12.010

[58] Majewski S, Pryliński M. Materiały i technologie współczesnej protetyki stomatologicznej. Wydawnictwo Czelej: Lublin; 2013. 188 p 978-83-7563-188-3

[59] Papaspyridakos P, White GS, Lal K. Flapless CAD/CAM-guided surgery for staged transition from failing dentition to complete arch implant rehabilitation: A 3-year clinical report. Journal of Prosthetic Dentistry. 2012;**107**:143-150. DOI: 10.1016/S0022-3913 (12)00025-x

[60] Malara P, Dobrzański LB. Designing and manufacturing of implantoprosthetic fixed suprastructures in edentulous patients on the basis of digital impressions. Archives of Materials Science and Engineering. 2015;**76**:163-171

[61] Dobrzański LA. Metaloznawstwo opisowe stopów metali nieżelaznych. Wydawnictwo Politechniki Śląskiej: Gliwice; 2008. 406 p 978-83-7335-516-3

[62] Al Jabbari YS. Physico-mechanical properties and prosthodontic applications of co-Cr dental alloys: A review of the literature. Journal of Advanced Prosthodontics. 2014;**6**: 138-145. DOI: 10.4047/jap.2014.6.2.138

[63] Augustyn-Pieniążek J, Łukaszczyk A, Szczurek A, Sowińska K. Struktura i własności stopów dentystycznych na bazie kobaltu stosowanych do wykonywania protez szkieletowych. Inżynieria Materiałowa. 2013;**34**:116-120

[64] Malkondu Ö, Tinastepe N, Akan E, Kazazoğlu E. An overview of monolithic zirconia in dentistry. Biotechnology & Biotechnological Equipment. 2016;**30**:644-652. DOI: 10.1080/13102818.2016.1177470

[65] Madfa AA, Al-Sanabani FA, Al-Qudami NH, Al-Sanabani JS, Amran AG. Use of zirconia in dentistry: An overview. The Open Biomaterials Journal. 2014;**5**:1-9

[66] Craig RG. Materiały stomatologiczne. Wrocław: Elsevier Urban & Partner; 2008. 602 p. ISBN: 978-83-7609-072-6

[67] Conrad HJ, Seong WJ, Pesun IJ. Current ceramic materials and systems with clinical recommendations: A systematic review. Journal of Prosthetic Dentistry. 2007;**98**:389-404. DOI: 10.1016/S0022-3913(07)60124-3

[68] Miyazaki T, Nakamura T, Matsumura H, Ban S, Kobayashi T. Current status of zirconia restoration. Journal of Prosthodontic Research. 2013;**57**:236-261. DOI: 10.1016/j.jpor.2013.09.001

[69] Komine F, Blatz MB, Matsumura H. Current status of zirconia-based fixed restorations. Journal of Oral Science. 2010;**52**:531-539

[70] Dejak B, Kacprzak M, Suliborski B, Śmielak B. Struktura i niektóre właściwości ceramik dentystycznych stosowanych w uzupełnieniach pełnoceramicznych w świetle literatury. Protetyka Stomatologiczna. 2006;**56**:471-477

[71] Bachhav VC, Aras MA. Zirconia-based fixed partial dentures: A clinical review. Quintessence International. 2011;**42**:173-182

[72] Vagkopoulou T, Koutayas SO, Koidis P, Strub JR. Zirconia in dentistry: Part 1. Discovering the nature of an upcoming bioceramic. The European Journal of Esthetic Dentistry. 2009;**4**:130-151

[73] Dobrzański LA, Matula G, Dobrzańska-Danikiewicz AD, Malara P, Kremzer M, Tomiczek B, Kujawa M, Hajduczek E, Achtelik-Franczak A, Dobrzański LB, Krzysteczko J. Composite materials infiltrated by aluminium alloys based on porous skeletons from alumina, mullite and titanium produced by powder metallurgy techniques. In: Dobrzański LA, editor. Powder Metallurgy – Fundamentals and Case Studies. Rijeka: InTech; 2017. p. 95-137. DOI: 10.5772/65377

[74] Dobrzańska J, Gołombek K, Dobrzański LB. Polymer materials used in endodontic treatment – In vitro testing. Archives of Materials Science and Engineering. 2012;**58**: 110-115

[75] Dobrzański LA. Applications of newly developed nanostructural and microporous materials in biomedical, tissue and mechanical engineering. Archives of Materials Science and Engineering. 2015;**76**:53-114

[76] Dobrzańska-Danikiewicz AD, Gaweł TG, Wolany W. Ti6Al4V Titanium alloy used as a modern biomimetic material. Archives of Materials Science and Engineering. 2015;**76**: 150-156

[77] Dobrzański LA, Dobrzańska-Danikiewicz AD, editors. Metalowe materiały mikroporowate do zastosowań medycznych. Vol. 7. Open Access Library; 2017 (in press)

[78] Malara P, Dobrzański LB. Computer aided manufacturing and design of fixed bridges restoring the lost dentition, soft tissue and the bone. Archives of Materials Science and Engineering. 2016;**81**:68-75. DOI: 10.5604/18972764.1230551

[79] Malara P, Dobrzański LB. Screw-retained full arch restorations – Methodology of computer aided design and manufacturing. Archives of Materials Science and Engineering. 2017;**83**:23-29

Chapter 5

Properties of Co-Cr Dental Alloys Fabricated Using Additive Technologies

Tsanka Dikova

Additional information is available at the end of the chapter

http://dx.doi.org/10.5772/intechopen.69718

Abstract

The aim of the present paper is to make a review of the properties of dental alloys, fabricated using Additive Technologies (AT). The microstructure and mechanical properties of Co-Cr alloys as well as the accuracy and surface roughness of dental constructions are discussed. In dentistry two different approaches can be applied for production of metal frameworks using AT. According to the first one the wax/polymeric cast patterns are fabricated by 3D printing, than the constructions are cast from dental alloy with as-printed patterns. Through the second one the metal framework is manufactured form powder alloy directly from 3D virtual model by Selective Electron Beam Melting (SEBM) or Selective Laser Melting (SLM). The microstructure and mechanical properties of Co-Cr dental alloys, cast using 3D printed patterns, are typical for cast alloys. Their dimensional and adjustment accuracy is higher comparing to constructions, produced by traditional lost-wax casting or by SLM. The surface roughness is higher than that of the samples, cast by conventional technology, but lower comparing to the SLM objects. The microstructure of SLM Co-Cr dental alloys is fine grained and more homogeneous comparing that of the cast alloys, which defines higher hardness and mechanical properties, higher wear and corrosion resistance.

Keywords: materials science and engineering, biomaterials, regenerative stomatology, additive technologies, Co-Cr dental alloys, microstructure and properties

1. Introduction

Dental alloys on the basis of cobalt and chromium are one of the most preferred for production of metal frameworks of dental constructions because of their high strength, high corrosion and wear resistance, high biocompatibility, and a relatively low cost [1, 2]. The chemical composition of Co-Cr dental alloys consists of 53–67% of Co, 25–32% of Cr, 2–6% of Mo, and small quantities of W, Si, Al, and others [3]. Cr, Mo, and W are added for strengthening of the solid solution. Due to the relatively large amount of Cr, dense passive layer of Cr_2O_3 with 1–4 nm thickness on the surface as well as carbides in the microstructure of the details is formed, determining the high hardness, high corrosion, and wear resistance [4, 5]. According to the phase diagram, Co-Cr dental alloys are characterized with face-centered cubic (fcc) lattice, γ phase in high temperatures, and with hexagonal close packed (hcp) lattice, ε phase in room temperature [4, 6]. The γ phase defines the ductility, while the ε phase defines the corrosion and wear resistance of the alloy [7]. In proper alloying, the microstructure of the Co-Cr dental alloys consists mainly of γ phase and carbides of the $M_{23}C_6$ type [4]. Therefore, the properties of Co-Cr dental alloys depend on the ratio between γ/ε phases and the type, quantity and distribution of the carbide phase in the microstructure.

The microstructure and properties of dental alloys depend on the manufacturing process and the technological regimes. Since the beginning of the last century, casting is the most common technology for production of metal constructions in dentistry. During this conventional technology, the metal frameworks are fabricated by centrifugal casting using hand-built wax patterns. The technological process is characterized with large amount of manual work, and although it is performed by a qualified dental technician, it can lead to low accuracy and satisfactory quality. The advent of the modern CAD/CAM systems and the additive technologies in the last 30 years of the last century allow to decrease the amount of manual work and production time, to improve the quality of dental constructions and as a consequence to reduce their price [8, 9].

The additive technologies (AT) are developed in the late 1980s as an alternative of the subtractive technologies. They are characterized with building of one layer at a time from a powder or liquid that is bonded by means of melting, fusing, or polymerization. The American Society for Testing and Materials (ASTM) defined additive manufacturing as 'the process of joining materials to make objects from 3D model data, usually layer upon layer, as opposed to subtractive manufacturing methodologies' [8–10]. These processes are also known as 'three-dimensional printing', 'layered manufacturing', 'free-form fabrication', 'rapid prototyping' and 'rapid manufacturing' [9–11]. The ASTM international committee, intended for specification of standards for additive manufacturing—ASTM F42—created a categorization of all 3D printing technologies into seven major groups [11]. According to it, the following 3D printing technologies are used in work with biomaterials: 3D plotting/direct ink writing, laser-assisted bioprinting, selective laser sintering, stereolithography, fused deposition modeling, and robot-assisted deposition/robocasting. The stereolithography (SLA), fused deposition modeling (FDM), selective electron beam melting (SEBM), selective laser sintering (SLS), selective laser melting (SLM), and ink-jet printing (IJP) are among the AT, mostly used in the dental medicine [8, 9, 12–15].

During the SLA process, a concentrated beam of UV light is focused on the surface of a tank filled with liquid photopolymer. As the light beam draws the object on the surface of the liquid, each time a layer of resin is polymerized or cross-linked until the real object is obtained [8, 9, 12, 14, 16]. The FDM characterizes with extruding the thermoplastic materials through heated nozzle, or the material is fed from a reservoir through a syringe [9, 12, 16]. By SEBM, the parts are manufactured by melting metal powder layer per layer with an electron beam in a high vacuum [8, 9, 14, 16]. In SLS/SLM technology, layers of particular powder material (mainly polymers and porcelains in SLS and pure metals or alloys in SLM) are fused into a real detail by a computer-directed laser [8, 9, 12, 17]. During IJP process, an extremely small ink droplet is ejected toward the substrate. Different substances can be used as ink aqueous solution of colouring agents and binders to a ceramic suspension to produce zirconia dental restorations [8, 9, 12, 18]. Another variant of the IJP is by depositing droplets of a polymer, and each formed layer is cured by UV light, which allows polymer material with characteristics similar to the technical wax to be used [8, 9, 13].

The main advantages of AT include production of complex objects from different materials—polymers, composites, metals and alloys; manufacturing of parts with dense/porous structure and predetermined surface roughness and controllable, is an easy and relatively quick process [9]. Due to the various materials used and the great variety of additive manufacturing processes, AT can be successfully applied in many fields of dentistry for the production of different types of dental constructions. AT give even opportunities to fabricate structures of hard-to-handle materials such as cobalt and chromium alloys [19]. Polymeric study models instead of dental plaster casts, surgical guides for placement of dental implants, temporary crowns and bridges as well as resin models for lost-wax casting can be fabricated by SLA and FDM [8, 9, 12, 20–22]. SEBM and SLM are used for manufacturing of customized implants for maxillofacial surgery, dense or porous dental implants, dental crowns and bridges as well as partial denture frameworks [23–31]. IJP is suitable for printing of zirconia dental restorations, polymeric dental models, orthodontic bracket guides, wax patterns of complete dentures, or wax models for casting [8, 9, 32, 33].

There are two main approaches for the production of metal constructions from Co-Cr dental alloys using AT. The first one concerns to the fabrication of wax/polymeric cast patterns by 3D printing, and the second one is casting of the framework from dental alloy. Through the second one, the metal object is manufactured directly from the 3D virtual model by SEBM or SLM. As the additive technologies are relatively new, develop extremely fast and with their indisputable advantages enter in many areas, so their implementation into the dentistry is faster than the research and data about the quality of the details produced by them. The aim of the present chapter is to make a review of the properties of dental alloys, fabricated using additive technologies. The microstructure and mechanical properties of Co-Cr alloys as well as the accuracy and surface roughness of dental constructions are discussed.

2. Properties of Co-Cr dental alloys cast with 3D-printed patterns

The wax/polymeric cast patterns can be produced from generated 3D virtual models by laser-assisted or digital light projection (DLP) stereolithography, fused deposition modeling, and ink-jet printing. The type of the additive manufacturing process, the parameters of its technological regimes, and the properties of the used materials influence mainly on the geometrical characteristics, adjustment accuracy, and surface roughness of the cast patterns [33–36], thus defining the accuracy and surface quality of the cast dental constructions. As with any AT, the layer's thickness, the position of the object toward the print direction, and the optical properties of the polymers (in SLA) have a decisive effect on the object's accuracy. The thinner the layer, the lower the angle to the print direction, and the lower the surface roughness, the higher the resolution and the dimensional accuracy, but the longer the production time [37].

The accuracy of dental constructions, produced by the modern CAD/CAM systems, is defined in the standard ISO 12836:2015 'Dentistry—Digitizing devices for CAD/CAM systems for indirect dental restorations—Test methods for assessing accuracy' [38]. Three standardized geometrical figures for accuracy evaluation are described in its Annexes A, B, and C [38]. Braian et al. [39] used the figures in the Annexes A and C, specifying the measurement of an inlay-shaped object and a multiunit specimen to simulate a four-unit bridge, for determining the production tolerances of four commercially available additive manufacturing systems, working on the SLS, multi-jet, or poly-jet principles. The samples were printed with different layer's thicknesses, optimal for each system, and ensuring the highest product's accuracy (as per the equipment manufacturer). According to the ISO/IEC GUIDE 99:2007(E/F) [40], the 'accuracy' is the closeness of agreement between a measured quantity value and a true quantity value of a measured object. The authors [39] defined the terms 'resolution' and 'repeatability'. The resolution refers to the smallest feature that the system can produce, while the reproducibility is described as the system ability to produce consistent output time after time. The researchers established that the accuracy of both types of samples, produced with four printers, is different in the three directions. The multi-jet printing ensures the highest accuracy of the linear as well as angular dimensions, followed by poly-jet process and SLS. It was concluded that the suitable type of printer should be chosen according to the intended dental application.

In evaluation of restorations, fabricated by digital technologies, subtractive (milling of wax and zirconia) and additive (SLA of photopolymer and SLS of Co-Cr alloy), Bae et al. [41] found out that the dimensions of the samples of all four groups are smaller than the reference data. Concerning to the repeatability of AT, they established the smallest difference from the reference data in the specimens, fabricated using SLA, and no significant differences between the SLA and SLS methods. Therefore, they concluded that the accuracy of additive manufacturing methods is better than the subtractive ones, as the mean accuracy discrepancies are the smallest for SLA, followed by the SLS, wax, and zirconia milling. But they pointed out the following as disadvantages: the print layers, clearly seen on the surface of the SLA sample, and circular, sunken forms of approximately 80 μm in diameter on the SLS specimen surface. The investigations of Mai et al. [42] proved that the crowns fabricated by CAD/CAM systems (milling and polymer-jet 3D printing) have higher fitting accuracy than that of the molded ones. Among them, the polymer-jet 3D printing significantly enhanced the fit of the interim crowns, particularly in the occlusal region. According to Kim et al. [43], the fitting accuracy of dental crowns is affected by the number of copings fabricated by micro-stereolithography. Based on the marginal discrepancy, the most precise copings are printed when three arrays are used on a single-built platform. Therefore, there are limitations concerning to the reproducibility and accuracy of polymeric dental constructions, manufactured by 3D printing.

Ishida and Miyasaka [44] investigated the accuracy of the patterns for casting of all metal crowns, manufactured by four 3D printers, using different manufacturing processes: laser-assisted SLA, SLA with concentrated UV beam, FDM, and IJP. They established that in all types of manufacturing process, the outer and the inner diameters of the crowns are smaller than that of the virtual 3D model, which could be compensated with 3–5% increasing the dimensions of the virtual model. The FDM-printed crowns are characterized with the highest surface roughness, while the lowest is typical for the crowns, fabricated by laser-assisted SLA. Additionally, the difference of the roughness values along the 3D axes exists. They established that the roughness along the tooth axis of the crown created by laser-assisted SLA was the smallest, while that along the horizontal direction of the crown printed by multi-jet modeling (MJM) was the smallest. This can be explained with the specific features of the additive manufacturing process [8, 12, 45].

The higher roughness of the surfaces, parallel, or inclined to the print direction (*Z* axes), compared to that along *X* and *Y* axes is also established in the research of Dikova et al. [46]. Their investigations of cubic samples, printed with different polymers by DLP SLA, show that the roughness of the surfaces of the cubes with horizontal position toward the basis is less than those of the cubes, printed inclined in almost all types of polymers. All sizes of the cubes in both positions are larger than the size of the virtual model. In their subsequent investigations [47], it is observed that the decrease of the layer's thickness from 50 (recommended by the equipment manufacturer) to 35 μm leads to nearly twice decrease of the surface roughness of the samples, made of resin NextDent Cast (developed especially for manufacturing of cast patterns), in both printing positions. It also leads to the highest dimensional accuracy and the least interval of deviation. The research team has shown that the dimensions, parallel to the basis, axes *X* and *Y*, are the most precise, while those, parallel or inclined to the printing direction, axis Z, are the most deviating. The dimensional and adjustment accuracy as well as the surface roughness of four-part dental bridges, made of polymers NextDent Cast and NextDent C+B (for temporary crowns and bridges) by DLP SLA, are investigated in the work of Dikova et al. [48]. They established that the dimensions in both directions of the bridges from both polymers, fabricated with less layer's thickness of 35 μm, are smaller with 0.29–1.10%, compared to those of the virtual model, while the sizes of the bridges with larger layer's thickness of 50 μm are 1.51–3.45% more than the virtual model. Dimensions of the samples of both polymers, situated in the building direction, are larger with 1.51–3.45%, and those, which are not in the print direction, are 0.49–0.53% smaller than those of the virtual 3D model. The inaccuracy in the geometrical dimensions causes inaccuracy of adjusting and absence of gap between the bridge constructions and the gypsum model. The surface roughness of the cast patterns, manufactured with regime, recommended by the producer (50 μm layer's thickness), is relatively high with average arithmetic deviation of the surface roughness $Ra = 3.24$ μm. Decreasing the layer's thickness to 35 μm leads to the lower surface roughness of $Ra = 2.18$ μm. The surface observations with optical microscopy proved the layered structure, which is a specific feature of the objects built up via stereolithography. The higher roughness of the 3D-printed cast patterns can complicate the cast process and cause higher roughness of the cast itself.

The geometrical accuracy of fixed dental prostheses, manufactured by additive technologies, is investigated in Ref. [49]. Four-part dental bridges are produced of Co-Cr alloy by three technological processes: conventional lost-wax casting with wax models, manufactured in silicon mold; lost-wax casting with 3D-printed (multi-jet modeling (MJM) cast patterns, and direct fabricating by SLM. It was established that the surface roughness of Co-Cr fixed partial dentures, cast with 3D-printed wax patterns, is more than three times higher than the roughness of dentures, cast by conventional lost-wax technology (Ra = 3.39 μm and Ra = 1.11 μm, respectively). The surface observation of Co-Cr dental bridges proves the higher roughness as well as the traces of the layered manufacturing of the cast patterns, left even after sandblasting. The fixed partial dentures, cast with 3D-printed cast patterns, possess higher accuracy of the shape, sizes, and adjustment compared to the dentures, produced by conventional lost-wax casting. This is mainly due to the minimal manual work, because 3D-printed cast patterns are produced directly from the virtual model.

The microstructure of Co-Cr sample, cast with 3D-printed pattern (MJM), is a typical cast microstructure—inhomogeneous, consisting of large grains with dendrite morphology [50]. A large amount of lamellar and blocky carbides of different sizes are located mainly along the grain boundaries. The dendrites consist of γ phase, while the inter-dendritic regions consist of microeutectic (Co solid solution with intermetallic precipitations) and carbides of the $(Cr, Mo)_{23}C_6$ type. The hardness of the Co-Cr samples, cast with 3D-printed patterns, is in the range 327HV–343HV with uneven distribution due to the inhomogeneous microstructure. The dendritic grains with intermetallic phases, precipitated along the grain boundaries, were observed in the microstructure of the samples, cast with 3D-printed patterns from Co-Cr alloy remanium star CL by Kim et al. [51]. They characterized with yield strength 540 ± 20 MPa, mean percent elongation 10 ± 2 and Young's modulus 260 ± 20 GPa.

The microstructure and strength properties of Co-Cr dental alloys, cast with 3D-printed patterns, are typical for castings. Their dimensional and fitting accuracy is higher than that of the constructions, manufactured by conventional casting, while their surface roughness is more than three times higher. Therefore, the dimensional and fitting accuracy and the surface roughness of Co-Cr dental substructures are strongly influenced by the quality of the cast patterns fabricated via AT. The data about the dimensional accuracy of wax/polymeric dental constructions, produced by different additive manufacturing processes, are contrary. In some cases, the dimensions of the printed objects are smaller than the virtual model; in another case, they are larger. It depends not only on the type of the manufacturing process but also on the scale factor—the sizes of the printed object. As a consequence of the low dimensional accuracy, the fitting accuracy is lower. The strategy of printing should be chosen very carefully, because too much objects in the same built platform can cause lower dimensional and fitting accuracy. Due to the features of the 3D printing processes, the sizes of the detail can be different in the three directions X, Y, and Z; therefore, it should be paid attention to the position of the object, especially toward the print direction. This will influence the surface roughness, too. The parameters of the technological regimes also influence the dimensional accuracy. Decreasing the layer's thickness leads to smaller sizes and lower roughness, which in some cases defines the higher accuracy. The special requirements to the materials, used for 3D printing of cast patterns, exist—to have no burned out residue and null or minimal thermal expansion—otherwise, the casting with low quality will be the result.

For fabricating the high-quality castings from Co-Cr dental alloys using 3D-printed patterns, some steps concerning to the processes of 3D printing and casting should be observed. In 3D printing (1) the printer, ensuring the cast pattern with high dimensional and fitting accuracy and satisfactory surface roughness, has to be chosen. The present review shows that the systems, working on the principle of the laser-assisted SLA, DLP SLA, MJM, and polymer-jet printing, are good candidates; (2) As each printer works with its specific materials, the right material, meeting the requirements for manufacturing the cast patterns, should be used; (3) The optimal technological regime has to be chosen, including the layer's thickness; (4) The strategy of printing has to be developed—positions, and the number of the objects; (5) Preliminary tests, calibration on the three axes *X*, *Y*, and *Z* of the printer and compensation (if needed), have to be done to guarantee dimensions with high accuracy; and (6) 3D printing, post-curing (in some processes), and final surface treatment to enhance the smoothness. In the second part, the casting processes are (1) selection of the investment material, relevant to the material of the 3D-printed cast pattern, should be done for manufacturing casting mold; (2) heating of the casting mold with temperature regime concerning to the material of the 3D-printed cast pattern and the Co-Cr alloy and (3) casting with the given Co-Cr alloy, keeping the manufacturer's requirement.

3. Properties of Co-Cr dental alloys fabricated via SLM

The properties of the Co-Cr dental alloys depend on the microstructure, its morphology, and composition, which are defined by the manufacturing process and the technological regimes. The SLM is a complex thermophysical process, depending on a number of important parameters. During SLM, layers of metal powder are fused into a real object by a computer-directed laser. The process characterizes with high heating and cooling rates, leading to fine-grained microstructure of the solidified layer. As the heat is led away through the solid body, phase transformations run in the underneath layers heated above the transition temperatures [50]. Due to the high-temperature gradients during the SLM process, high residual stresses are generated in the details, which can cause subsequent deformations [17, 52–54]. These characteristics determine the specific microstructure and properties of the objects, produced by SLM as compared with that, manufactured by casting. The main technological parameters that are crucial for the production of high-quality construction are the laser power, scanning speed, laser beam diameter and distance between the traces, layer's thickness, and the working area [55–58]. In the development of any process for production of an object by SLM, it is necessary to evaluate the density, accuracy, surface roughness, hardness, and strength properties.

In SLM process, the volume of the molten metal depends on the volume energy density E_v, which is directly proportional to the power density N_s and inversely proportional to the scanning speed V. If the volume energy density E_v is insufficient, the incomplete melting of the deposited layer will occur. Therefore, the lower volume energy density E_v is the main reason for the porous structure [50]. This can be avoided by optimization of the technological parameters—increasing the laser power or decreasing the scanning speed. Varying with the input energy and scan spaces, Takaichi et al. [59] established that dense structure of SLM Co-29Cr-6Mo alloy can be obtained when the input energy of the laser scan increased more than 400 J/mm^3 and porous, in input energy less than 150 J/mm^3. Vandenbroucke and Kruth [60] ensured 99.9% density of the SLM Co-Cr-Mo alloy, working with optimized technological regime. In the investigation of the possibility for manufacturing of three-part dental bridge from Co-Cr alloy by SLM, Averyanova et al. [29] stated that all samples are densed with porosity less than 1%.

The specific features of the SLM process, characterizing with high heating and cooling rates, define unique microstructure of Co-Cr dental alloys and mechanical properties higher than that of the cast alloys. Meacock and Vilar [61] reported that the microstructure of biomedical Co-Cr-Mo alloy, produced by laser powder micro-deposition, is homogeneously composed of fine cellular dendrites. The average hardness was 460 HV0.2, which is higher than the values obtained by the other fabrication process. In investigation of Co-Cr-Mo parts, produced by direct metal laser sintering, Barucca et al. [62] established that the microstructure consists of γ and ε phases. The ε phase is formed by athermal martensitic transformation, and it is distributed as network of thin lamellae inside the γ phase. The higher hardness (47 HRC) is attributed to the presence of the ε-lamellae grown on the $\{111\}_\gamma$ planes that restricts the dislocation movement in the γ phase. Lu et al. [63] investigated the microstructure, hardness, mechanical properties, electrochemical behaviour, and metal release of Co-Cr-W alloy fabricated by SLM in two different scanning strategies—line and island. They established the coexistence of the γ and ε phases in the microstructure and nearly the same hardness, 570 HV for line-formed alloy and 564 HV for island-formed alloy. Their research shows that the results of tensile strength (1158.22 ± 21 MPa for line scheme and 1115.56 ± 19 MPa for island scheme), hardness, density, electrochemical, and metal release tests are independent of the scanning strategy, and the yield strength of both samples meets the ISO 22764:2006 standard for dental restorations. The tensile strength of two Co-Cr dental alloys—cast remanium GM and SLM F75— was investigated in the work of Jevremovic et al. [64]. They established more than 1.5 times higher tensile strength of the SLM samples compared to the cast ones (1363–1472 MPa and 900 MPa, respectively). Vandenbroucke and Kruth [60] did complex investigation of SLM titanium and Co-Cr-Mo alloys. The mechanical tests proved that the SLM Co-Cr details fulfil the requirements for hardness, strength, and stiffness. Concerning to the corrosion—the SLM Co-Cr samples showed lower emission than the cast ones due to the more homogeneous and finer microstructure of the laser molten material.

In investigation of the influence of the object's position to the building direction, some researchers established anisotropy of the mechanical properties and especially of fatigue strength. The microstructure and mechanical properties of SLM Co-29Cr-6Mo alloy were studied out in the work of Takaichi et al. [59]. Unique microstructure was formed, consisting of the fine cellular dendrites in the elongated grains, parallel to the building direction. The cellular boundaries were enriched with Cr and Mo, and the γ phase was dominant in the SLM building. Due to the unique microstructure, the mechanical anisotropy was confirmed in the samples, but the yield strength, Ultimate Tensile Strength (UTS), and elongation were higher than that of the cast alloy. The research of Kajima et al. [65] shows that the microstructure of SLM specimens of Co-Cr-Mo alloys is quite different from those of the cast samples, which consists of coarse dendrites with visible precipitates in the inter-dendritic regions. The fine cellular dendrites with diameter about 0.5 μm were observed in the SLM samples parallel to the building direction, while in directions, perpendicular or inclined at 45° to the building direction, fine columnar structures with diameter about 0.5 μm, elongated along the building direction, were found. In lower magnification, gradual arch-shaped molten pool boundaries, typical for the SLM process, were observed. The tensile strength of the three groups of specimens is in the range 1170–1274 MPa, which is higher than that of the cast samples. Concerning to the fatigue strength, the results confirmed the anisotropy of the SLM alloy. The samples, parallel to the building direction, exhibit significantly longer fatigue life than the cast specimens, while the fatigue life of the two groups is significantly shorter than that of the cast specimens.

The research of Kim et al. [51] shows that the SLM Co-Cr alloy remanium star CL clearly exhibits the laser scan traces, as in higher magnification, the presence of fine grains with sizes about 35 μm can be recognized. The SLM samples showed the highest mean ultimate tensile strength, followed by milled/post-sintered, cast, and milled samples. The yield strength and mean percent elongation of SLM alloy were higher than the cast alloy ($R_{0.2}$ = 580 ± 50 MPa and $R_{0.2}$ = 540 ± 20 MPa, respectively), while Young's modulus was lower (200 ± 10 GPa for SLM and 260 ± 20 GPa for cast). A high yield strength and relatively low but sufficient modulus of elasticity of the SLM samples allow this technology to be used for manufacturing dental constructions, such as removable partial dentures, clasps, thin-veneered crowns, and wide-span bridges.

Averyanova et al. [29] investigated three-part dental bridge of Co-Cr alloy, manufactured by SLM. The hardness of the SLM Co-Cr alloy is in the range of 400 ± 14 HV10 and 462 ± 22 HV0.05, while the average tensile strength is 1157 MPa, which is equal to that of the wrought alloy and about twice higher than the cast Co-Cr alloy (655 MPa). The microstructure of SLM dental bridges is nonequilibrium, consisting mainly of 98.7 ± 1.8% fcc Co-rich solid solution and 1.3 ± 0.5% of hcp ε phase. The investigation of Dikova et al. [50] established similar hardness of Co-Cr four-part dental bridges, produced by SLM (407–460 HV), which is higher than the cast alloy and has nearly even distribution along the depth of each crown. Their subsequent investigations [66] show that the tensile strength and the yield strength of the SLM Co-Cr alloy are higher than the cast alloy ($R_{0.2}$ = 720 MPa and $R_{0.2}$ = 410 MPa, respectively), while the elastic modulus is comparable (213 and 209 GPa, accordingly). The higher hardness and more homogeneous microstructure of SLM Co-Cr dental alloys determine their higher wear and corrosion resistance [67].

The specific features of the SLM process and the parameters of the powder materials used determine the high surface roughness of the SLM Co-Cr dental alloys. The object position and orientation to the building direction as well as the choice and modeling of the supports are also of great importance [58]. The accuracy and the relation between the surface roughness and the sloping angle were researched in Ref. [60] using special benchmark models. It was proven that the surface roughness depends on the layer's thickness and sloping angle to the basis. The average arithmetic deviation of the surface roughness *Ra* varies between 6–18 μm for 20 μm layer's thickness and 13–33 μm for 50 μm layer's thickness. The higher values concern to the lower sloping angle to the basis of 8°, while the lower values for the larger angle of 70°. The research of Kajima et al. [65] confirmed that the surface roughness depends on the building direction. It is the lowest in the samples, perpendicular to the building direction,

Ra = 10.22 μm, followed by that of the samples, inclined at 45° with *Ra* = 13.67 μm, and the highest in the samples, parallel to the building direction, *Ra* = 18.17 μm. It was established in Ref. [49] that the roughness of the vestibular surface of the second premolar of four-part dental bridge, manufactured by SLM, is nearly four times higher than the roughness of the conventional cast bridge (*Ra* = 4.24 μm and *Ra* = 1.11 μm, respectively) and 25% higher compared to that cast with 3D-printed patterns (*Ra* = 4.24 μm and *Ra* = 3.39 μm, respectively). It is proposed that the considerably higher roughness and partially melted powder on the surface of the SLM samples could lead to the increase of the mechanical as well as the chemical components of the adhesion of the porcelain to the Co-Cr alloys, thus promoting higher adhesion strength of the porcelain coating.

Concerning to the standard ISO 9693-1:2012 [68], the minimum acceptable bond strength of metal-ceramic is 25 MPa. Kaleli and Sarac [69] compared porcelain bond strength of Co-Cr frameworks manufactured by conventional lost-wax technique, milling, direct metal laser sintering (DMLS), and direct process powder-bed method. There was no significant difference between the values of porcelain bond strength to the samples, produced by different methods. The mean bond strength was 38.08 MPa for cast samples, while that of the DMLS samples was 40.73 MPa. The type of failure of the cast samples was adhesive/mixed, while that of the DMLS samples was cohesive/mixed. Li et al. [70] confirmed that there are no significant differences between the bond strength of the cast, milled, and SLM Co-Cr alloys. The milled and SLM groups showed significantly more porcelain adherence than the cast group. Akova et al. [71] also revealed no statistically significant difference of the shear bond strength of porcelain to the cast and SLM Co-Cr dental alloys (72.9 and 67.0 MPa, respectively). The failure type of the porcelain of the cast samples was of the mixed adhesion-cohesion type, while the failure of the SLM samples was of the mixed/adhesive type. The similar porcelain bond strength to the cast and SLM Co-Cr dental alloys was confirmed in the work of Wu et al. [72]—54.17 and 55.78 MPa accordingly. They established that the SLM alloy had an intermediate layer with elemental interpretation between the alloy and the porcelain, resulting in an improved bonding strength. Xiang et al. [73] established no significant difference for the mean bond strength of the SLM (44 MPa) and traditional cast (43 MPa) Co-Cr samples. A mixed fracture mode on the debonding interface of both the SLM and the cast groups was observed, but the SLM group showed significantly more porcelain adherence. Only Wang et al. [74] stated that there are statistically significant differences of the porcelain bond strength to the cast, CNC milled and SLM Co-Cr samples (37.7 ± 6.5 MPa, 43.3 ± 9.2 MPa and 46.8 ± 5.1 MPa, respectively), as the debonding surface of the all samples was of the cohesive failure mode. It should be noticed out that the surface roughness of the SLM samples decreases two–three times after sandblasting before porcelain firing (from *Ra* = 15 μm of as-received SLM Co-Cr samples to *Ra* = 8 μm of glass blasted and to *Ra*-5 μm in ultrasonic ceramic field) [60]; as in the most cases, the physical appearance and the surface quality are similar to the conventionally manufactured by investment casting [58]. Current investigations show that the SLM metal-ceramic system exhibits a bonding strength that exceeds the requirement of ISO 9691: 1:2012. It even shows a better behaviour in porcelain adherence comparable to the traditional cast methods.

The higher roughness of SLM Co-Cr alloys can cause lower dimensional accuracy and comparatively satisfactory adjustment accuracy in comparison to the objects, cast by conventional technology or with 3D-printed patterns. In SLM of Co-Cr-Mo alloy, Vandenbroucke and Kruth [60] established the process accuracy below 40 μm, which fulfil requirements of most medical and dental applications. According to Bibb et al. [30], SLM-manufactured Co-Cr frameworks for removable partial denture possess accuracy and quality of fit comparable to the existing traditional methods used in the dental laboratories. Averyanova et al. [29] confirmed that the geometrical accuracy of three-part dental bridge of Co-Cr alloy, produced by SLM, is comparable to that of the substructures, produced by conventional technology. The good repeatability of the SLM process in manufacturing Co-Cr four-part dental bridges was reported by Dzhendov et al. [49]. The maximal deviations of the dimensions of SLM bridges were the lowest compared to the conventional casting and casting with 3D-printed patterns. But the dimensions of the SLM dentures were lower than that of the base model with −0.07/−0.23 mm. The adjustment accuracy of the SLM bridges is comparable to that of the bridges, cast with 3D-printed patterns, and is higher than that of the conventionally cast dentures.

There is no consensus regarding the clinically acceptable limits of marginal fit of dental restorations. Most researchers agree on an acceptable marginal discrepancy (distance from the abutment margin to the metal coping in a straight line) below the range of 100–120 μm [75], as values, greater than 120 μm, are considered not clinically acceptable [76–78]. The research of Kim et al. [75] stated that the marginal fit values of the Co-Cr alloy greatly depended on the fabrication methods and, occasionally, on the alloy systems. They found out that the marginal discrepancy of the SLM crowns (98.7 μm for 20-μm-thick layer and 128.8 μm for 30-μm-thick layer) is larger than the cast crowns (65.3–70.4 μm). In SLM samples, the marginal discrepancy increases with the increase of the layer's thickness. Kaleli and Sarac [76] compared the marginal adaptation after fabrication of the framework, porcelain application, and cementation of metal-ceramic restorations prepared by conventional lost-wax technique, milling, DMLS, and a direct powder-bed process. They observed the lowest marginal discrepancy values in the crowns, prepared by direct process powder-bed method, followed by the DMLS, milling, and casting. The research of Pompa et al. [77] concerns to the differences of marginal fit of laser-fused and conventional technologies for production of fixed dental prostheses. They established that the copings, manufactured by SLM, have better marginal adaptation within an acceptable range. But the cement gap characterized with irregular distribution was wider in the region of the shoulder than at the point of closure. The marginal discrepancy increased after porcelain application and cementation. In comparison of the marginal fit of metal laser sintered (MLS) Co-Cr alloy copings and conventional cast Ni-Cr alloy copings, Sundar et al. [79] concluded that the MLS copings had a better marginal fit and a decrease in micro-leakage compared to the copings, manufactured by conventional lost-wax technique. Huang et al. [80] compared the marginal and internal fit of SLM metal-ceramic crowns with lost-wax cast ones. They established that the SLM Co-Cr metal-ceramic crowns were better in marginal fit, not significantly different in axial fit and less accurate in occlusal fit than that of the cast samples. Concerning to the gap distribution, Tamac et al. [81] reported nearly the same results in comparing the clinical marginal and internal adaptation of metal-ceramic crowns, fabricated by CAD/CAM milling, DMLS, and traditional casting. They established that mean marginal gap values were 86.64 μm for milling, 96.23 μm for DMLS, and 75.92 μm for casting. The gap values in the axial wall region were the higher for the three groups of samples, followed by the gap values of axio-occlusal and occlusal surface regions. The cement film thickness at the occlusal region and axio-occlusal region was higher for the DMLS crowns. Consequently, the laser-assisted technologies for direct production of metal dental restorations, such as SLM, SLS, and DMLS, ensure improved or at least clinically acceptable fitting values compared with that of the conventional casting.

The Co-Cr dental alloys, fabricated by SLM, characterize with homogeneous fine-grained microstructure, consisting mainly of γ and ε phases in different ratios. Their unique microstructure defines higher mechanical properties,—hardness, yield and tensile strength, fatigue strength as well as higher wear and corrosion resistance—compared to the cast alloys. As a consequence of the peculiarities of the SLM manufacturing and the position of the object toward the building direction, anisotropy in mechanical properties, especially in the fatigue life, can be observed. The work with optimized technological regimes can guarantee the constructions with density higher than 99% and comparatively low roughness. But as a whole, the surface roughness of the SLM Co-Cr alloys is higher than the alloys, cast conventionally or with 3D-printed patterns, due to the specific features of the manufacturing process and the use of metal powder as raw material. It was expected that the higher surface roughness could decrease the dimensional and fitting accuracy and promote the higher adhesion strength of the porcelain coating. But the current review shows that the dimensional accuracy of the SLM details is higher than the cast samples and the fitting accuracy is improved or clinically accepted, most probably due to the CAD/CAM nature of the manufacturing process, which additionally enables high repeatability. Concerning to the adhesion strength of the porcelain to the SLM dental alloys, more authors reported that it is comparable with the adhesion strength to the cast alloys, which may be due to the decreased roughness of the SLM samples after sandblasting before porcelain firing. The Co-Cr dental alloys, fabricated by SLM, comply with the standards and requirements concerning the dimensional and fitting accuracy as well as strength properties and can be successfully used in production of dental constructions.

4. Conclusion

In dentistry, two different approaches can be applied for the production of metal frameworks using AT. According to the first one, the wax/polymeric cast patterns are fabricated by 3D printing, and the constructions are cast from dental alloy with as-printed patterns. Through the second one, the metal framework is a manufactured form of powder alloy directly from the 3D virtual model by SEBM or SLM.

The specific features of the manufacturing processes, the parameters of the technological regimes, and the properties of the materials influence on the accuracy and properties of Co-Cr dental constructions.

The microstructure and mechanical properties of Co-Cr dental alloys, cast using 3D-printed patterns, are typical for cast alloys. Their dimensional and adjustment accuracy is higher compared to the constructions, produced by traditional lost-wax casting or by SLM. The surface roughness is higher than that of the samples, cast by conventional technology, but lower compared to the SLM objects.

The microstructure of SLM Co-Cr dental alloys is fine grained and more homogeneous comparing to that of the cast alloys, which defines higher hardness and mechanical properties, higher wear and corrosion resistance. The surface roughness of SLM Co-Cr dental alloys is higher than that of the alloys, cast conventionally or with 3D-printed patterns. The dimensional accuracy of SLM Co-Cr details is higher than the cast samples, while the fitting accuracy is improved or clinically accepted.

Author details

Tsanka Dikova

Address all correspondence to: tsanka_dikova@abv.bg

Medical University "Prof. d-r Paraskev Stoyanov", Varna, Bulgaria

References

[1] Youssef S, Jabbari Al. Physico-mechanical properties and prosthodontic applications of Co-Cr dental alloys: a review of the literature. Journal of Advanced Prosthodontics. 2014;**6**(2):138-145

[2] Eliasson A, Arnelund CF, Johansson A. A clinical evaluation of cobalt-chromium metal-ceramic fixed partial dentures and crowns: A three- to seven-year retrospective study. The Journal of Prosthetic Dentistry. 2007;**98**:6-16

[3] Kissov H. Stomatologichna keramika, chast I, Osnovni principi, materiali i instrumentarium [in Bulgarian]. 1st ed. Sofia: Index; 1997. 432 p

[4] Podrrez-Radziszewska M, Haimann K, Dudzinski W, Morawska-Soltysik M. Characteristic of intermetallic phases in cast dental CoCrMo alloy. Archives of Foundry Engineering. 2010;**10**(3):51-59

[5] Bellefontaine G. The Corrosion of CoCrMo Alloys for Biomedical Applications [thesis]. Birmingham: University of Birmingham; 2010. 88 p

[6] Gupta P. The Co-Cr-Mo (Cobalt-Chromium-Molybdenum) system. Journal of Phase Equilibria and Diffusion. 2005;**26**(1):87-92

[7] Kurosu Sh, Nomura N, Chiba A. Effect of sigma phase in Co-29Cr-6Mo alloy on corrosion behavior in saline solution. Materials Transaction. 2006;**47**(8):1961-1964

[8] van Noort R. The future of dental devices is digital. Dental Materials. 2012;**28**(3-12)

[9] Dikova T, Dzhendov D, Simov M, Katreva-Bozukova I, Angelova S, Pavlova D, Abadzhiev M, Tonchev T. Modern trends in the development of the technologies for production of dental constructions. Journal of IMAB. 2015;**21**(4):974-981

[10] Dovbish VM, Zabednov PV, Zlenko MA. Additivnie tehnologii I izdelia iz metala [in Russian]. [Internet] 2013. Available from: http://nami.ru/uploads/docs/centr_technology_docs/55a62fc89524bAT_metall.pdf [Accessed: 06.04.2017]

[11] Bandyopadhyay A, Bose S, Das S. 3D printing of biomaterials. MRS Bulletin. 2015; **40**:108-115

[12] Torabi K, Farjood E, Hamedani Sh. Rapid prototyping technologies and their applications in prosthodontics, a review of literature. Journal of Dentistry Shiraz University of Medical Sciences. 2015;**16**(1):1-9

[13] Sofronov Y, Nikolov N, Todorov G. Analysis of technologies for rapid prototyping of dental constructions. Scripta Scientifica Medicinae Dentalis. 2016;**2**(1):32-38

[14] Önoral O, Ulusoy M. New approaches in computer aided printing technologies. Cumhuriet Dental Journal. **19**(3):256-266. DOI: 10.7126/cumudj.298920

[15] Bilgin MS, Baytaroglu EN, Erdem A, Dilber E. A review of computer-aided design/ computer-aided manufacture techniques for removable denture fabrication. European Journal of Dentistry. 2016;**10**:286-291. DOI: 10.4103/1305-7456.178304

[16] Konidena A. 3D printing: Future of dentistry?. Journal of Indian Academy of Oral Medicinea and Radiology. 2016;**28**:109-110

[17] Thomas D. The Development of Design Rules for Selective Laser Melting [dissertation]. Cardiff: University of Wales Institute; 2009. 318 p

[18] Ebert J, Ozkol E, Zeichner A, et al. Direct inkjet printing of dental prostheses made of zirconia. Journal of Dental Research. 2009;**88**:673-679

[19] Barazanchi A, Li KC, Al-Amleh B, Lyons K and Waddell JN. Additive technology: Update on current materials and applications in dentistry. Journal of Prosthodontics. 2017;**26**(2):156-163. DOI: 10.1111/jopr.12510

[20] Kasparova M, et al. Possibility of reconstruction of dental plaster cast from 3D digital study models. BioMedical Engineering Online [Internet]. 2013;**12**:49. Available from: http://www.biomedical-engineering-online.com/content/12/1/49 [Accessed: 11-04-2017]

[21] Whitley D, Eidson RS, Rudek I, Bencharit S. In-office fabrication of dental implant surgical guides using desktop stereolithographic printing and implant treatment planning software: A clinical report. Journal of Prosthetic Dentistry. Forthcoming. DOI: 10.1016/j.prosdent.2016.10.017

[22] Digholkar S, Madhav V, Palaskar J. Evaluation of the flexural strength and microhardness of provisional crown and bridge materials fabricated by different methods. The Journal of Indian Prosthodontic Society. 2016;**16**:328-334

[23] Dobrzański LA, Achtelik-Franczak A, Król M. Computer aided design in Selective Laser Sintering (SLS)—application in medicine. Journal of Achievements of Materials and Manufacturing Engineering. 2013;**60**(2):66-75

[24] Dobrzański LA, Dobrzanska-Danikiewicz AD, Gawel TG. Ti6Al4V porous elements coated by polymeric surface layer for biomedical applications. Journal of Achievements of Materials and Manufacturing Engineering. 2015;**71**(2):53-59

[25] Dobrzański LA. Applications of newly developed nanostructural and microporous materials in biomedical, tissue and mechanical engineering. Archives of Material Science & Engineering. 2015;**76**(2):53-114

[26] Dobrzański LA. The concept of biologically active microporous engineering materials and composite biological-engineering materials for regenerative medicine and dentistry. Archives of Material Science & Engineering. 2016;**80**(2):64-85

[27] Traini T, Mangano C, Sammons RL, et al. Direct laser metal sintering as a new approach to fabrication of an isoelastic functionally graded material for manufacture of porous titanium dental implants. Dental Materials. 2008;**24**:1525-1533

[28] Tara MA, Eschbach S, Bohlsen F, Kern M. Clinical outcome of metal–ceramic crowns fabricated with laser-sintering technology. International Journal of Prosthodontics. 2011;**24**:46-54

[29] Averyanoiva M, Bertrand P, Verquin B. Manufacture of Co-Cr dental crowns and bridges by selective laser melting technology. Virtual and Physical Prototyping. 2011; **6**(3):179-185

[30] Bibb R, Eggbeer D, Williams R. Rapid manufacture of removable partial denture frameworks. Rapid Prototyping Journal. 2006;**12**:95-99

[31] Kruth J-P, Vandenbroucke B, Van Vaerenbergh J, Naert I. Rapid Manufacturing of Dental Prostheses by Means of Selective Lsser Sintering/Melting. http://doc.utwente.nl/52914/1/Wa1025.pdf

[32] Chen H, Wang H, Lv P, Wang Y, Sun Y. Quantitative evaluation of tissue surface adaption of CAD-designed and 3D printed wax pattern of maxillary complete denture. Hindawi Publishing Corporation, BioMed Research International. 2015;**2015**:ID 453968, 5 p. http://dx.doi.org/10.1155/2015/453968

[33] Dikova T, Dzhendov D, Bliznakova K, Ivanov D. Application of 3D printing in manufacturing of cast patterns. In: Sveto C and Goran N, editors. VII-th International Metallurgical Congress; 9-12.06.2016; Ohrid, Macedonia. Skopje: Macedonian union of metallurgists; 2016. CD-ROM

[34] Minev R, Minev E. Technologies for rapid prototyping (RP)–basic concepts, quality issues and modern trends. Scripta Scientifica Medicinae Dentalis. 2016;**2**(1):29-39

[35] Bliznakova K. The use of 3D printing in manufacturing anthropomorphic phantoms for biomedical applications. Scripta Scientifica Medicinae Dentalis. 2016;**2**(1):40-48

[36] Kuo RF, Fang KM, Su FC. Open-source technologies and workflows in digital dentistry. In: Sasaki K, Suzuki O, Takahashi N, editors. Interface Oral Health Science 2016. 1st ed. Singapore: Springer; 2017. pp. 165-170. DOI: 10.1007/978-981-10-1560-1_14

[37] Jacobs PF. Rapid Prototyping & Manufacturing. 1st edition, second printing ed. Dearborn, USA: Society of Manufacturing Engineers; 1992

[38] ISO. ISO 12836:2015(en). Dentistry—Digitizing Devices for CAD/CAM Systems for Indirect Dental Restorations—Test Methods for Assessing Accuracy" [Internet]. 2015. Available from: https://www.iso.org/obp/ui/#iso:std:iso:12836:ed-2:v1:en. [Accessed: 06.04.2017]

[39] Braian M, Jimbo R, Wennerberg A. Production tolerance of additive manufactured polymeric objects for clinical applications. Dental Materials. Forthcoming. http://dx.doi.org/10.1016/j.dental.2016.03.020

[40] ISO. ISO/IEC Guide 99:2007(en) International Vocabulary of Metrology — Basic and General Concepts and Associated Terms (VIM); [Internet]. 2007-12.Available from: https://www.iso.org/standard/45324.html [Accessed: 06.04.2017]

[41] Bae EJ, Jeong ID, Kim WC, Kim JH. A comparative study of additive and subtractive manufacturing for dental restorations. The Journal of Prosthetic Dentistry. Forthcoming. DOI: 10.1016/j.prosdent.2016.11.004

[42] Mai HN, Lee KB, Lee DH. Fit of interim crowns fabricated using photopolymer-jetting 3D printing. The Journal of Prosthetic Dentistry. Forthcoming. DOI: 10.1016/j.prosdent.2016.10.030

[43] Kim DY, Jeon JH, Kim JH, Kim HY, Kim WC. Reproducibility of different arrangement of resin copings by dental microstereolithography: Evaluating the marginal discrepancy of resin copings. The Journal of Prosthetic Dentistry . 2017;**117**:260-265

[44] Ishida Y, Miyasaka T. Dimensional accuracy of dental casting patterns created by 3D printers. Dental Materials Journal. 2016;**35**(2):250-256

[45] Formlabs. 3D Printing Technology Comparison: SLA vs. DLP. [Internet]. 2016 Nov 1. Available from: https://formlabs.com/blog/3d-printing-technology-comparison-sla-dlp/; [Accessed: 06-04-2017]

[46] Dikova T, Dzhendov D, Katreva I, Pavlova D, Simov M, Angelova S, Abadzhiev M, Tonchev T. Possibilities of 3D printer rapidshape D30 for manufacturing of cubic samples. Scripta Scientifica Medicinae Dentalis. 2016;**2**(1):9-15

[47] Dikova T, Dzhendov D, Katreva I, Pavlova D, Tonchev T, Doychinova M. Geometry and Surface Roughness of Polymeric Samples Produced by Stereolithography. International Journal of Machines, Technologies, Materials. 2017;**4**:201-205

[48] Dikova T, Dzhendov D, Katreva I, Pavlova D. Accuracy of polymeric dental bridges manufactured by stereolithography. Archives of Materials Science and Engineering. 2016;**78**(1):29-36

[49] Dzhendov D, Pavlova D, Simov M, Marinov N, Sofronov Y, Dikova T, et al. [Geometrical accuracy of fixed dental constructions, manufactured by additive technologies] [in Bulgarian]. In: 8th International Conference "Technical Science and Industrial Management"; 2014 Sep; Varna, Bulgaria. Sofia: Scientific-Technical Union of Mechanical Engineering; 2014. pp. 13-17

[50] Dikova T, Dzhendov D, Simov M. Microstructure and hardness of fixed dental prostheses manufactured by additive technologies. Journal of Achievements in Mechanical and Materials Engineering. 2015;**71**(2):60-69

[51] Kim HR, Jang SH, Kim YK, Son JS, Min BK, Kim KH, Kwon TY. Microstructures and mechanical properties of Co-Cr dental alloys fabricated by three CAD/CAM-based processing techniques. Materials. 2016;**9**(956):1-4. DOI: 10.3390/ma9070596

[52] Rehme O, Emmelmann C. Reproducibility for properties of selective laser melting. In: Lasers in Manufacturing 2005, LIM, International WLT-Conference on Lasers in Manufacturing, by AT-Verlag, Stuttgart; 2005;**3**:227-232

[53] Shiomi M, Osakada K, Nakamura K, Yamashita T, Abe F. Residual stress within metallic model made by selective laser melting process. CIRP Annals—Manufacturing Technology. 2004;**53**:195-198

[54] Vranken B. Study of Residual Stresses in Selective Laser Melting [dissertation]. Leuven: KU Leuven-Faculty of Engineering Science; 2016. 255 p

[55] Yadroitsev I, Bertrand P, Smurov I. Parametric analysis of the selective laser melting process. Applied Surface Science. 2007;**253**(19):8064-8069

[56] Yadroitsev I, Krakhmalev P, Yadroitsava I, Johansson S, Smurov I. Energy input effect on morphology and microstructure of selective laser melting single track from metallic powder. Journal of Materials Processing Technology. 2013;**213**:606-613

[57] Yadroitsev I, Smurov I. Surface morphology in selective laser melting of metal powders. Physics Procedia. 2011;**12**:264-270

[58] Averyanova M. Quality control of dental bridges and removable prostheses manufactured using Phenix systems equipment. In: AEPR'12, 17th European Forum on Rapid Prototyping and Manufacturing; 12-14 June 2012; Paris, France. Ecole Centrale Paris; 2012

[59] Takaichi A, Suyalatu, Nakamoto T et al. Microstructures and mechanical properties of Co–29Cr–6Mo alloy fabricated by selective laser melting process for dental applications. Journal of the Mechanical Behavior of Biomedical Materials. 2013;**21**:67-76

[60] Vandenbroucke B, Kruth JP. Selective laser melting of biocompatible metals for rapid manufacturing of medical parts. Rapid Prototyping Journal. 2007;**13**(196-203)

[61] Meacock CG, Vilar R. Structure and properties of a biomedical Co–Cr–Mo alloy produced by laser powder microdeposition. Journal of Laser Applications 2009;**21**:88-95

[62] Barucca G, Santecchia E, Majni G et al. Structural characterization of biomedical Co-Cr-Mo components produced by direct metal laser sintering. Materials Science and Engineering C. 2015;**48**:263-269

[63] Lu Y, Wu S, Gan Y et al. Investigation on the microstructure, mechanical property and corrosion behavior of the selective laser melted CoCrW alloy for dental application. Materials Science and Engineering C. 2015;**49**:517-525

[64] Jevremovic D, Puskar T, Kosec B, Vukelic D, Budak I, Aleksandrovic S, Egebeer D, Williams R. The analysis of the mechanical properties of F75 Co-Cr alloy for use in selective laser melting (SLM) manufacturing of removable partial dentures (RPD). Metalurgija. 2012;**51**(2):171-174

[65] Kajima Y, Takaichia A, Nakamoto T, et al. Fatigue strength of Co–Cr–Mo alloy clasps prepared by selective laser melting. Journal of the Mechanical Behavior of Biomedical Materials. 2016;**59**:446-458

[66] Dolgov NA, Dikova T, Dzhendov D, Pavlova D, Simov M. Mechanical properties of dental Co-Cr alloys fabricated via casting and selective laser melting. Materials Science. Non-Equilibrium Phase Trasformations. 2016;**3**:3-7

[67] Atapek H, Dikova T, Aktaş G, Polat Ş, Dzhendov D, Pavlova D. Tribo-corrosion behavior of cast and selective laser melted Co-Cr alloy for dental applications. International Journal of Machines, Technologies, Materials. 2016;**12**:61-64

[68] ISO. ISO 9693-1:2012. Dentistry—Compatibility Testing—Part 1: Metal-Ceramic Systems [Internet]. 2012. Available from: https://www.iso.org/obp/ui/#iso:std:iso:9693:-1:ed-1:v1:en [Accessed: 06-04-2017]

[69] Kaleli N, Sarac D. Comparison of porcelain bond strength of different metal frameworks prepared by using conventional and recently introduced fabrication methods. The Journal of Prosthetic Dentistry. Forthcoming. DOI: 10.1016/j.prosdent.2016.12.002

[70] Li J, Chen C, Liao J, Liu L, Ye X, Lin S, Ye J. Bond strengths of porcelain to cobalt-chromium alloys made by casting, milling, and selective laser melting. The Journal of Prosthetic Dentistry. Forthcoming. DOI: 10.1016/j.prosdent.2016.11.001

[71] Akova T, Ucar Y, Tukay A, Balkaya MC, Brantley WA. Comparison of the bond strength of laser-sintered and cast base metal dental alloys to porcelain. Dentistry Materials. 2008;**24**:1400-1404

[72] Wu L, Zhu H, Gai X, Wang Y. Evaluation of the mechanical properties and porcelain bond strength of cobalt-chromium dental alloy fabricated by selective laser melting. The Journal of Prosthetic Dentistry. 2014;**111**:51-55

[73] Xiang N, Xin XZ, Chen J, Wei B. Metal–ceramic bond strength of Co–Cr alloy fabricated by selective laser melting. Journal of Dentistry. 2012;**40**:453-457

[74] Wang H, Feng Q, Li N, Xu S. Evaluation of metal-ceramic bond characteristics of three dental Co-Cr alloys prepared with different fabrication techniques. The Journal of Prosthetic Dentistry. 2016;**116**:916-923

[75] Kim EH, Lee DH, Kwon SM, Kwon TY. A microcomputed tomography evaluation of the marginal fit of cobalt-chromium alloy copings fabricated by new manufacturing techniques and alloy systems. The Journal of Prosthetic Dentistry. 2017;**117**(3):393-399

[76] Kaleli N, Sarac D. Influence of porcelain firing and cementation on the marginal adaptation of metal ceramic restorations prepared by different methods. Prosthet Dent. 2017;**117**(5):656-661

[77] Pompa G, Di Carlo S, De Angelis F, Cristalli MP, Annibali S. Comparison of conventional methods and laser-assisted rapid prototyping for manufacturing fixed dental prostheses: An in vitro study. BioMed Research International. 2015. 1-7. http://www.hindawi.com

[78] McLean JW and von Fraunhofer JA. The estimation of cement film thickness by an in vivo technique. British Dental Journal. 1971;**131**(3):107-111

[79] Sundar MK, Chikmagalur SB, Pasha F. Marginal fit and microleakage of cast and metal laser sintered copings—An in vitro study. Journal of Prosthodontic Research. 2014;**58**:252-258

[80] Huang Z, Zhang L, Zhu J, Zhang X. Clinical marginal and internal fit of metal ceramic crowns fabricated with a selective laser melting technology. Journal of Prosthetic Dentistry. 2015;**113**:623-627

[81] Tamac E, Toksavul S, Toman M. Clinical marginal and internal adaptation of CAD/CAM milling, laser sintering, and cast metal ceramic crowns. Journal of Prosthetic Dentistry. 2014;**112**:909-913

Chapter 6

Perspective of Additive Manufacturing Selective Laser Melting in Co-Cr-Mo Alloy in the Consolidation of Dental Prosthesis

Marcello Vertamatti Mergulhão,
Carlos Eduardo Podestá and
Maurício David Martins das Neves

Additional information is available at the end of the chapter

http://dx.doi.org/10.5772/intechopen.69720

Abstract

This chapter seeks to compare the properties of samples manufactured by additive manufacturing (AM) by the selective laser melting (SLM) technology and compare with the precision casting (PC) processes using the Co-Cr-Mo (ASTM F75) alloy to manufacture of dental prosthesis. This AM process can be manufactured three-dimensional models by means of a laser beam that completely melts particles of powder deposited layer by layer. However, it is still relevant to know the properties of: performance, dimensional, mechanical and microstructural of this laser melting process and compare with a convencional process. The results of mechanical evaluation showed that the SLM technique provides superior mechanical properties compared to those obtained by the PC technique. It is possible to verify that the consolidation by SLM technique results in lower presence of porosity than PC technique. In addition, PC samples presented a gross dendritic microstructure of casting process. Microstructural analysis of SLM samples results in a characteristic morphology of layer manufacturing with ultrafine grains and a high chemical homogeneity. In this way, the development of the present study evidenced to improve the manufacture of customized components (copings) using the SLM technology.

Keywords: Co-Cr-Mo alloy, biomaterial, additive manufacturing, selective laser melting, precision casting

1. Introduction

Metal powders of cobalt-chromium (Co-Cr) alloy are widely used in various sectors of the automotive, aeronautics, and aerospace industry, because of its high wear resistance and adequate corrosion resistance also being used in surface coating to increase performance components [1, 2]. In addition, the biocompatibility properties are suitable and are being used in the manufacture of medical and dental prosthetics [3–7]. The use of Co-Cr alloys is widely discussed to manufacture medical and dental implants or prostheses [7–9] presenting positive aspects in relation to biocompatibility analysis. The necessity for characterization and biological evaluations, physical-chemical, and mechanical are basic requirements for the development of new biomaterials applied in medical devices. In general, biomaterials need to present a final clinical characteristic (bio-functionality) and biocompatibility [10, 11].

Since 1930, Cobalt-Chromium-Molybdenum (Co-Cr-Mo) alloys processed by casting were used as dental alloys and later adapted for use in orthopedic implants [12, 13]. According to Jabbari et al., Co-Cr alloys are used almost exclusively in the manufacture of metal structures prostheses and recently is replacing Ni-Cr alloy or alternatively for the production of restorations in porcelain fused to metal (PFM), because Co-Cr alloy is Ni-free and does not have allergic responses or toxic effects related to Nickel [14].

The coefficient of thermal expansion (CTE) is a thermal property of the alloy, is of great interest in cases of applications in dental components, that requires ceramic coating, such as the dental crown. In this case, it is shown by Refs. [6, 15, 16] that Co-Cr alloys should have a CTE value in the range of 14.0–14.6 × 10^{-6} $°C^{-1}$ at temperatures from 500 to 600°C to the correct ceramic firing process, as coating of the metal component. The CTE of ceramic materials for coating applications in metallic materials should be close, providing a good adjustment due to contraction and expansion during heating, thus avoiding the possibility of voids or cracks occurring during the firing process [15, 17].

Currently, the lost wax casting method is the most widely used, but has faced competition from other manufacturing processes [17]. Several authors [18–20] describe the development of AM technologies providing the creation of final customized implants. Techniques such as stereolithography were implemented to manufacture resin models for posterior manufacture of dental prostheses (crowns and bridges) by conventional process of lost wax castings. Mechanical, chemical, and microstructural properties are evaluated in comparison to new AM technologies, for example, the selective laser melting (SLM) in relation to conventional techniques as lost wax casting [21–23]. In this way, the preparation of medical and dental components provides customized final components with high mechanical properties, compared to conventional techniques (see **Figure 1**) [24].

Notably in health area, this technology is competitive over other traditional manufacturing processes by advances occurred in the area of processes using powder metallurgy techniques [25, 26]. Selective laser melting technology is one of the innovative technologies in additive manufacturing development in the middle of the 1980s after the creation of selective laser sintering (SLS) process. SLM is a process based on the 3D construction in which it is possible by laser beam to completely melt the metallic powder particles on a previous layer [27–31].

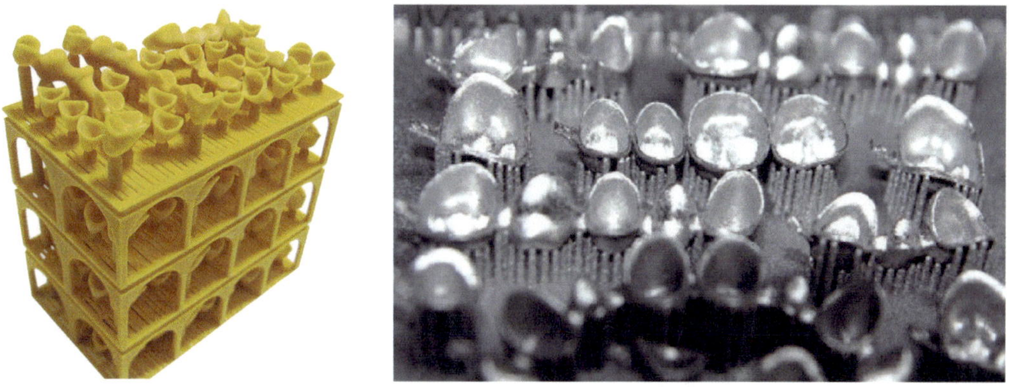

(a) (b)

Figure 1. Models of dental components manufactured by AM techniques in (a) resin model by stereolithography for posterior precision casting process and (b) copings manufactured by SLM technique [17].

In SLM technique, the raw material is in metallic powder form and the thermal energy required for the complete melting of the powder layer comes from a laser beam, usually is used a source of Ytterbium (Yb) fiber [29]. The maximum laser power on SLM machines is approximately 400 W and the laser focus may have a diameter of approximately 100 μm [30]. In turn, the laser beam is commanded by an interface that transmits it to the optical assembly, which selectively directs (X–Y plane) the laser beam, causing the powders to melt [28–33]. The metal powder is stored in a container, which may or may not be the distributor of powder (deposited by gravity), which in turn uniformizes the powder layer (between 50 and 100 μm) on an object consolidation platform. At each consolidated layer, the platform moves on the Z–axis, according to the next layer until the component is completely consolidated. The process of component consolidation occurs in a consolidation chamber (internal environment of the SLM equipment) that is under inert atmosphere protection (argon gas) [28–33]. The basic scheme of the components present in the consolidation process by SLM can be observed in **Figure 2**, and other machine parameters and working conditions are presented in **Table 1**.

Details of the main parameters of the SLM technique are shown schematically in **Figure 3**.

The SLM technique has several process parameters and can be grouped in five families, being these related to laser, scanning, material (powder), temperature, and consolidation chamber [28, 29, 31].

- Laser: power "*P*", beam diameter, pulse duration, pulse frequency;
- Scanning: speed "*v*", track distance, strategy;
- Powder (material): material properties, particle size, distribution, bed density of powder, layer thickness;
- Temperature: consolidation layer, powder feeder, uniformity;
- Compounding chamber: composition of protective atmosphere.

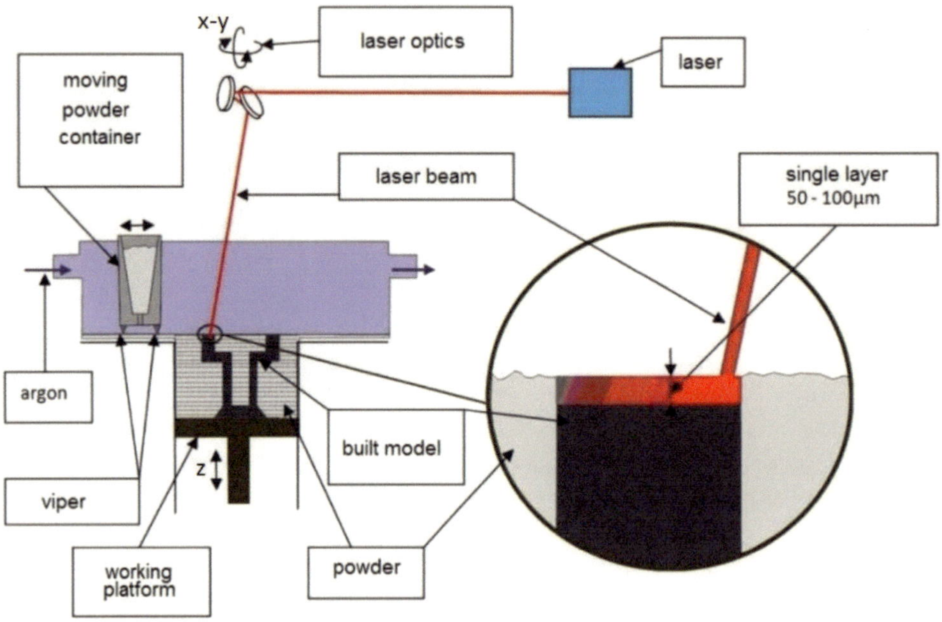

Figure 2. Basic schematic of components present in a SLM machine [33].

The most common parameters to be adjusted in the SLM process to optimize the manufacture of components are: laser power (*P*), scan speed (v), track (hatch) distance, and layer thickness. According to Refs. [27, 34, 35], the volumetric energy density of the laser (ψ, given in $J.mm^{-3}$) relates the main parameters of consolidation, in relation to the laser as shown in Eq. (1), being: laser power (*P*), track distance (*t*), scan speed (v), and layer thickness (*L*).

$$\psi = \frac{P}{t \cdot v \cdot L} \tag{1}$$

System parameters	
Laser power	400 W Yb-Fiber-laser
Build speed	20 ccm/h
Pract. layer thickness	20–75 µm min
Scan line/wall thickness	150 µm
Operational beam focus	80–120 µm
Scan speed	15 m/s
Inert gas consumption in operation	Ar/N_2, 2.5–3.0 l/min
Inert gas consumption venting	Ar/N_2, 1700 l @ 100 l/min
Compressed air requirement	18 l/min @ 1.5 bar

Table 1. Typical technical parameters of the SLM®280^{HL} machine.

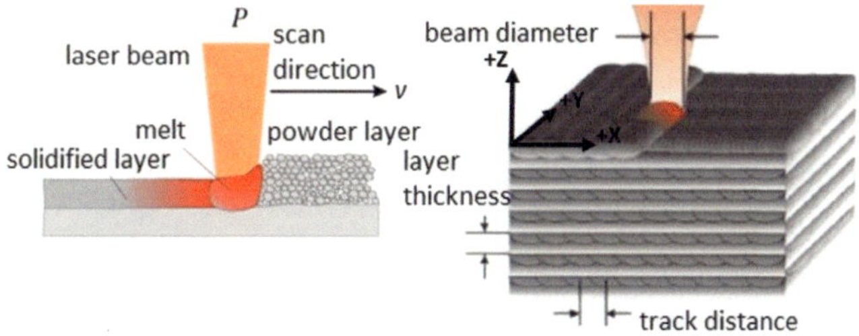

Figure 3. Schematic of parameters involved in SLM consolidation [31].

Several authors in Refs. [27, 34, 35] report that these parameters affect the volumetric energy density, determinant in the powder melting and that in turn influences the mechanical properties and roughness of the surface of the consolidated parts. The combination of these variables can generate excess (or insufficient) energy during the consolidation process, which can lead to the balling phenomenon at consolidated specimen, which corresponds to the dissimilar or noncontinuous scan tracks [30, 36, 37]. Additionally, the balling phenomenon can generate uniform deposition of next powder layer, can cause uncontrollable porosity and delamination by the absence of inter-fusion between layers [37, 38].

As observed, the parameters of the SLM process involve a certain complexity, in order to obtain the fabrication of components of complete density. In order to optimize the mechanical and physical-chemical properties of the final components manufactured the consolidation strategies are the subject of discussion and study [39]. The consolidation strategies refer to the consolidation parameters already presented, as well as to the direction and orientation of laser beam scanning, angle of rotation between the layers, and the number passes of laser beam (per layer), as seen in **Figure 4a–h** [40–42]. Also, the physical properties of the components may be associated with the manufacturing anisotropy of the samples, see **Figure 4i** [43–45].

An interesting point to consider for AM processes is the feedstock (or raw material–metal powder). These new technologies demanded a characteristic powder size distribution, format and physical properties (flowability and packing) [36]. In this case, to produce spherical metal powder the most common process is gas atomization [29, 30]. However, it is remarkable that the use of gas-atomized powders in the SLM process by the better physical properties is compared to water-atomized powders. The characteristic format of powders (gas and water atomized) is possible to observe in **Figure 5**.

In addition, the capability to reusability of feedstock material in AM processes is a significant issue to promote economic and environmental manufacture processes [28, 46]. However, the effects and influence of the powder reuse on manufactured parts are the subject of much discussion [47–50]. The conclusions of Tang et al. [48] study, about reuse of Ti-6Al-4V powders of electron beam melting (EBM) process, appoint the increased oxygen content and particles became less spherical. Although the reuse powder improved the flowability (by little presence of satellite particles), increased the yield strength and the ultimate tensile on the AM process of Ti-6Al-4V [48].

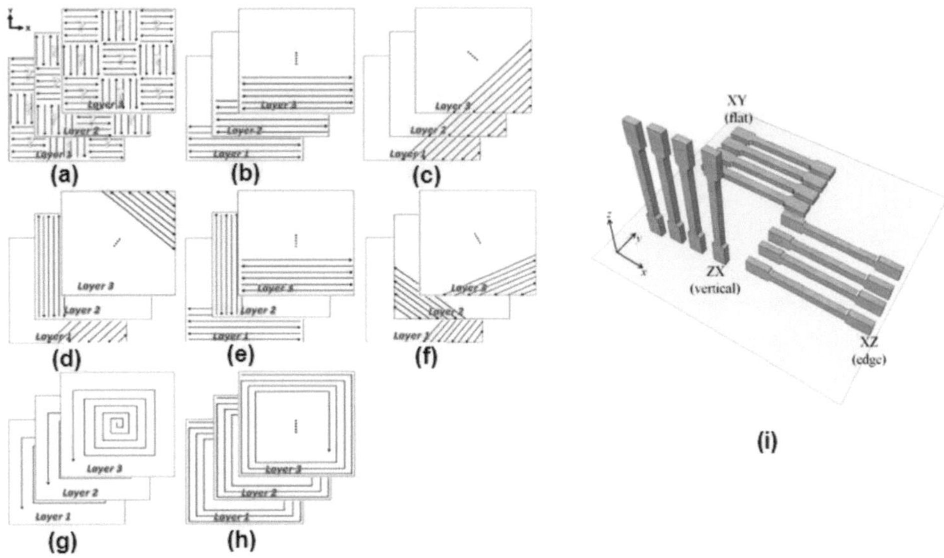

Figure 4. Representation of laser scanning strategies for sample consolidation via SLM, scanning in: (a) island, (b) line, (c) line at 45°, (d) line and rotation line at 45°, (e) line and rotation line at 90°, (f) line rotate at 67°, (g) internal spiral, (h) external spiral [42], and (i) building orientation of specimens [45].

Considering this important field in expansion, this chapter is part of this scenario with a focus on the dental sector, more specifically on the evaluation of mechanical properties and microstructural analysis of Co-Cr-Mo alloy to manufacture dental prostheses (copings). The aim of this chapter is to evaluate the mechanical properties and microstructures of standardized specimens made by powder metallurgy techniques using SLM from powdered gas powder of the Co-Cr-Mo alloy. The results obtained by SLM will be compared with the results of samples manufactured by precision casting.

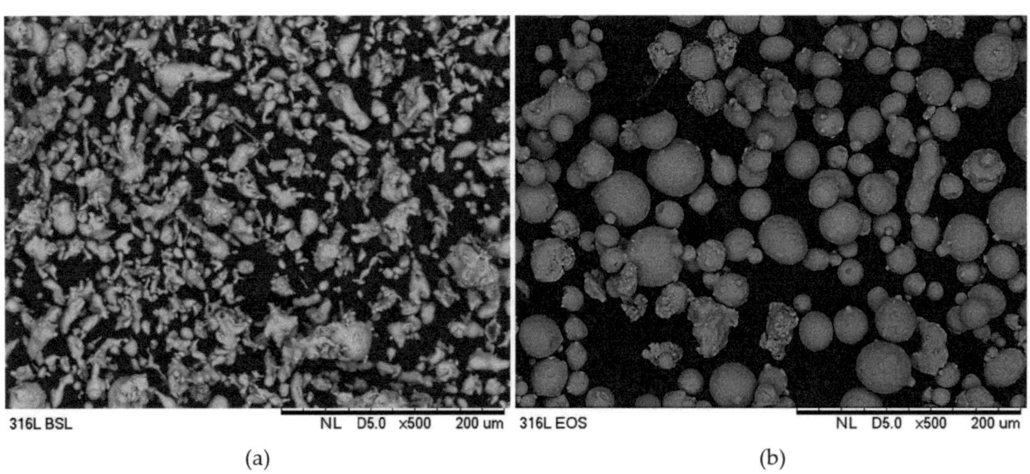

Figure 5. Characteristic format of powders (316L alloy) produced by water atomization (a) and gas atomization (b). Magnitude ×500.

2. Experimental procedure

2.1. Powder characterization

Co-Cr-Mo alloy gas atomized (H.C Starck®, Lübeck, Germany) was provided by the HighBond® (Indaiatuba, Brazil) in the particle size (granulometric range) of 15–45 μm. The confirmation of the chemical composition was performed by energy dispersive X-ray (Shimadzu EDX-720 equipment and by LECO). This study was based on alloy/powder with certification of ANVISA (Brazilian agency) for use in health care segment.

Several physical properties of gas-atomized powder were obtained such as flow time (ASTM B212 [51]), apparent density (ASTM B213 [52]), and tap density (ASTM B527 [53]). The particle size distribution was performed using a particle analyzer by laser scattering (Cilas–Model 1064). The particle format and microstructural characterization of powders were performed via optical and scanning electron microscopy (OM–Olympus BX51M and SEM-EDS Philips XL30).

To evaluate the internal porosity of powders sample were measured by the pycnometer density in comparison to theoretical density. The density by Helium pycnometry considered only the internal porosity (excluding the open porosity) and was performed using the Micromeritics equipment (Model Accu PYC 1330 Pycnometer).

Differential scanning calorimetry (DSC) analysis was performed using a sample of gas-atomized Co-Cr-Mo powder. Three runs of heating curves at rates of 10, 20, and 30°C/min were performed and under static atmosphere constituted in argon (99.999%) for minimizing the oxidation of the samples. In all experiments, both the crucible (sample holder and the reference–empty during all tests) were composed of alumina (Al_2O_3) with a volume of approximately 100 μL. The equipment used was Setsys 16/18, from Setaram with a thermocouple rod of Pt/Pt Rh 10%.

2.2. Manufacturing specimens using precision casting and selective laser melting

Precision casting and selective laser melting techniques performed the manufacture of gas-atomized Co-Cr-Mo powders. The tensile and three-point bending specimens were manufactured in standard dimensions according to ISO 22674-06 [54] and ASTM B528-12 [55]. **Figure 6** shows specimens manufactured.

The precision casting (PC) samples were performed according to ASTM F75-12 [13] by HighBond® (Indaiatuba, Brazil). The PC fabrication process method satisfied the following steps: machining of wax disks in the standard dimensions of tensile and flexural test specimens, assembly and shell building, dewaxing and pouring the Co-Cr-Mo alloy was by an induction furnace at a temperature of 1489°C.

The consolidation of SLM samples was carried out by SLM Solutions™ using a selective laser melting machine SLM®280HL with a single Ytterbium laser beam (maximum power 400 W). The building consolidation of specimens was parallel to laser beam and performed using parameters such as: layer thickness of 30 μm and diameter of laser beam of 76 μm.

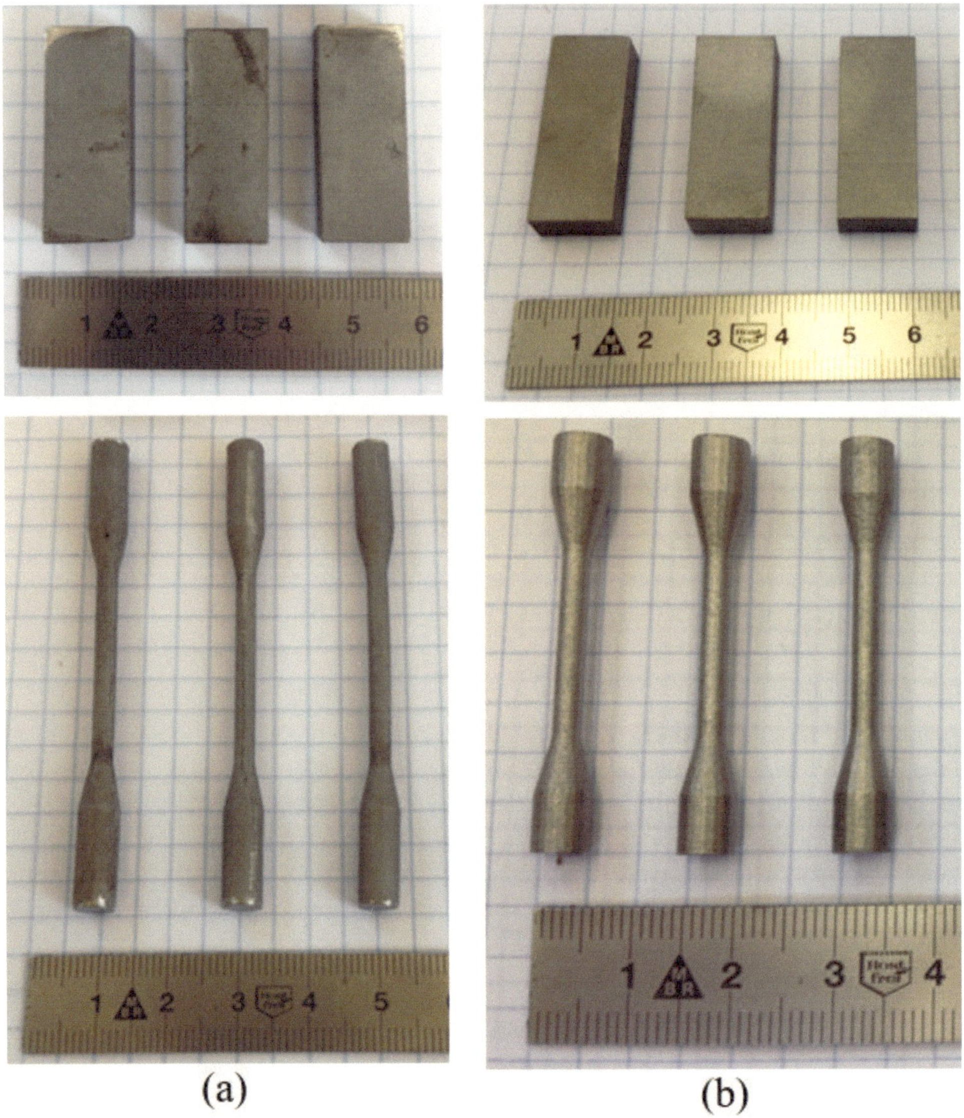

Figure 6. Specimens manufactured of Co-Cr-Mo alloy: (a) specimens made by FP technique and (b) specimens made by SLM technique.

2.3. Characterization of samples manufactured

To evaluate the susceptibility to cell growth in the Co-Cr-Mo alloy after the consolidation processes (PC and SLM) was performed by the cytotoxicity analysis, according to ISO 10993-5 [56]. The determination of the cytotoxicity was obtained by the quantitative evaluation method, which is carried out by the measurement of cell death, cell proliferation or formation of cellular colonies.

To evaluate the internal porosity of cast, and SLM samples were measured by the pycnometry density in comparison to the theoretical density. The density by Helium pycnometry, considered only the internal porosity (excluding the open porosity) was measured using the Micromeritics equipment–Model Accu PYC 1330 Pycnometer.

The thermomechanical analysis (TMA) was performed on samples consolidated by SLM and PC of the Co-Cr-Mo alloy. The purpose of the technique was to obtain the coefficient of thermal expansion (CTE). In addition, the SLM samples were analyzed in the parallel and transversal building direction (SLM 1—parallel direction and SLM 2—transversal direction). The routine of the TMA remained the heating rate was from 10°C/min until the temperature of 1300°C. The equipment used was a Setaram–Setsys 16/18, using a thermocouple rod of Pt/Pt Rh 10% under a static atmosphere (argon—99.999%) to exclude the sample oxidation.

2.4. Mechanical characterization

Mechanical characterization of consolidated samples by PC and SLM techniques was held in five samples of each test (tensile and three-point bending), respectively, according to ISO 22674-06 [54] and ASTM B528 [55]. The three-point bending test determined the transversal rupture strength (TRS) of specimens. The TRS relates to the applied load (P) and the distance between the supports (L), over the cross area of the sample (thickness "*t*" and width "*w*"), as show in Eq. (2). Mechanical tests were performed using a universal testing machine (Instron 3366) under a crosshead speed of 0.2 mm/min at room temperature.

$$TRS = \frac{3 \cdot P \cdot L}{2 \cdot t2 \cdot w} \tag{2}$$

2.5. Microstructural evaluation

The microstructural characterization of consolidated Co-Cr-Mo and the fracture analysis were evaluated after tensile test. Metallography preparation consisted of mechanical grinding in SiC paper #1200 and final chemical polishing with OP-S 0.02 μm with addition of 10% HCl. The specimens were etching in solution: 100 ml HCl and 2 ml H_2O_2 (1–2 min at room temperature). The microstructural characterization was performed in both building directions using an optical microscope (OM) Olympus–BX51M and scanning electron microscope (SEM) with energy dispersive X-ray (EDS) Philips XL30 and JEOL–JSM6701F.

3. Results and discussion

3.1. Powder characterization

The confirmation of chemical composition was performed in Co-Cr-Mo powder alloy, as also in the samples manufactured (PC and SLM). The chemical composition is presented in **Table 2** comparing with the standard ASTM F75-12 [13].

Elements (%)	Powder	PC	SLM	ASTM F75
Co	63.93 ± 0.16	66.38 ± 0.15	65.38 ± 0.32	Balanço
Cr	28.83 ± 0.19	26.76 ± 0.21	27.68 ± 0.13	27.00 – 30.00 ± 0.30
Mo	7.07 ± 0.31	6.68 ± 0.03	6.61 ± 0.16	5.00 – 7.00 ± 0.15
Fe	0.17 ± 0.01	0.18 ± 0.08	0.33 ± 0.06	0.75 ± 0.03
C	0.03 ± 0.01	0.02 ± 0.01	0.03 ± 0.01	0.350 ± 0.020
S	0.01 ± 0.01	0.01 ± 0.01	0.01 ± 0.01	0.010 ± 0.003
N_2	0.0820 ± 0.0011	0.0416 ± 0.0015	0.1330 ± 0.0015	0.250 ± 0.020
O_2	0.0940 ± 0.0015	0.0187 ± 0.0016	0.0240 ± 0.0010	–

Table 2. Chemical composition (weight %) of Co-Cr-Mo samples (powder, PC and SLM) in accordance with standard ASTM F75.

The characteristic format of the powder process fabrication by gas atomization is observed in **Figure 7**. The analysis in SEM shows that the powders are spherical and presented satellites (appointed by arrows—**Figure 7a,b**). The satellites can be formed in the surface particles during the cooling process of the spherical powder particles during gas atomization. It is noteworthy that the shape of the particle influences on packing properties, flow hate, and compressibility, as well as reports on the powder metallurgy process [2, 57, 58]. The cross-sectioned powder (**Figure 7d,e.**) shows the dendritic morphology with the primarily arms and ramifications, characterizing the rapid solidification of gas atomization process.

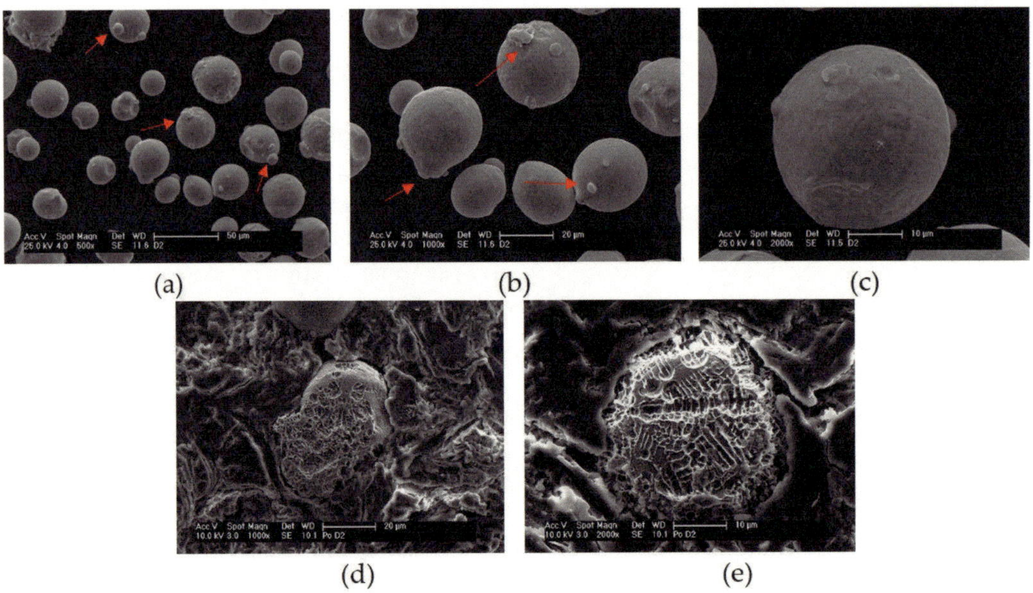

Figure 7. SEM images of Co-Cr-Mo powder: (a) magnitude ×500, (b) magnitude ×1000, (c) magnitude ×2000, (d) and (e) cross-section powder after chemical etch (etch solution: HCl, H_2SO_4, and HNO_3 for 60–240 s at 45°C, respectively, magnitude ×1000 and ×2000).

Physical properties	Co-Cr-Mo powder			
Granulometric distribution (μm)	D10	D50	D90	D mean
	20.88	31.11	46.10	32.36
Flow rate (s/50g)	15.85 ± 0.11			
Apparent density (g/cm³)	4.51 ± 0.01			
Tap density (g/cm³)	5.26 ± 0.05			
Theoretical density (g/cm³)	8.38			
Helium pycnometry (g/cm³)	8.30 ± 0.001			

Table 3. Results of physical properties for Co-Cr-Mo powder in the SLM range.

The powders to manufacture samples via SLM technique have a mean diameter less than 50 μm to improve the physical properties such as flow time, apparent density, and tap density [33]. The results of physical powder properties are summarized in **Table 3**.

According to Haan et al. [59], Co-Cr-Mo powders with diameter D90 equals to 39 μm, the flowability was 18.60 s/50 g. The results were similar to those obtained for the present study, such as 15.86 s/50 g for D90 equals to 46.10 μm.

The result of tap density tends to be higher than the result of the apparent density, because of the particle's accommodation there is a decrease in the amount of voids between the particles [2]. Also, the smaller the apparent density, the greater the percentage of increase the tap density.

It is possible to verify the presence of closed porosity that is not considered as a measure of the volume of Helium and consequently reduces the value of pycnometry density (8.30 g/cm³). The presence of internal porosity calculated in relation to theoretical density (8.38 g/cm³) is approximately 1.3%.

To investigate and confirm the thermal events present in the Co-Cr-Mo alloy, the heating curves (different rates: 10, 20, and 30°C/min) of the DSC analysis obtained using the Co-Cr-Mo powder are shown in **Figure 8**. The presence of three events occurring in the heating curves of DSC is observed, being the first exothermic, the second endothermic, and the final event corresponding to the fusion of the Co-Cr-Mo alloy. It is possible to verify the temperature variation of the events between the different temperature rates.

In relation to the first event (exothermic), occurring around 582.81°C, it is related to the phase transformation of the alloy (precisely from Co), from the cubic face (αCo) phase to the compact hexagonal phase (εCo). In a similar analysis, Santos [16] obtained a slight peak at 600°C in the thermal analysis (DTA), however the author does not approach the occurrence. Facchini [60] describes this occurrence, the event occurs at approximately 650°C, but is described by an endothermic peak, diverging from the present analysis, in which the curve of 20°C/min occurs at approximately 600°C and which describes an exothermic peak.

The second event (endothermic) occurs around 944.52°C, may be related to allotropic transformation of element Co, by the transition of the phase of compact hexagonal structure (εCo) to the phase of cubic structure of face centered (αCo). This transformation can be confirmed

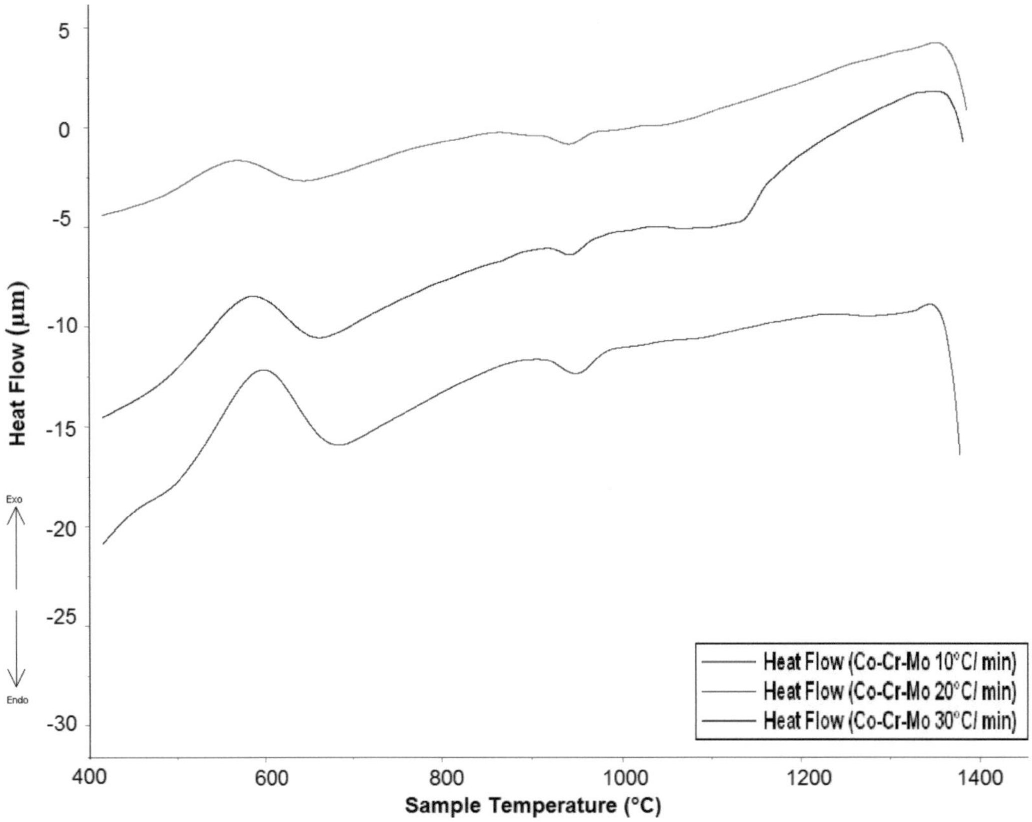

Figure 8. Heating curves of DSC analysis for Co-Cr-Mo powder in different rates (10, 20, and 30°C/min).

in the Co-Cr binary diagram occurring at about 950°C). This occurrence is similar to that described in Ref. [16, 60], which obtains endothermic peaks, respectively, at approximately 970 and 1000°C, relative to that obtained in the present study of 944.52°C. This temperature difference is associated with the chemical composition of the alloy (64Co-29Cr-7Mo of the present study), which represent alloys according to ASTM F75-12 (stoichiometry is 66Co-28Cr-6Mo) and therefore there are temperature difference of 26 and 56°C relative to the cited references. This difference can be associated to different calibrations, among the equipment used in the analysis.

In interpreting the DSC curves, it is possible to verify the melting temperature of the Co-Cr-Mo alloy. By means of the average value of the three heating rates, the melting temperature is approximately 1354.5°C, with a variation of 4°C between rates.

3.2. Evaluation of samples manufactured

The result of the cytotoxicity analysis for the processed samples is shown in **Figure 9**. According to the cytotoxicity assay with respect to the pure extract, without dilution, the samples processed by precision casting and selective laser melting showed no toxicity.

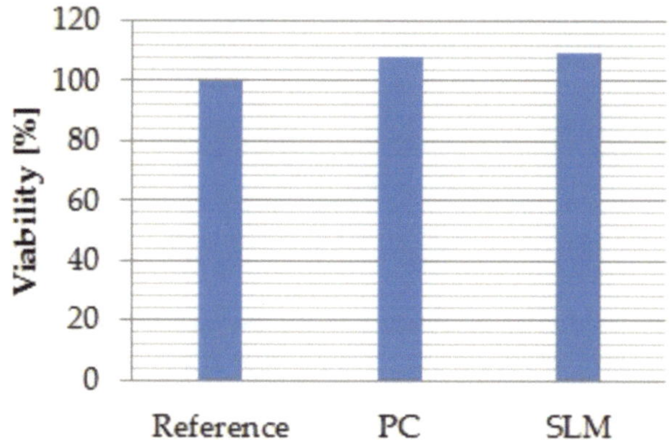

Figure 9. Result of viability of cell growth (cytotoxicity analysis) for Co-Cr-Mo specimens manufactured by PC and SLM techniques.

The results showed the expected results for the Co-Cr-Mo alloy, because as there is a need for specific mechanical properties to be reached, the use of Co-Cr alloys for the manufacture of medical and dental implants or prostheses did not show any toxicity with the medium biological [7, 8, 61].

The porosity of the samples was determined and a comparison was made with average densities: theoretical, volumetric, and by helium pycnometry. The results of the mean densities obtained for the samples consolidated by PC and SLM can be seen in **Table 4**.

Analyzing the results of volumetric density, it is possible to identify that PC samples present a lower result than the one obtained in the SLM samples. This premise is confirmed by the result obtained, evidencing that the PC process presents superior open porosity than SLM technology. When correlating with the density by Helium pycnometry, it can be verified that the open porosity results in 0.24% for PC sample and 0.12% for SLM sample.

Analyzing the Helium pycnometry, it is possible to verify that both consolidation processes have the same theoretical density (8.24 g/cm^3). Relating the Helium pycnometry density to the theoretical density is possible to check the internal porosity, that results, respectively, for the PC and SLM process of 2.14 and 1.80%. It can be concluded that the SLM consolidation process produces samples with lower internal porosity, and can obtain components with densities around 98.20% of theoretical, in contrast to the PC process that obtains components with density of 97.86% of theoretical.

The heating curves obtained from TMA in the form of PC and SLM sample are presented in **Figure 10**. To understand the events occurred at TMA, the heating curves of DSC analysis were juxtaposed. The CTE for the consolidated samples has a different behavior between the processes of the Co-Cr-Mo alloy in a similar analysis to the present study. The CTE for the consolidated samples has a distinct behavior between the processes of the Co-Cr-Mo alloy. As it is possible to verify the CTE at temperature of 500°C of samples (SLM 1, SLM 2, and PC) is, respectively, 15.0/19.5/22.0 × 10^{-6} °C^{-1}. At the temperature of 600°C, the coefficient value decreases to the values of 12.5/14.5/18.5 × 10^{-6} °C^{-1}. This difference is greater for the

Sample	Densities (g/cm³)		
	Theoretical	Helium pycnometry	Volumetric
PC	8.42	8.24 ± 0.01	8.22 ± 0.10
SLM	8.39	8.24 ± 0.01	8.23 < 0.01

Table 4. Results of densities (medium values): theoretical, Helium pycnometry and volumetric for samples PC and SLM.

alloy processed by PC, which is associated to the dendritic microstructure formed and in relation to the casting process because it characterizes a fine grain microstructure and more packaging.

As can be seen in the TMA and DSC curves are similar in the temperature ranges of events. Facchini [60] shows the DSC and TMA curves (heating rate of 20°C/min) performed in an ASTM F75-12 composition alloy and processed by electron beam fusion (EBM). EBM has an effect similar to the SLM because both processes have a concentrated heat source and is possible to relate the DSC analysis, in which, the first peak (565–900°C) is associated with the transition from the FCC (αCo) phase to the HCP (εCo) phase, and the second peak (900–1000°C) reduces the HCP phase (εCo) and reappearance of the FCC phase (αCo).

Is possible observed two events, the first event occurs in the range of 514–614°C and the second event at 923–961°C. In the case of the PC sample (TMA curve), it is possible to verify that the events occur in a higher temperature in relation to the samples processed by SLM. This occurrence is associated to the microstructure samples of analysis performed. The powder

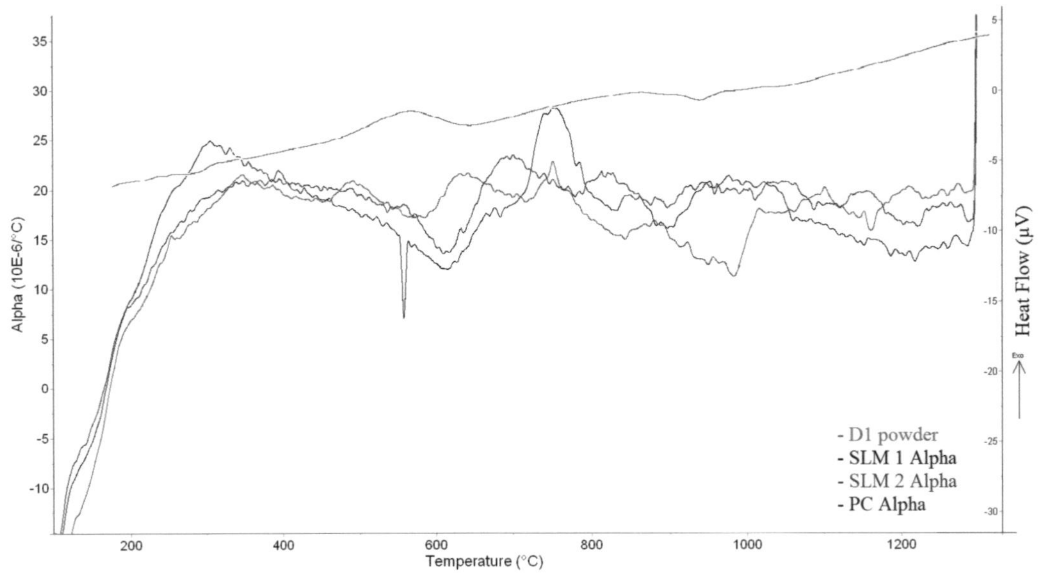

Figure 10. Heat curve of thermal analysis for Co-Cr-Mo samples: DSC curve of Co-Cr-Mo powder sample and TMA curves of consolidated samples PC and SLM (SLM 1 and SLM 2).

sample (at DSC analysis) has a dendritic microstructure, as the same of the PC sample (at TMA). TMA curves of SLM samples (SLM 1 and SLM 2) show more evident peaks, but the transition of these events set at increased temperatures. Thus, the laser fusion process has a refined and more homogeneous microstructure what hinders phase transitions and requires higher temperatures.

3.3. Mechanical behavior

The mechanical results of the tests for the PC and SLM specimens are present in **Table 5**. Analyzing the values is possible to verify that in all properties the SLM technique results in higher properties than PC technique. According to standard ISO22674:06 [54], the SLM and PC specimens satisfied the type 5 criteria in all mechanical properties.

The result of TRS samples (SLM = 2501.2 ± 9.7 MPa and PC = 1072.3 ± 4.6 MPa) was satisfactory. However, there was no rupture of PC sample (test interrupted) evidenced the ductility of the precision casting process, that was confirmed by the value of higher elongation. According to Mengucci et al. [62], the TRS result for a similar composition of Co-Cr-Mo alloy, after the shoot-peened treatment followed by heat treatment for strain relief, resulted a TRS equal to 2700 ± 25 MPa. Therefore, the present results are acceptable comparing the data obtained with the study by Mengucci et al. [62].

The heat treatment, as hot isostatic pressing (HIP), after the additive manufacturing process (by laser melting process–SLM and EBM) of parts has been used successfully by medical and aeronautic manufactures. The HIP process is effective to obtain better results of mechanical properties (ductility and fatigue resistance) and decrease the porosity [60, 63]. Although this present study evaluated the mechanical properties of SLM and PC samples without any post-process of heat treatment is possible, check the relevant mechanical properties obtained by manufacture process. According to the results present evaluation is presented in **Table 6** to compare the mechanical properties to those presented in the literature [14, 41, 44, 62, 64].

Mechanical properties	Consolidation technique		Standards	
	PC	SLM	ISO 22674 "type 5"	ASTM F75 "casting"
Yield strength (MPa)	646.7 ± 44.4	731.5 ± 40.3	500	450
Rupture strength (MPa)	742.2 ± 106.8	1127.9 ± 0.1	–	–
Ultimate tensile strength (MPa)	771.7 ± 103.3	1136.9 ± 1.0	–	655
Elongation (%)	14.20 ± 2.8	13.7 ± 5.3	2	8
Elastic modulus "E" (GPa)	223.42 ± 15.7	225.2 ± 14.4	150	–
Hardness Vickers (HV)	272.2 ± 20.5	334.8 ± 16.0	–	266–345
TRS (MPa)	1072.3 ± 4.6	2501.2 ± 9.7	–	–

Table 5. Mechanical properties of Co-Cr-Mo alloy manufactured by PC and SLM compared to the minimum properties required by standards.

References		Alloy (wt%)	σ_{YS} (MPa)	El (%)	σ_{UTS} (MPa)	TRS (MPa)	Hardness (HV)
Present work Co-Cr-Mo	PC	64Co-29Cr-7Mo	646.76 ± 44.36	14.20 ± 2.76	771.70 ± 103.32	1072.3 ± 4.6	256.7 ± 12.9
	SLM		731.50 ± 40.31	13.73 ± 5.32	1136.95 ± 0.92	2501.2 ± 9.7	358.1 ± 9.8
Takaichi et al. [64]	C	Co-28Cr-6Mo	296 ± 25	9.6 ± 2.5	912 ± 39	–	–
	SLM		516 ± 28	10.7 ± 2.9	591 ± 37	–	–
Qian et al. [41]	C	60-65Co 26-30Cr 5-7Mo	610	–	741	–	–
	SLM		873 ± 76	–	1303 ± 73	–	–
Kajima et al. [44]	C	63Co-29Cr-6Mo	571 ± 23	11.2 ± 2	775 ± 67	–	–
	SLM	60-65Co 26-30Cr 5-7Mo	877 ± 37	12.3 ± 3	1170 ± 29	–	–
Mengucci et al. [62]	SLM	63.8Co-24.7Cr-5.1Mo-5.4W	–	–	1340 ± 20	2700 ± 25	434 ± 22
Jabbari et al. [14]	C	61.6Co-30Cr-6.5Mo	–	–	–	–	320 ± 12
	SLM	Co-29Cr-5.5Mo	–	–	–	–	371 ± 10
Liverani et al. [40]	SLM	Co 27-30Cr 5-7Mo	677	–	–	–	361 ± 31

Table 6. Comparative results for mechanical properties obtained in the present study with Co-Cr alloys manufactured by selective laser melting (SLM) and casting (C) process presented in the literature.

3.4. Microstructural analysis

To understand the mechanical properties improved in the SLM specimens in relation to the casting process technique carried out the microstructural analysis by OM and SEM-EDS.

The microstructural analysis by OM of PC samples (**Figure 11a,b**) describe dendritic arms and ramifications with different solidification orientations [14]. In addition, PC sample present porous (microporous) as the SLM samples, but are uneven (a little larger but in small quantity). This occurrence is possible to form by problems of dispersing the powder in the bed layer and the presence of satellites/porous in the powder particles. SLM specimens show a characteristic morphology (weld-like structure) of laser beam melting. Is possible to check the layers formed during the manufacture process (**Figure 11c,d**)? The vertical section of SLM sample is to observe the building direction of specimen (indicated by arrow—**Figure 11d**) characterized by the overlapping of each layer and the morphology formation by the action of the laser beam such as the weld pool.

Figure 12 represents the SEM images of PC samples and the semi-quantitative analysis of interesting points by EDS. It is possible to identify the cast specimen with a second phase (white area) in the matrix. The semi-quantitative analysis with the EDS and the respective spectrums (**Figure 12c,d**) show that the composition of white area (point 1) is rich in Mo element, and the matrix (point 2) is composed by Co-Cr elements, with a small percentage of Mo. The phase (point 1) shows the confirmation of carbide ($M_{23}C_6$) presence, rich in chromium and

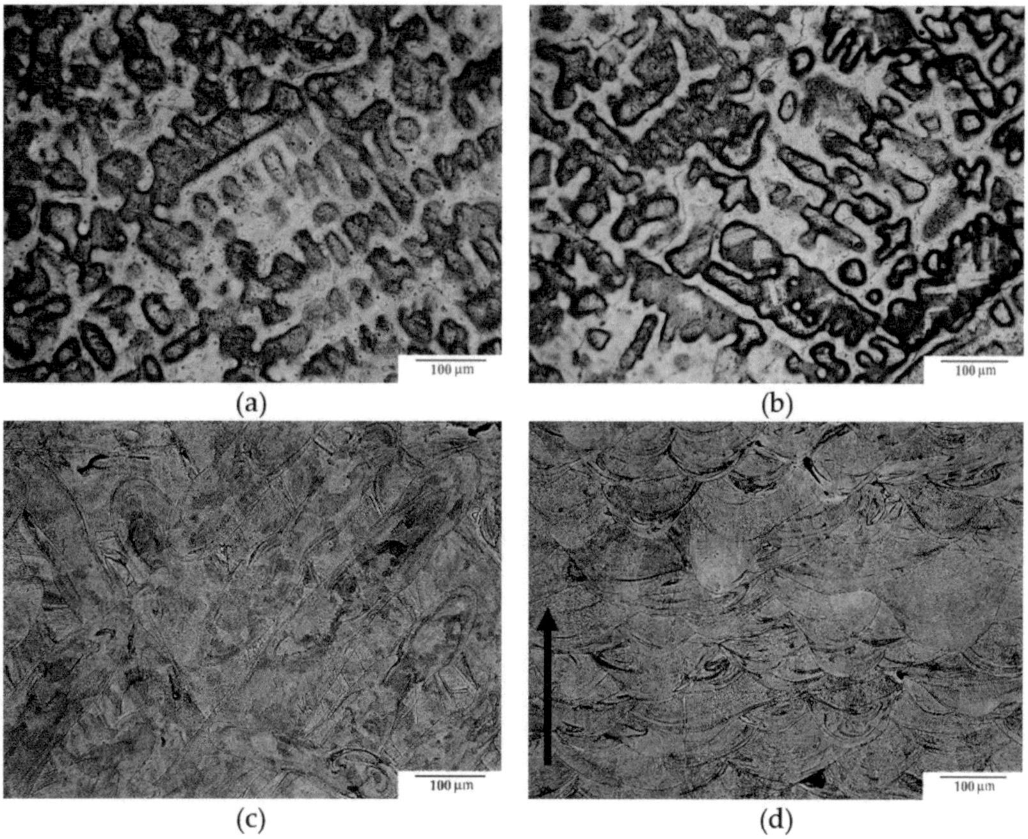

Figure 11. MO images of Co-Cr-Mo alloy consolidated: (a–b) transversal and longitudinal view of precision casting sample and (c–d) transversal and longitudinal (arrow indicates building consolidation) view of selective laser melting (Etch: 100 ml HCl and 2 ml H_2O_2. Magnitude: ×200).

molybdenum [14]. The $M_{23}C_6$ carbide results in a micro hardness of 699 ± 131 HV (1 mN/15 s) in opposition of 338 ± 14 HV (1 mN/15 s) to the micro hardness of matrix.

Figure 13 shows the microstructure of SLM specimen. It is observed that a microstructure is formed with small grains characterizing the rapid solidification during the SLM manufacturing process. The semi-quantitative analysis in the fine grains shows that it does not have different elements compositions. SLM specimen presents a homogeneous matrix with Co-Cr-Mo elements. The morphology formation of the laser melting sample was also observed after electrolytic attack [59, 61]. Also confirms that the fine grains are oriented in direction of the laser scanning. This characteristic microstructure of laser melting technique allows to achieve better mechanical properties than the cast technique.

The fractures of the tensile samples were SEM analyzed (**Figure 14**) observed the formation of dimples homogeneously distributed in the microfracture of both samples. Regions with the presence of dimples are ductless and with higher toughness. However, it is apparent that the dimple formations on the SLM sample extends completely by the fracture planes and are of

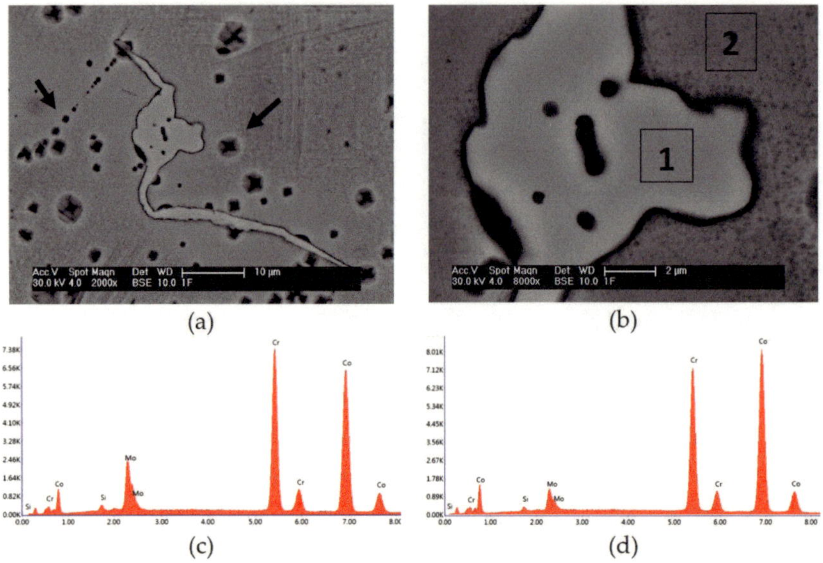

Figure 12. SEM images of PC sample: (a) magnitude ×2000, (b) magnitude ×8000, and (c–d) EDS spectrograms of analysis at point 1 and point 2.

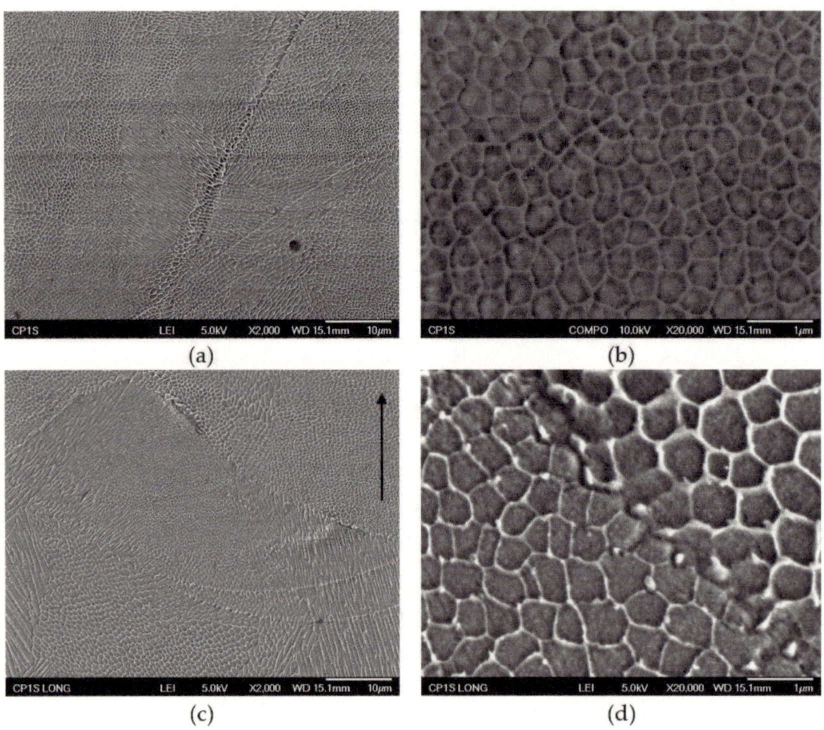

Figure 13. SEM images of SLM specimens: (a) horizontal section from backscattered electrons, and (b) from secondary electrons, (c–d) vertical section from backscattered electrons (black arrow indicates the building consolidation).

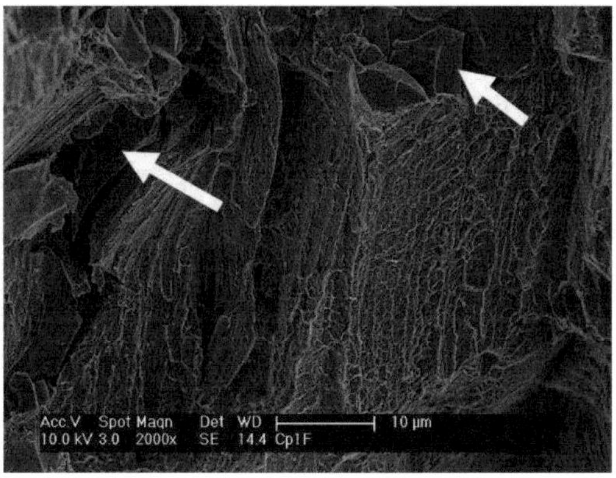

(a)

(b)

Figure 14. SEM images of tensile fracture: (a) PC sample and (b) SLM sample (magnitude: ×2000).

finer size, compared to the fracture with dimples geometrically larger of PC sample. It can be verified with SLM samples and confirmed with the mechanical results in relation to PC samples. In addition, some planar regions (indicated by arrows) show a semi-cleavage morphology. The type of fracture observed in the samples, according to Takaichi et al. [64], describes the formation of dimples along the fracture surface, as well as cracking of the wedge is appointed as a possible formation of cleavage fracture over favorable crystallographic planes.

4. Conclusions

In general, the results of powders characterization showed that the granulometric range of 20–50 μm is the one that best fits in the properties of packaging, for the consolidation by SLM.

The biocompatibility of the samples obtained a positive result for both processing techniques. In this way, the development of the present study evidenced to improve the manufacture of customized dental components (copings) using the SLM technique.

Microstructural analysis obtained for SLM samples results in a characteristic morphology of layer manufacturing with ultrafine grains and a high chemical homogeneity. The conventional technique presented a differentiated microstructure by the gross dendritic microstructure of casting process.

The mechanical evaluation showed that the SLM technique provides superior mechanical properties (as yield strength, rupture strength, ultimate tensile strength, TRS, and hardness) compared to those obtained by the precision casting technique.

The thermal analyses showed the present phase transitions of Co-Cr-Mo alloy, as well as being possible to correlate them (TMA to DSC curves). The coefficient of thermal expansion (CTE) resulted for both processes a similar value to alloys used in dental materials.

The processing using laser melting proved better mechanical and thermal properties to precision casting processing technique without post-processing (thermal treatment). SLM technique evidenced a promising use to manufacture prosthetics and dental implants. Nevertheless, it is still of great concern and promising further development of laser melting process (SLM and EBM) in relation to the parameters and variables of process, as also to the post-processing method apply to AM parts. Such characteristics should be addressed to new materials and investigate in relation to the performance and bio-functionality of specific application part.

Acknowledgements

This study was financially support by CNPq and FAPESP. The authors also thank to Ms. Amed Belaid and SLM® Solution for SLM specimens and collaboration.

Author details

Marcello Vertamatti Mergulhão*, Carlos Eduardo Podestá and
Maurício David Martins das Neves

*Address all correspondence to: marcellovertamatti@gmail.com

Nuclear and Energy Research Institute (IPEN/CNEN-SP), CCTM, São Paulo, Brazil

References

[1] Davis JR, editor. ASM Specialty Handbook: Nickel, Cobalt, and Their Alloys. 1st ed. ASM International; Materials Park, OH, 2000. p. 442

[2] Lee PW, editor. Powder Metal Technologies and Applications. Vol. 7. 9th ed. ASM International; Materials Park, OH, 1998. p. 1147

[3] Misch CE. Dental Implant Prosthetics. 2nd ed. Elsevier; Missouri, 2015. p. 1008

[4] Ivanova EP, Bazaka K, Crawford RJ. Cytotoxicity and biocompatibility of metallic biomaterials. In: New Funct. Biomater. Med. Healthc. Elsevier; Cambridge, 2014. pp. 148-172. DOI: 10.1533/9781782422662.148

[5] Ratner BD, Hoffman AS, Schoen FJ, Lemons JE. Biomaterials Science: An Introduction to Materials in Medicine. 2nd ed. Academic Press; San Diego, California, 2004. p. 864. ISBN: 9780080470368

[6] McCabe JF, Walls AWG, editors. Applied Dental Materials. 9th ed. Blackwell Publishing Ltd.; Oxford, 2008. p. 312. ISBN: 978-1-118-69712-2

[7] Wataha JC. Biocompatibility of dental casting alloys: A review. The Journal of Prosthetic Dentistry. 2000;**83**:223-234. DOI: 10.1016/S0022-3913(00)80016-5

[8] Kim HR, Kim YK, Son JS, Min BK, Kim KH, Kwon T-Y. Comparison of in vitro biocompatibility of a Co-Cr dental alloy produced by new milling/post-sintering or traditional casting technique. Materials Letters. 2016;**178**:300-303. DOI: 10.1016/j.matlet.2016.05.053

[9] Hedberg YS, Qian B, Shen Z, Virtanen S, Wallinder IO. In vitro biocompatibility of CoCrMo dental alloys fabricated by selective laser melting. Dental Materials. 2014;**30**:525-534. DOI: 10.1016/j.dental.2014.02.008

[10] Borelli V. Pesquisa e desenvolvimento de biomateriais: estudo das inter-relações científicas, tecnológicas e normativas. [Master]. São Paulo: Instituto de Pesquisas Energéticas e Nucleares (IPEN/USP); 2011

[11] Williams DF. On the mechanisms of biocompatibility. Biomaterials. 2008;**29**:2941-2953. DOI: 10.1016/j.biomaterials.2008.04.023

[12] Niinomi M, Narushima T, Nakai M, editors. Advances in Metallic Biomaterials. Vol. 3. Berlin, Heidelberg: Springer Berlin Heidelberg; 2015. p. 348. DOI: 10.1007/978-3-662-46836-4

[13] ASTM F75-12. Standard Specification for Cobalt-28 Chromium-6 Molybdenum Alloy Castings and Casting Alloy for Surgical Implants (UNS R30075). West Conshohocken, PA: ASTM International; 2012

[14] Jabbari YSA, Koutsoukis T, Barmpagadaki X, Zinelis S. Metallurgical and interfacial characterization of PFM Co-Cr dental alloys fabricated via casting, milling or selective laser melting. Dental Materials. 2014;**30**:e79-e88. DOI: 10.1016/j.dental.2014.01.008

[15] Oyagüe RC, Sánchez-Turrión A, López-Lozano JF, Montero J, Albaladejo A, Suárez-García MJ. Evaluation of fit of cement-retained implant-supported 3-unit structures fabricated with direct metal laser sintering and vacuum casting techniques. Odontology. 2012;**100**:249-253. DOI: 10.1007/s10266-011-0050-1

[16] Santos LA. Processamento e caracterização da liga 66Co-28Cr-6Mo (% peso) para implantes. [Master]. São Paulo: Universidade de São Paulo; 2012

[17] Craig RG. Craig's Restorative Dental Materials. 13th ed. Elsevier/Mosby; Philadelphia, 2012. p. 416

[18] Ren L, Memarzadeh K, Zhang S, Sun Z, Yang C, Ren G, et al. A novel coping metal material CoCrCu alloy fabricated by selective laser melting with antimicrobial and antibiofilm properties. Materials Science and Engineering C. 2016;**67**:461-467. DOI: 10.1016/j.msec.2016.05.069

[19] van Noort R. The future of dental devices is digital. Dental Materials. 2012;**28**:3-12. DOI: 10.1016/j.dental.2011.10.014

[20] Utela B, Storti D, Anderson R, Ganter M. A review of process development steps for new material systems in three dimensional printing (3DP). Journal of Manufacturing Processes. 2008;**10**:96-104. DOI: 10.1016/j.jmapro.2009.03.002

[21] Örtorp A, Jönsson D, Mouhsen A, von Steyern PV. The fit of cobalt-chromium three-unit fixed dental prostheses fabricated with four different techniques: A comparative in vitro study. Dental Materials. 2011;**27**:356-363. DOI: 10.1016/j.dental.2010.11.015

[22] Li KC, Prior DJ, Waddell JN, Swain MV. Comparison of the microstructure and phase stability of as-cast, CAD/CAM and powder metallurgy manufactured Co-Cr dental alloys. Dental Materials. 2015;**31**:e306-e315. DOI: 10.1016/j.dental.2015.10.010

[23] Bilgin MS, Erdem A, Dilber E, Ersoy İ. Comparison of fracture resistance between cast, CAD/CAM milling, and direct metal laser sintering metal post systems. Journal of Prosthodontic Research. 2016;**60**:23-28. DOI: 10.1016/j.jpor.2015.08.001

[24] Kale PJ, Metkar RM, Hiwase SD. Development and optimization of dental crown using rapid prototyping integrated with CAD. In: Wimpenny DI, Pandey PM, Kumar LJ, editors. In: Advanced 3D Printing and Additive Manufacturing Technology. Singapore: Springer Singapore; 2017. pp. 169-182. DOI: 10.1007/978-981-10-0812-2_15

[25] Shah P, Racasan R, Bills P. Comparison of different additive manufacturing methods using computed tomography. Case Studies in Nondestructive Testing and Evaluation. 2016;**6**:69-78. DOI: 10.1016/j.csndt.2016.05.008

[26] Gibson I. Advanced Manufacturing Technology for Medical Applications: Reverse Engineering, Software Conversion and Rapid Prototyping. John Wiley & Sons Ltd, Chichester, UK, 2005

[27] Grzesiak D, Krawczyk M. Effects of the selective laser melting process parameters on the functional properties of the Co-Cr alloy. International Journal of Recent Contributions from Engineering, Science & IT (iJES). 2015;**3**:39-42. DOI: 10.3991/ijes.v3i1.4291

[28] Germanovix AA. Establishing a benchmark part to analyze the capabilities of selective laser melting systems. [Bacharel]. Universidade Federal de Santa Catarina: Santa Catarina, Brasil, 2011

[29] Bremen S, Meiners W, Diatlov A. Selective laser melting: A manufacturing technology for the future? Laser Technik Journal. 2012;**9**:33-38. DOI: 10.1002/latj.201290018

[30] Yap CY, Chua CK, Dong ZL, Liu ZH, Zhang DQ, Loh LE, et al. Review of selective laser melting: Materials and applications. Applied Physics Reviews. 2015;**2**:041101. DOI: 10.1063/1.4935926

[31] Meiners W. Selective Laser Melting: Generative Fertigung fur die Produktion der Zukunft Optische Technologien in der Produktionstechnik. Aachen: Fraunhofer Institut fur Lasertechnik: 2012

[32] Gu DD, Meiners W, Wissenbach K, Poprawe R. Laser additive manufacturing of metallic components: Materials, processes and mechanisms. International Materials Reviews. 2012;**57**:133-164. DOI: 10.1179/1743280411Y.0000000014

[33] Kurzynowski T, Chlebus E, Kuźnicka B, Reiner J. Parameters in selective laser melting for processing metallic powders. In: Beyer E, Morris T, editors; 2012. pp. 823-914. DOI: 10.1117/12.907292

[34] Sallica-Leva E, Jardini AL, Fogagnolo JB. Microstructure and mechanical behavior of porous Ti-6Al-4V parts obtained by selective laser melting. Journal of the Mechanical Behavior of Biomedical Materials. 2013;**26**:98-108. DOI: 10.1016/j.jmbbm.2013.05.011

[35] Simchi A, Pohl H. Effects of laser sintering processing parameters on the microstructure and densification of iron powder. Materials Science and Engineering A. 2003;**359**:119-128. DOI: 10.1016/S0921-5093(03)00341-1

[36] Calignano F, Manfredi D, Ambrosio EP, Biamino S, Lombardi M, Atzeni E, et al. Overview on additive manufacturing technologies. Proceedings of IEEE. 2017:1-20. DOI: 10.1109/JPROC.2016.2625098

[37] Zhou X, Liu X, Zhang D, Shen Z, Liu W. Balling phenomena in selective laser melted tungsten. Journal of Material Processing Technology. 2015;**222**:33-42. DOI: 10.1016/j.jmatprotec.2015.02.032

[38] Gu D, Shen Y. Balling phenomena in direct laser sintering of stainless steel powder: Metallurgical mechanisms and control methods. Materials & Design. 2009;**30**:2903-2910. DOI: 10.1016/j.matdes.2009.01.013

[39] Senthilkumaran K, Pandey PM, Rao PVM. Influence of building strategies on the accuracy of parts in selective laser sintering. Materials & Design. 2009;**30**:2946-2954. DOI: 10.1016/j.matdes.2009.01.009

[40] Liverani E, Fortunato A, Leardini A, Belvedere C, Siegler S, Ceschini L, et al. Fabrication of Co-Cr-Mo endoprosthetic ankle devices by means of Selective Laser Melting (SLM). Materials & Design. 2016;**106**:60-68. DOI: 10.1016/j.matdes.2016.05.083

[41] Qian B, Saeidi K, Kvetková L, Lofaj F, Xiao C, Shen Z. Defects-tolerant Co-Cr-Mo dental alloys prepared by selective laser melting. Dental Materials. 2015;**31**:1435-1444. DOI: 10.1016/j.dental.2015.09.003

[42] Cheng B, Shrestha S, Chou K. Stress and deformation evaluations of scanning strategy effect in selective laser melting. Additive Manufacturing. 2016;**12**:240-251. DOI: 10.1016/j.addma.2016.05.007

[43] Alsalla H, Hao L, Smith C. Fracture toughness and tensile strength of 316L stainless steel cellular lattice structures manufactured using the selective laser melting technique. Materials Science and Engineering A. 2016;**669**:1-6. DOI: 10.1016/j.msea.2016.05.075

[44] Kajima Y, Takaichi A, Nakamoto T, Kimura T, Yogo Y, Ashida M, et al. Fatigue strength of Co-Cr-Mo alloy clasps prepared by selective laser melting. Journal of the Mechanical Behavior of Biomedical Materials. 2016;**59**:446-458. DOI: 10.1016/j.jmbbm.2016.02.032

[45] Simonelli M, Tse YY, Tuck C. Effect of the build orientation on the mechanical properties and fracture modes of SLM Ti-6Al-4V. Materials Science and Engineering A. 2014;**616**:1-11. DOI: 10.1016/j.msea.2014.07.086

[46] Džugan J, Nový Z. Powder Application in Additive Manufacturing of Metallic Parts. In: Dobrzanski LA, editor. Powder Metallurgy - Fundamentals and Case Studies [Internet]. InTech; Rijeka, Croatia, 2017

[47] Sun YY, Gulizia S, Oh CH, Doblin C, Yang YF, Qian M. Manipulation and characterization of a novel titanium powder precursor for additive manufacturing applications. JOM. 2015;**67**:564-572. DOI: 10.1007/s11837-015-1301-3

[48] Tang HP, Qian M, Liu N, Zhang XZ, Yang GY, Wang J. Effect of powder reuse times on additive manufacturing of Ti-6Al-4V by selective electron beam melting. JOM. 2015;**67**:555-563. DOI: 10.1007/s11837-015-1300-4

[49] Slotwinski JA, Garboczi EJ, Stutzman PE, Ferraris CF, Watson SS, Peltz MA. Characterization of metal powders used for additive manufacturing. Journal of Research of the National Institute of Standards and Technology. 2014;**119**:460. DOI: 10.6028/jres.119.018

[50] Gaytan SM, Murr LE, Medina F, Martinez E, Lopez MI, Wicker RB. Advanced metal powder based manufacturing of complex components by electron beam melting. Materials Technology. 2009;**24**:180-190. DOI: 10.1179/106678509X12475882446133

[51] ASTM International. B212-13—Standard Test Method for Apparent Density of Free-Flowing Metal Powders Using the Hall Flowmeter Funnel. Pensilvânia: ASTM; 2013

[52] ASTM International. B213-13—Standard Test Methods for Flow Rate of Metal Powders Using the Hall Flowmeter Funnel. Pensilvânia: ASTM; 2013

[53] ASTM International. B527-14—Standard Test Method for Determination of Tap Density of Metal Powders and Compounds. Pensilvânia: ASTM; 2014

[54] ISO. 22674-06—Dentistry—Metallic Materials for Fixed and Removable Restorations and Appliances. Geneva: ISO, 2006

[55] ASTM International. B528-12—Standard Test Method for Transverse Rupture Strength of Powder Metallurgy (PM) Specimens. Pensilvânia: ASTM; 2012

[56] ISO. 10933-5—Biological Evaluation of Medical Devices—Part 5: Tests for Cytotoxicity: In Vitro Methods; 1995

[57] German RM. Powder Metallurgy Science. 2nd ed. Princeton: Metal Powder Industry Federation; 1994

[58] Gessinger GH. Powder Metallurgy of Superalloys. Baden, Switzerland: Butterworth & Co.; 1984

[59] Haan J, Asseln M, Zivcec M, Eschweiler J, Radermacher R, Broeckmann C. Effect of subsequent hot isostatic pressing on mechanical properties of ASTM F75 alloy produced by Selective Laser Melting. Powder Metallurgy. 2015;**58**:161-165. DOI: 10.1179/0032589915Z.000000000236

[60] Facchini L. Microstructure and Mechanical Properties of Biomedical Alloys produced by Rapid Manufacturing Techniques. Doutorado: University of Trento; 2010

[61] Xin XZ, Xiang N, Chen J, Wei B. In vitro biocompatibility of Co-Cr alloy fabricated by selective laser melting or traditional casting techniques. Materials Letters. 2012;**88**:101-103. DOI: 10.1016/j.matlet.2012.08.032

[62] Mengucci P, Barucca G, Gatto A, Bassoli E, Denti L, Fiori F, et al. Effects of thermal treatments on microstructure and mechanical properties of a Co-Cr-Mo-W biomedical alloy produced by laser sintering. Journal of the Mechanical Behavior of Biomedical Materials. 2016;**60**:106-117. DOI: 10.1016/j.jmbbm.2015.12.045

[63] Benedetti M, Torresani E, Leoni M, Fontanari V, Bandini M, Pederzolli C, et al. The effect of post-sintering treatments on the fatigue and biological behavior of Ti-6Al-4V ELI parts made by selective laser melting. Journal of the Mechanical Behavior of Biomedical Materials. 2017;**71**:295-306. DOI: 10.1016/j.jmbbm.2017.03.024

[64] Takaichi A, Suyalatu, Nakamoto T, Joko N, Nomura N, Tsutsumi Y, et al. Microstructures and mechanical properties of Co-29Cr-6Mo alloy fabricated by selective laser melting process for dental applications. Journal of the Mechanical Behavior of Biomedical Materials. 2013;**21**:67-76. DOI: 10.1016/j.jmbbm.2013.01.021

Application of 3-D Printing for Tissue Regeneration in Oral and Maxillofacial Surgery: What is Upcoming?

Seied Omid Keyhan, Hamidreza Fallahi,
Alireza Jahangirnia,
Mohammad Taher Amirzade-Iranaq and
Mohammad Hosein Amirzade-Iranaq

Additional information is available at the end of the chapter

http://dx.doi.org/10.5772/intechopen.70323

Abstract

The ultimate goal of any surgical procedure is to improve perioperative form and function and to minimize operative and postoperative morbidity. In recent years, many exciting and novel technological advances have been introduced in the field of oral and maxillofacial surgery. One example of such technology that is continuing to increase in prevalence is the use of 3-dimensional (3-D) printing techniques with special properties, which seems hopeful for practitioners in the field of regenerative medicine. Tissue engineering is a critical and important area in biomedical engineering for creating biological alternatives for grafts, implants, and prostheses. One of the main triad bases for tissue engineering is scaffolds, which play a great role for determining growth directions of stem cells in a 3-dimensional aspect. Mechanical strength of these scaffolds is critical as well as interconnected channels and controlled porosity or pores distribution. However, existing 3-D scaffolds proved less than ideal for actual clinical applications. In this chapter, we review the application and advancement of rapid prototyping (RP) techniques in the design and creation of synthetic scaffolds for use in tissue engineering. Also, we survey through new and novel merging era of "bioprinting."

Keywords: 3-D printing, prototyping, tissue engineering, scaffolds, bioprinting, stem cells, regenerative medicine, oral surgery, maxillofacial surgery

1. Introduction

Three-dimensional printing—also known as rapid prototyping—was first introduced in 1980s; during past three decades, enormous changes and development have been performed by scientists through modifying this technology by uses, material, and also accuracy.

With increasing attention of scientific societies, recently, scientific literature bolded feasibility of 3-D-printed tissues and organs and its usage within laborious clinical situations. Also, this technology was used largely in accurate and highly customized devices, such as tracheobronchial splints, bionic ears, and even more. Within the field of craniofacial surgery, 3-D surgical models have been used as templates to create bone grafts, tailoring bioprosthetic implants, plate bending, cutting guides for osteotomies, and intraoperative oral splints. Using 3-D models and guides has been shown to shorten operative time and potentially reduce the complications associated with prolonged operative times.

The goal of surgical procedures for a clinician is to improve perioperative form, recovery of function, and also minimizing operative and postoperative morbidity. Many exciting and new technological advances have ushered in a new era in the field of oral and maxillofacial surgery over the last years, which within no exaggeration 3-D printing is the novelist and controversial one.

The aim of this chapter is to introduce 3-D printing method and its role in the contemporary oral and maxillofacial surgery and to review current advantages of its application in the field of regenerative medicine.

1.1. History of the technology

Three-dimensional (3-D) printing has been utilized in diverse aspects of manufacturing to produce different objects from guns, boats, and food to models of unborn babies. From over 1450 articles related to 3-D printing listed in PubMed, nearly a third of them were solely published in the last 2 years [1].

Three-dimensional (3-D) printing is a manufacturing process that objects are fabricated in a layering method during fusing or depositing different materials such as plastic, metal, ceramics, powders, liquids, or even living cells to build a 3-D matter [2, 3]. It is a process of generating physical models from digital layouts [4, 5]. This technology demonstrates a technique that a product designed through a computer-aided scheme is manufactured in a layer-by-layer system [6]. This process is also cited as rapid prototyping (RP), solid freeform technology (SFF), or additive manufacturing (AM) [7].

3-D printing techniques are not brand new and have been existed since 30 years ago [8–10]. This technology is first introduced and invented by Charles Hull in 1986, and at first, it was utilized in the engineering and automobile industry for manufacturing polyurethane frameworks for different models, pieces, and instruments [11]. Originally, Hull employed the phrase "stereolithography" in his US Patent 4,575,330, termed "Apparatus for Production of Three—Dimensional Objects by Stereolithography" published in 1986. Stereolithography technique included subjoining layers over the top of each other, by curing photopolymers with UV lasers [12, 13].

Since then, 3-D models have been used for a diversity of different objectives. Since 1986, this process has started to accelerate and has honored recognition globally and has influenced different arenas, such as medicine.

The developing agora for 3-D desktop printers encourages wide-ranging experimentations in that subject. Generally, medical indications of these printers are such as treatment planning, prosthesis, implant fabrications, medical training, and other usages [4].

Having being used in military, food industry, and art, rapid prototyping is receiving a lot of attention in the field of surgery in the last 10 years [6, 14].

The pioneering usage of stereolithography in oral and maxillofacial surgery was by Brix and Lambrecht in 1985. Later this technique was used by them for treatment planning in craniofacial surgery [15].

In 1990, stereolithography was used by Mankovich et al. for treating patients having craniofacial deformities [16, 17]. They used it to simulate bony anatomy of the cranium using computed tomography with complete internal components [17, 18].

By aiding in complex craniofacial reconstructions, 3-D printing has recently earned reputation in medicine and surgical fields [19–21].

Today, maxillofacial surgery can benefit from additive manufacturing in various aspects and different clinical cases [22]. This technique can help with bending plates, manufacturing templates for bone grafts, tailoring implants, osteotomy guides, and intraoperative occlusal splints [23–27]. Rapid prototyping can shorten surgery duration and simplify pre and intraoperative decisions. It has enhanced efficacy and preciseness of surgeries [10].

2. Current 3-D printing techniques used in oral and maxillofacial surgery

From first innovation till nowadays, there are different kind of technologies introduced for 3-D printing. Binder jetting (BJ), electron beam melting (EBM), fused deposition modeling (FDM), indirect processes, laser melting (LM), laser sintering (LS), material jetting (MJ), photopolymer jetting (PJ), and stereolithography (SL) are well-known technologies of 3-D printing [14, 28, 29].

There are many different 3-D printing techniques. Benefits and disadvantages are factors to differ each technology system [14]. Among this variety of different techniques, there is a huge discussion and usage in oral and maxillofacial region for SL, FDM, and PJ [1, 28, 30].

Each technology has its own characteristics, properties, and advantages which **Table 1** summarizes some different three dimensional printing technologies.

3. Biomaterials available for 3-D printing

As researchers aim to investigate new materials for 3-D printing in last decade, it is obvious to see variety of biomaterials with different properties and also different applications. As **Table 2** summarizes all biomaterials used within studies all over the world for generating scaffolds for bone tissue engineering, it has to be noticed that from this large spectrum of biomaterials

Techniques	Advantages	Disadvantages
Light cured resin		
1. Stereolithography (SLA) Light sensitive polymer cured layer by layer by a scanning laser in a vat of liquid polymer.	Rapid fabrication. Able to create complex shapes with high feature resolution. Lower cost materials if used in bulk.	Only available with light curable liquid polymers. Support materials must be removed. Resin is messy and can cause skin sensitization and may irritate by contact and inhalation. Limited shelf life and vat life. Cannot be heat sterilized. High cost technology.
2. Photojet—light sensitive polymer is jetted onto a build platform from an inkjet type print head, and cured layer by layer on an incrementally descending platform.	Relatively fast. High resolution, high-quality finish possible. Multiple materials available various colors and physical properties including elastic materials. Lower cost technology.	Tenacious support material can be difficult to remove completely. Support material may cause skin irritation. Cannot be heat sterilized. High cost materials.
3. DLP (digital light processing) Liquid resin is cured layer by layer by a projector light source. The object is built upside down on an incrementally elevating platform.	Good accuracy, smooth surfaces, relatively fast. Lower cost technology.	Light curable liquid polymers and wax-like materials for casting. Support materials must be removed. Resin is messy and can cause skin sensitization, and may be irritant by contact Limited shelf life and vat life. Cannot be heat sterilized. Higher cost materials.
Powder binder		
Plaster or cementaceous material set by drops of (colored) water from 'inkjet' print head. Object built layer by layer in a powder bed, on an incrementally descending platform.	Lower cost materials and technology. Can print in color. Un-set material provides support Relatively fast process. Safe materials.	Low resolution. Messy powder. Low strength. Cannot be soaked or heat sterilized.
Sintered powder		
Selective laser sintering (SLS) for polymers. Object built layer by layer in powder bed. Heated build chamber raises temperature of material to just below melting point. Scanning laser then sinters powder layer by layer in a descending bed.	Range of polymeric materials including nylon, elastomers, and composites. Strong and accurate parts. Self-supported process. Polymeric materials—commonly nylon may be autoclaved. Printed object may have full mechanical functionality. Lower cost materials if used in large volume.	Significant infrastructure required, e.g., Compressed air, climate control. Messy powders. Lower cost in bulk. Inhalation risk. High cost technology. Rough surface.
Selective laser sintering (SLS)—for metals and metal alloys. Also described as selective laser melting (SLM) or direct metal laser sintering (DMLS). Scanning laser sinters metal powder layer by layer in a cold build chamber as the build platform descends. Support structure used to tether objects to build platform.	High strength objects can control porosity. Variety of materials including titanium, titanium alloys, cobalt chrome, stainless steel. Metal alloy may be recycled. Fine detail possible.	Elaborate infrastructure requirements. Extremely costly technology moderately costly materials. Dust and nanoparticle condensate may be hazardous to health. Explosive risk. Rough surface. Elaborate post-processing is required: Heat treatment to relieve internal stresses in printed objects. Hard to remove support materials. Relatively slow process.
Electron beam melting (EBM, Arcam). Heated build chamber. Powder sintered layer by layer by scanning electron beam on descending build platform.	High temperature process, so no support or heat treatment needed afterwards. High speed. Dense parts with controlled porosity.	Extremely costly technology moderately costly materials. Dust may be hazardous to health. Explosive risk. Rough surface. Less post-processing required. Lower resolution.

Techniques	Advantages	Disadvantages
Thermoplastic		
Fused deposition modeling (FDM) First 3-DP technology, most used in 'home' printers. Thermoplastic material extruded through nozzle onto build platform.	High porosity. Variable mechanical strength. Low- to mid-range cost materials and equipment. Low accuracy in low cost equipment. Some materials may be heat sterilized.	Low cost but limited materials—only thermoplastics. Limited shape complexity for biological materials. Support material must be removed.

Table 1. 3-D printing modalities and materials [14, 31].

Composed scaffolds	Synthetic scaffolds		Natural scaffolds	
Nano-hydroxyapatite/collagen/PLLA	Ceramic	Polymeric	Inorganic	Organic
Octacalcium phosphate/collagen	Calcium Magnesium Phosphate cement (CMPC)	PLGA	Silver	Collagen sponge
Nano-hydroxyapatite/polyamide 6	βTCP	PLG	Coral	PRP
Nano-hydroxyapatite/polyamide66	HA/TCP	PLLA	Silk fibroin protein	Gelatin sponge
Hydroxyapatite-coated PLGA	Flurohydroxyapatite	PGA	Premineralized silk fibroin protein	Gelatin Hydrogel
HA/PLGA	Ca deficient hydroxyapatite (CDHA)	PLA	ABB	PuraMatrix
βTCP/collagen		PLA-PEG	Deer horn	Alginate
DBM/PLA		Fibronectin-coated PLA		Partially demineralized bone matrix
Nano-hydroxyapatite/polyamide		PEG-DA		Bio-Oss
OsteoSet		PEG-MMP		Allograft
Octacalcium phosphate precipitated (OCP) alginate		PVDC		Fibrin sealant
Demineralized bone powders/PLA		Polycaprolactone		Gelatin foam
Apatite-coated PLGA				Collagen gel
				Hyaluronic acid based hydrogel

TCP, tri-calcium phosphate; HA, hydroxyapatite; DBM, demineralized bone matrix; PLGA, poly(lactic-co-glycolic acid); PLA, poly(D,L-lactic acid); PGA, poly(glycolic acid); PLLA, poly(L-lactic acid); PVDC, polyvinylidene chloride; PEG, polyethylene glycol; DA, diacrylate; MMP, matrix metalloproteinases; ABB, anorganic bovine bone; Puramatrix, a self-assembling peptide nanomaterial.

Table 2. Types of scaffolds used in bone tissue engineering in maxillo-craniofacial region [51].

just a whole bit of them are available for application in 3-D printing. As follows, we discuss four large categories of materials for 3-D printing of scaffolds and craniofacial tissues, which researches still aim to determine these materials complete properties and advantages.

3.1. Polymers and hydrogels

Polymer hydrogels are ideal candidates for the development of printable materials for tissue engineering. Hydrogels are known for remarkable tunability of rheological also presenting great mechanical, chemical, and biological properties; high biocompatibility; and similarity to native extracellular matrix (ECM) [32]. For three-dimensional printing of polymers and hydrogels, the use of materials with controlled viscosity should been noticed. This defines the range of printability of the ink. Polymer inks, which are typically printed in the prepolymer phase, need enough viscosity allowing structural support of subsequent printed layers, also enough fluidity to prevent nozzle clogging. For avoiding these difficulties, alginate hydrogels have been cross-linked with calcium ions immediately before the ink leaves the printing head or just after extrusions [33].

In recent researches, for providing suitable ink for bioprinting applications, prepolymerized cell-laden methacrylated gelatin hydrogels have been used successfully [34, 35]. Synthetic hydrogels used for cell encapsulation may limit cell-cell interactions. These interactions are critical for efficient cell proliferation, differentiation, and finally, tissue development. This can represent one of the limitations of bioprinting cell-laden hydrogels which is not present in 3-D printed scaffolds with cells seeded onto or in bioprinting of dense cell aggregates, which will discuss as follow. Hence, the requirement for the development of ECM-derived hydrogels that have tunable physical and chemical properties, are compatible with high cell viability, and provide the adequate binding sites (RGDs) for cell attachment and matrix remodeling during their early proliferative stage [32].

Synthetic polymers are most commonly used materials for 3-D printing in biomedical applications [36, 37]. However, since high temperature is usually involved during the printing of these materials, the direct incorporation of cells or growth factors in the polymer mixture is generally avoided as the cell viability or bioactivity [37] cannot be maintained throughout the manufacturing process.

Although hydrogels provide great advantages for tissue engineering applications, such as the ability of exposing cells to highly hydrated 3-D microenvironments that is similar to the natural ECM [32]. In contrast, they generally present very low stiffness (in the kPa range) compared with the majority of load-bearing tissues in the craniofacial complex (in the GPa range). Therefore, reconstruction of tissues subjected to higher mechanical loads, such as bones and teeth, usually requires the use of ceramic materials or composite scaffolds which provide more mechanical advantages, where polymers are commonly combined with inorganic fillers to increase scaffold stiffness [38].

3.2. Ceramics

Ceramic scaffolds are usually composed of calcium and phosphate mineral phases, such as hydroxyapatite [39] or b-tricalcium phosphate [40]. The noticeable ability of these scaffolds to upregulate osteogenesis due to inherent properties of the formation of a bioactive ion-rich cellular microenvironment, also as mentioned before their ability to mechanically provide space maintenance, makes these materials interesting choice for 3-D scaffold fabrication for craniofacial applications. In contrast, ceramic scaffolds are not compatible with cell encapsulation for bioprinting. In 3-D printed ceramic scaffolds, cells quickly populate the scaffold surface, which establishing close cell-cell interactions lead to promotion of cell proliferation and differentiation. On the other hand, ceramics with properties lead to lower rates of degradation than hydrogels, which aids in prolonged guided tissue remodeling and structural support. In contrast, ceramic scaffolds are too brittle for implantation in load-bearing defect sites. Ideal scaffolds would combine the high calcium content of calcium and phosphate ceramics with the outstanding toughness of natural bone, which perhaps can only be obtained by creating scaffolds that are biomimetically mineralized and hierarchically structured, as recent researches demonstrated that in [41].

Fused deposition of ceramics (FDC) in a direct printing mode generally consists of extruding a slurry including a high content (>50% w/v) of inorganic components [42]. The manufacturing of such scaffolds follows 3 steps:

1. Mixture phase, which involves the preparation of the slurry. The bioceramic particles are mixed in a solvent (aqueous or nonaqueous) with a low concentration of organic polymers/surfactants, called the binder, to obtain adequate flowability.
2. Green ceramic and binder burnout phase involving the deposition of filaments of slurry following a predetermined pattern prior to drying and exposure to high temperature to burn out the organic component of the mixture.
3. Sintering phase, which involves the exposure of the green form to elevated temperature (above 1000°C) to initiate the migration of atoms between adjacent ceramic particles, hence creating physical bonds called "necks."

It is critical for reproducible manufacturing of 3-D rapid prototyped bioceramics to have shape retention, a challenge that can be reached by adjusting the viscosity of the slurry and the evaporation rate of the solvent [43].

3.3. Composite materials

Printable composites, which are usually in the form of copolymers, polymer-polymer mixtures, or polymer-ceramic mixtures [44], allow ability for the combination of variety of advantageous properties of their included components, which provide a remarkable candidate as "bioink". Considering the advantages of polymer composite hydrogels, such as interpenetrating polymer networks (IPNs) or hybrid hydrogels [45], the incorporation of synthetic fillers to printable materials recently discussed in researches [33]. The addition of silicate fillers [38] and a range of nanoparticles have been used to synthesize different types of composite scaffolds [46] to promote greater control over viscosity and stiffness of polymer hydrogels. In addition, several of silica-containing hydrogels with higher expression of genes encoding morphogenetic cytokines, such as bone morphogenetic proteins (BMPs) seems promising [47]. The combination and manufacturing mixture of hydrogels with filler materials and/or natural peptides with morphogenetic capacity demonstrate great future for application in 3-D printing in aim to reach ultimate goal in regenerative craniofacial repair.

3.4. Cell aggregates and spheroids

Over recent years, many of researches aimed to evaluate and study cell aggregates and spheroids for use in tissue engineering and regenerative medicine [48]. As this method cited correctly and appropriately as "scaffold-free printing," in fact small quantities of hydrogel are used to facilitate cell aggregation. In this method for 3-D printing, or in an appropriate

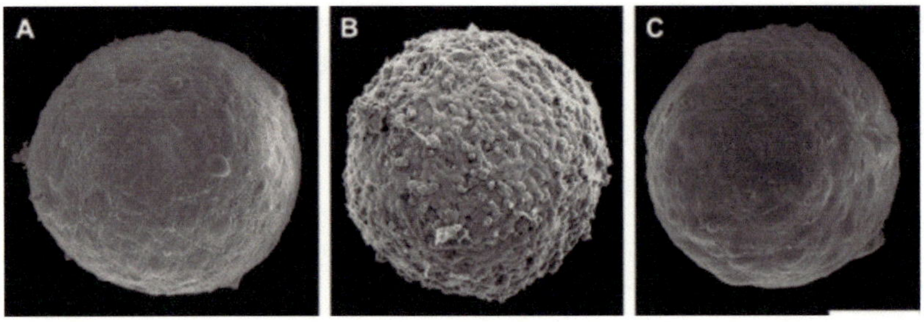

Figure 1. SEM view of multicellular spheroids of HUSMCs (A), CHO cells (B) and HFBs (C) (adapted from Norotte et al. [50]).

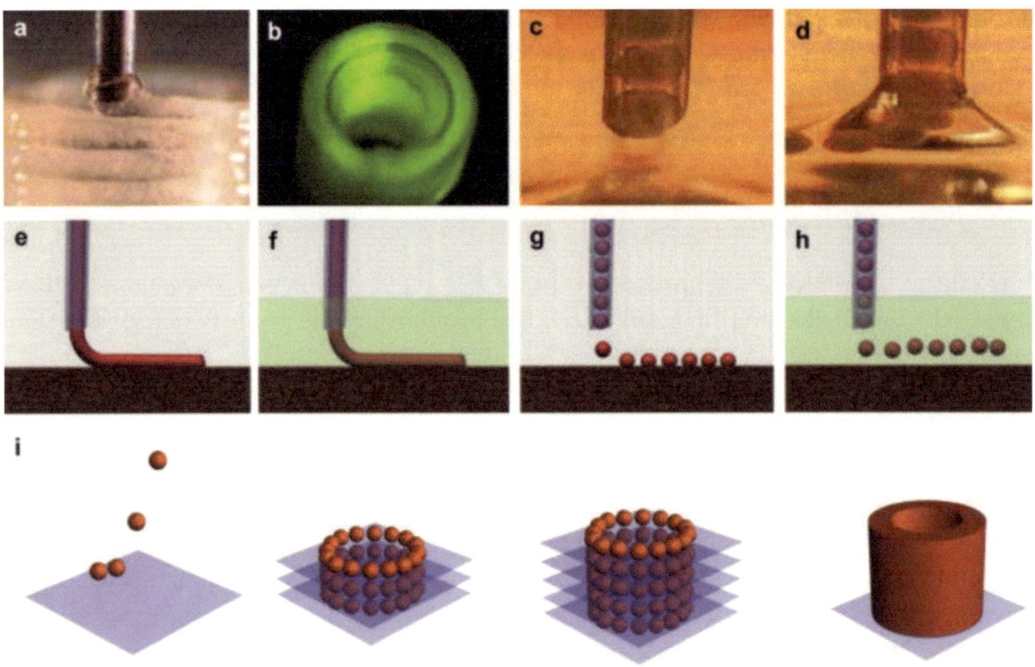

Figure 2. Principles of spheroids bioprinting technology: (a) bioprinter (general view); (b) multiple bioprinter nozzles; (c) tissue spheroids before dispensing; (d) tissue spheroids during dispensing; (e) schematic view of continuous dispensing in air; (f) schematic view of continuous dispensing in fluid; (g) schematic view of digital dispensing in air; (h) schematic view of digital dispensing in fluid; (i) schematic view of bioassembly of tubular tissue construct using bioprinting of self-assembled tissue spheroids illustrating sequential steps of layer-by-layer tissue spheroid deposition and tissue fusion process (adopted from Mironov et al. [48]).

way called "bioprinting," multicellular spheroids are deposited using extrusion printers and allowed to self-assemble into the desired 3-D structure (**Figure 1**). As it is clear, these systems allow direct fabrication of tissue constructs which in contrast to other methods have extremely high cell densities. Although in load-bearing tissues with high amount of mineral components and noticeable mechanical properties use of this methods still looks uncertain, the ability to position aggregates of heterotypic cells with microscale precision (**Figure 2**) seems promising as an excellent alternative to bioprint complex tissues consisting variety of cells [49].

4. Manufacturing of scaffolds with 3-D printing technology

Researches aimed to investigate novel technologies for 3-D printing and introduced some novel methods including phase-separation, self-assembly, electrospinning, freeze drying, solvent casting/particulate leaching, gas foaming, and melt molding [52]. Using scaffolds, the architecture of native extracellular matrices can be mimicked at the nanoscale level and therefore provide the primary base for the regeneration of new tissue [53]. Originally, a "top-down" approach was used as a tissue engineering method for scaffold fabrication. In this method, cells are seeded onto a biodegradable and biocompatible scaffold and are predicted to migrate and fill the scaffold hence creating their own matrix. By using this technique, several avascular tissues such as bladder [54] and skin [55] have been engineered effectively. However, due to the limited diffusion properties of these scaffolds, this technique faces several challenges for fabrication of more complex tissues such as heart and liver [56]. Therefore, "bottom-up" methods have been developed to overcome this problem [57]. Bottom-up approaches include cell-encapsulation with microscale hydrogels, cell aggregation by self-assembly, generation of cell sheets, and direct printing of cells [58]. These complex tissue blocks can be assembled using various methods including microfluidics [59], magnetic fields [60], acoustic fields [61], and surface tension [62]. These methods are relatively easy and have provided a solid foundation for the fabrication of scaffolds. However, as mentioned previously, these conventional methods suffer from several limitations including inadequate control over scaffold properties such as pore size, pore geometry, distribution of high levels of interconnectivity, and mechanical strength. As such, it is necessary to develop technologies with sufficient control so as to design more intricate tissue-specific scaffolds. In addition, scaffolds can be coated using surface modification techniques (such as introducing functional groups) to enhance cell migration, attachment and proliferation. Three-dimensional printing allows scaffolds to become more precisely fabricated (similar to that of the computer-aided design (CAD)) with higher flexibility in the type of materials used to make such scaffolds. Three-dimensional printing uses an additive manufacturing process where a structure is fabricated using a layer-by-layer process. Materials deposited for the formation of the scaffold may be cross-linked or polymerized through heat, ultraviolet light, or binder solutions. Using this technology, 3-D printed scaffolds can be prepared for optimized tissue engineering [52].

For appropriate formation of tissue architecture, the seeding cells (often stem cells) require a 3-D environment/matrix similar to that of the ECM. The ECM acts as a medium to provide proteins and proteoglycans among other nutrients for cellular growth. The

ECM also provides structural support to allow for cellular functionality such as regulating cellular communication, growth, and assembly [63]. With this in mind, scientists and engineers originally attempted to replicate the ECM through conventional techniques, which consequently established a framework for using more advanced techniques, such as 3-D printing, to yield higher quality scaffolds. The 3-D printing technique can create defined scaffold structures with controlled pore size and interconnectivity and the ability to support cell growth and tissue formation [64–66]. The current methods for 3-D printing involve a CAD, which is then relayed to each 3-D printing system to "print" the desired scaffold structure. Through various 3-D printing technologies, discussed below, researchers are trying to fabricate biocompatible scaffolds that efficiently support tissue formation (**Table 3**).

5. Bioprinting advantages aiming for clinical use

The goal of tissue engineering is to create functional tissues and organs for regenerative therapies and ultimately organ transplantation/replacement. Trial and error was the long and tedious process mainly used to advance the field of regenerative medicine by clarifying the success of techniques.

Researchers needed to come up with a list of requirements in order to measure their successes or failures in tissue fabrication [48, 67]. This list was generated from the observations of natural human tissue.

As gold standard of fabricated tissues is to be as similar as possible to natural tissues in the human body in different parameters, then these fabricated tissues must:

1. Be able to integrate with naturally occurring tissue, and attach via microsutures, glues [68], or through cell adhesion [69–71].

2. Be capable of essential functions in vivo [48].

3. Become fully vascularized in order to sustain its functionality [68, 71].

Also, the printers used for tissue fabrication required standardization as well [67, 69].

1. The bioprinting machines required set extreme sterilization methods to eliminate unwarranted contamination with previously used materials or foreign matter from the environment.

2. The conditions for printing must be ideal for tissue fabrication, so factors such as humidity and temperature must be closely monitored.

3. Nozzle size and methods of delivery affect the viability of the materials being printed; therefore, there must be set ideals for delivery methods in relation to the various printing materials.

http://dx.doi.org/10.5772/intechopen.70323

Printing method	Advantages	Disadvantages	Preclinical progress
Direct 3-D printing/ inkjet	• Versatile in terms of usable materials • No support is necessary for overhang or complex structures	• Potential toxicity (incompletely removed binders) • Low mechanical strength prints compared to laser sintering • Time consuming (post-processing)	• (Rat/bone) • (Rabbit/bone) • (Mouse/bone)
W/electrospinning			• (Mouse/cartilage)
Bioplotting	• Prints viable cells • Soft tissue applications	• Limitation on nozzle size* (*Must not be cytotoxic during processing) • Requires support structure for printing complex shapes	• (Rabbit/trachea) • (Rabbit/cartilage) • (Rat/cartilage) • (Mouse/cartilage) • (Mouse/tooth regeneration) • (Mouse/skin)
Fused deposition modeling	• Low cytotoxicity vs direct 3-D printing • Relatively inexpensive (printers and materials)	• Limitation on materials (often requires thermoplastics) • Materials used are nonbiodegradable • Requires support structure for overhangs and complex shapes • Post-processing may be necessary • Low Resolution	• (Swine/bone) • (Rat/bone)
Selective laser sintering	• Provides scaffolds with high mechanical strength • Powder bed provides support for complex structures • Fine resolution	• Limitation on materials (must be shrinkage and heat resistant) • Very high temp required (up to 1400°C) • Expensive and time consuming (processing and post processing)	• (Mouse/bone) • (Rat/heart) • (Rat/bone) • (Mouse/skin) • (Mouse/heart)

Printing method	Advantages	Disadvantages	Preclinical progress
Stereolithography	• Very high resolution • Speed of fabrication • Smooth surface finish	• Materials must be photopolymers • Expensive (two photon printers) • Support system is necessary for overhang and intricate objects	• (Rat/bone) • (Rabbit/trachea) • (Pig/tendon)
Electrospinning	• Speed of fabrication • Cell printing • Soft tissue engineering • Low shear stress (bioelectrospraying)	• Random orientation of fibers • Nonuniform pore sizes • High voltage (1–30 kV) requirements	• (Mouse/ biocompatibility) • (Rat/bone) • (Rabbit/vascular tissue)
Indirect 3-D printing	• Good for prototyping/preproduction • Material versatility casting once mold is obtained	• Requires proprietary waxes for biocompatibility (wax printing) • Low accuracies/resolution • Mold required for casting • Long production times (mold → cast → processing → product)	• (Rat/bone) • (Mouse/tooth regeneration)

Table 3. Preclinical researches on various 3-D printing techniques for manufacturing scaffolds for tissue engineering [52].

http://dx.doi.org/10.5772/intechopen.70323

As a result, researchers created a few methods of printing with the goal of finding a solution to the given problems for optimal tissue biofabrication [48, 68, 69]. Thermal inkjet bioprinting with bioink and direct-write bioprinting both make use of modified inkjet printers but with varied application techniques. Organ printing with tissue spheroids is the recent achievement of researches which seems promising to fabricate tissues directly. **Table 4** review advantages and disadvantages of all three common methods "Thermal Inkjet Bioprinting," "Direct-Write Bioprinting," and "Spheroid Organ Printing." Organ printing, otherwise known as the biomedical

Type of bioprinting	Method	Tissue characteristics	Note
Thermal inkjet bioprinting	• Bottom up • Layer-by-layer	• Avascular • Aneural • Alymphatic • Thin • Only nourishable via diffusion	"Bioink," which is a water-based liquid consisting of proteins, enzymes, and cells suspended in a media or saline.
Direct-write bioprinting	• Digital control of print. • Several printing units simultaneously. • Application of variety of materials simultaneously. • Faster turnaround time for printed products.	Possibility of printing tissues with different compositions.	• The bioink of direct-write printers may consist of hydrogels of varying consistencies that are composed of agarose, alginate, collagen type I, and Pluronic F127. • This method categorized in pneumatic, mechanical, and a pneumatic-mechanical hybrid. It was concluded that the pneumatic systems work better with high viscosity materials, while mechanical systems are better suited in working with materials of low viscosity.
Spheroid organ printing	Spheroids are punched into "biopaper" which is a sprayed layer of hydrogel. Each spheroid is made of living cells, thereby creating a ball of "living materials" capable of self-assembly and self-fusion. Alternatively, the spheroids can be digitally placed, undergo self-assembly, and fuse without the use of hydrogel.	• Self-organization is defined as, "a process in which patterning at the global level of a system emerges solely from numerous interactions among the lower-level components of the system." • Self-assembly is defined to be, "the autonomous organization of components into patterns or structures without human intervention."	Researchers fabricated three types of spheroids to create a vascular tree: solid or nonlumenized spheroids, spheroids with one big lumen (mono-lumenized spheroids), and microvascularized tissue spheroids.

Table 4. 3-D bioprinting technique advantages and properties[67].

application of rapid prototyping, may be defined as additive layer-by-layer biomanufacturing of cells. Advantages of organ printing include its automated approach offering a pathway for a scalable and reproducible mass production of tissue-engineered products. This also allows the precise simultaneous 3-D positioning of several cell types, hence enabling the creation of tissue with a high level of cell density. Organ printing may be used to solve the problem of vascularization in thick tissue constructs, and moreover, this technology may be done in situ. Therefore, this emerging transforming technology has potential for surpassing traditional solid scaffold-based tissue engineering [72].

6. Current limitations

6.1. Vascularization

In order to create a complete and functional organ, the researchers must be able to create thick complex tissues with full vascularization containing lumens of various sizes, large vascular structures to microstructures, in order to sustain the surrounding organ tissue. The best way to achieve this type of vascularization is to fabricate the vascular system and tissue simultaneously, of which is easier said than done [48]. Thorough vascularization remains a common theme for current bioprinting limitations. Without a functional circulatory system, tissue constructs are limited to a means of diffusion for nutrition, which in itself is limited to just a few hundred microns [69].

Current methods of vascularization call for the infiltration of host microvessels into an implanted construct [67, 73, 74].

Yet, this strategy is lacking in control and specificity for the developing microvessels. The invading microvessels have a limited penetration depth which prevents the successful incorporation of the microvessels into larger layered constructs. Additionally, the penetration of the vascular system itself may result in a distortion of the region penetrated or in the destruction of the fabricated tissue altogether. For these reasons, it would be ideal to construct tissues with direct vessel in-growth, or vascularization created within the tissue itself, all before implantation.

6.2. Tissue components and costs

In addition to vascularization, native tissues contain unique cellular combinations and organizations. There is a need to develop techniques that mimic the complexity of native tissues in order to drive tissue recovery and replacement for medical applications [69]. With the production of organs such as kidneys, for example, at least one million glomeruli and nephrons would need to be generated. Not only would the fabrication be a massive undertaking but also the fabricated tissue would need to be scalable. Scalability of biofabricated tissues is not presently a reality. Yet, spheroids have shown promise toward being scalable with further development. Finally, another major limitation for the development of natural-like, fully functioning fabricated human tissue is economic [68]. This challenge must definitely be overcome if biofabrication technology is to allow the creation of a functional living human organ.

7. Future aspects of 3-D printing for regenerative medicine

In this chapter, we have illustrated current guiding principles for 3-D bioprinting in tissue fabrication, as well as recent advances and technological developments. The speed at which our knowledge has advanced with additive manufacturing and automated printing systems shows a promise to expand our basic science and engineering capabilities toward addressing health care problems. One of the significant developments in 3-D bioprinting is to manufacture cell microenvironments from molecular to macroscopic scales, which are requested and suitable for tissue engineering and regenerative medicine. As novel methods and technologies introduced in recent years for 3-D printing of biomaterials, promising overview of future appears to manufacture scaffolds for tissue engineering that reach the gold standards and also better comprehensions of stem cells microenvironments and interactions. By aid of various novel technologies, such as microfluidic systems [75, 76], biopatterning [77], and layer-by-layer assembly [76, 78], researchers are now able to biomanufacture microtissue constructs within scaffolds and even also within scaffold-free environments. Considering the great and enormous improvements of biomaterial for tissue engineering, in contrast, there are still certain challenges and difficulties that need more attention. Vascularization is one of the limitations which receive most of attentions [79, 80] due to the fact that this challenge leads to hypoxia, apoptosis, and immediate cell death. For resolving this issue and providing sufficient space for vascularization, researchers attempts to fabricate porous scaffolds [81], to provide sufficient space for vascularization. However, this approach cannot overcome the vascularization challenge completely due to the diffusion of cells and other materials into these porous structures [82]. Forming interconnected, well-defined vascular structures during biomanufacturing process seems to lead to resolving this difficulty and providing better results during process. Other issues that have to be noticed are mechanical strength and stability in 3-D tissue engineering which is one of the key requirements [83]. To be clear in regeneration of hard (e.g., bone) and soft (e.g., vascular grafts) tissues, modulus of elasticity is a crucial parameter that desires improvement [84–86]. Furthermore, the development of a totally closed bioprinting system that integrates printing and post-printing processes such as in-vitro culture and maturation of tissue constructs continues to be a challenge.

With advances in near future, which help finding solutions for the challenges mentioned above, bioprinting technologies will potentially help improvements of rapid clinical solutions and advances in medical implants. Further, we envision that the integration of cells and biomaterials through bioprinting with microfluidic technologies are likely to create unique microenvironments for various applications in cancer biology, tissue engineering, and regenerative medicine [87–91]. Additionally, developments on high-throughput biomanufacturing of 3-D architectures will pave the way for further advancements of in vitro screening and diagnostic applications, potentially enabling complex organ constructs. In the meantime, it is only the effective interplay of engineering concepts in combination with the well-established fundamentals of biology that will realize the true potential of this exciting area.

Author details

Seied Omid Keyhan[1,2,3,4], Hamidreza Fallahi[5], Alireza Jahangirnia[6], Mohammad Taher Amirzade-Iranaq[7] and Mohammad Hosein Amirzade-Iranaq[8,9,10,11]*

*Address all correspondence to: h.amirzade@gmail.com

1 Department of Oral & Maxillofacial Surgery, Faculty of Dentistry, Birjand University of Medical Science, Birjand, Iran

2 Vice Presidential Organization of Technology of the Islamic Republic of Iran, Iran

3 Stem Cell & Regenerative Medicine Network, Shahid Beheshti University of Medical Sciences, Tehran, Iran

4 Cranio Maxillofacial Research Center, Tehran Dental Branch, Islamic Azad University, Tehran, Iran

5 Oral and Maxillofacial Surgery, Jundishapur University of Medical Sciences, Ahvaz, Iran

6 Oral and Maxillofacial Surgeon, Private Practice, Tehran, Iran

7 Biomaterial Medical Engineer, Department of Materials, Faculty of Materials Engineering, Islamic Azad University Najaf Abad Branch, Isfahan, Iran

8 Student Research Committee (SRC), Baqiyatallah University of Medical Sciences, Tehran, Iran

9 Universal Network of Interdisciplinary Research in Oral and Maxillofacial Surgery (UNIROMS), Universal Scientific Education and Research Network (USERN), Tehran, Iran

10 Student Research Committee, Shahid Sadoughi University of Medical Sciences, Yazd, Iran

11 International Otorhinolaryngology Research Association (IORA), Universal Scientific Education and Research Network (USERN), Tehran, Iran

References

[1] Gibbs DM, Vaezi M, Yang S, Oreffo RO. Hope versus hype: What can additive manufacturing realistically offer trauma and orthopedic surgery? Regenerative Medicine. 2014;**9**(4):535-549

[2] Canstein C, Cachot P, Faust A, Stalder A, Bock J, Frydrychowicz A, et al. 3D MR flow analysis in realistic rapid-prototyping model systems of the thoracic aorta: Comparison with in vivo data and computational fluid dynamics in identical vessel geometries. Magnetic Resonance in Medicine. 2008;**59**(3):535-546

[3] Müller A, Krishnan KG, Uhl E, Mast G. The application of rapid prototyping techniques in cranial reconstruction and preoperative planning in neurosurgery. Journal of Craniofacial Surgery. 2003;**14**(6):899-914

[4] Hoy MB. 3D printing: Making things at the library. Medical Reference Services Quarterly. 2013;**32**(1):93-99

[5] Rengier F, Mehndiratta A, von Tengg-Kobligk H, Zechmann CM, Unterhinninghofen R, Kauczor H-U, et al. 3D printing based on imaging data: Review of medical applications. International Journal of Computer Assisted Radiology and Surgery. 2010;**5**(4):335-341

[6] Chae MP, Rozen WM, McMenamin PG, Findlay MW, Spychal RT and Hunter-Smith DJ (2015) Emerging applications of bedside 3D printing in plastic surgery. Front. Surg. 2:25. DOI: 10.3389/fsurg.2015.00025

[7] Mertz L. New world of 3-d printing offers "completely new ways of thinking": Q&A with author, engineer, and 3-d printing expert hod lipson. Pulse, IEEE. 2013;**4**(6):12-14

[8] Ibrahim AMS, Jose RR, Rabie AN, Gerstle TL, Lee BT, Lin SJ. Three-dimensional Printing in Developing Countries. Plastic and Reconstructive Surgery – Global Open. 2015;**3**(7):e443

[9] Chan HH, Siewerdsen JH, Vescan A, Daly MJ, Prisman E, Irish JC. 3D rapid prototyping for otolaryngology—head and neck surgery: Applications in image-guidance, surgical simulation and patient-specific modeling. PLoS One. 2015;**10**(9):e0136370

[10] Mendez BM, Chiodo MV, Patel PA. Customized "In-Office" three-dimensional printing for virtual surgical planning in craniofacial surgery. Journal of Craniofacial Surgery. 2015;**26**(5):1584-1586

[11] Cunningham LL, Madsen MJ, Peterson G. Stereolithographic modeling technology applied to tumor resection. Journal of Oral and Maxillofacial Surgery. 2005;**63**(6):873-878

[12] AlAli AB, Griffin MF, Butler PE. Three-Dimensional Printing Surgical Applications. Eplasty. 2015;**15**:e37

[13] Hull CW. Apparatus for Production of Three-Dimensional Objects by Stereolithography. Google Patents; 1986

[14] Dawood A, Marti BM, Sauret-Jackson V, Darwood A. 3D printing in dentistry. British Dental Journal. 2015;**219**(11):521-529

[15] Brix F, Hebbinghaus D, Meyer W. Verfahren und Vorrichtung für den Modellbau im Rahmen der orthopädischen und traumatologischen Operationsplanung. Röntgenpraxis. 1985;**38**:290-292

[16] Sinn DP, Cillo Jr JE, Miles BA. Stereolithography for craniofacial surgery. Journal of Craniofacial Surgery. 2006;**17**(5):869-875

[17] Mankovich NJ, Cheeseman AM, Stoker NG. The display of three-dimensional anatomy with stereolithographic models. Journal of Digital Imaging. 1990;**3**(3):200-203

[18] Suomalainen A, Stoor P, Mesimäki K, Kontio RK. Rapid prototyping modelling in oral and maxillofacial surgery: A two year retrospective study. Journal of Clinical and Experimental Dentistry. 2015;**7**(5):e605

[19] Barker T, Earwaker W, Lisle D. Accuracy of stereolithographic models of human anatomy. Australasian Radiology. 1994;**38**(2):106-111

[20] Frühwald J, Schicho KA, Figl M, Benesch T, Watzinger F, Kainberger F. Accuracy of craniofacial measurements: Computed tomography and three-dimensional computed tomography compared with stereolithographic models. Journal of Craniofacial Surgery. 2008;**19**(1):22-26

[21] Mazzoli A, Germani M, Moriconi G. Application of optical digitizing techniques to evaluate the shape accuracy of anatomical models derived from computed tomography data. Journal of Oral and Maxillofacial Surgery. 2007;**65**(7):1410-1418

[22] Mehra P, Miner J, D'Innocenzo R, Nadershah M. Use of 3-d stereolithographic models in oral and maxillofacial surgery. Journal of Maxillofacial and Oral Surgery. 2011;**10**(1):6-13

[23] Cohen A, Laviv A, Berman P, Nashef R, Abu-Tair J. Mandibular reconstruction using stereolithographic 3-dimensional printing modeling technology. Oral Surgery, Oral Medicine, Oral Pathology, Oral Radiology, and Endodontology. 2009;**108**(5):661-666

[24] Mazzoni S, Marchetti C, Sgarzani R, Cipriani R, Scotti R, Ciocca L. Prosthetically guided maxillofacial surgery: Evaluation of the accuracy of a surgical guide and custom-made bone plate in oncology patients after mandibular reconstruction. Plastic and Reconstructive Surgery. 2013;**131**(6):1376-1385

[25] Eppley BL, Sadove AM. Computer-generated patient models for reconstruction of cranial and facial deformities. Journal of Craniofacial Surgery. 1998;**9**(6):548-556

[26] Gerstle TL, Ibrahim AM, Kim PS, Lee BT, Lin SJ. A plastic surgery application in evolution: Three-dimensional printing. Plastic and Reconstructive Surgery. 2014;**133**(2):446-451

[27] Chopra K, Gastman BR, Manson PN. Stereolithographic modeling in reconstructive surgery of the craniofacial skeleton after tumor resection. Plastic and Reconstructive Surgery. 2012;**129**(4):743e-745e

[28] Melchels FP, Feijen J, Grijpma DW. A review on stereolithography and its applications in biomedical engineering. Biomaterials. 2010;**31**(24):6121-6130

[29] Yan X, Gu P. A review of rapid prototyping technologies and systems. Computer-Aided Design. 1996;**28**(4):307-318

[30] Choi JW, Kim N. Clinical application of three-dimensional printing technology in craniofacial plastic surgery. Archives of Plastic Surgery. 2015;**42**(3):267-277

[31] Seied Omid Keyhan, Sina Ghanean, Alireza Navabazam, Arash Khojasteh and Mohammad Hosein Amirzade Iranaq (2016). Three-Dimensional Printing: A Novel Technology for Use in Oral and Maxillofacial Operations, A Textbook of Advanced Oral and Maxillofacial Surgery Volume 3, Prof. Mohammad Hosein Kalantar Motamedi (Ed.), InTech, DOI: 10.5772/63315

[32] Annabi N, Tamayol A, Uquillas JA, Akbari M, Bertassoni LE, Cha C, et al. 25th anniversary article: Rational design and applications of hydrogels in regenerative medicine. Advanced Materials. 2014;**26**(1):85-124

[33] Bakarich SE, Gorkin III R, in het Panhuis M, Spinks GM. Three-dimensional printing fiber reinforced hydrogel composites. ACS Applied Materials & Interfaces. 2014;**6**(18): 15998-16006

[34] Bertassoni LE. Bioprinting of human organs. Australasian Science. 2015;**36**(3):34

[35] Bertassoni LE, Cardoso JC, Manoharan V, Cristino AL, Bhise NS, Araujo WA, et al. Direct-write bioprinting of cell-laden methacrylated gelatin hydrogels. Biofabrication. 2014;**6**(2):024105

[36] Woodruff MA, Hutmacher DW. The return of a forgotten polymer—polycaprolactone in the 21st century. Progress in Polymer Science. 2010;**35**(10):1217-1256

[37] Hutmacher DW, Sittinger M, Risbud MV. Scaffold-based tissue engineering: Rationale for computer-aided design and solid free-form fabrication systems. TRENDS in Biotechnology. 2004;**22**(7):354-362

[38] Xavier JR, Thakur T, Desai P, Jaiswal MK, Sears N, Cosgriff-Hernandez E, et al. Bioactive nanoengineered hydrogels for bone tissue engineering: A growth-factor-free approach. ACS Nano. 2015;**9**(3):3109-3118

[39] Michna S, Wu W, Lewis JA. Concentrated hydroxyapatite inks for direct-write assembly of 3-D periodic scaffolds. Biomaterials. 2005;**26**(28):5632-5639

[40] Tarafder S, Dernell WS, Bandyopadhyay A, Bose S. SrO- and MgO-doped microwave sintered 3D printed tricalcium phosphate scaffolds: Mechanical properties and in vivo osteogenesis in a rabbit model. Journal of Biomedical Materials Research Part B: Applied Biomaterials. 2015;**103**(3):679-690

[41] Wang Y, Azaïs T, Robin M, Vallée A, Catania C, Legriel P, et al. The predominant role of collagen in the nucleation, growth, structure and orientation of bone apatite. Nature Materials. 2012;**11**(8):724-733

[42] Sousa F, Evans JR. Sintered hydroxyapatite latticework for bone substitute. Journal of the American Ceramic Society. 2003;**86**(3):517-519

[43] Morissette SL, Lewis JA, Cesarano J, Dimos DB, Baer T. Solid freeform fabrication of aqueous alumina–poly (vinyl alcohol) gelcasting suspensions. Journal of the American Ceramic Society. 2000;**83**(10):2409-2416

[44] Tevlin R, McArdle A, Atashroo D, Walmsley G, Senarath-Yapa K, Zielins E, et al. Biomaterials for craniofacial bone engineering. Journal of Dental Research. 2014;**93** (12):1187-1195

[45] Hutson CB, Nichol JW, Aubin H, Bae H, Yamanlar S, Al-Haque S, et al. Synthesis and characterization of tunable poly (ethylene glycol): Gelatin methacrylate composite hydrogels. Tissue Engineering Part A. 2011;**17**(13-14):1713-1723

[46] Gao G, Schilling AF, Yonezawa T, Wang J, Dai G, Cui X. Bioactive nanoparticles stimulate bone tissue formation in bioprinted three-dimensional scaffold and human mesenchymal stem cells. Biotechnology Journal. 2014;**9**(10):1304-1311

[47] Müller WE, Schröder HC, Feng Q, Schlossmacher U, Link T, Wang X. Development of a morphogenetically active scaffold for three-dimensional growth of bone cells: biosilica–alginate hydrogel for SaOS-2 cell cultivation. Journal of Tissue Engineering and Regenerative Medicine. 2015;**9**(11):E39-E50

[48] Mironov V, Visconti RP, Kasyanov V, Forgacs G, Drake CJ, Markwald RR. Organ printing: Tissue spheroids as building blocks. Biomaterials. 2009;**30**(12):2164-2174

[49] Ikeda E, Morita R, Nakao K, Ishida K, Nakamura T, Takano-Yamamoto T, et al. Fully functional bioengineered tooth replacement as an organ replacement therapy. Proceedings of the National Academy of Sciences. 2009;**106**(32):13475-13480

[50] Norotte C, Marga FS, Niklason LE, Forgacs G. Scaffold-free vascular tissue engineering using bioprinting. Biomaterials. 2009;**30**(30):5910-5917

[51] Tabatabaee FS MS, Gholipour F, Khosraviani K, Khojasteh A..Craniomaxillofacial bone engineering by scaffolds loaded with stem cells: A Systematic review. Journal of Dental School. 2012;**30**(2):115-131

[52] Do A-V, Khorsand B, Geary SM, Salem AK. 3D Printing of scaffolds for tissue regeneration applications. Advanced Healthcare Materials. 2015;**4**(12):1742-1762

[53] Wei G, Ma PX. Nanostructured biomaterials for regeneration. Advanced Functional Materials. 2008;**18**(22):3568-3582

[54] Korossis S, Bolland F, Southgate J, Ingham E, Fisher J. Regional biomechanical and histological characterisation of the passive porcine urinary bladder: Implications for augmentation and tissue engineering strategies. Biomaterials. 2009;**30**(2):266-275

[55] Groeber F, Holeiter M, Hampel M, Hinderer S, Schenke-Layland K. Skin tissue engineering — In vivo and in vitro applications. Advanced Drug Delivery Reviews. 2011;**63**(4-5): 352-366

[56] Huang G, Wang L, Wang S, Han Y, Wu J, Zhang Q, et al. Engineering three-dimensional cell mechanical microenvironment with hydrogels. Biofabrication. 2012;**4**(4):042001

[57] Nichol JW, Khademhosseini A. Modular tissue engineering: Engineering biological tissues from the bottom up. Soft Matter. 2009;**5**(7):1312-1319

[58] Napolitano AP, Chai P, Dean DM, Morgan JR. Dynamics of the self-assembly of complex cellular aggregates on micromolded nonadhesive hydrogels. Tissue Engineering. 2007;**13**(8):2087-2094

[59] Chung SE, Park W, Shin S, Lee SA, Kwon S. Guided and fluidic self-assembly of microstructures using railed microfluidic channels. Nature Materials. 2008;**7**(7):581-587

[60] Xu F, Wu CaM, Rengarajan V, Finley TD, Keles HO, Sung Y, et al. Three-dimensional magnetic assembly of microscale hydrogels. Advanced Materials. 2011;**23**(37):4254-4260

[61] Xu F, Finley TD, Turkaydin M, Sung Y, Gurkan UA, Yavuz AS, et al. The assembly of cell-encapsulating microscale hydrogels using acoustic waves. Biomaterials. 2011;**32**(31):7847-7855

[62] Kachouie NN, Du Y, Bae H, Khabiry M, Ahari AF, Zamanian B, et al. Directed assembly of cell-laden hydrogels for engineering functional tissues. Organogenesis. 2010;**6**(4):234-244

[63] Geckil H, Xu F, Zhang X, Moon S, Demirci U. Engineering hydrogels as extracellular matrix mimics. Nanomedicine. 2010;**5**(3):469-484

[64] Lee K-W, Wang S, Lu L, Jabbari E, Currier BL, Yaszemski MJ. Fabrication and characterization of poly (propylene fumarate) scaffolds with controlled pore structures using 3-dimensional printing and injection molding. Tissue Engineering. 2006;**12**(10):2801-2811

[65] Xiaoming Li, Rongrong Cui, Lianwen Sun, et al., "3D-Printed Biopolymers for Tissue Engineering Application," International Journal of Polymer Science, vol. 2014, Article ID 829145, 13 pages, 2014. DOI:10.1155/2014/829145

[66] Bose S, Vahabzadeh S, Bandyopadhyay A. Bone tissue engineering using 3D printing. Materials Today. 2013;**16**(12):496-504

[67] Patra S, Young V. A review of 3D printing techniques and the future in biofabrication of bioprinted tissue. Cell Biochemistry and Biophysics. 2016;**74**(2):93-98

[68] Cui X, Boland T, D D'Lima D, K Lotz M. Thermal inkjet printing in tissue engineering and regenerative medicine. Recent Patents on Drug Delivery & Formulation. 2012;**6**(2): 149-155

[69] Chang CC, Boland ED, Williams SK, Hoying JB. Direct-write bioprinting three-dimensional biohybrid systems for future regenerative therapies. Journal of Biomedical Materials Research Part B: Applied Biomaterials. 2011;**98**(1):160-170

[70] Jakab K, Damon B, Marga F, Doaga O, Mironov V, Kosztin I, et al. Relating cell and tissue mechanics: Implications and applications. Developmental Dynamics. 2008;**237**(9): 2438-2449

[71] Dean DM, Morgan JR. Cytoskeletal-mediated tension modulates the directed self-assembly of microtissues. Tissue Engineering Part A. 2008;**14**(12):1989-1997

[72] Ozbolat IT, Yu Y. Bioprinting toward organ fabrication: Challenges and future trends. IEEE Transactions on Biomedical Engineering. 2013;**60**(3):691-699

[73] Barralet J, Gbureck U, Habibovic P, Vorndran E, Gerard C, Doillon CJ. Angiogenesis in calcium phosphate scaffolds by inorganic copper ion release. Tissue Engineering Part A. 2009;**15**(7):1601-1609

[74] Borselli C, Ungaro F, Oliviero O, d'Angelo I, Quaglia F, La Rotonda MI, et al. Bioactivation of collagen matrices through sustained VEGF release from PLGA microspheres. Journal of Biomedical Materials Research Part A. 2010;**92**(1):94-102

[75] Nakao Y, Kimura H, Sakai Y, Fujii T. Bile canaliculi formation by aligning rat primary hepatocytes in a microfluidic device. Biomicrofluidics. 2011;**5**(2):022212

[76] Chang R, Emami K, Wu H, Sun W. Biofabrication of a three-dimensional liver micro-organ as an in vitro drug metabolism model. Biofabrication. 2010;**2**(4):045004

[77] Gurkan UA, El Assal R, Yildiz SE, Sung Y, Trachtenberg AJ, Kuo WP, et al. Engineering anisotropic biomimetic fibrocartilage microenvironment by bioprinting mesenchymal stem cells in nanoliter gel droplets. Molecular Pharmaceutics. 2014;**11**(7):2151-2159

[78] Snyder J, Hamid Q, Wang C, Chang R, Emami K, Wu H, et al. Bioprinting cell-laden matrigel for radioprotection study of liver by pro-drug conversion in a dual-tissue microfluidic chip. Biofabrication. 2011;**3**(3):034112

[79] Rupnick MA, Panigrahy D, Zhang C-Y, Dallabrida SM, Lowell BB, Langer R, et al. Adipose tissue mass can be regulated through the vasculature. Proceedings of the National Academy of Sciences. 2002;**99**(16):10730-10735

[80] Langer RS, Vacanti JP. Tissue engineering: The challenges ahead. Scientific American. 1999;**280**:86-89

[81] Hutmacher DW. Scaffold design and fabrication technologies for engineering tissues—State of the art and future perspectives. Journal of Biomaterials Science, Polymer Edition. 2001;**12**(1):107-124

[82] Henze U, Kaufmann M, Klein B, Handt S, Klosterhalfen B. Endothelium and biomaterials: Morpho-functional assessments. Biomedicine & Pharmacotherapy. 1996;**50**(8):388

[83] Hockaday L, Kang K, Colangelo N, Cheung P, Duan B, Malone E, et al. Rapid 3D printing of anatomically accurate and mechanically heterogeneous aortic valve hydrogel scaffolds. Biofabrication. 2012;**4**(3):035005

[84] Mironov V, Boland T, Trusk T, Forgacs G, Markwald RR. Organ printing: Computer-aided jet-based 3D tissue engineering. TRENDS in Biotechnology. 2003;**21**(4):157-161

[85] Prestwich GD. Hyaluronic acid-based clinical biomaterials derived for cell and molecule delivery in regenerative medicine. Journal of Controlled Release. 2011;**155**(2):193-199

[86] Guven S, Lindsey JS, Poudel I, Chinthala S, Nickerson MD, Gerami-Naini B, et al. Functional maintenance of differentiated embryoid bodies in microfluidic systems: A platform for personalized medicine. Stem Cells Translational Medicine. 2015;**4**(3):261-268

[87] Rizvi I, Gurkan UA, Tasoglu S, Alagic N, Celli JP, Mensah LB, et al. Flow induces epithelial-mesenchymal transition, cellular heterogeneity and biomarker modulation in 3D ovarian cancer nodules. Proceedings of the National Academy of Sciences. 2013;**110**(22):E1974-E1983

[88] Luo Z, Güven S, Gozen I, Chen P, Tasoglu S, Anchan RM, et al. Deformation of a single mouse oocyte in a constricted microfluidic channel. Microfluidics and Nanofluidics. 2015;**19**(4):883-890

[89] Anchan R, Lindsey J, Ng N, Parasar P, Guven S, El Assal R, et al. Human IPSC-derived steroidogenic cells maintain endocrine function with extended culture in a microfluidic chip system. Fertility and Sterility. 2015;**104**(3):e73

[90] Gurkan UA, Tasoglu S, Akkaynak D, Avci O, Unluisler S, Canikyan S, et al. Smart interface materials integrated with microfluidics for on-demand local capture and release of cells. Advanced Healthcare Materials. 2012;**1**(5):661-668

[91] Gurkan UA, Anand T, Tas H, Elkan D, Akay A, Keles HO, et al. Controlled viable release of selectively captured label-free cells in microchannels. Lab On a Chip. 2011;**11**(23):3979-3989

Chapter 8

Tissue Engineering: Use of Growth Factors in Bone Regeneration

Carmen Mortellaro and Massimo Del Fabbro

Additional information is available at the end of the chapter

http://dx.doi.org/10.5772/intechopen.69875

Abstract

Tissue healing is a complex process involving a cascade of cellular and molecular events that are mostly shared by the different tissues of the body. Interestingly, the tissue repair process initiates immediately after a traumatic injury and is mediated and controlled by a wide range of cytokines, proteins, and growth factors released from platelets upon activation. Consequently, many growth factors have been considered as therapeutic molecules for the repair or regeneration of a wide range of tissues. Although their role has been only partially elucidated, the potential benefit of most growth factors has been demonstrated. In the last few years, the development of platelet-rich preparations has revolutionized the field of regenerative medicine, due to the repair capacities of the platelet-released growth factors that stimulate and accelerate both soft and hard tissue healing and regeneration. Today, autologous platelet concentrates (APCs) are used in a wide range of disciplines such as dentistry, oral surgery, orthopedics, sport medicine, dermatology, and plastic and reconstructive surgery. The purpose of this chapter is to describe the current evidence regarding the benefits of using autologous platelet concentrates in various oral surgery procedures, using a systematic review approach.

Keywords: platelet concentrates, growth factors, tissue regeneration, oral surgery

1. Introduction

Autologous platelet concentrates (APCs) have been widely used in many different clinical situations that require a rapid tissue healing and regeneration as it is especially the case in oral and maxillofacial surgery, orthopedics, sports medicine, ophthalmology, and in the treatment of skin ulcers. APCs are hemocomponents, obtained through centrifugation of patient's own blood, in order to collect the most active blood components: platelets, fibrin, and in certain cases also leukocytes. The final product has a platelet concentration higher than the basal level, consequently has an increased number of platelet-derived growth factors [1]. The rationale of the clinical use of such platelet-rich preparations is based upon the concept of exploiting their contents enriched of numerous mitogenic platelet-derived growth factors (including platelet-derived growth factor (PDGF), transforming growth factor-β (TGF-β), endothelial growth factor (EGF), vascular endothelial growth factor (VEGF), insulin-like growth factor-1 (IGF-1), basic fibroblast growth factor (FGF), and hepatocyte growth factor (HGF)) as well as other key molecules in promoting tissue healing (as adhesive proteins, pro-coagulant factors, cytokines, chemokines, and anti-microbial proteins [2–5]) to stimulate many biological functions, such as chemotaxis, angiogenesis, proliferation, and differentiation (**Table 1**) in order to enhance hard and soft tissue healing.

Category	Term	Biological activities
Adhesive proteins	VWF + propeptide, Fg, Fn, Vn, TSP-1, and laminin-8	Cell contact interactions, clotting, and extracellular matrix composition
Clotting factors and associated proteins	F V/Va, F XI, Gas6, protein S, HMWK, AT, and TFPI	Thrombin production and regulation and angiogenesis
Fibrinolytic factors and associated proteins	Pgn, PAI-I, u-PAm, OSN, α2-AP, HRGP, TAFI, and α2-M	Plasmin production and vascular modeling
Proteases and antiproteases	TIMP-4, MMP-4, inhibitor of FIX, PN-II, C1-INH, and A1AT	Angiogenesis, vascular modeling, regulation of coagulation, and regulation of cellular behavior
Growth factors, cytokines, and chemokines	PDGF, TGFβ-1 and -2, EGF, IGF-1,VEGF, bFGF, FGF-2, HGF, CCL5, IL-8, MIP-1α, CXCL5, MCP-3, ANG-1, and IL-1β, neutrophil chemotactic protein	Chemotaxis, cell proliferation and differentiation, and angiogenesis
Basic proteins and others	PF4, β-TG, PBP, CTAP III, NAP-2, and ES	Regulation of angiogenesis, vascular modeling, and cellular interactions
Antimicrobial proteins	TC	Bactericidal and fungicidal properties
Others	CS-4, AB, and Ig	Diverse
Membrane glycoproteins	αIIbβ3, αvβ3, GPIb, PECAM-1, most plasma membrane constituents, receptors for primary agonists, CD40L, TF, and P-selection	Platelet aggregation and adhesion, endocytosis of proteins, inflammation, thrombin generation, and platelet-leukocyte interactions

Extracted and readapted from Anitua et al. [2].

Table 1. Platelet α-granule contents and their functional categories.

2. Brief history of platelet concentrates

2.1. Fibrin adhesives

More than 40 years ago, these technologies were originally used as sealant-adhesive agents in the treatment of hemorrhage with the aim of blocking the blood leakage [6]. Subsequently, other molecules involved in the coagulation process were combined to such fibrin preparations to improve their adhesive properties. These preparations were referred as "platelet-fibrinogen-thrombin mixtures" and were successfully used in ophthalmology [7, 8], general surgery [9], and neurosurgery [10]. Other authors termed them as "gel foam" [11].

It is noteworthy that the application of these preparations was essentially related to their adhesive properties and the platelets were served only to reinforce the fibrin matrix architecture. A few years later, it was developed the concept that these preparations could have healing and regenerative properties. In the late 1980s, Knighton et al. [12–14] used the autologous "Platelet-Derived Wound Healing Factors (PDWHF)," which were prepared through two-step centrifugation process, in the treatment of chronic nonhealing cutaneous ulcers. In 1997, Whitman et al. used platelet concentrate referred as "platelet gel" in oral and maxillofacial surgery [15].

2.2. Platelet-rich plasma

The term "Platelet-Rich Plasma" (PRP) was, for the first time, introduced by Kingsley et al. to describe a thrombocyte concentrate [16] used for the treatment of severe thrombopenia. However, the use of PRP term really started with Marx in 1998 [1] when he published a comparative clinical study in which the PRP regenerative potential was demonstrated in a series of patients undergoing mandibular reconstruction. Afterward, the PRP product was associated with the concept of platelet growth factors and their potential contribution to the enhancement of tissue healing.

According to the PRP protocol, the blood is collected in tubes containing anticoagulants and processed by two centrifugation steps. **Figure 1** illustrates schematically the specific protocol [17]. The final PRP product can be applied to the surgical site with a syringe or be activated by thrombin and/or calcium chloride to trigger platelet activation and to stimulate the fibrin polymerization.

After blood collection into tubes with anticoagulant, the first centrifugation at low force (soft spin) allows the separation of blood into three distinct layers: red blood cells at the bottom, a cellular plasma (platelet-poor plasma (PPP)) in the upper portion, and a whitish layer called buffy coat located between them containing the highest concentration of platelets and leukocytes. For the production of Pure-PRP (P-PRP), PPP and the superficial buffy coat layer are transferred into another tube and centrifuged at high forces (hard spin), after which most of the PPP and leukocytes are discarded and the final P-PRP can be collected. For the production of PRP rich in leukocytes (L-PRP), PPP, the entire buffy coat layer and some residual red blood cells are collected and transferred in another tube to be hard spin centrifuged. To obtain

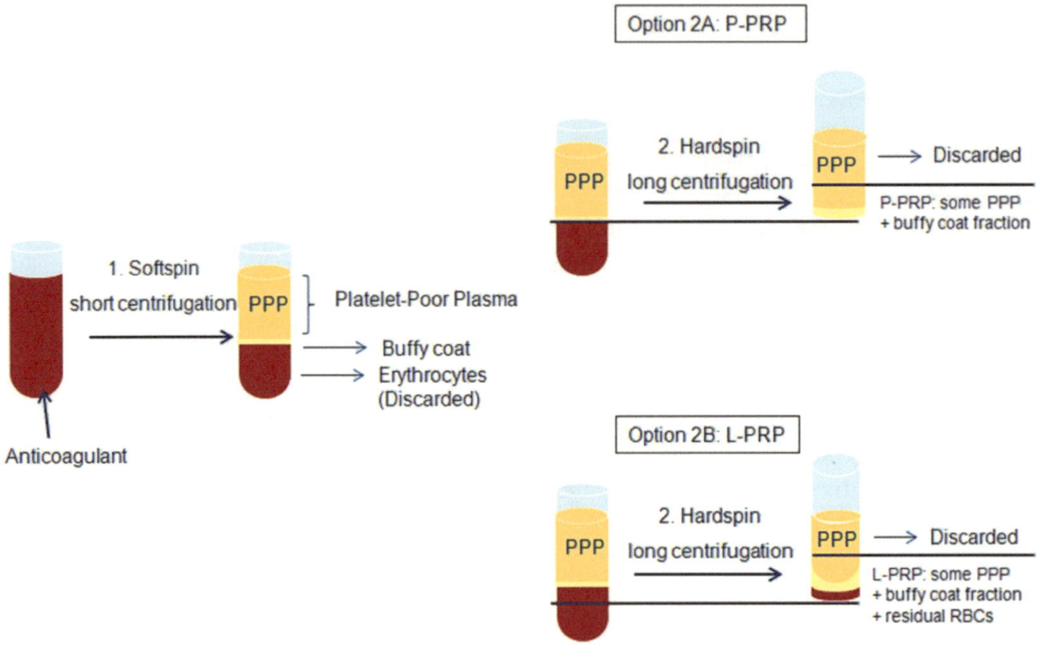

Figure 1. Protocol for PRP production.

the final L-PRP, PPP is discarded leading to an L-PRP that contains the buffy coat with most of the platelets and leukocytes, some residual red blood cells and PPP (adapted from Dohan Ehrenfest et al. [17]).

Currently, more than 20 different commercial systems for PRP preparation exist, which may lead to products with different features, especially regarding the composition and the cell concentration rate respect to baseline. On average, a 5–8× concentration is achieved though a ratio of up to 11× has been reported with PRP.

2.3. Platelet-rich fibrin

In 2001, Choukroun et al. developed a protocol for producing a hemocomponent named platelet-rich fibrin (PRF) [18]. Here, the blood is collected in tubes in the absence of anticoagulant and centrifuged with moderate forces (3000 rpm in a dedicated centrifuge) for 12 min. Afterward, three layers are formed: red blood cells and acellular plasma are located, respectively, at the bottom and at the top of the tube, and the fibrin clot, positioned between them, is PRF (**Figure 2**). Since the formation of the PRF clot naturally occurs within the tube, it has a strong fibrin matrix in which most of the platelets and leukocytes are embedded [19]. Since its introduction, PRF has undergone some development: the advanced PRF (a-PRF) was launched a few years ago, characterized by a reduced centrifugation speed and time (2700 rpm, 8–10 min), which allows a more even cell distribution within the clot [20]. Recently, the injectable PRF (i-PRF) has also been developed, which may be obtained with a further softer spin (1500 rpm, 3 min), consisting of a liquid form, very rich in white cells that can be used by infiltration into tissues and joints.

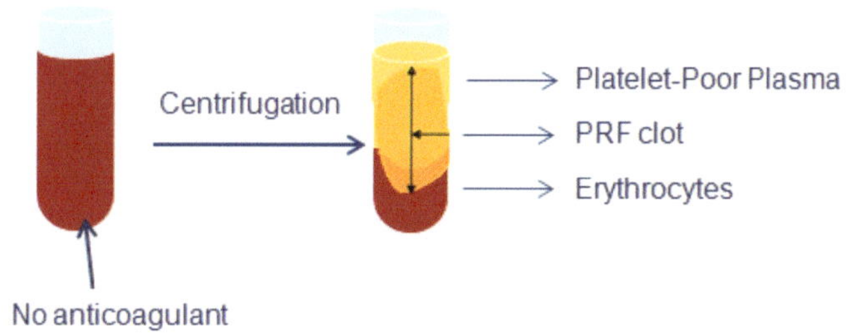

Figure 2. Protocol for PRF production. (Adapted from Dohan Ehrenfest et al. [19]).

2.4. Plasma rich in growth factors

In parallel to the introduction of PRP and PRF, Anitua in 1999 proposed another platelet concentrate protocol, denominated plasma rich in growth factors (PRGF) [21]. Briefly, blood is collected in tubes containing 0.2 ml of 3.8% trisodium citrate as anticoagulant. After a centrifugation at 580 g for 8 min, red blood cells and buffy coat layer are deposited at the bottom of the tube and the plasma component above. The latter is then manually separated into two fractions. The lower portion of about 2 ml, above the buffy coat, is the PRGF, whereas the upper portion is the plasma poor in growth factors (PPGF) (**Figure 3**). The final PRGF product may be applied as a liquid fraction to the target site or may be preactivated by adding 0.2 ml of 10% $CaCl_2$ to induce the clot formation [22].

2.5. Technical differences between PRGF, PRP, and PRF

PRGF differs from PRP for the following technical aspects:

1. Blood volume drawn is minimal (5–40 ml).
2. Requires a single centrifugation for the preparation.
3. Does not contain leukocytes.
4. Does not contain proinflammatory cytokines.
5. Platelet concentration is reduced (2–3 fold the baseline, respect to 5–8× for PRP).

In addition, PRGF also differs from PRF for these features:

1. Different products can be obtained (liquid, gel, membrane, and fibrin clot).
2. PRGF liquid can be combined with bone graft materials for bone regeneration procedure.

Differences between PRP and PRF are as follows:

1. PRP preparation requires two centrifugations.
2. Different products can be obtained (liquid and fibrin clot).
3. PRP liquid can be mixed with bone graft materials for bone regeneration procedures.

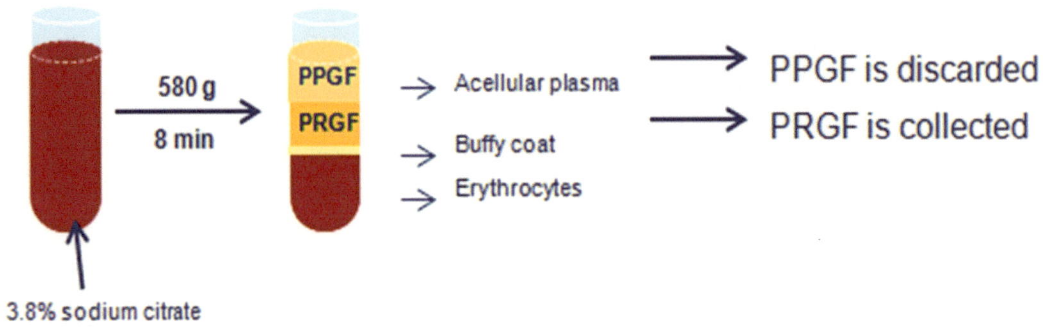

Figure 3. Process of PRGF production. Details are given in the text. (Adapted from Anitua [22]).

3. Clinical evidence of the efficacy of autologous platelet concentrates in oral regenerative surgical procedures

3.1. General aspects

In the last couple of decades, it has been observed a growing interest in the use of autologous platelet concentrates during oral regenerative surgical procedures as an adjunctive tool to enhance the hard and soft tissue healing. The following sections will summarize the recent evidence on the efficacy of autologous platelet concentrates in the dental field. The recent evidence has been obtained using a systematic review approach. The focused question was: "Does the adjunct of autologous platelet concentrate produce benefits to hard and soft tissue healing in oral surgery procedures in terms of tissue parameters, postoperative complications as well as patient's postoperative quality of life?" In order to address the aim of this chapter, electronic searches were performed on the main scientific databases (MEDLINE, Scopus, and Cochrane Central Register of Controlled Trials). Proper search terms were used, combined by Boolean operators. Only controlled clinical trials, randomized clinical trials, and existing systematic reviews and meta-analyses of the literature were included. The outcomes were complications and adverse events, treatment success, discomfort/quality of life, bone healing and remodeling assessed by histological and radiographic techniques, and soft tissue healing. The surgical procedures taken into consideration were: tooth extraction, periodontal surgery, endodontic surgery and treatment of immature necrotic teeth, maxillary sinus augmentation, and implant treatment. When possible, a quantitative analysis was undertaken by meta-analysis approach, using the software RevMan (RevMan, Version 5.3, The Nordic Cochrane Center, The Cochrane Collaboration, Copenhagen, Denmark, 2014).

3.2. Alveolar postextraction healing

Several recent systematic reviews evaluated the efficacy of autologous platelet concentrates in enhancing alveolar socket healing after tooth extraction [23, 24]. Beneficial effects were generally reported in terms of better soft tissue healing, better clinical and histological epithelialization of wound margins, and a faster wound closure, although the heterogeneity of the data could not allow sound meta-analyses. Regarding the bone formation, the qualitative synthesis of the histological analyses reported a better bone quality in biopsies retrieved from sites treated with platelet concentrate. Furthermore, the meta-analysis of the histomorphometric evaluation of the bone formation (including only few of the included studies) revealed that sites treated with platelet concentrate showed a statistically significant greater proportion of new bone than the controls, at 3 months of follow-up [23]. Even though the results of the meta-analysis suggest a beneficial effect of autologous platelet concentrates on bone formation, caution should be paid on interpreting such results, since the available evidence is scarce and of limited quality [23]. In spite of the relatively numerous randomized clinical studies assessing the value of APCs for enhancing postextraction socket healing, the main reason that prevents a wide meta-analysis is that the methods used for assessing bone regeneration and socket preservation are very different, providing different information that cannot be aggregated. In fact, different studies used different techniques like histological and immunohistological analysis, histomorphometric evaluation, scintigraphy, micro-TC, intraoral radiography, cone beam computed tomography, and clinical assessment, for the evaluation of different variables as percentage of new bone formation, osteoblasts activity, bone density, crestal bone level changes, ridge width and height, and soft tissues health status [23, 24]. All studies that evaluated postoperative symptoms like pain, swelling, trismus, and adverse events like dry socket, alveolitis, and acute inflammation concluded that APCs are effective in reducing symptoms and the incidence of adverse events, with an overall improvement of patients' quality of life [24]. The most recent systematic review on this topic concludes that APCs should be used in postextraction sites in order to improve clinical and radiographic outcomes such as bone density and soft tissue healing, as well as to reduce postoperative symptoms [24]. The actual benefit of APCs on decreasing of pain in extraction sockets, however, though consistently reported, is still not quantifiable [24]. In **Figure 4**, a brief sequence of pictures documenting a

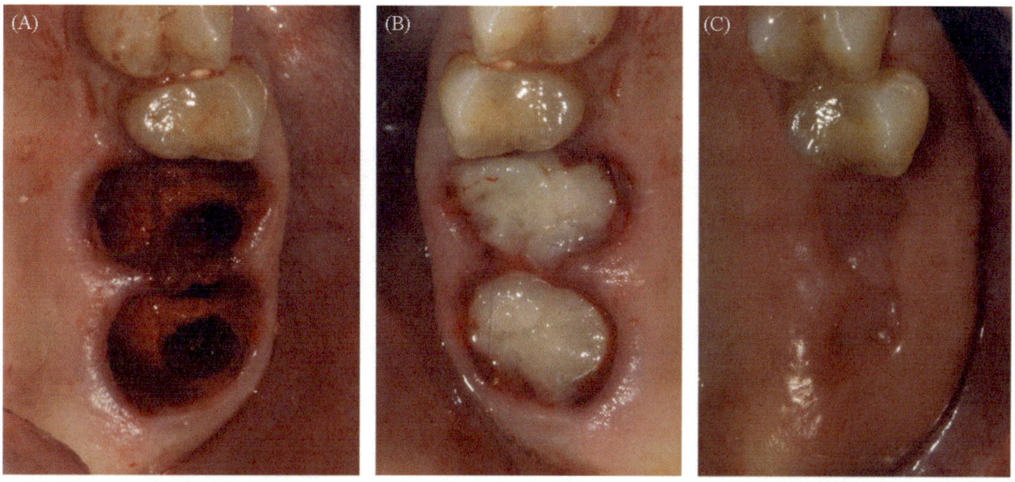

Figure 4. Double tooth extraction in the upper jaw and placement of PRGF in the extraction socket. (A) Fresh sockets after atraumatic extraction. (B) Sockets filled with PRGF clot; it can be sutured for a better stabilization. (C) 14 days after surgery: excellent healing of soft tissues is shown.

case of double extraction in the posterior upper jaw, subsequent positioning of platelet concentrate (PRGF) into extraction sockets, and the postsurgical clinical healing is shown.

3.3. Periodontal defects

Several systematic reviews and some meta-analyses evaluated the efficacy of autologous platelet concentrates in the treatment of periodontal defects, including intrabony defects, gingival recessions, and furcation defects [25–33]. Beneficial effects on clinical and radiographic outcomes in the treatment of infrabony defects were reported, although a high heterogeneity emerged among the clinical studies in terms of outcomes evaluated and bioactive agents/procedures combined with autologous platelet concentrates [25, 28, 30–32]. Furthermore, two meta-analyses [28, 30] concluded that PRP might exert positive adjunctive effects in the surgical treatment of such defects when combined with grafting materials, but no adjunctive effects were found in association with the guided tissue regeneration technique. Indeed, the latter is considered the gold standard for the treatment of periodontal intrabony defects and its use could probably mask the PRP effect. Regarding the autologous platelet concentrates' effect on gingival recessions, very few systematic reviews have been conducted, presumably due to the limited data about it. PRP or PRF did not show any clinical improvements in the treatment of gingival recessions or furcation defects [28, 33]. A Cochrane systematic review on this topic is still ongoing [34] and its results will certainly shed light on the actual evidence level regarding this topic, possibly confirming early indications of previous systematic reviews.

3.4. Endodontics and endodontic surgery

Platelet concentrates have been recently used in the clinical treatment of immature necrotic teeth, with the aim of regenerating the intracanal pulp and stimulating tooth development, as well as in the surgical treatment of teeth with apical periodontitis to enhance healing of periapical tissues. Clinical evidence on the benefits of the use of platelet concentrates in these pathologies exists but is still scarce. A recent systematic review concluded that periapical healing and apical closure were improved in those immature necrotic teeth treated with PRP compared with the control group without PRP, even though not statistically significant, and a significant better thickening of dentinal walls and root lengthening were also reported [35]. However, from the histological point of view, it seems that a true regeneration of necrotic pulp tissue of either mature or immature teeth was not achieved after using platelet concentrates. In fact, the neoformed intracanal tissues were mainly cemento-like, bone-like, and connective tissue. Root canals were repopulated with living tissue that only marginally resembled the original pulp. Despite this, the root maturation may be achieved and teeth function is not compromised [36].

Though the use of APCs in the management of periapical lesions could be considered a proper indication, very scarce studies are present regarding this topic. One randomized clinical study evaluated the postoperative quality of life in patients undergoing endodontic surgery [37]. The test group of 18 patients was treated with the adjunct of P-PRP and the control group was treated conventionally, without P-PRP. The test group showed significantly less pain and swelling, fewer analgesics taken, and improved functional activities as compared with the control group [37]. This suggests that the adjunct of P-PRP to the endodontic surgery procedure may produce significant beneficial effect to patients' quality of life during the early postoperative period.

Another pilot clinical study on endodontic surgery compared a group of seven patients treated with the adjunct of PRF versus four control patients [38]. In addition to confirming the beneficial effects in the early postoperative period regarding significant reduction of pain and swelling, a significantly better healing of the lesion was observed after 2 and 3 months but not after 1-year follow-up [38]. This preliminary study had a very small sample size, so results should be interpreted cautiously. The latter two studies have been included in a recently published Cochrane systematic review on endodontic procedures for the retreatment of periapical defects, which concluded that there is evidence that adjunctive use of a gel of plasma rich in growth factors reduces postoperative pain compared with no grafting [39]. Regarding other possible beneficial effects of APCs in endodontic surgery, further evidence is needed.

3.5. Maxillary sinus augmentation

The use of platelet concentrates in association with grafting material during maxillary sinus augmentation procedure provided conflicting results in both preclinical and clinical studies [40, 41]. A recent meta-analysis documented that PRP combined to graft materials, in this type of surgical procedure, had no adjunctive effect on bone formation, on implant survival and implant stability as well as it did not show any statistically significant differences on marginal bone loss or alveolar bone height, compared to the bone graft alone [42]. Similar conclusions were also reported in other systematic reviews [43, 44]. However, another meta-analysis reported opposite conclusions concerning the bone formation supporting the use of PRP for sinus bone graft [45]. Furthermore, beneficial effects on soft tissue healing as well as reduction of postoperative discomfort were often reported [43]. Such variability in results could be ascribed to a number of factors. First of all, different techniques have been adopted for the preparation of platelet concentrates, leading to the products with different characteristics (final concentration of platelets and white cells, presence or absence of leukocytes, use of anticoagulants and activators, different mechanical consistence of the product, and association with different graft types) and, presumably, different biological activities. Secondly, different studies may also differ in experimental design, objectives, outcome variables, inclusion criteria, and follow-up duration. Furthermore, the sinus augmentation technique, though representing a very popular model for the assessment of bone substitutes for bone regeneration, suffers from a number of confounding factors that make standardization difficult, like patient age, residual bone quality and quantity, smoking habits, volume of graft used, porosity and general features of the graft material (e.g., intraoral or extraoral autografts, allografts, xenografts, and alloplasts), graft resorbability over time, graft healing time, and intra- and postsurgical complications like membrane perforation, infection, expertise of the clinician, techniques adopted for evaluating bone formation, including the position of the biopsy (crestal or lateral). Finally, if one aims at evaluating the effect of APCs on implant survival and success in the maxillary sinus augmentation, a number of additional factors concerning the implants and the prosthetic reconstruction must be considered, e.g., implant length and width, shape, surface micro- and nano-geometry, type of implant-abutment connection, implant primary stability, and number and position of the implants. In fact, it is well known that implant survival in the augmented maxillary sinus is more variable than that of implants placed in the posterior maxilla.

A randomized clinical trial evaluating the effect of P-PRP adjunct on postoperative quality of life of patients undergoing maxillary sinus augmentation procedure, found a beneficial effect of P-PRP regarding pain, swelling, hematoma, and other postoperative symptoms, improving patient's acceptance of this often demanding procedure [46].

3.6. Implant dentistry

As it is claimed that platelet concentrates may promote bone regeneration, several animal studies have been conducted to assess the PRP effect on the osseointegration process, through histological and histomorphometrical evaluation, but controversial results have been reported. In fact, some studies did not demonstrate any advantages of PRP over non-PRP control groups at stimulating faster bone formation or higher bone-implant contact [47–49]. By contrast, histomorphometric analyses of the bone-implant interface in the early healing phase after implantation (6 or 8 weeks) revealed a significantly higher percentage of bone-implant contact in implants coated with liquid PRP formulation compared to those not PRP-bioactivated [50–52]. In addition to being time dependent, PRP effect is also site dependent since its effect has been reported to decrease with increasing distance from the site of application [52]. Similarly, liquid-PRP showed a tendency to increase the bone apposition to roughened titanium implants during early healing phase [53, 54].

Clinical studies reported a higher bone formation around the implants [55] and a good preservation of the alveolar crest around postextraction implants [56, 57] when APCs were used.

APCs have been also combined with several different types of grafting materials during regenerative procedures associated with implantoprosthesis rehabilitations, showing satisfying results and positive patient-related outcomes [58–61]. A long-term clinical study (10–12 years) on short implant placement in association with PRGF reported an implant survival rate of 98.9% and marginal bone loss inferior to 1 mm [62].

3.7. Systemically compromised patients

Patients affected by chronic systemic conditions like osteoporosis, cancer, diabetes type I or II, immunodeficiency, hematological/coagulation defects ,and other conditions, often present difficulty in healing even after simple surgical procedures, like the extraction of a tooth. Therefore, the use of a safe tool that may enhance the healing process in a natural way may represent a remarkable benefit for these patients.

A few studies have been published on the use of platelet concentrates in systemically compromised patients, most of them are regarding postextraction healing. Some examples will be reported.

3.8. Diabetes

In a split-mouth study in 34 patients with diabetes mellitus candidate to bilateral tooth extraction of a total of 127 teeth, alveolar sockets on the test side were treated with P-PRP and on the control side they were left to heal in a natural way [63]. The P-PRP group showed a significantly better healing (evaluated through the Healing Index) and a faster closure of the alveolus (by evaluation of the residual socket volume) at 3 and 7 days after extraction, as compared to the control group [63]. The authors concluded that platelet growth factors stimulate a faster epithelialization, protecting the alveolus in early healing steps. Hence, it is avoided occurrence of alveolitis, very common in diabetic patients following tooth extraction.

3.9. Irradiated patients

In a split-mouth study on 20 patients that underwent radiotherapy for head and neck cancer, and candidate to bilateral tooth extraction, alveolar sockets on the directly irradiated side were treated with P-PRP (test group) and on the untreated side they were left to heal in a natural way (control group) [64]. Twenty-four bilateral extractions were performed in the mandible and 33 in the maxilla for a total of 114 extractions. The P-PRP group showed a significantly better healing, in terms of Healing Index and residual socket volume at each follow-up (7, 14, and 21 days), as compared to the control group. Patients were followed up to 24 months after surgery. In the control group, two cases with bone exposure were retreated with P-PRP and subsequently healed [64].

3.10. Osteoradionecrosis (ORN)

In a series of 10 patients who developed osteoradionecrosis, debridement of necrotic bone (performed with ultrasonic instruments) was associated with the adjunct of P-PRP. All patients successfully healed, with no intraoperative or postoperative complications up to 12 months of follow-up [65]. Tissue regeneration and closure was excellent, and postoperative pain, assessed through visual analogue score (VAS) scores, was low. In spite of the absence of a control group, this study suggested that P-PRP may be beneficial as an adjunct to surgical treatment of ORN, for predictable enhancement of tissue vascularization and epithelialization in patients with a history of head and neck radiotherapy.

3.11. Coagulation defects

A case-control study was performed on 66 patients affected by severe thrombocytopenia (<50,000 platelets/μL) and in need for at least two tooth extraction each [66]. For these patients, postoperative bleeding represents an important issue. Teeth were extracted in two consecutive interventions. In one intervention, the patients received a platelet transfusion before extraction (systemic treatment) and in the other session, the postalveolar sockets were filled with P-PRP (local treatment). Patients were evaluated frequently in the first 7 days after extraction. The group treated locally with P-PRP showed a statistically significant reduction in postoperative bleeding, hematoma, and need for reintervention, as compared to the group receiving a systemic infusion of platelets [66]. Therefore, P-PRP in postalveolar sockets may strongly reduce the risk of hemorrhage and related complications in thrombocytopenic patients, due to the ability of stimulating healing and its hemostatic properties.

3.12. Bisphosphonate-related osteonecrosis of the jaws

Recently renamed medication-related osteonecrosis of the jaw (MRONJ), it is an adverse drug reaction consisting of progressive bone destruction in the maxillofacial region of patients under current or previous treatment with a bisphosphonate or another antiresorptive or antiangiogenic drug. Since APCs demonstrated to enhance bone and soft tissue healing in oral surgery procedures, it is reasonable to believe that they might provide benefits to these patients. A recent systematic review included 18 studies, reporting on 362 patients undergoing oral surgery in combination with APCs [67]. The adjunct of APC in the surgical treatment of necrotic bone removal significantly reduced osteonecrosis recurrence as compared to control (**Figure 5**). APC was associated with a reduced, though not significant, incidence of BRONJ after tooth extraction. Heterogeneity among studies was found regarding bisphosphonate type, clinical indication, administration route, treatment duration, triggering factors, study design, follow-up duration, type of APC, and outcomes adopted to evaluate treatment success [67]. Though the results of this review must be cautiously interpreted, since they are based on low evidence level studies, with limited sample size, they are suggestive of possible beneficial effects of APC when associated with surgical procedures for treatment or prevention of BRONJ. To confirm such indication, prospective comparative studies with a large sample size are urgently needed. Another subsequent systematic review on the same topic substantially confirmed these results, highlighting the need for well-done evidence-based comparative studies [68].

3.13. Implants in patients assuming bisphosphonates

In a multicenter study on a cohort of 235 middle-aged osteoporotic women under bisphosphonates therapy, the outcome of a total of 1267 implants was evaluated after a minimum follow-up of 24 months [69]. The implants were always placed in combination with P-PRP, used as a coating over implant surface at insertion (**Figure 6**). The main outcomes (adverse

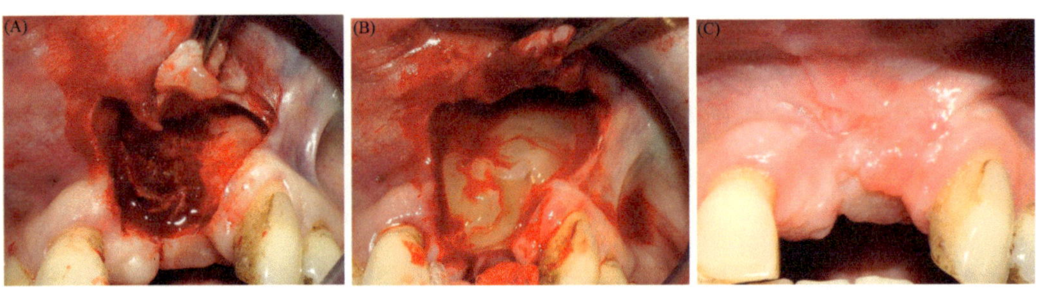

Figure 5. (A) A case of an oncologic patient under bisphosphonate affected by osteonecrosis of the upper jaw, undergoing respective surgery for removal of the necrotic bone. (B) PRGF is placed within the region involved from resection. (C) One year after surgery, the region is completely healed.

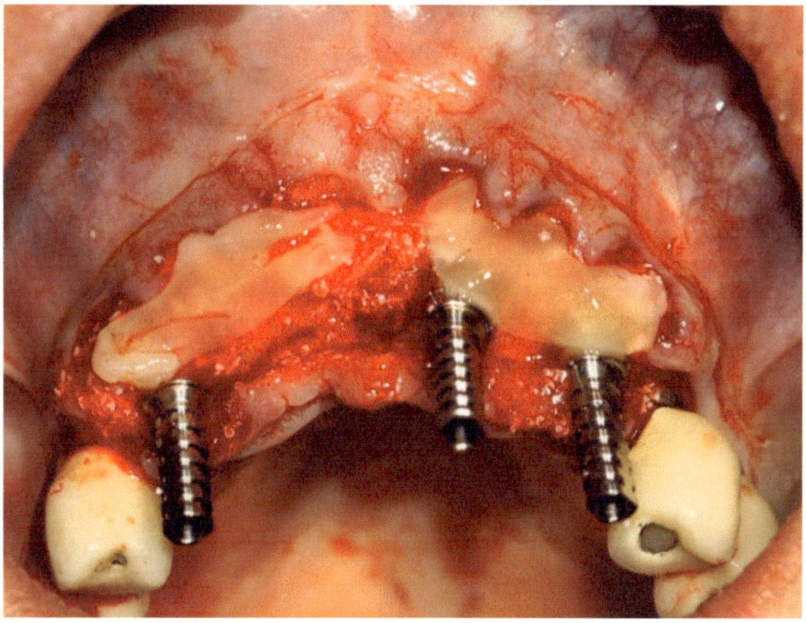

Figure 6. Multiple implant placements in an osteoporotic woman under bisphosphonates. PRGF membrane was used to cover the region involved from implant surgery.

events) were implant failure and incidence of osteonecrosis of the jaw (ONJ). Sixteen implants were lost in 16 patients up to 120 months after placement, representing a survival of 98.7% on an implant basis and 93.2% on a patient basis [69] No cases of ONJ occurred. The results are perfectly in line with those of healthy patients.

Overall, the above results suggest that the use of procedures aimed at enhancing tissue healing, such as autologous platelet concentrates, may produce relevant benefits in patients at risk due to their compromised systemic condition and should be recommended.

4. Conclusion

The use of autologous platelet concentrates generally produced beneficial effects, though the level of evidence differs among various surgical procedures. In postextraction sockets and periodontal intrabony defects, the advantage of using APCs, both alone and combined with bone substitutes, is well documented by a number of randomized clinical studies. In maxillary sinus augmentation, controversial outcomes exist, due to few published controlled studies. Also, in endodontic surgery and implant treatment, there is a paucity of evidence-based studies, even though all show beneficial effects of APC. The variability among protocols and outcomes in different studies often prevents the possibility of performing meta-analysis and is thought to be related to the controversial results sometimes observed. Better soft tissue healing, improved patients' quality of life and reduced incidence of adverse events and complications are systematically reported when using APC. In conclusion, the use of such autologous products is recommended for improving predictability and patients' acceptance of treatment in oral surgery procedures.

Author details

Carmen Mortellaro[1]* and Massimo Del Fabbro[2,3]

*Address all correspondence to: carmen.mortellaro@med.uniupo.it

1 Department of Health Sciences, Avogadro University of Eastern Piedmont, Novara, Italy

2 Department of Biomedical, Surgical and Dental Sciences, Università degli Studi di Milano, Milano, Italy

3 IRCCS Istituto Ortopedico Galeazzi, Milano, Italy

References

[1] Marx RE, Carlson ER, Eichstaedt RM, Schimmele SR, Strauss JE, Georgeff KR. Platelet-rich plasma: Growth factor enhancement for bone grafts. Oral Surgery, Oral Medicine, Oral Pathology, Oral Radiology, and Endodontology. 1998;**85**:638-646

[2] Anitua E, Andia I, Ardanza B, et al. Autologous platelets as a source of proteins for healing and tissue regeneration. Thrombosis and Haemostasis. 2004;**91**:4-15

[3] Alsousou J, Thompson M, Hulley P, et al. The biology of platelet-rich plasma and its application in trauma and orthopaedic surgery: A review of the literature. Journal of Bone and Joint Surgery. British Volume. 2009;**91**:987-996

[4] Boswell SG, Cole BJ, Sundman EA, et al. Platelet-rich plasma: A milieu of bioactive factors. Arthroscopy. 2012;**28**:429-439

[5] Pelletier MH, Malhotra A, Brighton T, et al. Platelet function and constituents of platelet rich plasma. International Journal of Sports Medicine. 2013;**34**:74-80

[6] Matras H. Effect of various fibrin preparations on reimplantations in the rat skin. Osterreichische Zeitschrift fur Stomatologie. 1970;**67**:338-359

[7] Rosenthal AR, Harbury C, Egbert PR, Rubenstein E. Use of a platelet-fibrinogen-thrombin mixture as a corneal adhesive: Experiments with sutureless lamellar keratoplasty in the rabbit. Investigative Ophthalmology. 1975;**14**:872-875

[8] Rosenthal AR, Egbert PR, Harbury C, Hopkins JL, Rubenstein E. Use of platelet-fibrinogen-thrombin mixture to seal experimental penetrating corneal wounds. Albrecht von Graefe's archive for clinical and experimental ophthalmology. 1978;**207**:111-115

[9] Pearl RM, Wustrack KO, Harbury C, Rubenstein E, Kaplan EN. Microvascular anastomosis using a blood product sealant-adhesive. Surgery, Gynecology & Obstetrics. 1977;**144**: 227-231

[10] Silverberg GD, Harbury CB, Rubenstein E. A physiological sealant for cerebrospinal fluid leaks. Journal of Neurosurgery. 1977;**46**:215-219

[11] Fischer H. A method of suture-free anastomosis of nerve transplantation is being reported, using facial nerve as the example (author's transl). Laryngologie, Rhinologie, Otologie. 1979;**58**:154-156

[12] Knighton DR, Ciresi KF, Fiegel VD, Austin LL, Butler EL. Classification and treatment of chronic nonhealing wounds. Successful treatment with autologous platelet-derived wound healing factors (PDWHF). Annals of Surgery. 1986;**204**:322-330

[13] Knighton DR, Doucette M, Fiegel VD, Ciresi K, Butler E, Austin L. The use of platelet derived wound healing formula in human clinical trials. Progress in Clinical and Biological Research. 1988;**266**:319-329

[14] Knighton DR, Ciresi K, Fiegel VD, Schumerth S, Butler E, Cerra F. Stimulation of repair in chronic, nonhealing, cutaneous ulcers using platelet-derived wound healing formula. Surgery, Gynecology & Obstetrics. 1990;**170**:56-60

[15] Whitman DH, Berry RL, Green DM. Platelet gel: An autologous alternative to fibrin glue with applications in oral and maxillofacial surgery. Journal of Oral and Maxillofacial Surgery. 1997;**55**:1294-1299

[16] Kingsley CS. Blood coagulation; evidence of an antagonist to factor VI in platelet-rich human plasma. Nature. 1954;**173**:723-734

[17] Dohan Ehrenfest DM, Rasmusson L, Albrektsson T. Classification of platelet concentrates: From pure platelet-rich plasma (P-PRP) to leucocyte- and platelet-rich fibrin (L-PRF). Trends in Biotechnology. 2009;**27**:158-167

[18] Choukroun J, Adda F, Schoeffler C, Vervelle A. Une opportunité en paro-implantologie: Le PRF. Implantodontie. 2001;**42**:55-62

[19] Dohan Ehrenfest DM, Del Corso M, Diss A, Mouhyi J, Charrier JB. Three-dimensional architecture and cell composition of a Choukroun's platelet-rich fibrin clot and membrane. Journal of Periodontology. 2010;**81**:546-555

[20] Ghanaati S, Booms P, Orlowska A, Kubesch A, Lorenz J, Rutkowski J, Landes C, Sader R, Kirkpatrick CJ, Choukroun J. Advanced platelet-rich fibrin: A new concept for cell-based tissue engineering by means of inflammatory cells. Journal of Oral Implantology. 2014;**40**:679-689

[21] Anitua E. Plasma rich in growth factors: Preliminary results of use in the preparation of sites for implants. The International Journal of Oral & Maxillofacial Implants. 1999;**14**:529-535

[22] Anitua E. The use of plasma-rich growth factors (PRGF) in oral surgery. Practical Procedures & Aesthetic Dentistry. 2001;**13**:487-493

[23] Del Fabbro M, Corbella S, Taschieri S, Francetti L, Weinstein R. Autologous platelet concentrate for post-extraction socket healing: A systematic review. European Journal of Oral Implantology. 2014;**7**:333-344

[24] Del Fabbro M, Bucchi C, Lolato A, Corbella S, Testori T, Taschieri S. Healing of postextraction sockets preserved with autologous platelet concentrates. A systematic review and meta-analysis. Journal of Oral and Maxillofacial Surgery. 2017. http://dx.doi.org/10.1016/j.joms.2017.02.009

[25] Plachokova AS, Nikolidakis D, Mulder J, Jansen JA, Creugers NH. Effect of platelet-rich plasma on bone regeneration in dentistry: A systematic review. Clinical Oral Implants Research. 2008;**19**:539-545

[26] Martínez-Zapata MJ, Martí-Carvajal A, Solà I, Bolibar I, Angel Expósito J, RodriguezL, García J. Efficacy and safety of the use of autologous plasma rich in platelets for tissue regeneration: A systematic review. Transfusion. 2009;**49**:44-56

[27] Kotsovilis S, Markou N, Pepelassi E, Nikolidakis D. The adjunctive use of platelet-rich plasma in the therapy of periodontal intraosseous defects: A systematic review. Journal of Periodontal Research. 2010;**45**:428-443

[28] Del Fabbro M, Bortolin M, Taschieri S. Weinstein R. Is platelet concentrate advantageous for the surgical treatment of periodontal diseases? A systematic review and meta-analysis. Journal of Periodontology. 2011;**82**:1100-1111

[29] Del Fabbro M, Ceci C, Taschieri T. Systematic review on the effect of platelet concentrates for the surgical treatment of periodontal defects. Dental Cadmos. 2013;**81**:138-145

[30] Panda S, Doraiswamy J, Malaiappan S, Varghese SS, Del Fabbro M. Additive effect of autologous platelet concentrates in treatment of intrabony defects: A systematic review and meta-analysis. Journal of Investigative and Clinical Dentistry. 2014;**5**:1-14

[31] Shah M, Deshpande N, Bharwani A, Nadig P, Doshi V, Dave D. Effectiveness of autologous platelet-rich fibrin in the treatment of intra-bony defects: A systematic review and meta-analysis. Journal of Indian Society of Periodontology. 2014;**18**:698-704

[32] Roselló-Camps À, Monje A, Lin GH, Khoshkam V, Chávez-Gatty M, Wang HL, Gargallo-Albiol J, Hernandez-Alfaro F. Platelet-rich plasma for periodontal regeneration in the treatment of intrabony defects: A meta-analysis on prospective clinical trials. Oral Surgery, Oral Medicine, Oral Pathology and Oral Radiology. 2015;**120**:562-574

[33] Moraschini V, Barboza Edos S. Use of platelet-rich fibrin membrane in the treatment of gingival recession: A systematic review and meta-analysis. Journal of Periodontology. 2016;**87**:281-290

[34] Del Fabbro M, Panda S, Jayakumar ND, Sankari M, Varghese S, Ramamoorthi S, Ceci C, Ceresoli V, Taschieri S. Autologous platelet concentrates for treatment of periodontal

defects (Protocol). Cochrane Database of Systematic Reviews. 2014;(12). Art.No.: CD011423. DOI: 10.1002/14651858.CD011423

[35] Lolato A, Bucchi C, Taschieri S, Kabbaney AE, Fabbro MD. Platelet concentrates for revitalization of immature necrotic teeth: A systematic review of the literature of the clinical studies. Platelets. 2016;**27**:383-392

[36] Del Fabbro M, Lolato A, Bucchi C, Taschieri S, Weinstein RL. Autologous platelet concentrates for pulp and dentin regeneration: A literature review of animal studies. Journal of Endodontics. 2016;**42**:250-257

[37] Del Fabbro M, Ceresoli V, Lolato A, Taschieri S. Effect of platelet concentrate on quality of life after periradicular surgery: A randomized clinical study. Journal of Endodontics. 2012;**38**:733-739

[38] Angerame D, De Biasi M, Kastrioti I, Franco V, Castaldo A, Maglione M. Application of platelet-rich fibrin in endodontic surgery: A pilot study. Giornale Italiano di Endodonzia. 2015;**29**:51-57

[39] Del Fabbro M, Corbella S, Sequeira-Byron P, Tsesis I, Rosen E, Lolato A, Taschieri S. Endodontic procedures for retreatment of periapical lesions. Cochrane Database of Systematic Reviews. 2016;(10). Art. No.: CD005511. DOI: 10.1002/14651858.CD005511.pub3

[40] Choi BH, Im CJ, Huh JY, Suh JJ, Lee SH. Effect of platelet-rich plasma on bone regeneration in autogenous bone graft. International Journal of Oral and Maxillofacial Surgery. 2004;**33**:56-59

[41] Thor A, Franke-Stenport V, Johansson CB, Rasmusson L. Early bone formation in human bone grafts treated with platelet-rich plasma: Preliminary histomorphometric results. International Journal of Oral and Maxillofacial Surgery. 2007;**36**:1164-1171

[42] Lemos CA, Mello CC, dos Santos DM, Verri FR, Goiato MC, Pellizzer EP. Effects of platelet-rich plasma in association with bone grafts in maxillary sinus augmentation: A systematic review and meta-analysis. International Journal of Oral and Maxillofacial Surgery. 2016;**45**:517-525

[43] Del Fabbro M, Bortolin M, Taschieri S, Weinstein RL. Effect of autologous growth factors in maxillary sinus augmentation: A systematic review. Clinical Implant Dentistry and Related Research. 2013;**15**:205-216

[44] Rickert D, Slater JJ, Meijer HJ, Vissink A, Raghoebar GM. Maxillary sinus lift with solely autogenous bone compared to a combination of autogenous bone and growth factors or (solely) bone substitutes. A systematic review. International Journal of Oral and Maxillofacial Surgery. 2012;**41**:160-167

[45] Bae JH, Kim YK, Myung SK. Effects of platelet-rich plasma on sinus bone graft: Meta-analysis. Journal of Periodontology. 2011;**82**:660-667

[46] Del Fabbro M, Corbella S, Ceresoli V, Ceci C, Taschieri S. Plasma rich in growth factors improves patients' postoperative quality of life in maxillary sinus floor augmentation:

Preliminary results of a randomized clinical study. Clinical Implant Dentistry and Related Research. 2015;**17**:708-716

[47] Streckbein P, Kleis W, Buch RS, Hansen T, Weibrich G. Bone healing with or without platelet-rich plasma around four different dental implant surfaces in beagle dogs. Clinical Implant Dentistry and Related Research. 2014;**16**:479-486

[48] Thor AL, Hong J, Kjeller G, Sennerby L, Rasmusson L. Correlation of platelet growth factor release in jawbone defect repair-a study in the dog mandible. Clinical Implant Dentistry and Related Research. 2013;**15**:759-768

[49] Weibrich G, Hansen T, Kleis W, Buch R, Hitzler WE. Effect of platelet concentration in platelet-rich plasma on peri-implant bone regeneration. Bone. 2004;**34**:665-671

[50] Anitua EA. Enhancement of osseointegration by generating a dynamic implant surface. Journal of Oral Implantology. 2006;**32**:72-76

[51] Anitua E, Orive G, Pla R, Roman P, Serrano V, Andía I. The effects of PRGF on bone regeneration and on titanium implant osseointegration in goats: A histologic and histomorphometric study. Journal of Biomedical Materials Research Part A. 2009;**91**:158-165

[52] Zechner W, Tangl S, Tepper G, Fürst G, Bernhart T, Haas R, Mailath G, Watzek G. Influence of platelet-rich plasma on osseous healing of dental implants: A histologic and histomorphometric study in mini pigs. The International Journal of Oral & Maxillofacial Implants. 2003;**18**:15-22

[53] Nikolidakis D, van den Dolder J, Wolke JG, Stoelinga PJ, Jansen JA. The effect of platelet-rich plasma on the bone healing around calcium phosphate-coated and non-coated oral implants in trabecular bone. Tissue Engineering. 2006;**12**:2555-2563

[54] Nikolidakis D, van den Dolder J, Wolke JG, Jansen JA. Effect of platelet-rich plasma on the early bone formation around Ca-P-coated and non-coated oral implants in cortical bone. Clinical Oral Implants Research. 2008;**19**:207-213

[55] Georgakopoulos I, Tsantis S, Georgakopoulos P, Korfiatis P, Fanti E, Martelli M, Costaridou L, Petsas T, Panayiotakis G, Martelli FS. The impact of platelet rich plasma (PRP) in osseointegration of oral implants in dental panoramic radiography: Texture based evaluation. Clinical Cases in Mineral and Bone Metabolism. 2014;**11**:59-66

[56] Kutkut A, Andreana S, Monaco Jr E. Clinical and radiographic evaluation of single-tooth dental implants placed in grafted extraction sites: A one-year report. Journal of the International Academy of Periodontology. 2013;**15**:113-124

[57] Rosano G, Taschieri S, Del Fabbro M. Immediate postextraction implant placement using plasma rich in growth factors technology in maxillary premolar region: A new strategy for soft tissue management. Journal of Oral Implantology. 2013;**39**:98-102

[58] Anitua E, Alkhraisat MH, Miguel-Sánchez A, Orive G. Surgical correction of horizontal bone defect using the lateral maxillary wall: Outcomes of a retrospective study. Journal of Oral and Maxillofacial Surgery. 2014;**72**:683-693

[59] Jeong SM, Lee CU, Son JS, Oh JH, Fang Y, Choi BH. Simultaneous sinus lift and implantation using platelet-rich fibrin as sole grafting material. Journal of Cranio-Maxillofacial Surgery. 2014;**42**:990-994

[60] Tajima N, Ohba S, Sawase T, Asahina I. Evaluation of sinus floor augmentation with simultaneous implant placement using platelet-rich fibrin as sole grafting material. Journal of Oral and Maxillofacial Surgery. 2013;**28**:77-83

[61] Taschieri S, Corbella S, Del Fabbro M. Mini-invasive osteotome sinus floor elevation in partially edentulous atrophic maxilla using reduced length dental implants: Interim results of a prospective study. Clinical Implant Dentistry and Related Research. 2014;**16**:185-193

[62] Anitua E, Piñas L, Begoña L, Orive G. Long-term retrospective evaluation of short implants in the posterior areas: Clinical results after 10-12 years. Journal of Clinical Periodontology. 2014;**41**:404-411

[63] Mozzati M, Gallesio G, di Romana S, Bergamasco L, Pol R. Efficacy of plasma-rich growth factor in the healing of postextraction sockets in patients affected by insulin-dependent diabetes mellitus. Journal of Oral and Maxillofacial Surgery. 2014;**72**:456-462

[64] Mozzati M, Gallesio G, Gassino G, Palomba A, Bergamasco L. Can plasma rich in growth factors improve healing in patients who underwent radiotherapy for head and neck cancer? A split-mouth study. Journal of Craniofacial Surgery. 2014;**25**:938-943

[65] Gallesio G, Del Fabbro M, Pol R, Mortellaro C, Mozzati M. Conservative treatment with plasma rich in growth factors-Endoret for osteoradionecrosis. Journal of Craniofacial Surgery. 2015;**26**:731-736

[66] Cocero N, Mozzati M, Bergamasco L. Oral surgery in severe thrombocytopenia patients: A case-control comparison of platelet concentrate versus platelet transfusion. Indian Journal of Dental Research. 2012;**2**:27-32

[67] Del Fabbro M, Gallesio G, Mozzati M. Autologous platelet concentrates for bisphosphonate-related osteonecrosis of the jaw treatment and prevention. A systematic review of the literature. European Journal of Cancer. 2015;**51**:62-74

[68] Lopez-Jornet P, Sanchez-Perez A, Amaral Mendes R, Tobias A. Medication-related osteonecrosis of the jaw: Is autologous platelet concentrate application effective for prevention and treatment? A systematic review. Journal of Cranio-Maxillofacial Surgery. 2016;**44**:1067-1072

[69] Mozzati M, Arata V, Giacomello M, Del Fabbro M, Gallesio G, Mortellaro C, Bergamasco L. Failure risk estimates after dental implants placement associated with plasma rich in growth factor-Endoret in osteoporotic women under bisphosphonate therapy. Journal of Craniofacial Surgery. 2015;**26**:749-755

Chapter 9

Laser Processing of Silicon for Synthesis of Better Biomaterials

Candace Colpitts and Amirkianoosh Kiani

Additional information is available at the end of the chapter

http://dx.doi.org/10.5772/intechopen.69856

Abstract

The increasing demand for new biomaterials and fabrication methods provides an opportunity for silicon to solve current challenges in the field. Laser processing is becoming more common as the public begins to understand its simplicity and value. When an abundant material is paired with a reliable and economic fabrication method, biomedical devices can be created and improved. In this chapter, different laser parameters of the Nd:YAG laser are investigated and the topographic and physical trends are analyzed. The biocompatibility is assessed for scanning speed, line spacing, overlap number, pulse frequency, and laser power with the use of simulated body fluid (SBF) and fibroblast culturing (NIH 3T3). Not only can nanosecond pulses increase the biocompatibility of silicon by generating silicon oxide nanofibers, but the substrate becomes bioactive with the manipulation of cell interactions.

Keywords: silicon, nanofibers, Nd:YAG laser, fibroblasts

1. Introduction

Science fiction has motivated intelligent minds to enhance the quality of living for the last century. A well-known example in fantasy is bionic limbs controlled by the mind. Individuals who have lost or permanently damaged limbs can benefit from procuring an aesthetically pleasing and fully functioning bionic replacement to restore or improve their quality of life. The field of biomaterials engineering has been making monumental advances by producing devices such as biosensors, bioMEMS, and artificial hearts [1–3]. There is a continuous growth in population today, demanding the attention from biomedical fields to improve lifestyles and create better body functionality. Although current devices' interfaces with the human body have come a long way, there is still a long way to go in the fabrication methods of the required scaffolds.

This chapter outlines one single pathway of research done to broaden the opportunities in the biomaterials industry. The bioactivity of laser-treated silicon is investigated through the use of *in vitro* testing. This research investigates of the trends of different laser parameters, including power, frequency, scanning speed, line spacing, and overlap number.

1.1. Challenges in the biomaterials industry

The current challenges in the biomedical field include finding biocompatible and bioactive organic or inorganic materials and simplifying manufacturing processes. Devices that are implanted inside the body require materials that are biocompatible; the behavior of this material when interacting with the human body must not have any toxic effects and must perform a specific task. For a material to be bioactive, it requires to be biocompatible and have a biological effect and provoke a positive and controlled biological response. Current biocompatible materials that are commonly used are gold, titanium, polymers, bioceramics, and composites [4–8].

Silicon was chosen as the material for this research due to its abundance and semiconductor abilities. Microfabricated silicon is widely used today in the microelectronics and photovoltaics industry [9, 10]. Silicon in its pure form is not biocompatible [11, 12]. However, silicon can be packaged in a biocompatible material such as titanium [13, 14]. It has been found that porous silicon is biocompatible [11]. The current method used for creating a porous layer is etching with hydrochloric acid. Acid etching is a long process that requires the use of dangerous chemicals and is consequently environmentally friendly. The challenge that silicon faces in the biomaterials industry is to find a superior surface alteration method.

1.2. Laser processing

Technology that easily controls and creates an accurate pattern on a microscale is required in the microelectronics industry. A good solution to this criterion is a laser, which has been commonly used for surface texturing of steel [15, 16]. It was found that this method of surface treatment allowed the generation of micropores with different characteristics. Unlike acid etching, a laser is great for the modification of silicon since it is very clean, high resolution, and controllability of intensity and depth of penetration. The Nd:YAG pulsed laser is a particularly good solution since it is cost-effective, stable, and has the required high power range. Another advantage to this method is that there are no chemicals involved, which eliminates the complex processes of preparation and environmental concerns. Above all, using a laser is a single-step process. The economic and simplistic benefits that are associated with this approach are valuable to the biomedical industry.

High-end picosecond and femtosecond pulsed laser systems have also been used for generation of porous silicon particles [17, 18]. In this research, it is found that a Nd:YAG nanosecond pulse laser can achieve the desired biocompatible silicon. The nanosecond laser is much more economical and commercially available than the faster pulse lasers. The nanosecond laser is also currently used in the medical industry for procedures such as eye and dental surgeries [19, 20].

2. Laser processing and surface characterization

Microfabrication with lasers is becoming increasingly popular in many industries including biomaterials [21]. Laser irradiation introduces surface irregularities and chemical changes to the silicon surface. The laser irradiates a simple line pattern onto pure silicon with <100> orientation. There are a number of methods used to analyze the condition of the laser-processed substrat. Images of the samples are taken with field emission scanning electron microscopy (FESEM) and 3D optical microscopy.

2.1. Laser system

The laser used in this research is a SOL-20 1064 nm Nd:YAG nanosecond laser by Bright Solutions Inc. The JD2204 Sino Galvo two-axis Galvo scanner has an input of 10 mm and a beam displacement of 13.4 mm. The theoretically determined spot size diameter is 19 μm. The laser pulses can range from lengths of 6 to 35 ns, frequencies of 10 to 100 kHz, powers up to 20 W, and scanning speeds up to 3000 mm/s. For this research, scanning speeds of 100–1000 mm/s, powers of 7, 10, and 15 W, and frequencies of 25, 70, and 100 kHz are used. Line spacing varies from 0.025, 0.05, and 0.10 mm. Overlap number, or number of times the laser repeats the same pattern, varies from 1, 2, and 3. The manipulation of these parameters is easily executed through the laser-operating software.

2.2. Biocompatibility evaluation

The biocompatibility of a material is influenced by surface roughness, reflectivity, and chemical content of the substrate. The chemical content is assessed using micro-Raman and energy-dispersive X-ray (EDX) analysis. The surface roughness is determined with the use of 3D optical microscopy, and the reflectivity is determined with light spectroscopy. The biocompatibility is also determined with the use of simulated body fluid, which is a form of *in vitro* assessment—testing done outside of the body. Simulated body fluid is a solution that mimics the ion concentration of human blood plasma. When a biocompatible material is submerged in the liquid, hydroxyapatite forms on the surface [22, 23]. The submerged samples in the SBF are kept in an incubator at 36.5°C for up to 6 weeks.

Cell interactions with the laser-processed silicon substrate are also evaluated with cultured mouse embryonic fibroblasts (NIH 3T3). Cells are seeded at 2400 cells/cm^2 in triplicate and incubated for 72 h at 37°C. The samples are incubated under 5% CO_2 in Dulbecco's modified Eagle medium (DMEM) supplemented with 10% heat-inactivated calf serum. Phosphate-buffered saline (PBS) was then used to rinse the nonadherent cells from the samples overnight at 4°C. Staining was then done to the samples with phalloidin (1:2000 dilution) and draq5 (1:10,000 dilution) overnight for fluorescence imaging.

2.3. Temperature evaluation

Different frequencies, powers, and pulse widths change the behavior of the laser pulses. Determining the temperature will help investigate the pulse energy and how it affects the surface topographic properties and chemical structure. The surface temperature is modeled using the two-dimensional heat equation in cylindrical coordinates. The heat equation in cylindrical coordinates is found in Eq. (1).

$$\frac{1}{r}\frac{\partial}{\partial r}\left(\kappa r\frac{\partial T}{\partial r}\right)+\frac{1}{r^2}\frac{\partial}{\partial \phi}\left(\kappa\frac{\partial T}{\partial \phi}\right)+\frac{\partial}{\partial z}\left(\kappa\frac{\partial T}{\partial z}\right)+\dot{q}=\rho c_p\frac{\partial T}{\partial t} \quad (1)$$

where κ is the heat conduction coefficient, ρ is the material density (kg/m^3), c_p is the specific heat (J/kgK), and $\dot{q}$ is the rate at which energy is generated per unit volume of the medium (W/m^3) [24]. Since it is assumed that κ, ρ, and c_p are constant, and there is no energy generation within the silicon ($\dot{q}=0$), Eq. (1) can be simplified into Eq. (2).

$$\frac{1}{r}\frac{\partial}{\partial r}\left(r\frac{\partial T}{\partial r}\right)+\frac{\partial}{\partial z}\left(\frac{\partial T}{\partial z}\right)=\frac{1}{a}\frac{\partial T}{\partial t} \quad (2)$$

where $a = \kappa/\rho c_p$, which is the thermal diffusivity (m^2/s). The boundary conditions for this equation include the initial temperature being room temperature, the pulse intensity is at its maximum during the pulse at $z = 0$, the intensity is zero between pulses, and the temperature change is zero at $r = \infty$, and $z = \infty$. With these conditions, the single-pulse temperature change of a high absorption material, ΔT, for a square pulse, can be obtained (Eq. (3)) [25].

$$\Delta T(r,z,\tau)=\frac{I_{max}\gamma\sqrt{\kappa}}{\sqrt{\pi}K}\int_0^{\tau}\frac{1}{\sqrt{t}\left[1+\frac{8\kappa t}{W^2}\right]}e^{-\left[\frac{z^2}{4\kappa t}+\frac{r^2}{4\kappa t+\frac{1}{2}W^2}\right]}dt \quad (3)$$

$$\text{where } I_{max}=\frac{P_{peak}}{A}=\frac{4P_{measured}}{\pi d^2 f\tau} \quad (4)$$

where I_{max} (Eq. (4)) is the peak intensity which is the peak power divided by the spot area, $P_{measured}$ is the experimental measured power, f is the pulse frequency, τ is the pulse duration, γ is equal to the Fresnel energy reflectivity (R) subtracted from 1 (1–R) with a R value of 0.325 and a γ value of 0.675, κ is Silicon's diffusivity with a value of 9.07×10^{-3} m^2/s, K is Silicon's conductivity with a value of 155 W/mK, τ is the laser pulse duration, W is the beam's filed radius (1/e) with a value of 1.94×10^{-5} m, z is the ablation pit depth, and r is the ablation pit radius. Using this equation with the assumption that the pulse is square-shaped, the temperature can be determined at the center of ablation (r = 0) and at the surface (z = 0). From Eq. (3), an analytical expression can be made to determine the depth of the ablated groove at the center of the ablation with respect to radius by using the mean value theorem.

$$h(r)=\sqrt{-4\kappa\tau\ln\left\{\frac{\beta K\Delta T_B}{\gamma I_{max}}\sqrt{\frac{\pi}{\kappa\beta\tau}}\left(1+\frac{8\beta\kappa\tau}{W^2}\right)\right\}-\frac{r^2}{1+\frac{W^2}{8\beta\kappa\tau}}}\leq h(0) \quad (5)$$

where β is an experimentally determined correction factor of 0.5 and ΔT_B is silicon's boiling temperature of 2972 K. A more accurate representation of the experimental results found in

Eqs. (6) and (7) will determine the ablation depth after a train of pulses. Each consecutive pulse adds to the penetration of the preceding pulse, S, resulting in a deeper groove.

$$h_{scan}(r) = h(r) + h(S - r) \tag{6}$$

$$S = \frac{v\_scan}{f} \tag{7}$$

where S is the spatial separation between each consecutive pulse, v_scan is the scanning speed in m/s, and f is the pulse frequency.

3. Effect of scanning parameters

The three scanning parameters discussed in this section are overlap number, line spacing, and scanning speed. Each of these is easily set through the laser-operating software. These parameters have a direct effect of the surface topography and oxidation levels of the silicon substrates.

3.1. Topography analysis

The field emission scanning electron microscope images are an effective way of investigating the physical results from the laser ablation on the silicon samples. **Figure 1** shows the effect of different overlap numbers. At 1 overlap (OL), the line pattern is distinct and relatively clean. When the OL number is increased to 2, the line pattern is less definite and contains more irregularities. Finally, increasing the overlap number once again to 3, the line pattern is almost unrecognizable with a substantial amount of imperfections.

The effect of different line spacing was then investigated with FESEM. The overlap number was kept constant at 1. At the largest line spacing of 0.10 mm, the line pattern is discrete. At the smallest line spacing of 0.025 mm, the amount of imperfections is high with no distinctive line separation. The effect of line spacing can be seen in **Figure 2**.

Knowing that a higher overlap number and a smaller line spacing made for the highest level of laser-ablated substrate, a sample with 0.025 mm line spacing and three overlaps was created to observe the surface characteristics. The high magnification FESEM image in **Figure 3** of this sample shows a nanofibrous substrate. These nanofibers are the result of a high-energy reactive plume that forms on the surface during laser ablation [26]. The plume generates a heat-affected zone that causes the silicon ions to react with the oxygen ions, creating these nanoscale SiO_2 fibers [27–29].

The effect of scanning speed was then investigated with 3D optical microscopy. The data of the results from scanning speeds of 100, 200, 500, 800, and 1000 mm/s are mapped and compared in **Figure 4**. It is expected that the lower scanning speeds have larger depths due to a higher number of pulses ablating the surface area. However, with a closer look at **Figure 4**, the lower scanning speeds have shallow depths and a relatively high wall of built-up material along the

Figure 1. FESEM images of pattern overlaps of 1, 2, and 3 with a line spacing of 0.10 mm, a laser power of 10.5 W, a frequency of 100 kHz, and a scanning speed of 400 mm/s.

Figure 2. FESEM images of line spacing of 0.025, 0.05, and 0.10 mm with an overlap number of 1, a laser power of 13.3 W, a frequency of 100 kHz, and a scanning speed of 400 mm/s.

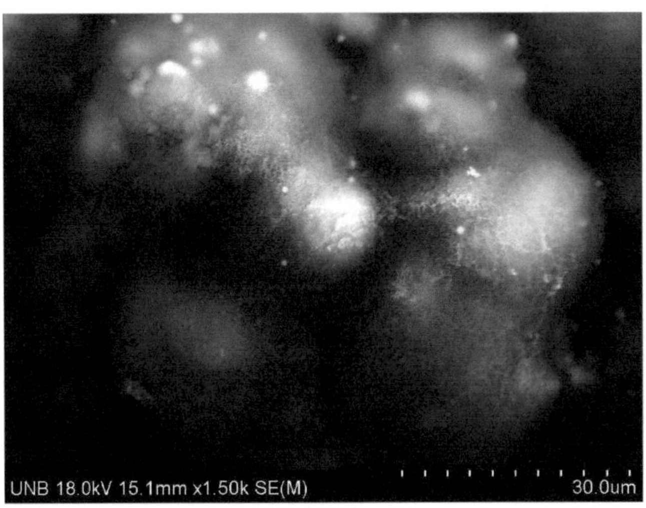

Figure 3. Presence of nanofibers detected on FESEM image of sample with three overlaps and 0.025 mm line spacing with power of 13.3 W, frequency of 100 kHz, and a scanning speed of 400 mm/s.

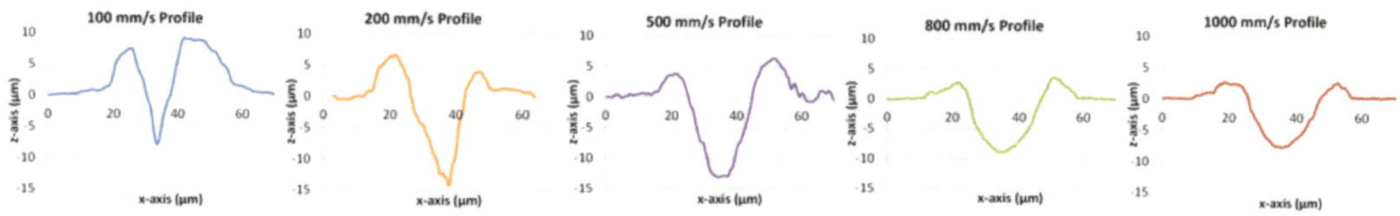

Figure 4. Profile data from the 3D optical microscope for scanning speeds of 100, 200, 500, 800, and 1000 mm/s at an overlap number of 1, a power of 15 W, and a frequency of 100 kHz.

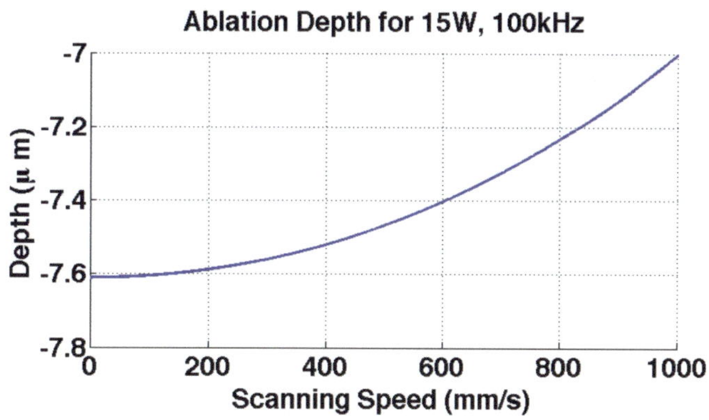

Figure 5. Ablation depth after a train of pulses at difference scanning speeds at a power of 15 W and a frequency of 100 kHz.

sides of the groove. At higher scanning speeds, when the pulses are farther apart, there is less penetration on the surface, leading to a shallower groove. At lower scanning speeds, the high-temperature ablated material from the walls caves back into the deep groove and solidifies into a much shallower groove than initially dredged. Scanning speeds of 200 and 800 mm/s show a deep groove with a smaller amount of material built up than the 100 mm/s sample.

Eqs. (6) and (7) are then used to find the theoretical ablation depths at various scanning speeds. Observing the trend in **Figure 5**, it is clear that the ablation depth decreases with increasing scanning speed. Both the experimental data and theoretical results are in close agreement.

3.2. Bioactivity assessment

Each sample was submerged in simulated body fluid (SBF) for 6 weeks and kept at a constant temperature of 36.5°C. The samples were then emerged from the fluid and assessed with energy-dispersive X-ray (EDX). The SBF-soaked samples were found to contain a traces of phosphorous and calcium, which is indicative of the presence of a bone-like apatite. Hydroxy-apatite is formed by the nucleation of calcium phosphate ions [5, 27]. The silicon oxide layer created by the laser plume has a negative charge, which attracts the positively charged calcium phosphate. The resulting substrate contains this bone-like apatite, which was seen on the laser-treated silicon samples. The EDX results from the sample with three overlap and 0.025 mm line spacing can be seen in **Figure 6**.

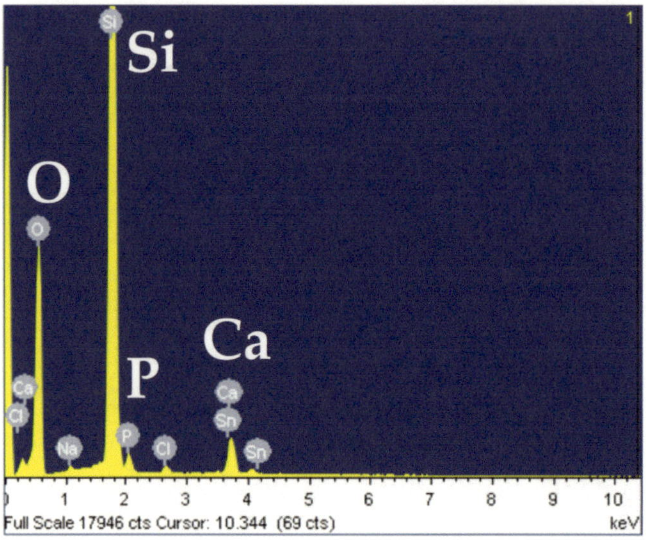

Figure 6. EDX results of sample with three overlaps, 0.025 mm line spacing, 400 mm/s, a power of 13.3 W, and a frequency of 400 mm/s.

This presence of bone-like apatite confirms that the biocompatibility of pure silicon was enhanced with nanosecond laser pulses. A smaller line spacing and higher overlap number generates more SiO_2 nanofibers, which provides a favorable site for the nucleation of apatite.

4. Effect of frequency

The range of frequencies used in this section is 25, 70, and 100 kHz. For these experiments, the scanning speed was kept constant at 100 mm/s, the power at 15W, and the overlap number at 1.

4.1. Topography analysis

Figures 7 and **8** show the topography changes in the frequency samples. A lower frequency produces a wider and shallower groove, while the higher frequencies yield a thinner yet deeper ablated groove. The theoretical results in for a single pulse in **Figure 9** show that the groove increases in depth and decreases in width as frequency increases, which is in close agreement with the experimental results.

4.2. Temperature analysis

The temperature is determined for each frequency with Eq. (3) and can be found in **Figure 10**. The higher temperatures are associated with the lower frequencies on both z-axis and r-axis. By increasing the frequency, the pulse energy decreases which results in lower temperatures and a smaller heat-affected zone [30]. Due to reduced size of the heat-affected zone, the shape of the groove consequently changes size as well. This results in the thinner grooves at higher

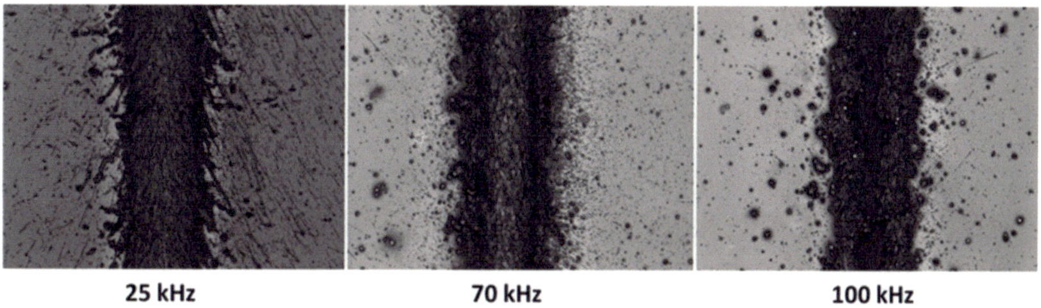

Figure 7. 3D optical microscopy images of samples with frequencies of 25, 70, and 100 kHz at a power of 15 W, a scanning speed of 100 mm/s, and 1 overlap.

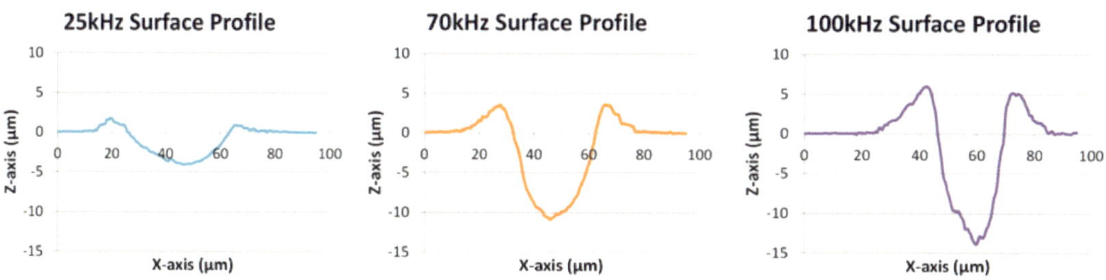

Figure 8. Experimental profile data from the 3D optical microscope for frequencies of 25, 70, and 100 kHz at an overlap number of 1, a power of 15 W, and a scanning speed of 100 mm/s.

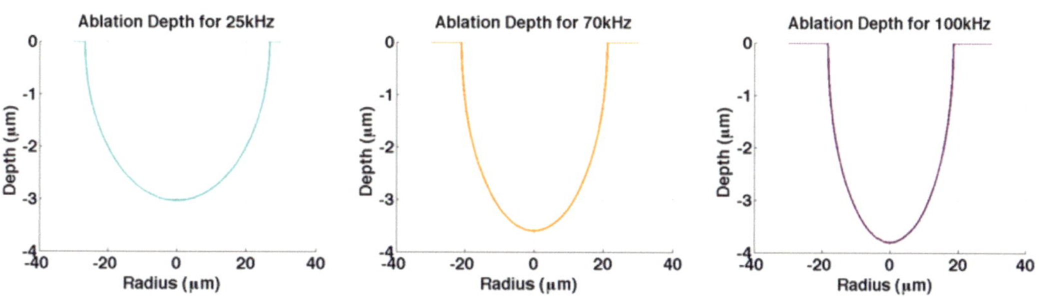

Figure 9. Theoretical profile data for a single pulse for frequencies of 25, 70, and 100 kHz at an overlap number of 1, a power of 15 W, and a scanning speed of 100 mm/s.

frequencies. Each recurring pulse adds to the penetration of preceding pulse. Higher frequencies have more pulses, resulting in a deeper penetration of the surface, which develops a trench with a larger depth.

4.3. Bioactivity assessment

Mouse embryonic fibroblast cell interactions were examined for each frequency. The cell count in **Figure 11** establishes that there are a higher number of cells in the higher frequency grooves.

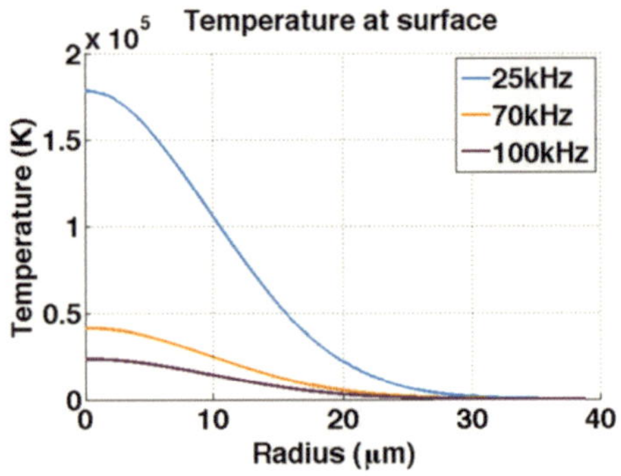

Figure 10. The single-pulse temperature on the surface of the silicon with respect to radius.

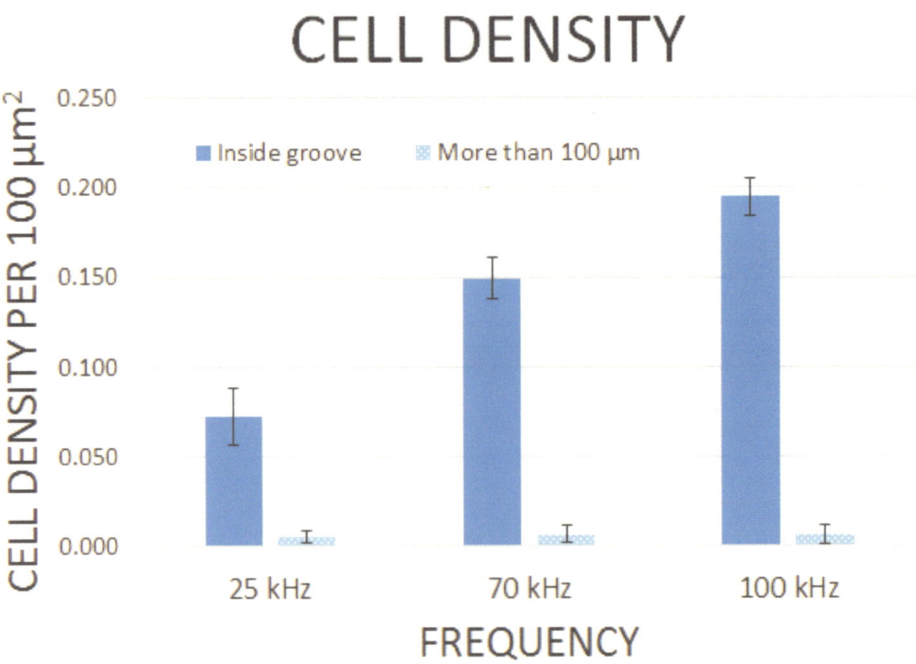

Figure 11. The number of cells within the laser-treated groove for each frequency as well as the amount of cells outside the groove within 100 μm from the edge of the groove [30].

The cells show a strong preference for the treated areas and show avoidance in the untreated areas. There is also a presence of fibronectin within the cells, which is an ECM protein secreted during embryonic development and wound healing, potentially leading to collagen deposition and tissue morphogenesis [30, 31]. These results confirm that the biocompatibility is enhanced with higher frequencies.

5. Effect of power

Laser power immensely influences the surface properties when treatment is done to a material. The power for this section varies from 7, 10 to 15 W. For these experiments, the frequency was kept constant at 100 kHz, the scanning speed was set to 400 mm/s, and the overlap number was 1.

5.1. Topography analysis

The FESEM images of each power sample are shown in **Figure 12**. The experimental 3D optical microscopy profile data is shown in **Figure 13**. These results show that at higher powers, the groove will increase in both width and depth. Unlike the frequency trends, the size of the heat-affected zone increases with power. The theoretical single-pulse depths found with Eq. (5) are shown in **Figure 14**. These results also show that the groove width and depth increase with power and are in close agreement with the experimental results.

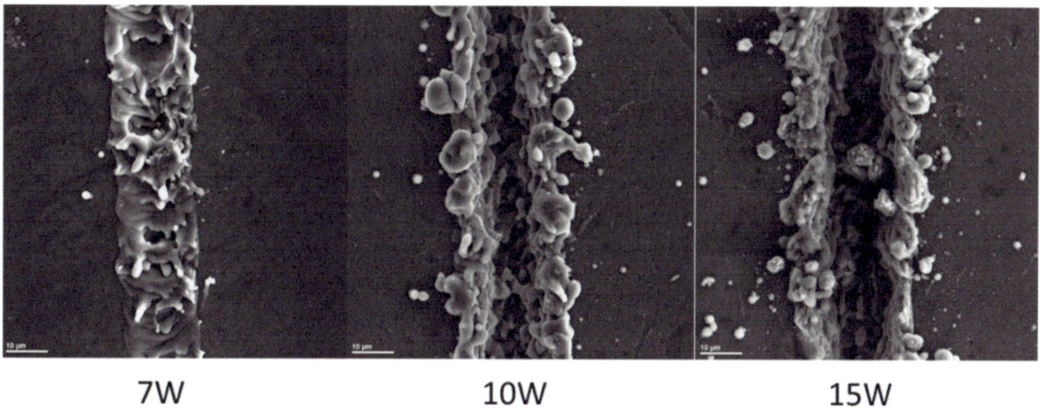

Figure 12. FESEM images of samples with powers of 7, 10, and 15 W at a frequency of 100 kHz, a scanning speed of 400 mm/s, and 1 overlap.

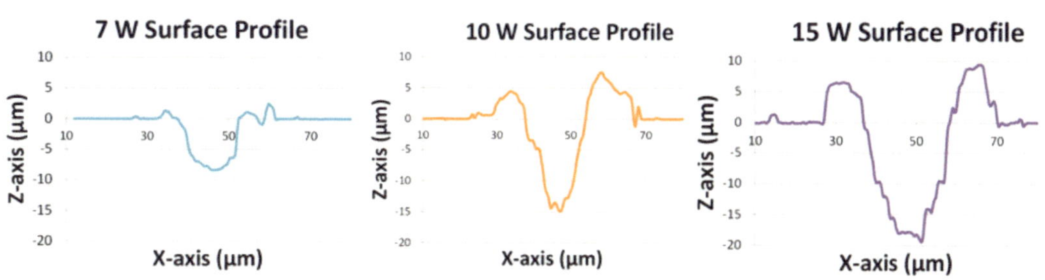

Figure 13. Experimental profile data from the 3D optical microscope for powers of 7, 10, and 15 W at an overlap number of 1, a frequency of 100 kHz, and a scanning speed of 400 mm/s.

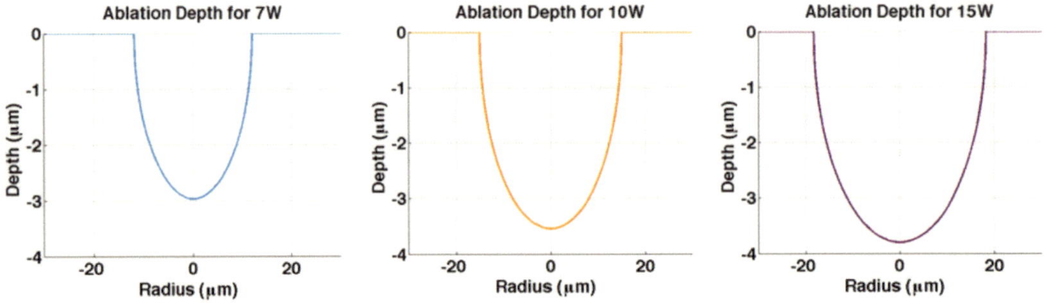

Figure 14. Theoretical profile data for a single pulse for powers of 7, 10, and 15 W with an overlap number of 1, a frequency of 100 kHz, and a scanning speed of 400 mm/s.

5.2. Temperature analysis

The temperature is determined for each power with Eq. (3) and can be found in **Figure 15**. As expected, the higher temperatures are found with higher powers. At lower powers, the heat-affected zone is smaller, allowing for both a thinner and shallower groove. When the pulse power is increased, the temperatures in the high-density plume are increased, causing more generation of the SiO_2 nanofibers [26].

5.3. Bioactivity assessment

Samples were once again assessed with fibroblast culturing for each power. When viewing the cell interactions under the microscope, cells were accumulated inside the laser-treated area as expected. Interestingly, the cell count was low directly beside the grooves and began to become more concentrated farther away from the edge of the groove. The fibroblasts avoided the zones immediately beside the grooves on each side. This can be seen in **Figure 16**.

This phenomenon is a result from the shockwave that is generated from the high-energy plume during laser ablation [26]. The shockwave transfers energy to the surface with results in high intensity thermal stress. The thermal shock causes a small zone directly beside the

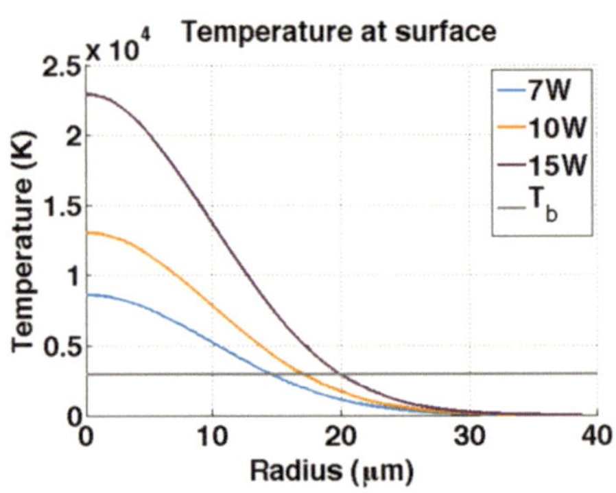

Figure 15. The single-pulse temperature on the surface of the silicon with respect to radius. T_b is the boiling temperature of silicon at 3538 K.

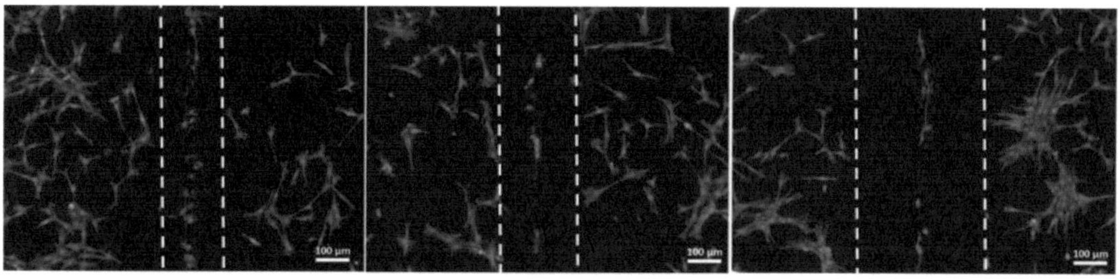

Figure 16. Fibroblast cells avoiding the zones beside the grooves for powers of 7, 10, and 15 W.

ablated areas, which contains residual stress. When the power is increased, there is a larger transfer energy, resulting in a larger stress zone. These residual stress zones contain mismatched crystal orientations due to tensile stresses causing crystal distortion.

Knowing that the power level of the laser pulse can control the residual stress zone size and cell behavior, this research can provide opportunities in cell manipulation and cell programming.

6. Summary

This chapter aims to introduce nanosecond laser processing for the enhancement of biocompatibility of pure silicon for various biomedical technologies. These results can contribute to the design of manufacturing processes of innovative biomedical devices to enhance the quality of living for a number of individuals. This research investigates the trends of various laser parameters including three scanning parameters (line spacing, overlap number, and scanning speed), pulse frequency, and laser power. Biocompatible *in vitro* assessment was conducted through the use of simulated body fluid (SBF) and cell culturing with NIH 3T3 fibroblasts. Samples with smaller line spacing and higher overlap numbers showed more generation of SiO_2 nanofibers, which were shown to be biocompatible under SBF assessment. Biocompatibility increased with frequency due to the SiO_2 being more prominent on high frequency samples and containing more fibroblast cell proliferation. Fibroblasts also showed preference to higher powers. However, the heat-affected zone immediately outside the ablated areas showed a mismatch of crystal orientations causing residual stress. These stress zones were avoided by cells, which led to promising results for the potential in cell programming and manipulation.

Author details

Candace Colpitts and Amirkianoosh Kiani*

*Address all correspondence to: a.kiani@unb.ca

Silicon Hall: Laser Micro/Nanofabrication Facility, Department of Mechanical Engineering, University of New Brunswick, NB, Canada

References

[1] Turner AP, Pickup JC. Diabetes mellitus: Biosensors for research and management. Biosensors. 1985;**1**(1):85-115

[2] Grayson AR, et al. A BioMEMS review: MEMS technology for physiologically integrated devices. Proceedings of the IEEE. 2004;**92**(1):6-21

[3] Copeland JG, et al. Cardiac replacement with a total artificial heart as a bridge to transplantation. The New England Journal of Medicine. 2004;**351**(9):859-867

[4] Li X, et al. The use of nanoscaled fibers or tubes to improve biocompatibility and bioactivity of biomedical materials. Journal of Nanomaterials. 2013;**2013**:14

[5] Radmanesh M, Kiani A. Effects of laser pulse numbers on surface biocompatibility of titanium for implant fabrication. Journal of Biomaterials and Nanobiotechnology. 2015; **6**:168

[6] Swetha M, et al. Biocomposites containing natural polymers and hydroxyapatite for bone tissue engineering. International Journal of Biological Macromolecules. 2010;**47**(1):1-4

[7] Yue Z, et al. Controlled delivery for neuro-bionic devices. Advanced Drug Delivery Reviews. 2013;**65**(4):559-569

[8] Vallet-Regí M, Colilla M, González B. Medical applications of organic–inorganic hybrid materials within the field of silica-based bioceramics. Chemical Society Reviews. 2011;**40**(2): 596-607

[9] Green ML, et al. Ultrathin (<4 nm) SiO_2 and Si–O–N gate dielectric layers for silicon microelectronics: Understanding the processing, structure, and physical and electrical limits. The Journal of Applied Physics. 2001;**90**(5):2057-2121. DOI: http://dx.doi.org/10.1063/1.1385803

[10] Jeong S, et al. Hybrid silicon nanocone-polymer solar cells. Nano Letters. 2012;**12**(6): 2971-2976. Available: http://dx.doi.org/10.1021/nl300713x. DOI: 10.1021/nl300713x

[11] Buckberry L, Baylis S. Porous silicon as a biomaterial. Materials World. 1999;**7**:213-215

[12] Shaoqiang C, et al. Hydroxyapatite coating on porous silicon substrate obtained by precipitation process. Applied Surface Science. 2004;**230**(1):418-424

[13] Mwenifumbo S, et al. Cell/surface interactions on laser micro-textured titanium-coated silicon surfaces. The Journal of Materials Science: Materials in Medicine. 2007;**18**(1):9-23

[14] Myllymaa S, et al. Adhesion, spreading and osteogenic differentiation of mesenchymal stem cells cultured on micropatterned amorphous diamond, titanium, tantalum and chromium coatings on silicon. Journal of Materials Science: Materials in Medicine. 2010;**21**:329-341

[15] Vilhena LM, et al. Surface texturing by pulsed Nd:YAG laser. Tribology International. 2009;**42**(10):1496-1504. DOI: dx.doi.org/10.1016/j.triboint.2009.06.003

[16] Radmanesh M, Kiani A. Nd: YAG laser pulses ablation threshold of stainless steel 304. Materials Sciences and Applications. 2015;**6**(07):634

[17] Bonse J, et al. Femtosecond laser ablation of silicon–modification thresholds and morphology. Applied Physics A. 2002;**74**(1):19-25

[18] Kiani A, et al. Leaf-like nanotips synthesized on femtosecond laser-irradiated dielectric material. The Journal of Applied Physics. 2015;**117**(7):074306

[19] Wood JP, et al. Nanosecond pulse lasers for retinal applications. Lasers in Surgery and Medicine.2011;**43**(6):499-510

[20] Al-Hadeethi Y, et al. Data fitting to study ablated hard dental tissues by nanosecond laser irradiation. PLoS One. 2016;**11**(5):e0156093

[21] Khademhosseini A, et al. Microscale technologies for tissue engineering and biology. Proceedings of the National Academy of Sciences of the United States of America. 2006;**103**(8):2480-2487. DOI: 0507681102 [pii]

[22] Canham LT. Bioactive silicon structure fabrication through nanoetching techniques. Advanced Materials. 1995;**7**(12):1033-1037

[23] Kamitakahara M, et al. Bioactivity and mechanical properties of polydimethylsiloxane (PDMS)-CaO-SiO_2 hybrids with different PDMS contents. The Journal of Sol-Gel Science and Technology. 2001;**21**(1-2):75-81

[24] Incropera FP, et al. Introduction to convection. In: Anonymous, editors. Introduction to Heat Transfer. 5th ed. Danvers, MA, USA: John Wiley & Sons; 2007. pp. 57

[25] Hendow ST, Shakir SA. Structuring materials with nanosecond laser pulses. Optics Express. 2010;**18**(10):10188-10199

[26] Tavangar A, Tan B, Venkatakrishnan K. Study of the formation of 3-D titania nanofibrous structure by MHz femtosecond laser in ambient air. The Journal of Applied Physics. 2013;**113**(2):023102

[27] Tavangar A, Tan B, Venkatakrishnan K. The influence of laser-induced 3-D titania nanofibrous platforms on cell behavior. Journal of biomedical nanotechnology. 2013;**9**(11):1837-1846

[28] Kiani A, Venkatakrishnan K, Tan B. Optical absorption enhancement in 3D nanofibers coated on polymer substrate for photovoltaic devices. Optics Express. 2015;**23**(11):A569-A575

[29] Colpitts C, Kiani A. Synthesis of bioactive three-dimensional silicon-oxide nanofibrous structures on the silicon substrate for bionic devices' fabrication. Nanotechnology and Nanomaterials. 2016;**6**:1-7

[30] Colpitts C, et al. Mammalian fibroblast cells show strong preference for laser-generated hybrid amorphous silicon-SiO_2 textures. Journal of Applied Biomaterials and Functional Materials. 2016. DOI: D2673E51-9D76-4504-A673-13AAC5CE7AD3 [pii]

[31] Vega ME, Schwarzbauer JE. Collaboration of fibronectin matrix with other extracellular signals in morphogenesis and differentiation. Current Opinion in Cell Biology. 2016;**42**:1-6. DOI: S0955-0674(16)30055-2 [pii]

Chapter 10

Measurement and Simulation of Permeation and Diffusion in Native and Cultivated Tissue Constructs

Hao-Hsiang Hsu, Katharina Schimek, Uwe Marx and Ralf Pörtner

Additional information is available at the end of the chapter

http://dx.doi.org/10.5772/intechopen.72904

Abstract

Characterization of native skin or cultured 3D skin models with respect to permeability plays an important role for the development and testing of pharmaceuticals and cosmetics. Extensive efforts have been dedicated to determining the key parameters describing permeability and diffusion. Whereas respective methods are well established for native skin biopsies, only few are available for 3D skin models, as these have usually much lower dimensions. In this chapter, some fundamentals about permeation and diffusion as well as state of the art of measurement methods used for skin biopsies are summarized. An alternative method for the determination of the permeation in a membrane insert system and the use of a modular simulation to support permeability studies is presented and discussed.

Keywords: skin models, permeation, diffusion, membrane insert system

1. Introduction

Permeability studies are indispensable to characterize the transport of substances through the skin, either natural skin or cultivated 3D skin models. This is evident for dermal drug delivery systems, where drugs can be applied in the form of creams or patches on the skin. Furthermore, permeability studies play an important role in toxicity tests applied in drug development as well as substance testing. In this respect, human three-dimensional skin models have become an interesting tool.

Animal experiments are still a common method of testing drugs and also for skin, which is ethically controversial, cost-intensive and time-consuming. The fact is that drug testing on human skin is more efficient in comparison to animal skin just like rat, mouse and guinea pig [1, 2]. Furthermore, in 2013 the European Regulation (EC) No 1223/2009 entered into force, which prohibits animal experiments for cosmetics products. Artificial skin models based on human cells are intended to replace animal experiments. Especially the barrier function between artificial and human skin can differ, so permeation and diffusion investigation in this area is necessary.

Most methods for determination of diffusion and permeability parameters have been developed for large biopsies and can hardly be applied for small-scale 3D skin tissue cultures. But this is indispensable, if multi-well test systems with several samples run in parallel or multi-organ-chips are applied. Therefore, this chapter will first give a brief overview of standard methods used to investigate permeation and diffusion on the skin. Then an alternative method based on skin tissue cultures in a membrane insert system is introduced. By this, the permeation coefficient of substances through a skin-tissue barrier can be determined. The diffusion coefficient is estimated via parameter optimization performed in COMSOL Multiphysics. This software tool helps to describe the physical effects of the experimental set-ups more precisely and can further be used to reduce the required amount of experiments significantly.

2. Penetration, diffusion and permeation through the skin

Penetration describes the entering of a substance into the skin. The entering process and depth of substance penetration through the skin depends on the physical and chemical character of the substance and the skin. The permeation of the skin is the pathway of a substance from the surface to the blood vessel. From a scientific point of view, it is the permeation of a substance through the skin layers. Diffusion is the physical process of randomized particle movement. If a concentration gradient exists, the particles move in the direction of lower concentration.

The human skin consists of three layers, the epidermis, dermis and subcutis. The stratum corneum (horny layer) is the upper layer of the epidermis and forms the main barrier of the skin. Substances such as drugs and chemicals penetrate through the skin barrier in three possible routes: the transcellular, the intercellular and the appendageal route (see **Figure 1**) [3–5].

The transcellular route leads the permeating substances directly through the cells. Here, the substances have to pass alternating lipophilic and hydrophilic layers. This is probably the

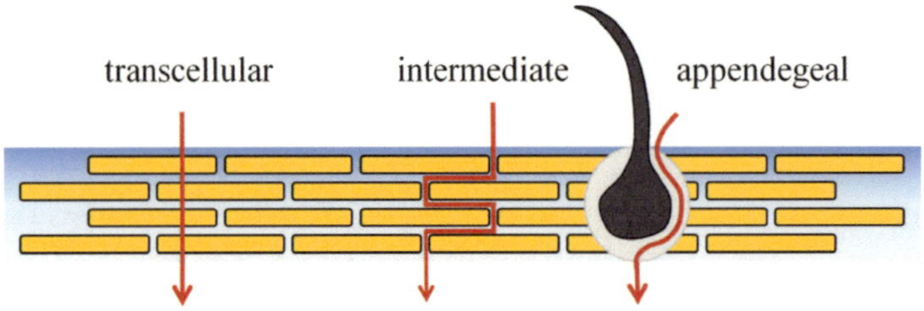

Figure 1. Schematic illustration of the three possible pathways of a permeating substance through the stratum corneum.

most difficult way for substances, because they should have lipophilic and hydrophilic properties. Until now, it is not clear if hydrophilic substances choose this pathway [6].

Alternatively, there is a way through the intercellular spaces between the cells. This is called the intercellular route. The intermediate space consists of cholesterols, ceramides and free fatty acids [7]. Because of the fatty acids, lipophilic substances can pass easier through the intercellular route in contrast to lipophobic substances [8]. Another barrier in this intercellular space is the tight junctions [9]. These are networks of strands which are formed by membrane proteins connecting cells. They are located between the keratinocytes in the stratum granulosum of the epidermis. An important function of the tight junctions, which are formed during the differentiation of keratinocytes, is to protect the skin from water loss [10].

The appendageal route describes the penetration of a substance through skin appendages like hair follicles and glands. Since hair follicles and glands build only a small part of the human skin, their relevance for skin permeation was neglected for a long time. Its importance was shown recently by a researcher as permeation is better in a skin area containing hair follicles and glands in comparison to an area without them. A specific characteristic of the hair follicle is it reservoir function. In follicles, substances can be stored up to 10 days and can penetrate gradually into the skin. This aspect is interesting for drug delivery over the skin. For example, alcohol prefers the appendageal penetration route, as it would otherwise evaporate quite fast on the surface of the skin. [11–14]

So far the penetration and permeation in the skin was described. To get more in detail, the physical aspect of diffusion and permeation will be explained. Diffusion is a transport process where molecules move via Brownian motion in a volume or area. It is driven by the concentration gradient in the direction from higher to lower concentration [15]. Adolf Fick (1829–1901) verified the coherence between heat transfer and diffusion, which led to the Fick's first law of diffusion:

$$F = -D\frac{\partial C}{\partial x} \tag{1}$$

According to Eq. (1), the flux F in the one partial direction x is proportional to the gradient of concentration C. D is the diffusion coefficient or diffusivity. [16, 17]

Permeation is an aspect of diffusion. Whereas diffusion is related to the movement of molecules in a system, permeation describes how fast molecules move through a system. An example is the permeation of a substance within a volume V_A and a donor concentration c_D through a membrane with a surface A. The acceptor concentration c_A of the permeating substance on the other side can be detected over the time t. Eq. (2) for the permeation coefficient P can be derived from the Fick's law of diffusion [18, 19]:

$$\frac{dc_A}{dt} = P \cdot A\frac{c_D}{V_A} \tag{2}$$

This equation can only be used for $c_D \gg c_A$. With respect to skin, the permeation coefficient is the preferred parameter, as it is easier to measure compared to the diffusion coefficient. Due to the different layers of the skin, permeation and diffusion coefficient changes all the time from layer to layer.

3. State of the art for investigation of penetration, diffusion and permeation within the skin

To understand and investigate diffusion and permeation of the skin biopsies, several methods have been established. Some are summarized in the following.

The Franz diffusion cell is a well-known device to measure the permeation of a substance through a skin biopsy. This device consists of two chambers where the skin biopsy (or any other barrier) is fixed in between. The test substance can be applied to the top chamber (donor) of the skin; it permeates through the barrier into the bottom chamber (acceptor). The fluid in the bottom chamber is mixed by means of a magnetic stirrer. On this side, samples can be taken and the concentration can be analyzed. The concentration of the acceptor is plotted over the time and the permeation coefficient can be calculated according to Eq. (2). The whole system can be temperature-controlled. The usual size (height) of a Franz diffusion cell is in the range of 19–179 mm. Besides permeation investigation, this system is also used to test the quality of skin models and the effects of pharmaceutical substances on the skin. [18, 20–26]

Fluorescence recovery after photobleaching (FRAP) is a method to measure molecular diffusion in tissues or gels, mainly for high molecular weight compounds. For this, the substance to be analyzed must be labeled with a fluorochrome. Mostly fluorescence-labeled proteins or FITC-dextranes (fluorescein isothiocyanate-dextranes) with different molecular sizes are used. The tissues or gels have to be soaked with this substance. This can be realized by storing the material in the fluorescence substance for some days or in case of a gel, to directly polymerize in the fluorescent substance. Then, a confocal laser is used to bleach out a certain area (mostly a line or a point) in the material. Because of diffusion, bleached molecules will move and change their position with fluorescent particles and the fluorescence recovers. After the bleaching process, the area will be scanned several times. The recovery time of fluorescence intensity is used to determine the diffusion coefficient of the substance in the material. For this, software for image analysis is used. [27–30]

Further examples for imaging methods for the determination of diffusion of molecules in skin are Fourier-transform-infrared (FTIR) spectroscopy [31, 32], two-photon fluorescence correlation spectroscopy in combination with fluorescence correlation spectroscopy (FCS) [33, 34] and optical coherence tomography [35]. These methods are noninvasive and nondestructive. Furthermore, some of them can detect molecules without fluorescence labeling. One big disadvantage is the equipment. For these imaging methods, special microscopes or also cost-intensive tomographs are needed.

A method to investigate the penetration process of substances into the skin is tape stripping. After treatment of the skin with the substance of interest the stratum corneum is ripped of layer by layer with an adhesive film. Then, the amount of the substance can be analyzed. For this, there are different methods to determine the concentration of the substance. One method is to detect the substances directly on the film, for example titanium dioxide can be analyzed with X-ray fluorescent measurement and fluorescent-labeled substances can be detected via laser scanning microscopy. Another possibility is to remove the skin layer from the film and apply standardized analytical methods to determine the substance concentration. With tape stripping it is possible to observe where the substance of interest is localized and how deep they can penetrate into the skin. It is minimal invasive and possible to investigate the penetration directly on human skin. A disadvantage of this method is the undefined thicknesses of the stripped layer. It varies from experiment to experiment and differs with the skin model or skin type. The thickness can be estimated by weighing. [36–39]

As mentioned before, the above methods all together provide a detailed characterization of diffusion and permeation effects within the skin. But most of them can hardly be adapted to skin tissue models used in drug and substance testing. Here usually small culture devices, e.g. 12- or 96-well plates are preferred, as they allow for handling of a large number of samples in parallel. Furthermore, most methods require treatment of the sample in one or the other way. Therefore, it is quite difficult to determine the time-depending changes of diffusion and permeation.

4. Skin tissue models

The need to evaluate skin permeation, test cosmetic products and toxicologically screen topically applied compounds is evident. Historically, several millions of animal experiments have been performed worldwide to address this purpose [40]. Since animal experiments are under massive debate, ethical and regulatory issues, but also severe differences between animal and human data pushed the development and commercialization of diverse *in vitro* skin models [41, 42]. Human skin equivalents (HSEs) can be categorized into two main groups: the epidermis-only and full-thickness models. For both, the differentiation of keratinocytes and hence development of the various layers of the epidermis is important to model actual skin barrier properties more closely. In this context, the direct exposure to air as well as the culture media that supply nutrients for cell growth and differentiation from below has been found to be beneficial [43]. Growing cells on a porous membrane is one of the most commonly used ways to accomplish this air-liquid interface culture. According to the Organization for Economic Co-operation and Development (OECD) test guideline 431 (skin corrosion) and 439 (skin irritation), currently validated skin models include EpiSkin™ (L'oreal, France), EpiDerm™ SIT (MatTek Corporation, USA), SkinEthic™ RHE (SkinEthic laboratories, France), EpiCS® (CellSystems, Germany) and LabCyte EPI-MODEL24 SIT (Japan Tissue Engineering Co., Japan). These 3D skin models are all composed of one cell type only, the keratinocytes, mimicking the epidermis of native human skin and are especially advantageous with respect to high reproducibility [44]. However still not validated, there are also commercially available full thickness skin models composed of an additional dermal layer (e.g., GraftSkin®, EpiDermFT®, and Pheninon®). The models described are nowadays widely used for animal-free tests in drug development as well as, the chemical and cosmetics industries.

5. Determination of permeation and diffusion coefficients in membrane insert systems via measurement and simulation

A common method for *in vitro* cultivation of skin models is the use of a membrane insert system. This system consists of a plastic vessel with a permeable membrane at the bottom. It enables the cultivation of skin tissue models in airlift on the membrane and guarantees the supply of nutrients from below. As the membrane insert system has two separate chambers, it can be used for permeation studies similar to the Franz diffusion cell. In the following, an experimental procedure for the determination of permeation coefficients in Transwell® systems (12 and 96 well) and simulations with COMSOL Multiphysics for estimation of diffusion coefficients will be discussed. Details can be found in [53].

5.1. Measurement and simulation.

The scheme of a permeation experiment in a membrane insert system is shown in **Figure 2**. The tissue barrier is established on the membrane. It is composed of agarose gel or 3D tissue to validate the method. The 3D tissues consist of a collagen matrix with human fibroblasts within and HaCaT cells on the top of the matrix. On top of the barrier, the donor is applied. The donor contains the substance to be analyzed, which permeates through the barrier. The acceptor, which collects the permeating substance, is located on the other side of the membrane in the receiver vessel. Temperature, humidity and mixing are parameters that influence the permeation and should be kept constant. It is recommended to perform the experiment on a shaker in an incubator with 37°C and ≥90% humidity (conditions for human cell culture). To avoid hydrodynamic pressure, the fluid surface in the insert (donor) and in the receiver vessel (acceptor) should be on the same level. The used volume for the experiment in 12 and 96 Transwell® systems is shown in **Table 1**. Sampling, like in the Franz diffusion cell, is difficult because of the small volume in the acceptor. A solution is the use of fluorescence-labeled substances, by which the permeate concentration can be detected via fluorescence measurement in the receiver vessel. A more elaborate possibility is to run several permeation experiments in parallel and take a sample from one vessel per time point. Then, the concentration of the substance can be measured analytically.

The permeation experiment was simulated with COMSOL Multiphysics. This program is based on the finite element method and offers different physic simulation modules. This

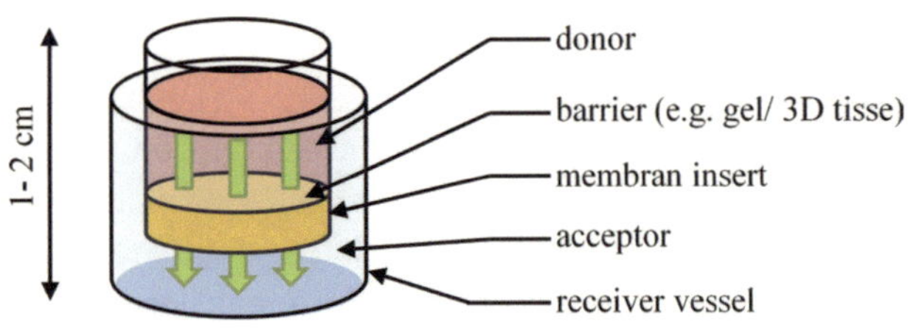

Figure 2. Schematic sketch of a membrane insert system.

Transwell System	Volume in acceptor	Volume in donor
96	75 µl	300 µl
12	590 µl	1845 µl

Table 1. Liquid volume in the acceptor and the donor in different membrane insert systems with a barrier of 2 mm thickness.

structure enables the computation of different physical problems in one simulation. The permeation experiment was simulated with the module "transport diluted species", which uses the Fick's law to simulate diffusion processes. In order to determine the diffusion coefficient, a parameter optimization was performed with the "optimization module". Some simplifications have to be done in order to simulate the experiment. In the experiment, the permeating substance passes through a barrier and a membrane. For the simulation, these two phases were resumed as one homogenous material. It is not possible to resolve the different phases, as investigations that are more detailed would be necessary. The diffusion coefficient of the permeating substance in the liquid phases (in donor and acceptor) was determined in preliminary mixing tests and was found to be 1×10^{-9} m^2/s. This parameter represents the molecular distribution in the mixing process. Furthermore, the geometry of the membrane insert system was simplified. In reality, the system is slightly conical. The simplified geometry is a cylinder. All boundaries of the geometry were set as "no slip". The concave surface of the agarose gel was approximated with a spherical shape. In **Figure 3**, the geometry and the mesh of the 96- and 12-well Transwell® systems are shown.

5.2. Influence of different settings and validation of the system.

Different membrane insert sizes influence the permeation within the system. Investigations with fluorescein sodium salt and 2% agarose gel in 96- and 12-well membrane insert systems are shown in **Figure 4a**. The time course of the acceptor concentration in the 12-well system was steeper in comparison to the 96-well system. Therefore, the fluorescein sodium salt permeates faster through the barrier in the 12-well system compared to the 96-well system. A reason for this is the concentration gradient in the gap between the membrane and the bottom of the receiver vessel. Because of the ratio "volume to permeation surface" and the gap size, the concentration will be balanced faster in case of the 12-well system. The gradient can be reduced by increasing the mixing frequency and amplitude. However, this is limited due to spillover of the liquid. By simulating these experiments, the different concentration distributions below the membrane can be visualized. In **Figure 4c** and **d** the concentration at different time points is plotted over the length below the membrane from the middle point to the edge of the receiver vessel (see red line on **Figure 4b**). The concentration difference between the middle and edge in 12-well systems is higher than that in 96-well systems. This indicates a better and faster concentration balance in the larger system, which explains the accelerated permeation.

The reproducibility of the suggested method is an important aspect. The permeation coefficient determined in different permeation experiments with fluorescein sodium salt through 2% agarose gel differed up to 40.9%. Although the value seems to be quite high, it is still within the range of deviation of permeability experiments with Franz diffusion cells reported in the literature [24]. It was found that small concentration variations during the preparation of the donor substance can cause large variation. By using one stock solution for different experiments, the deviation can be reduced to 29%. Further reasons for the deviation can also be small pipetting errors and variations of the barrier, for example, variation of the gel concentration or volume.

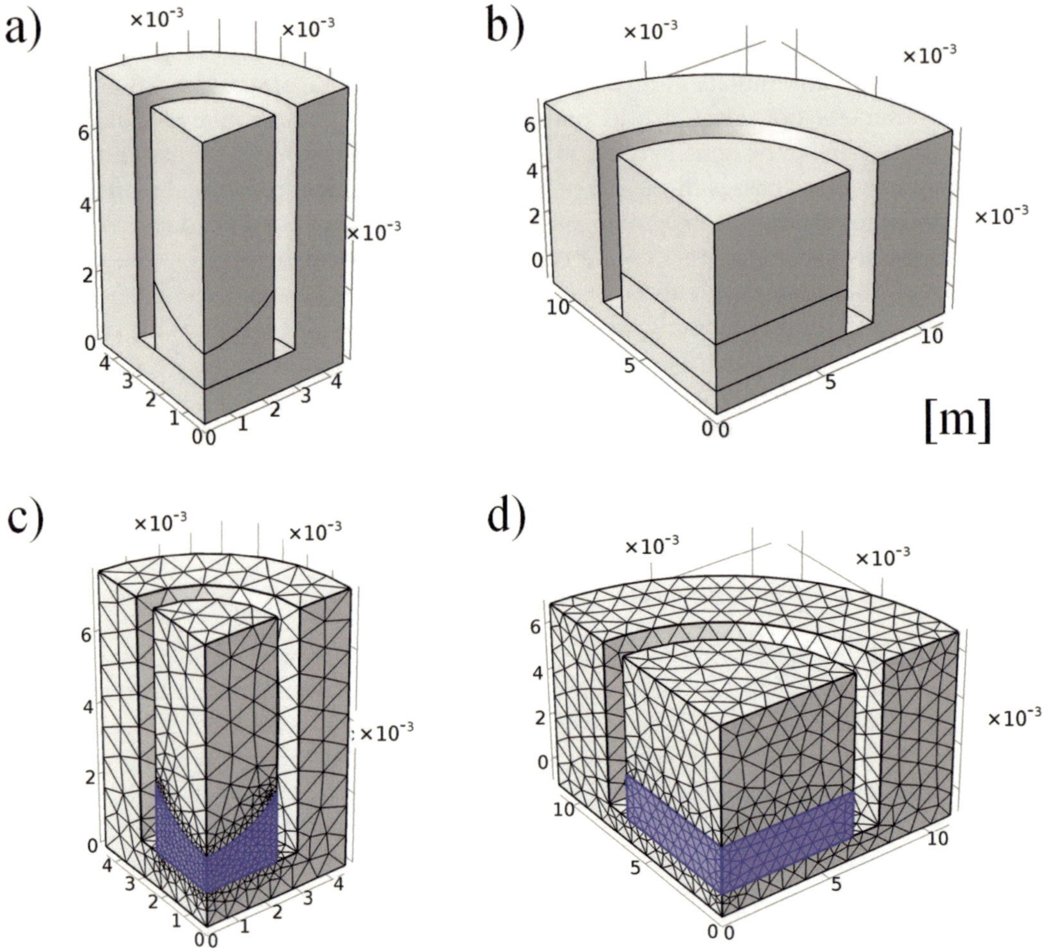

Figure 3. Geometry of the 96 a) and 12 b) Transwell® system and the used mesh c) and d) implemented in COMSOL **Multiphysics**.

The membrane itself also has an influence on the permeation. Same experiments as described above were carried out with different membranes in a 12 Transwell® system. Membranes consisting of polycarbonate (PC) with 0.4 and 3.0 μm pore size showed a quite similar permeation coefficient of 8.03 and 8.1 × 10^{-8} m/s. For polyethylene (PE) membranes with 0.4 μm pore size the mean value of the permeation coefficient was 5.94 × 10^{-8} m/s and for PE with 3.0 μm pore size it was 8.59 × 10^{-8} m/s (see **Figure 6a**). Except for the PE membrane with pore size of 3.0 μm there is no significant difference. The reason for this is the pore density of the membrane. In total, the pore surface of PE membranes with 0.4 μm pore size is 0.25 mm^2 per 1 cm^2 and for the other membranes 6.3–7.05 mm^2 per 1 cm^2. The material of the membrane seems not to influence the permeation.

The suggested method is sensitive enough to determine a cover layer of HaCaT cells of 3D skin tissue. To prove this, in a permeation experiment different 3D tissue model was tested in

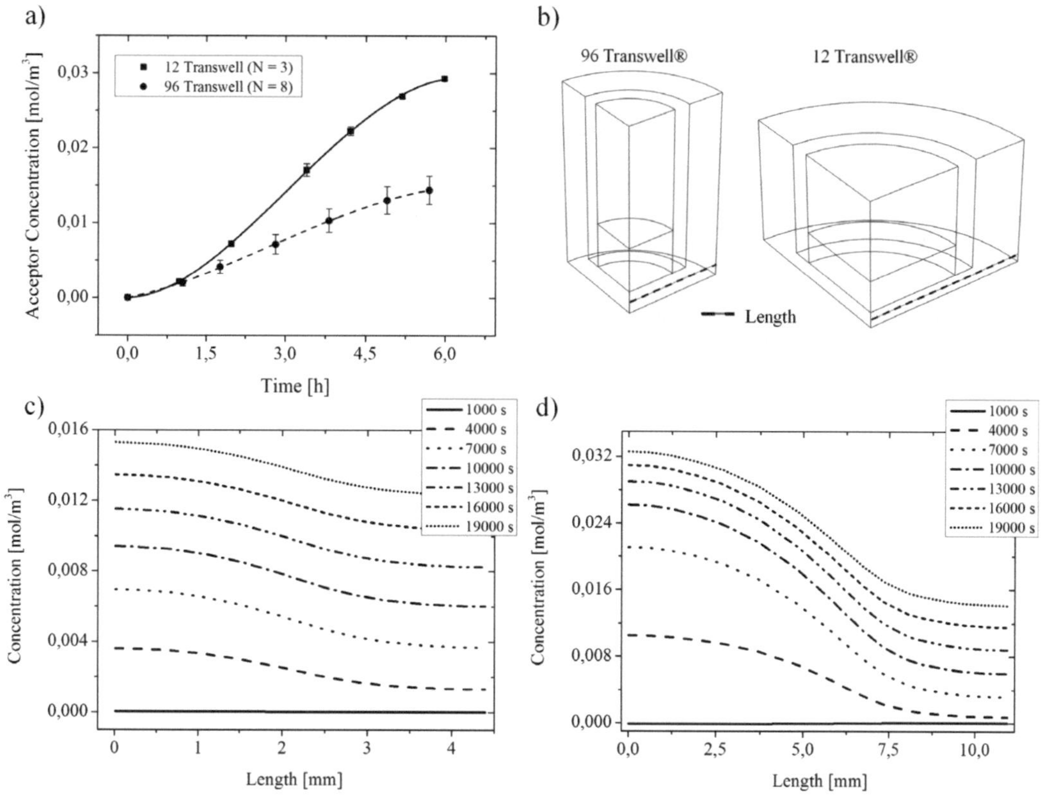

Figure 4. Permeation in 12- and 96-well membrane insert systems. (a) Experimental results of permeation of fluorescein sodium salt through 2% agarose. (b) Geometry of the 12 and 96 Transwell® system for the simulation and the position of the concentration measurement. (c) and (d) the concentration distribution at different time over the length at the red lined position on (b) ((c) 96 and (d) 12 Transwell® system).

a 12 Transwell® system. The 3D tissues consisted of a collagen matrix with different constellations of cell layers. Human primary fibroblasts were integrated into the collagen matrix and HaCaT cells were seeded on the top. A representative example of the 3D tissue is shown in **Figure 5**.

The results show that the permeation coefficient decreases when additional cell layers are added in the tissue model. The permeation coefficient of fluorescein sodium salt through collagen matrix (without cells) is 2.18×10^{-8} m/s, 1.85×10^{-8} m/s in tissue models with fibroblast and 1.67×10^{-8} m/s in models with HaCaT cells (see **Figure 6b**). These results represent very well the barrier function of keratinocytes of the skin [3].

The particle size influences the permeation behavior through gels and biomaterials. It is well known that smaller molecules permeate faster through a matrix mesh than larger particles. This was already observed for permeation experiments through sclera [19], human epidermal membrane [45], human skin [33] and rat skin [3]. Fluorescein sodium salt and fluorescein isothiocyanate-dextrans (FITC-dextran) were used to vary the molecular size from 376 g/mol up

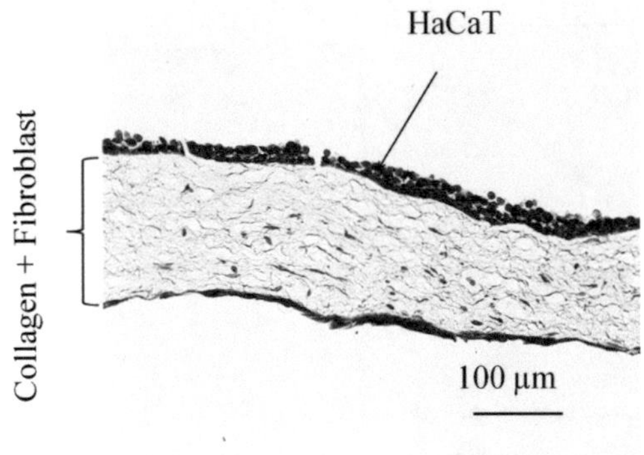

Figure 5. Hematoxylin and eosin stain of 3D tissue model consisting of collagen matrix with fibroblast and HaCaT seeded on the top.

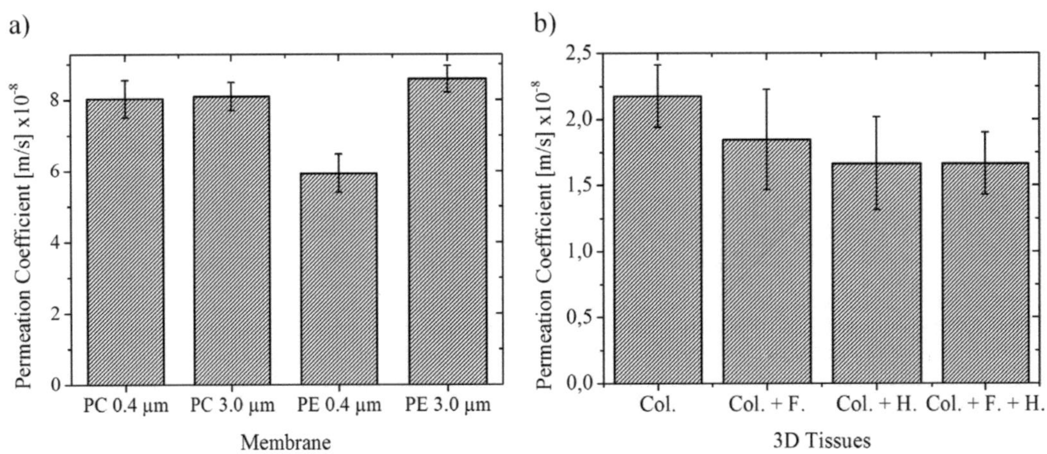

Figure 6. Permeation through different membranes and 3D skin tissues. (a) Results of permeation experiment with fluorescein sodium salt through 2% agarose gel on different membranes (n = 3). The 12 Transwell® system was used and membranes consisting of polycarbonate (PC) and polyethylene (PE) with pore sizes of 3.0 and 0.4 μm were tested. (b) Results of permeation experiments with fluorescein sodium salt through different 3D tissues in a 12 Transwell® system (n = 6). The 3D skin tissue consisted of collagen (Col.), collagen with primary human fibroblast (Col. + F.), collagen with HaCaTs (Col. + H.) and collagen with primary human fibroblast and HaCaTs (Col. F. + H.).

to 40,000 g/mol for permeation experiments in a 96 Transwell® system. The results show similar correlations between permeation coefficient and molecular size as the studies mentioned above. There is almost a linear relationship between these two factors, which is well described by the Navier–Stokes equation (see **Figure 7**). An exception is FITC-dextran 40,000 g/mol, which deviates from the linearity.

These experiments were simulated with COMSOL Multiphysics, where the diffusion coefficient is fitted on the experimental data. The simulation based on Fick's law is quite accurate for the permeation of substances with a small molecular size from 376 g/mol up to 4000 g/mol. The simulation shows good agreement with the experiment which is exemplary shown in **Figure 8a**. In the case of larger molecular size, the simulation is different from experimental results. A closer look shows that the experimental data increased and flattened faster compared to the simulation (see **Figure 8b)**. A possible reason for this effect could be the presence of particle size distribution in the substance. The migration of smaller particles reduces the lag time in the beginning of the permeation process, where large particles can increase the friction and slows down the diffusion. Especially the second effect leads to abnormal diffusion [46, 47]. This cannot be simulated with equations based on Fick's law.

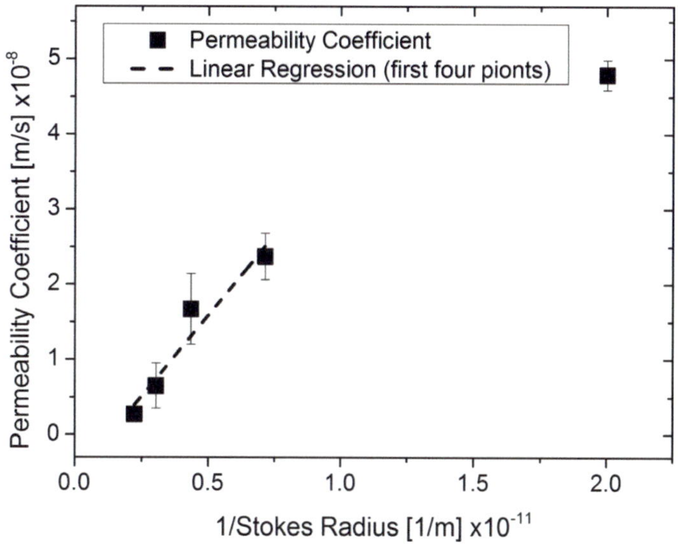

Figure 7. Permeability coefficient plotted over stokes radius. Results of permeation experiment in 96 Transwell® system with fluorescein sodium salt and FITC-dextran through 2% agarose gel.

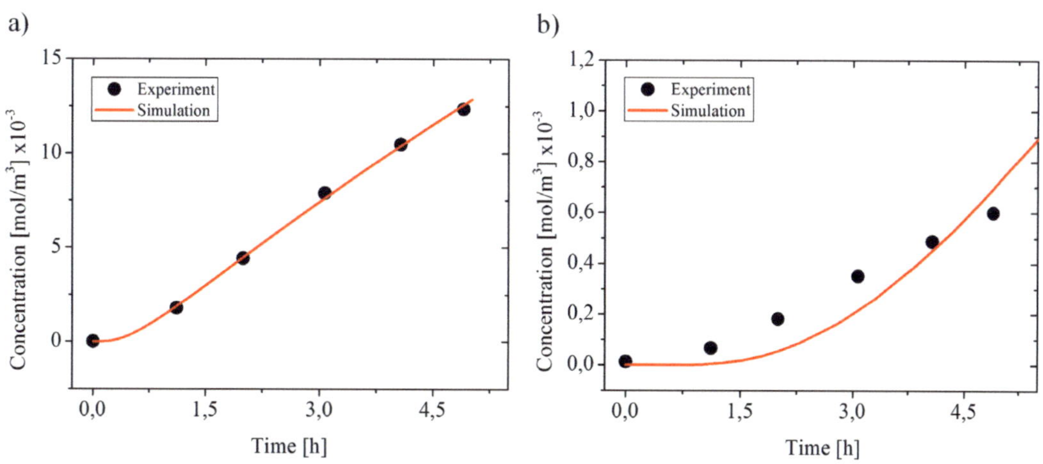

Figure 8. Simulation of permeation experiment with different molecular sizes. Results of permeation experiments and simulations with COMSOL **Multiphysics** of (a) fluorescein sodium salt and (b) FITC dextran 10,000 mg/mol in 96 Transwell systems through 2% agarose gel.

6. Conclusion and prospective

The studies have shown that the membrane insert system is a possible alternative for permeation studies. An advantage of the system is the small size. The membrane insert system of 96-well plates from CORNING has a cultivation surface of 0.143 cm^2 and a height of 2 cm. This reduces significantly the number of cells, materials and substances needed for the cultivation. In comparison to the Franz diffusion cell, the handling of such a system is easier and one experiment can be run with a large specimen number. A time-intensive mounting process of the samples (skin) is no longer required and the experiment can directly execute in the system. The sensitivity of this system is good enough to differ between 3D tissues as well as different cell layers and to detect different molecular sizes of the substance.

It should be considered that the permeation is detected through a membrane and the size of the system influences the permeation properties. The specimen has to cover up the whole membrane, otherwise the substance will pass by. Unlike the imaging and stripping method, it is not possible to measure the diffusion and penetration inside the membrane insert system. Alternatively, the diffusion can be calculated or estimated by simulation. Furthermore, this method can be used to investigate changes in the permeation behavior of the skin model during the cultivation or it can also be adapted for other systems, which use membrane insert systems. An example is the Two-Organ-on-a-Chip, a variant of TissUse's Multi-Organ-Chip platform [48–52]. This device enables the integration of skin models in a membrane insert system. Therefore this method can be used to investigate the permeation process into an organ-on-a-chip system in order to understand the substance distribution.

With the help of the simulation in COMSOL Multiphysics, it is possible to calculate the diffusion process in the membrane insert system. It is limited to small particle sizes and normal diffusion described by Fick's law. Otherwise, it is possible to optimize the simulation by integration of abnormal diffusion. Furthermore, the simulation is an attractive tool to support the experiments. On the one hand, it can be used to understand physical phenomena and to reduce experimental effort. On the other hand, it is modular and can be integrated into a more complex system to support permeation studies.

Acknowledgements

This work was created with financial support from Deutsche Forschungsgemeinschaft (DFG) under grant No. PO413/12-1 and LA 1028/7-1.

Appendices and nomenclatures

EC	European regulation
FITC dextran	fluorescein isothiocyanate-dextrans

FRAP	fluorescence recovery after photobleaching
FTIR	Fourier-transform-infrared
HaCaT	human adult low calcium high temperature
HSE	human skin equivalents
OECD	organization for economic co-operation and development

Author details

Hao-Hsiang Hsu[1], Katharina Schimek[2], Uwe Marx[2] and Ralf Pörtner[1]*

*Address all correspondence to: poertner@tuhh.de

1 Institute of Bioprocess and Biosystems Engineering, Hamburg University of Technology, Hamburg, Germany

2 TissUse GmbH, Berlin, Germany

References

[1] Ackermann K, Lombardi Borgia S, Korting H, Mewes K, Schäfer-Korting M. The Phenion® full-thickness skin model for percutaneous absorption testing. Skin Pharmacology and Physiology. 2010;**23**(2):105-112. DOI: 10.1159/000265681

[2] Asbill C. Evaluation of a human bio-engineered skin equivalent for drug permeation studies. Pharmaceutical Research. 2000;**17**(9):1092-1097. DOI: 10.1023/A:1026405712870

[3] Hadgraft J. Skin, the final frontier. International Journal of Pharmaceutics. 2001;**224**(1-2): 1-18. DOI: 10.1016/S0378-5173(01)00731-1

[4] Barry BW. Mode of action of penetration enhancers in human skin. Journal of Controlled Release. 1987;**6**(1):85-97. DOI: 10.1016/0168-3659(87)90066-6

[5] Valenzuela P, Simon JA. Nanoparticle delivery for transdermal HRT. Nanomedicine: Nanotechnology, Biology and Medicine. 2012;**8**:83-89. DOI: 10.1016/j.nano.2012.05.008

[6] Barbero AM, Frasch HF. Transcellular route of diffusion through stratum corneum: Results from finite element models. Journal of Pharmaceutical Sciences. 2006;**95**(10):2186-2194. DOI: 10.1002/jps.20695

[7] Bouwstra J, Pilgram G, Gooris G, Koerten H, Ponec M. New aspects of the skin barrier organization. Skin Pharmacology and Applied Skin Physiology. 2001;**14**:52-62. PubMed ID: 11509908

[8] Wertz PW. Lipids and barrier function of the skin. Acta Dermato-Venereologica. 2000; **208**:7-11. PubMed ID: 10884933

[9] Tsukita S, Furuse M, Itoh M. Multifunctional strands in tight junctions. Nature Reviews. Molecular Cell Biology. 2001;**2**(4):285-293. DOI: 10.1038/35067088

[10] Tsuruta D, Green KJ, Getsios S, Jonathan CRJ. The barrier function of skin: how to keep a tight lid on water loss. Trends in Cell Biology. 2002;**12**(8):355-357. PubMed ID: 12191905

[11] Knorr F et al. Follicular transport route – Research progress and future perspectives. European Journal of Pharmaceutics and Biopharmaceutics. 2009;**71**(2):173-180. DOI: 10.1016/j.ejpb.2008.11.001

[12] Schaefer H, Lademann J. The role of follicular penetration. Skin Pharmacology and Applied Skin Physiology. 2001;**14**:23-27. DOI: 10.1159/000056386

[13] Lademann J et al. Hair follicles—An efficient storage and penetration pathway for topically applied substances. Skin Pharmacology and Physiology. 2008;**21**(3):150-155. DOI: 10.1159/000131079

[14] Otberg N et al. The role of hair follicles in the percutaneous absorption of caffeine. British Journal of Clinical Pharmacology. 2008;**65**(4):488-492. DOI: 10.1111/j.1365-2125.2007.03065.x

[15] Wilkinson DS. Mass transport in solids and fluids. Cambridge University Press. 2000. ISBN: 9780521624947

[16] Cussler EL. Diffusion: Mass Transfer in Fluid Systems. 2nd ed. Cambridge University Press; 1997. ISBN: 0-521-45078-0

[17] Crank J. The Mathematics of diffusion. 2nd ed. Oxford, Clarendon Press; 1975. ISBN: 0-19-853411-6

[18] Netzlaff F et al. Permeability of the reconstructed human epidermis model Episkin® in comparison to various human skin preparations. European Journal of Pharmaceutics and Biopharmaceutics. 2007;**66**(1):127-134. DOI: 10.1016/j.ejpb.2006.08.012

[19] Jayakrishna A et al. Diffusion of high molecular weight compounds trough Sclera. IOVS. 2000;**41**(5):1181-1185. DOI: b, 19.02.2015

[20] Ackermann K, Lombardi Borgia S, Korting H, Mewes K, Schäfer-Korting M. The Phenion® full-thickness skin model for percutaneous absorption testing. Skin Pharmacology and Physiology. 2010;**23**(2):105-112. DOI: 10.1159/000265681

[21] Bran B et al. A new discriminative criterion for the development of Franz diffusion tests for transdermal pharmaceuticals. Journal of Pharmaceutical Sciences. 2010;**13**(2):218-230. DOI: 10.18433/J3WS33

[22] Pineau A, Guillard O, Favreau F, Marrauld A, Fauconneau B. In vitro study of percutaneous absorption of aluminum from antiperspirants through human skin in the Franz diffusion cell. Journal of Inorganic Biochemistry. 2012;**110**:21-26. DOI: 10.1016/j.jinorgbio.2012.02.013

[23] Larese Filon F et al. In vitro percutaneous absorption of cobalt. International Archives of Occupational and Environmental Health. 2004;(2):85-89. DOI: 10.1007/s00420-003-0455-4

[24] Ng S-F, Rouse JJ, Sanderson FD, Meidan V, Eccleston GM. Validation of a static Franz diffusion cell system for in vitro permeation studies. AAPS PharmSciTech. 2010;**11**(3):1432-1441. DOI: 10.1208/s12249-010-9522-9

[25] Bonferoni MC, Rossi S, Ferrari F, Caramella C. A modified Franz diffusion cell for simultaneous assessment of drug release and Washability of Mucoadhesive gels. Pharmaceutical Development and Technology. 1999;**4**(1):45-53. DOI: 10.1080/10837459908984223

[26] Bartosova L, Bajgar J. Transdermal drug delivery in vitro using diffusion cells. CMC. 2012;**19**(27):4671-4677. DOI: 10.2174/092986712803306358

[27] Pluen A, Netti PA, Jain RK, David A, Berk DA. Diffusion of macromolecules in agarose gels: Comparison of linear and globular configurations. Biophysical Journal. 1999;**77**(1):542-552. DOI: 10.1016/S0006-3495(99)76911-0

[28] Cornelissen LH, Bronneberg D, Oomens CWJ, Baaijens FPT. Diffusion measurements in epidermal tissues with fluorescent recovery after photobleaching. ISBS. 2008;**14**(4):462-467. DOI: 10.1111/j.1600-0846.2008.00313.x

[29] Leddy HA, Guilak F. Site-specific molecular diffusion in articular cartilage measured using fluorescence recovery after Photobleaching. Annals of Biomedical Engineering. 2003;**31**(7):753-760. DOI: 10.1114/1.1581879

[30] Seiffert S, Oppermann W. Systematic evaluation of FRAP experiments performed in a confocal laser scanning microscope. Journal of Microscopy. 2005;**220**(1):20-30. DOI: 10.1111/j.1365-2818.2005.01512.x

[31] Pirot F, Kalia YN, Stinchcomb AL, Keating G, Bunge A, Guy RH. Characterization of the permeability barrier of human skin in vivo. Proceedings of National Academy of Sciences USA. 1997;**94**:1562-1567. 0027-8424/97/941562-6$2.00/0

[32] Tetteh J et al. Local examination of skin diffusion using FTIR spectroscopic imaging and multivariate target factor analysis. Analytica Chimica Acta. 2009;**642**(1-2):246-256. DOI: 10.1016/j.aca.2009.03.002

[33] Guldbrand S et al. Two-photon fluorescence correlation spectroscopy as a tool for measuring molecular diffusion within human skin. European Journal of Pharmaceutics and Biopharmaceutics. 2013;**84**(2):430-436. DOI: 10.1016/j.ejpb.2012.10.001

[34] Berland KM, So PT, Gratton E. Two-photon fluorescence correlation spectroscopy: Method and application to the intracellular environment. Biophysical Journal. 1995;**68**(2):694-701. DOI: 10.1016/S0006-3495(95)80230-4

[35] Ghosn MG et al. Monitoring of glucose permeability in monkey skin in vivo using optical coherence tomography. Journal of Biophotonics. 2010;**3**(1-2):25-33. DOI: 10.1002/jbio.200910075

[36] Teichmann A et al. Differential stripping: Determination of the amount of topically applied substances penetrated into the hair follicles. The Journal of Investigative Dermatology. 2005;**125**(2):264-269. DOI: 10.1111/j.0022-202X.2005.23779.x

[37] Surber C, Schwarb FP, Smith EW. Tape-stripping technique. Cutaneous and Ocular Toxicology. 2002;**20**(4):461-474. DOI: 10.1081/CUS-120001870

[38] Weigmann H-J, Lademann MH, Schaefer H, Sterry W. Determination of the horny layer profile by tape stripping in combination with optical spectroscopy in the visible range as a prerequisite to quantify percutaneous absorption. Skin Pharmacology and Physiology. 1999;**12**(1-2):34-45. DOI: 10.1159/000029844

[39] Jacobi U, Weigmann H-J, Ulrich J, Sterry W, Lademann J. Estimation of the relative stratum corneum amount removed by tape stripping. ISBS. 2005;**11**(2):91-96. DOI: 10.1111/j.1600-0846.2005.00094.x

[40] Macfarlane M et al. A tiered approach to the use of alternatives to animal testing for the safety assessment of cosmetics: Skin irritation. Regulatory Toxicology and Pharmacology. 2009;**54**(2):188-196. DOI: 10.1016/j.yrtph.2009.04.003

[41] Basketter D, Jírova D, Kandárová H. Review of skin irritation/corrosion hazards on the basis of human data: A regulatory perspective. Interdisciplinary Toxicology. 2012;**5**(2):98-104. DOI: 10.2478/v10102-012-0017-2

[42] York M, Griffiths HA, Whittle E, Basketter DA. Evaluation of a human patch test for the identification and classification of skin irritation potential. Contact Dermatitis. 1996; **34**(3):204-212

[43] Prunieras M, Regnier M, Woodley D. Methods for cultivation of keratinocytes with an air-liquid Interface. The Journal of Investigative Dermatology. 1983;**81**:28-33. DOI: 10.1111/1523-1747.ep12540324

[44] Mathes SH, Ruffner H, Graf-Hausner U. The use of skin models in drug development. Advanced Drug Delivery Reviews. 2014;**69-70**:81-102. DOI: 10.1016/j.addr.2013.12.006

[45] Peck KD, Ghanem A, Higuchi WI. Hindered diffusion of polar molecules through and effective pore radii estimates of intact and ethanol treated human epidermal membrane. Pharmaceutical Research. 1994;**11**(9):1306-1314. DOI: 10.1023/A:1018998529283

[46] Laurent TC, Sundelöf L-O, Wik KO, Wärmegard B. Diffusion of dextran in concentrated solutions. European Journal of Biochemistry. 1976;**68**:95-102. DOI: 10.1111/j.1432-1033.1976.tb10767.x

[47] Metzler R, Klafter J. The random walk's guide to anomalous diffusion: A fractional dynamics approach. Physics Reports. 2000;**339**(1):1-77. DOI: 10.1016/S0370-1573(00)00070-3

[48] Ataç B et al. Skin and hair on-a-chip: In vitro skin models versus ex vivo tissue maintenance with dynamic perfusion. Lab on a Chip. 2013;**13**(18):3555-3561. DOI: 10.1039/c3lc50227a

[49] Maschmeyer I et al. Chip-based human liver–intestine and liver–skin co-cultures – A first step toward systemic repeated dose substance testing in vitro. European Journal of Pharmaceutics and Biopharmaceutics. 2015;**95**:77-87. DOI: 10.1016/j.ejpb.2015.03.002

[50] Materne E-M et al. The multi-organ Chip - a microfluidic platform for long-term multi-tissue coculture. JoVE. 2015;**98**:1-11. DOI: 10.3791/525264

[51] Schimek K et al. Integrating biological vasculature into a multi-organ-chip microsystem. Lab on a Chip. 2013;**13**(18):3588. DOI: 10.1039/c3lc50217a

[52] Wagner I et al. A dynamic multi-organ-chip for long-term cultivation and substance testing proven by 3D human liver and skin tissue co-culture. Lab on a Chip. 2013;**13**(18):3538. DOI: 10.1039/c3lc50234a

[53] Hsu H et al. A method for determination and simulation of permeability and diffusion in a 3D tissue model in a membrane insert system for multi-well plates. JOVE. 2017 https://www.jove.com/video/56412/a-method-for-determination-simulation-permeability-diffusion-3d

Chapter 11

Biomaterials for Tendon/Ligament and Skin Regeneration

Xingguo Cheng

Additional information is available at the end of the chapter

http://dx.doi.org/10.5772/intechopen.69716

Abstract

Tendon/ligament injury or skin injuries due to diseases, trauma, and surgery are common. Timely functional repair and tissue regeneration is a key to improve the quality of life of the patient while reducing health care cost. Tendon/ligament/skin is also enriched in a common extracellular matrix (ECM), collagen I, III, and elastin. Tissue engineering and regenerative medicine, the combination of (stem) cells, growth factors, and biomaterial scaffolds, is an emergent field, which has attracted substantial attention over the years. Biomaterials are considered the foundation of regenerative medicine. A key to find a new solution to tendon/ligament/skin healing is to synthesize new functional biomaterials, which have better biomechanical properties, biodegradability, and cell supporting properties. This chapter will review existing FDA-approved biomaterial-based therapy, as well as those in development.

Keywords: biomaterials, tendon, skin, regeneration

1. Introduction

From material science point of view, tendon/ligament and skin tissue are similar in that they are mainly composed of collagen and elastin. Up to 80% of the dry weight is collagen type I. However, their micro and hierarchical structures and functions are very different from each other. Tendon/ligament is composed of densely packed, aligned collagen fiber bundles, whereas skin is composed of a layered structure of collagen random naonfibril network. **Table 1** summarizes the key properties of tendon/ligament and skin tissue in human.

The regeneration of a large size, lost tendon/ligament, and skin tissue normally involves a type of stem cells/progenitor cells of endogenous origin or exogenous origin combined with a

	Tendon/ligament	Skin
Main composition	>80% Collagen, ~5% elastin elastin	>80% Collagen, ~5% elastin
Minor composition	Fibronectin, proteoglycan	Fibronectin, proteoglycan
Main cells	Fibroblast	Fibroblast and keratinocytes
Vasculature	Few	Abundant
Structure	1D Aligned fiber bundles (nano-macro)	2D random nanofibrils (weave basket pattern)
Mechanical properties	Strong and elastic	Weak and elastic

Table 1. A general comparison of two different connective tissues: tendon/ligament and skin.

biomaterial. Stem cells can be derived from tendon/ligament or skin, or from embryo, placenta, adipose tissue, bone marrow, umbilical cord, etc. This review focuses only on regeneration using exogenous stem cells coupled with a biomaterial matrix/carrier implanted to the wound site. Biomaterial can play a key role in protection of the cells from dehydration while it serves as a temporary substrate for stem cells to proliferate, or differentiate, and synthesize tissue-specific matrix. The morphology, topography, composition, stiffness of biomaterials may play a key role in controlling the differentiation of stem cells, in addition to biochemical factors, mechanical cues, or genetic/cellular cues. This chapter focuses on biomaterial explored for tendon/ligament/skin tissue regeneration applications.

1.1. Research methods

We performed a comprehensive search of PubMed using keywords "tendon," "ligament," "skin," "regeneration," "scaffold", over the years 1970–2016. All articles relevant to the subject were retrieved, and their bibliographies hand searched for further references in the context to biomaterials for tendon/ligament/skin regeneration

2. Results

2.1. Biomaterial directly derived from patients

2.1.1. PRP

Platelet-rich plasma (PRP) is derived from blood and PRP gel is widely used for tendon/ligament repair. Recently, PRP was combined with adipose-derived stem cells (ADSCs) and it was found that PRP combined with stem cells resulted in improved mechanical strength in a rabbit tendon model compared to PRP gel alone [1]. Similar results were also observed using PRP with tendon-derived stem cells (TDSCs) [2]. However, in a sheep model, no differences were observed between the PRP group and PRP-stem cell group [3]. This approach is highly translational, since both autologous stem cells and PRP can be obtained from the same patients. The concern may be the leucocytes-containing PRP (L-PRP) that have a catabolic effect, whereas pure PRP (P-PRP) without leucocytes have anabolic effects and results in over-proliferation and scar tissue formation. The complex interaction between PRP and stem cells may explain the different preclinical outcome and warrant a large random clinical trial.

2.1.2. *Fibrin*

Fibrin can be derived from human plasma. After addition of thrombin, it will form a gel, a process used for blood clotting. Stem cells can be added together with fibrin and thrombin and sprayed onto the wound for the promotion of wound closure and healing. The fibrin combined with bone marrow stem cells spray was successfully tested to prevent ulceration and accelerate wound closure in mice [4]. A small human clinical trial showed this approach accelerates wound closure and resurfacing without adverse effects [5]. Poly(ethylene glycol) PEG-modified fibrin combined with adipose-derived stem cells (ADSCs) also showed promising results in a pig burn model [6].

2.1.3. *Amniotic membrane*

Human placenta-derived biomaterial is unique in that it has immune privilege while it contains multiple growth factors. Human ADSCs (hADSCs) seeded onto radiosterilized human amnion are viable and can proliferate. These cells are able to migrate over these scaffolds as demonstrated by using time-lapse microscopy. In addition, the scaffolds induce hADSCs to secrete interleukin-10, an important negative regulator of inflammation [7]. This suggests that placenta-derived biomaterial may be a good substrate for stem cells and used for skin/tendon applications.

3D micronized (300–600 μm) amniotic membrane (mAM) was made by means of repeated freeze-thawing cycles to deplete cell components and homogenized with a macrohomogenizer in liquid nitrogen. These mAM loaded with epidermal stem cells (ESCs) (ESC-mAM) was further transplanted to full-thickness skin defects in nude mice. ESCs survived well and formed a new epidermis. Four weeks after transplantation, papilla-like structures were observed, and collagen fibers were well and regularly arranged in the newly formed dermal layer. In conclusion, the mAM as a novel natural microcarrier possesses an intact basement membrane structure and bioactivities [8].

2.2. FDA-approved ECM grafts for tendon augmentation and skin regeneration

A recent study showed that decellularized matrix from different tissues (tendon, bone, and skin) affect the differentiation of stem cell in a different way. Decellularized bone matrix may induce the undesirable osteogenic differentiation of stem cells, while tendon or skin matrix does not have such an effect [9]. The ECM components provide a niche for proper differentiation of stem cells. For example, ECM without biglycan (Bgn) and fibromodulin (Fmod) will affect the differentiation of tendon stem cells by modulating bone morphogenetic protein signaling and impairs tendon formation *in vivo* [10]. Autologous origin, but decelluarized dermal matrix using trypsin and Trion X-100, after combined with ADSCs, was found to enhance wound healing in a murine model of a full-thickness skin defect [11].

Many FDA-approved human or animal decelluarized tissue matrixes have been approved for direct tendon/ligament/skin repair without any (stem) cells (**Table 2**). The key advantages of decellularized tissue grafts are that they largely maintain the main architecture, composition, and mechanical properties of native tissues. These allografts/xenografts are processed to remove immunogenic cells, DNAs, and certain immunogenic molecules. Typical problems are that these grafts are slower to repair the tissue and some fail to restore the proper functions (e.g., scarring). For tissue regeneration using these ECM biological grafts, stem cells may need to be reseeded onto the grafts for recellularization. For a dense tendon allograft, direct cell seeding may be difficult. The recellularization onto the surface may be achieved using a cell-loaded gel coating [12]. However, it is highly desirable to get cells inside the grafts as well. Thus, ECM grafts may be processed to have a much higher porosity than the original tissue. Proration/incision into the ECM grafts may help the penetration of cells and nutrients. Instead of being coated with cell-laden gel, an interesting approach is to use a stem cell-sheet to wrap around a frozen tendon graft for implantation [13]. Interesting, a case report showed that a dermal allograft combined with PRP and autologous mesenchymal stem cells (MSCs) derived from peripheral blood (PB-MSCs) resulted in enhanced healing of human rotator cuff [14].

Some of the FDA-approved ECM biomatrices were combined with stem cells, and investigated for tendon/ligament/skin regeneration applications. Human acellular allograft (Alloderm) was investigated for direct cell seeding using ADSCs. It was concluded that human ADSCs can attach to Alloderm with the dermis side up in a petri dish [15]. ADSC seeded onto Alloderm was also implanted *in vivo* for skin regenerations with promising results [16]. Strattice was evaluated for seeding with rat MSCs [17]. Thus, the stem cell-seeded biologic graft can be used as a tendon wrap or a skin regeneration material. A study was performed to compare the survival and proliferation of stem cells via bioluminescent imaging. The use of biologic graft

Alloderm	Human skin (decelluarized)
GraftJacket	Human skin
Restore	10-layered porcine small intestine submucosa (SIS) treated with peracetic acid/ethanol (90% collagen, 5–10% lipids)
TissueMend	Noncrosslinked fetal biovine dermal matrix [19]
BioBlanket	Crosslinked porous bovine dermis
Permacol or Zimmer collagen repair patch	Porcine acelluar dermis treated with trypsin, solvent, and crosslinked with hexamethylene diisocyanate (HMDI)
Strattice	Porcine acelluar dermal matrix
Cuffpatch	EDC-crosslinked matrix from eight layers of treated SIS (97% collagen, 2% elastin)
OrthoADAPT	Crosslinked equine pericardium (90% collagen I, 10% collagen II)

Table 2. An example list of biologic ECM grafts with potential for tendon/skin regeneration.

patch (SIS) as carrier of ADSCs significantly increases the survival of stem cells as compared to direct injection of ADSCs into the skin wounds [18]. There may be a synergistic angiogenesis promoting effects of biologic graft with stem cells, which may be important for tendon/ skin wound healing [19].

Another interesting approach is to use ECM directly secreted by the cells as a carrier for stem cells. For example, stem cells can be incubated at 37°C in a temperature-responsive flask (e.g., upcell™, Cellseed, Japan) and they will product ECM after addition of ascorbic acid. The cell sheets can be lifted at room temperature since the Poly-n-isopropylacrylamide (Poly-NIPAM)-based substrate will become soluble at lower temperature. The cell-ECM sheet has been explored for promotion of tendon/ligament healing [20] as well as diabetic skin wound healing [21].

2.3. FDA-approved biomaterial

2.3.1. Collagen sponge scaffold

(5 mm × 2 mm) collagen sponge scaffolds (Zimmer Dental) were used for the culture of (BM) MSCs. Cell-seeded scaffolds were placed in culture dishes and incubated for 2 hours in a minimum volume of growth medium, after which more medium was applied to submerge the scaffolds. After an additional 24-hour culture, cells seeded in scaffolds were treated with 10 ng/mL of recombinant (BMP) 12 for 12 hours. The medium was then replaced with fresh growth medium and scaffolds were either cultured for an additional 7 days or immediately implanted into partial calcaneal tendon defects in rats. It was shown that after 21 days, the BMP12-treated, collagen-cell scaffold results in robust formation of tendon-like tissue [22]. Similarly, a collagen carrier combined with ASDCs showed they did not improve the biomechanical properties of the tendon-to-bone healing. However, the ADSCs group showed less inflammation, which may lead to a more elastic repair and less scarred healing in a rat model.

2.3.2. Integra bilayer

Integra® bilayer wound Matrix (LifeSciences Corp., Plainsboro, New Jersey) is a dermal acellular analog composed of bovine collagen type I crosslinked with glycosaminoglycans. Importantly, inclusion of WJ-MSC into Integra induced significant up-regulation of prototypical angiogenic and healing factors, stimulating pleiotropic aspects of neovascularization in experimental settings of angiogenesis, without altering the inflammatory response in the animals, thus demonstrating their potential benefit in therapeutic care of wounds and skin grafts [23]. Recently, Integra dermal matrix scaffold engineered with adult mesenchymal stem cells and platelet-rich plasma was investigated *in vitro* and demonstrated promising preliminary results [24].

2.4. Synthetic biopolymer matrix

Biopolymer matrix such as collagen, gelatin, hyaluronic acid, chitosan, silk, cellulose are the essential ECM components of human, animal, or plant cells. They have showed great promise for attachment, proliferation, and differentiation of stem cells and have been explored for tendon and skin regeneration.

2.4.1. *Collagen*

Collagen monomer solution can be extracted from skin of fetal bovine calf in close herd. Collagen-based biomaterial has been widely used for tendon/skin regeneration. For tendon application, it is highly desirable to prepare aligned collagen fiber scaffolds. Anisotropic collagen biomaterial can be prepared by directional freeze drying [25], electrospinning, or a novel process called electrochemical process. Electrochemically aligned collagen fiber and skin substrate have been coupled with stem cells for both tendon [26] and skin regeneration [27], respectively. Collagen carrier/gel [28] or collagen combined with stem cell accelerated the wound healing in healing-impaired db/db mice [29].

2.4.2. *Pullulan-collagen hydrogel*

A biomimetic pullulan-collagen hydrogel was used to create a functional biomaterial-based stem cell niche for the delivery of MSCs into wounds. Murine bone marrow-derived MSCs were seeded into hydrogels and compared to MSCs grown in standard culture conditions. Hydrogels induced MSC secretion of angiogenic cytokines and expression of transcription factors associated with maintenance of pluripotency and self-renewal. MSC-seeded hydrogels showed significantly accelerated healing and a return of skin appendages [30].

2.4.3. *Gelatin*

Gelatin is denatured collagen. Human ADSCs laden gelatin microcryogels (GMs) were evaluated *in vitro* as a stem cell carrier. The cell phenotype markers, stemness genes, differentiation, secretion of growth factors, cell apoptosis, and cell memory were compared against cells without a carrier. The priming effects of GMs on upregulation of stemness genes and improved secretion of growth factors of hASCs were demonstrated.

2.4.4. *Hyaluronic acid (HA)*

Hyaluronic acid (HA) is a nonsulfated, linear polysaccharide with the repeating disaccharide, β-1,4-D-glucuronic acid - β-1,3-N-acetyl-D-glucosamine (Mw: 100–8000 KDa). It is an ECM component. ADSCs combined with hyaluronic acid (ADSC-HA) dermal filler were implanted in rats and compared with HA alone. It was demonstrated that ADSC-HA has better filling effects than HA alone. A total of 70% of stem cells remain in the injection site after 3 months. These suggested stem cells have the potential to improve the esthetic effects and longevity of dermal fillers [31].

2.4.5. *Chitin/Chitosan*

Chitosan-hyaluron membrane: Hsu et al. investigated adult ADSCs spheroids combined with a chitosan-hyaluron membrane and showed biomaterial combined with stem cells promoted wound healing in a rat skin repair model [32]. Dense and porous chitosan-xanthan membranes seeded with multipotent mesenchymal stromal cells were evaluated for the treatment of skin wounds. The membranes showed to be nonmutagenic and allowed efficient adhesion and proliferation of the mesenchymal stromal cells *in vitro. In vivo* assays performed with mesenchymal stromal cells grown on the surface of the dense membranes showed acceleration of wound healing in Wistar rats [33].

2.4.6. Gelatin/PEG

A thiol-ene Michael-type addition was utilized for rapid encapsulation of MSCs within a gelatin/PEG biomatrix according to Eq. (1). The MSCs/gelatin/PEG biomatrix was applied as a provisional dressing to full-thickness wounds in Sprague-Dawley rats. Biomatrix resulted in attenuated immune cell infiltration, lack of foreign giant cell (FBGC) formation, accelerated wound closure and re-epithelialization, as well as enhanced neovascularization and granulation tissue formation by 7 days [34].

$$\text{Gelatin - PEG - SH} + \text{PEG-diacrylate} \rightarrow \text{Gelatin - PEG hydrogel} \tag{1}$$

2.4.7. Silk

Electrospun nanofiberous scaffolds prepared from silk fibroin protein were seeded with bone marrow-derived mesenchymal stem cells (MSCs) and epidermal stem cells (ESCs). The constructs were evaluated for wound re-epithelization, collagen synthesis, as well as the skin appendages regeneration. It was shown that both the transplantation of MSCs and ESCs could significantly accelerate the skin re-epithelization, stimulate the collagen synthesis. Furthermore, the regenerative features of MSCs and ESCs in activating the blood vessels and hair follicles formation, respectively, were suggested [35]. Combination of silk with collagen or poly(lactic-co-glycolic acid) (PLGA) [18] and stem cells were evaluated in a rabbit tendon defect model [36].

2.4.8. Fibrin-agarose

A stroma skin substitute was first generated by using a mixture of human fibrin obtained from frozen human plasma and 0.1% agarose. An average of 250,000 cultured skin fibroblasts were added to 5 mL of the mixture immediately before inducing the polymerization of the artificial stroma on Transwell (Corning Enterprises, Corning, NY, http://www.corning.com) porous inserts. Once the stromas jellified, human umbilical cord Wharton's jelly stem cells (HWJSCs) were seeded on top of the skin artificial stromas and cultured. It was demonstrated that this 3D bioactive scaffold can stratify and form epithelial cell-like layers and well-formed cell-cell junctions [37]. The authors also demonstrated similar strategy can be used for regeneration of oral mucosa using human Wharton's jelly stem cells (HWJSCs)-seeded fibrin-agarose-mucosal fibroblasts.

2.4.9. Sodium carboxymethylcellulose (CMC)

Sodium CMC was evaluated as a substrate for ADSCs and implanted in adult male Wistar rats. CMC at 10 mg/mL associated with ADSCs increased the rate of cell proliferation of the granulation tissue and epithelium thickness when compared to untreated lesions (Sham).

CMC is capable to allow the growth of ADSCs and is safe for this biological application up to the concentration of 20 mg/mL. These findings suggest that CMC is a promising biomaterial to be used in cell therapy [38].

2.5. Synthetic nondegradable polymer-based biomaterial

Table 3 summarizes common nondegradable polymer used as a biomaterial or as a modifier of ECM-based biopolymer for skin and tendon tissue engineering applications. The incorporation of such polymer allows biopolymer (collagen, gelatin, HA) to be able to gelled at physiological condition (e.g., under UV/blue light, ambient temperature, or room temperature aqueous free radical initiator. Stem cells can be directly encapsulated inside the substrate during gel formation.

2.5.1. Poly(NIPAM)-based

A biodegradable, multifunctional crosslinker and an n-isopropylacrylamide (NIPAM)-based, thermosensitive hydrogel was synthesized to carry BMSCs to treat diabetic skin ulcers. The crosslinker contains an arginylglycylaspartic acid(GRD)-like motif that promotes cell attachment and differentiation of BMSCs. After hydrogel association with BMSCs treated the diabetic skin wound in mice, significantly greater wound contraction was observed in the hydrogel + BMSCs group. Histology and immunohistochemistry results confirmed that this treatment contributed to the rapid healing of diabetic skin wounds by promoting granulation tissue formation, angiogenesis, extracellular matrix (ECM) secretion, wound contraction, and re-epithelialization. These results show that a hydrogel laden with BMSCs may be a promising therapeutic strategy for the management of diabetic ulcers [39].

Poly(NIPAM)	PEGDA	Poly(methyl acrylate) (PMA)
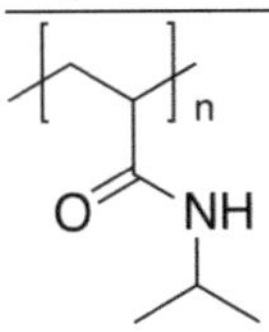	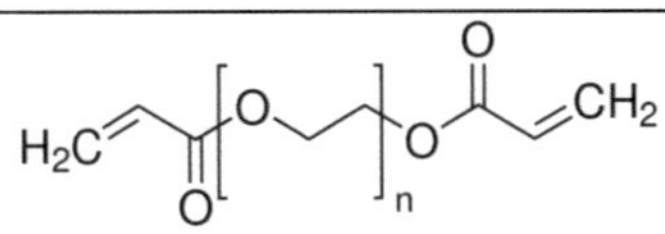	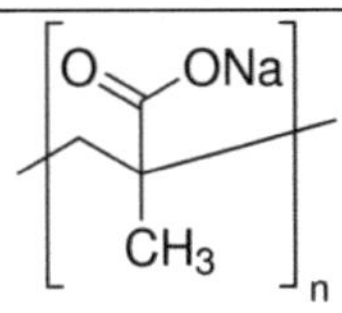

Table 3. Common synthetic, nondegradable polymer used as hydrogel or gel components.

2.6. Polyester-based degradable polymer biomaterial

The structure of common polyester is shown in **Table 4**. Stem cell-coated polyester suture was evaluated for tendon applications in a rat model [40]. Electrospun PLGA fiber may be more suitable for tendon regeneration than film [41]. (PGA/PLA) fiber combined with ADSCs improve tendon in a rabbit tendon model [42]. Open cell PLGA seeded with stem cells produced more collagen type I [43]. Knited PLGA encapsulated with stem cell/alginate gel [44]. PGA sheet with MSCs were able to regenerate tendon-bone insertions and the tendon in rabbit [45]. Electrospun polycaprolactone/gelatin (PCL/GT) membrane and human urine-derived stem cells (USCs) were evaluated for skin wound healing in a rabbit model [46]. USCs-PCL/GT-treated wounds closed much faster, with increased re-epithelialization, collagen formation, and

PLGA	PCL	PHBV
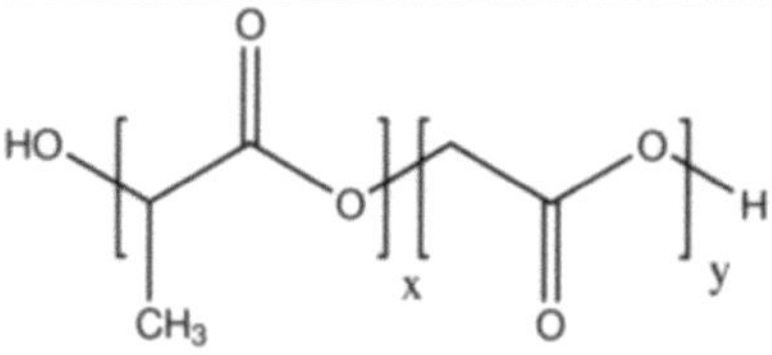	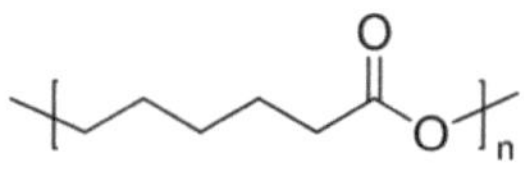	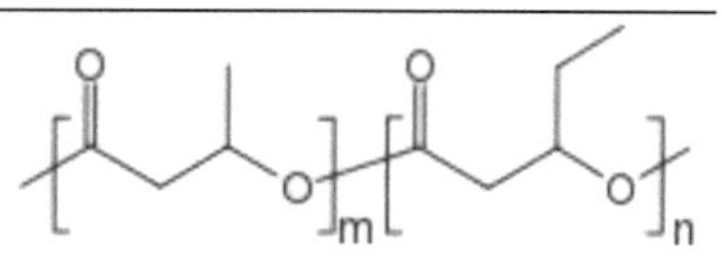

Table 4. Common synthetic, degradable polymer used for tendon/skin regeneration applications.

angiogenesis. Moreover, USCs could secrete vascular endothelial growth factor(VEGF) and transforming growth factor (TGF)-β1, and USC-conditioned medium enhanced the migration, proliferation, and tube formation of endothelial cells. This data suggested that USCs in combination with PCL/GT significantly prompted the healing of full-thickness skin wounds in rabbits. Similarly, electrospun poly (L-lactide-co-ε-caprolactone)/poloxamer (PLCL/P123) scaffolds combined with ADSCs enhanced skin wound healing in a rat model [47]. Chitosan-crosslinked poly (3-hydroxybutyrate-co-3-hydroxyvalerate) (PHBV) was used to load unrestricted somatic stem cells. The cell-laden scaffold showed better results during the healing process of skin defects in rat models [48, 49].

3. Conclusion

Biomaterials play an important role for attachment, survival, and function of stem cells. Many biomaterials are either used alone or as one component of the product for the regeneration of tendon/ligament/skin. Despite abundance of biomaterial developed, the optimal biomaterials that meet the structural, mechanical, functional requirement of tendon/ligament tissues to be regenerated remain a challenge. Novel biomaterial fabrication process, biomaterial design, and biomaterial synthesis toward tendon/ligament and skin regeneration are urgently needed.

Author details

Xingguo Cheng

Address all correspondence to: xcheng@swri.org

Southwest Research Institute, San Antonio, TX, USA

References

[1] Uysal CA, et al. Adipose-derived stem cells enhance primary tendon repair: Biomechanical and immunohistochemical evaluation. Journal of Plastic Reconstructive & Aesthetic Surgery. 2012:**65**(12):1712-1719

[2] Xu K, et al. Platelet-rich plasma activates tendon-derived stem cells to promote regeneration of Achilles tendon rupture in rats. Journal of Tissue Engineering and Regenerative Medicine. 2017;**11**(4):1173-1184

[3] Martinello T, et al. Effects of *in vivo* applications of peripheral blood-derived mesenchymal stromal cells (PB-MSCs) and platlet-rich plasma (PRP) on experimentally injured deep digital flexor tendons of sheep. Journal of Orthopaedic Research. 2013;**31**(2):306-314

[4] Falanga V, et al. Autologous bone marrow-derived cultured mesenchymal stem cells delivered in a fibrin spray accelerate healing in murine and human cutaneous wounds. Tissue Engineering. 2007;**13**(6):1299-1312

[5] Badiavas EV, Falanga V. Treatment of chronic wounds with bone marrow-derived cells. Archives of Dermatology. 2003;**139**(4):510-516

[6] Chung E, et al. Fibrin-based stem cell containing scaffold improves the dynamics of burn wound healing. Wound Repair and Regeneration. 2016;**24**(5):810-819

[7] Sanchez-Sanchez R, et al. Generation of two biological wound dressings as a potential delivery system of human adipose-derived mesenchymal stem cells. Asaio Journal. 2015;**61**(6):718-725

[8] Ji SZ, et al. An epidermal stem cells niche microenvironment created by engineered human amniotic membrane. Biomaterials. 2011;**32**(31):7801-7811

[9] Yin Z, et al. The effect of decellularized matrices on human tendon stem/progenitor cell differentiation and tendon repair. Acta Biomaterialia. 2013;**9**(12):9317-9329

[10] Bi Y, et al. Identification of tendon stem/progenitor cells and the role of the extracellular matrix in their niche. Nature Medicine. 2007;**13**(10):1219-1227

[11] Huang SP, et al. Adipose-derived stem cells seeded on acellular dermal matrix grafts enhance wound healing in a murine model of a full-thickness defect. Annals of Plastic Surgery. 2012;**69**(6):656-662

[12] Martinello T, et al. Successful recellularization of human tendon scaffolds using adipose-derived mesenchymal stem cells and collagen gel. Journal of Tissue Engineering and Regenerative Medicine. 2014;**8**(8):612-619

[13] Ouyang HW, et al. Mesenchymal stem cell sheets revitalize nonviable dense grafts: Implications for repair of large-bone and tendon defects. Transplantation. 2006;**82**(2): 170-174

[14] Protzman NM, Stopyra GA, Hoffman JK. Biologically enhanced healing of the human rotator cuff: 8-month postoperative histological evaluation. Orthopedics. 2013;**36**(1):38-41

[15] Altman AM, et al. Human adipose-derived stem cells adhere to acellular dermal matrix. Aesthetic Plastic Surgery. 2008;**32**(4):698-699

[16] Altman AM, et al. Dermal matrix as a carrier for *in vivo* delivery of human adipose-derived stem cells. Biomaterials. 2008;**29**(10):1431-1442

[17] Gao Y, et al. Methodology of fibroblast and mesenchymal stem cell coating of surgical meshes: A pilot analysis. Journal of Biomedical Materials Research Part B Applied Biomaterials. 2014;**102**(4):797-805

[18] Lam MT, et al. Effective delivery of stem cells using an extracellular matrix patch results in increased cell survival and proliferation and reduced scarring in skin wound healing. Tissue Engineering Part A. 2013;**19**(5-6):738-747

[19] Liu S, et al. Synergistic angiogenesis promoting effects of extracellular matrix scaffolds and adipose-derived stem cells during wound repair. Tissue Engineering Part A. 2011; **17**(5-6):725-739

[20] Mifune Y, et al. Tendon graft revitalization using adult anterior cruciate ligament (ACL)-derived CD34+ cell sheets for ACL reconstruction. Biomaterials. 2013;**34**(22):5476-5487

[21] Kato Y, et al. Allogeneic transplantation of an adipose-derived stem cell sheet combined with artificial skin accelerates wound healing in a rat wound model of type 2 diabetes and obesity. Diabetes. 2015;**64**(8):2723-2734

[22] Lee JY, et al. BMP-12 treatment of adult mesenchymal stem cells *in vitro* augments tendon-like tissue formation and defect repair *in vivo*. PLoS One. 2011;**6**(3):e17531

[23] Edwards SS, et al. Functional analysis reveals angiogenic potential of human mesenchymal stem cells from Wharton's jelly in dermal regeneration. Angiogenesis. 2014;**17**(4): 851-866

[24] Formigli L, et al. Dermal matrix scaffold engineered with adult mesenchymal stem cells and platelet-rich plasma as a potential tool for tissue repair and regeneration. Journal of Tissue Engineering and Regenerative Medicine. 2012;**6**(2):125-134

[25] Caliari SR, Harley BA. Structural and biochemical modification of a collagen scaffold to selectively enhance MSC tenogenic, chondrogenic, and osteogenic differentiation. Advanced Healthcare Materials. 2014;**3**(7):1086-1096

[26] Cheng XG, et al. Platelet-derived growth-factor-releasing aligned collagen–nanoparticle fibers promote the proliferation and tenogenic differentiation of adipose-derived stem cells. Acta Biomaterialia. 2014;**10**(3):1360-1369

[27] Cheng X, et al. Preparation and *in vitro* evaluation of electrochemically-aligned collagen matrix as a dermal substitute. MRS Advances. 2016;**1**(18):1295-1300

[28] Valencia Mora M, et al. Application of adipose tissue-derived stem cells in a rat rotator cuff repair model. Injury. 2014;**45**(Suppl 4):S22-27

[29] Nambu M, et al. Accelerated wound healing in healing-impaired db/db mice by autologous adipose tissue-derived stromal cells combined with atelocollagen matrix. Annals of Plastic Surgery. 2009;**62**(3):317-321

[30] Rustad KC, et al. Enhancement of mesenchymal stem cell angiogenic capacity and stemness by a biomimetic hydrogel scaffold. Biomaterials. 2012;**33**(1):80-90

[31] Nowacki M, et al. Filling effects, persistence, and safety of dermal fillers formulated with stem cells in an animal model. Aesthetic Surgery Journal. 2014;**34**(8):1261-1269

[32] Hsu SH, Hsieh PS. Self-assembled adult adipose-derived stem cell spheroids combined with biomaterials promote wound healing in a rat skin repair model. Wound Repair and Regeneration. 2015;**23**(1):57-64

[33] Bellini MZ, et al. Combining xanthan and chitosan membranes to multipotent mesenchymal stromal cells as bioactive dressings for dermo-epidermal wounds. Journal of Biomaterials Applications. 2015;**29**(8):1155-1166

[34] Xu K, et al. Thiol-ene Michael-type formation of gelatin/poly(ethylene glycol) biomatrices for three-dimensional mesenchymal stromal/stem cell administration to cutaneous wounds. Acta Biomaterialia. 2013;**9**(11):8802-8814

[35] Xie SY, et al. Adult stem cells seeded on electrospinning silk fibroin nanofiberous scaffold enhance wound repair and regeneration. Journal of Nanoscience and Nanotechnology. 2016;**16**(6):5498-5505

[36] Zhang W, et al. Weft-knitted silk-poly(lactide-co-glycolide) mesh scaffold combined with collagen matrix and seeded with mesenchymal stem cells for rabbit Achilles tendon repair. Connective Tissue Research. 2015;**56**(1):25-34

[37] Garzon I, et al. Wharton's jelly stem cells: a novel cell source for oral mucosa and skin epithelia regeneration. Stem Cells Translational Medicine. 2013;**2**(8):625-632

[38] Rodrigues C, et al. New therapy of skin repair combining adipose-derived mesenchymal stem cells with sodium carboxymethylcellulose scaffold in a pre-clinical rat model. PLoS One. 2014;**9**(5):e96241

[39] Chen S, et al. Mesenchymal stem cell-laden anti-inflammatory hydrogel enhances diabetic wound healing. Scientific Reports. 2015;**5**:18104

[40] Adams Jr SB, et al. Stem cell-bearing suture improves Achilles tendon healing in a rat model. Foot & Ankle International. 2014;**35**(3):293-299

[41] James R, et al. Tendon tissue engineering: Adipose-derived stem cell and GDF-5 mediated regeneration using electrospun matrix systems. Biomedical Materials. 2011;**6**(2):025011

[42] Deng D, et al. Repair of Achilles tendon defect with autologous ASCs engineered tendon in a rabbit model. Biomaterials. 2014;**35**(31):8801-8809

[43] Kim YS, et al. Survivorship of implanted bone marrow-derived mesenchymal stem cells in acute rotator cuff tear. Journal of Shoulder and Elbow Surgery. 2013;**22**(8):1037-1045

[44] Vaquette C, et al. A poly(lactic-co-glycolic acid) knitted scaffold for tendon tissue engineering: An *in vitro* and *in vivo* study. Journal of Biomaterials Science Polymer Edition. 2010;**21**(13):1737-1760

[45] Yokoya S, et al. Rotator cuff regeneration using a bioabsorbable material with bone marrow-derived mesenchymal stem cells in a rabbit model. The American Journal of Sports Medicine. 2012;**40**(6):1259-1268

[46] Fu Y, et al. Human urine-derived stem cells in combination with polycaprolactone/gelatin nanofibrous membranes enhance wound healing by promoting angiogenesis. Journal of Translational Medicine. 2014;**12**:274

[47] Gu J, et al. Adiposed-derived stem cells seeded on PLCL/P123 eletrospun nanofibrous scaffold enhance wound healing. Biomedical Materials. 2014;**9**(3):035012

[48] Biazar E, Keshel SH. The healing effect of stem cells loaded in nanofibrous scaffolds on full thickness skin defects. Journal of Biomedical Nanotechnology. 2013;**9**(9):1471-1482

[49] Zeinali R, et al. Regeneration of full-thickness skin defects using umbilical cord blood stem cells loaded into modified porous scaffolds. ASAIO Journal. 2014;**60**(1):106-114

Chapter 12

Hydrogels in Regenerative Medicine

Yasemin Budama-Kilinc, Rabia Cakir-Koc,
Bahar Aslan, Burcu Özkan, Hande Mutlu and
Eslin Üstün

Additional information is available at the end of the chapter

http://dx.doi.org/10.5772/intechopen.70409

Abstract

Polymer scaffolds have many various applications in the field of tissue engineering, drug delivery, and implantation. They are applied as dispensing devices for bioactive molecules and as three-dimensional (3D) structures that provide stimulants that organize cells and direct desired original tissue formation. Hydrogels are preferred scaffolding material because they are structurally similar to the extracellular matrix of many tissues, often processed under mild conditions, and can be delivered in a minimally invasive manner. Hydrogel materials formed a group of polymeric materials. The hydrophilic structure allows them to hold large amounts of water in their three-dimensional backbone. As a result, hydrogels are used as scaffolding material for drug and growth factor transmission, tissue engineering modifications, and many other applications. In this chapter, we describe the physical and chemical structure of hydrogels, side groups, cross-linkings, swelling properties, types of polymers and fabrication methods, and application fields.

Keywords: hydrogel, polymers, drug delivery, tissue engineering

1. Introduction

Hydrogels are polymer-based substances that contain high levels of water and show the physical characteristic of the extracellular matrix. In regenerative medicine, in terms of the material requirements for tissue scaffold or therapeutic transfer systems, the cell matrix is preferred due to its structural and compound-based similarities. Moreover, hydrogels are preferred as they promote cell attachment and proliferation.

In this chapter, compiling the studies about the preparation of hydrogels and application of hydrogels in regenerative medicine is aimed.

2. Physical and chemical properties of hydrogels

2.1. Crosslinking and general characteristic of hydrogels

Hydrogels are a three-dimensional (3D) polymeric network structure with the ability to absorb water and other biological fluids. The water content of hydrogels determines its specific physicochemical character: mildness, density, and low surface tension of hydrogels [1–4].

Hydrogels can be classified according to their preparation methods, ionic charges, physical structures, and cross-linking conditions. According to physical structures, hydrogels can be divided into some groups that are called amorphous hydrogels, semi-crystalline hydrogels, hydrogen-bonded hydrogels, physical hydrogels, and chemical hydrogels.

Various cross-linking approaches, including chemical and physical, have been used to create polymer networks and to protect 3D structures in aqueous media. Physical interactions between the polymer chains in the physically cross-linked gels prevent hydrogel breakdown, while the covalent bonds between the polymer chains in the chemically cross-linked gel form a constant hydrogel.

Physically cross-linked hydrogels are formed by changes in environmental conditions (pH, temperature, and ionic interactions), hydrogen bonds, and protein interactions. Chemically cross-linked gels are obtained by radical polymerization, chemical reactions, energy irradiation, and enzymatic cross-linking [5].

Hydrogels can also have many physical forms: solid mold forms (such as soft contact lenses), pressed powder matrices (pills or capsules taken orally), microparticles (bio-structured carriers or wound treatments), coatings (in implants or in catheters, in pills or capsules, or in capillary electrophoresis), membranes or disks (such as two-dimensional electrophoresis gels), encapsulated solute (in osmotic pumps), and liquids (gel forms that are heated or cooled) [6].

2.2. Hydrogels side groups

Hydrogels are also called "hydrophilic polymers" because of their water-like properties. Hydrogels have functional groups, such as $-SO_3H$, $-COOH$, $-CONH_2$, $-OH$, and $-NH_2$, which provide hydrophilic character to the polymer chains in their networks. Some of these functional groups ($-SO_3$ and $-COO$ hydrogel nets) are negatively charged in aqueous media and can adsorb metal ions through electrostatic interactions [7].

Depending on their chemical and physical structure, there are three types of hydrogels: neutral hydrogels, ionic hydrogels, and interconnected network structures. Neutral (nonionic) hydrogels are homopolymeric or copolymeric hydrogels which do not have charged groups in their structure. Ionic hydrogels, also known as polyelectrolytes, are prepared from ionically charged monomers. These hydrogels are termed cationic and anionic hydrogels, as their monomeric charge is positive and negative [8]. The presence of charged groups in the main chain of ionic hydrogels enhances their susceptibility to stimuli. The presence of ionizable functional groups such as carboxyl and sulfonic acid or ammonium salt increases the hydrophilicity of the polymer and thus increases the water absorption capacity. As a result of the electrostatic repulsive forces of these loads, the solubility and swelling ratios in the web structure are increased. In ionic hydrogels containing acid groups, swelling increases in basic medium, while hydrogels containing basic groups increase swelling in acidic media [9, 10].

2.3. Cross-linking methods in hydrogel fabrication

Hydrogels are polymeric webs that absorb and retain water in large quantities. In the polymeric cannula, there are hydrophilic groups or areas that provide a hydrated hydrogel structure in an aqueous environment. As is the case with the term "network," cross-links must be present to prevent the hydrophilic polymer chains from dissolving in aqueous phase.

2.3.1. Physically cross-linked hydrogel

Physical gel called hydrogels is the result of cross-linking of ionic interaction and physical interactions such as hydrogen bonds, coordination bonds, and hydrophobic interactions [11]. These hydrogels alter the temperature, pH, or solvent composition to form a homogeneous solution and re-gel when they return to their initial conditions. Because of these behaviors, physical gel is also known as reversible gel [12]. They are generally unstable and mechanically weak [13].

Generally, ionic interactions are used to produce physically cross-linked hydrogels. For example, alginate hydrogels are formed with calcium ions. The physical cross-links are found between ionic interactions [14], crystallization [14], amphiphilic block and graft copolymers [15], hydrogen bonds [16], and protein interactions [17].

2.3.2. Chemical cross-linked hydrogels

The hydrogels in this group are called as irreversible because they are cross-linked by strong chemical bonds between their chains and are not redissolved by changing temperature, pH, or solvent composition. Chemical cross-linked nets have permanent junctions [18, 19]. Chemical cross-linking involves permanent covalent bonds that result in better mechanical strength and more stable hydrogels than physical cross-linking. Incorporation of chemical cross-linkers, however, can cause toxicity problems. Cross-linking of hydrogel can occur with some other method such as radical polymerization, chemical reaction, and high-energy irradiation.

Chemically cross-linked gels can be obtained by radical polymerization of polymers of low molecular weight drugs in the presence of cross-linking agents; poly(2-hydroxyethyl methacrylate) (pHEMA) is a well-known and frequently studied hydrogel system. This hydrogel was first described by Wichterle and Lim [19] and was obtained by polymerization of HEMA in the presence of a suitable cross-linking agent (e.g., ethylene glycol dimethacrylate). A wide variety of hydrogel systems have been synthesized using similar procedures [20].

The covalent bonds between the polymer chains can be formed by reaction with complementary reactivity of the functional groups, formation of an amine carboxylic acid, or OH/NH_3 reaction such as isocyanate or Schiff base formation. Chemical cross-linking bonds can occur with aldehydes, with additional reactions and condensation reactions. Water-soluble polymers owe their solubility properties to the presence of functional groups (mainly —OH, COOH, NH) which can be used for hydrogel formation.

High-energy radiation, especially gamma and electron beams, can be used to polymerize unsaturated compounds. This means that water-soluble polymers derivatized with vinyl groups can be converted into hydrogels by high-energy irradiation [21]. Hydrogels may also be obtained by radiation-induced polymerization of a mixture consisting of a monofunctional acrylate (e.g., acryloyl-L-proline methyl ester) and a suitable cross-linker [22]. Moreover, high-energy irradiation cross-links water-soluble polymers without additional groups. During irradiation of aqueous solutions of polymers (gamma or electron transfer), the radicals are generally present in the polymer chain, e.g., homolytic cleavage of C–H bonds.

2.4. Water content and resources

Hydrogels can generally hold 20–100 times water and their own weight depends on their structures [23]. If hydrogels can hold higher than 100%, they are called superabsorbent hydrogels. According to the water contents, hydrogels can be categorized as low swelling grade hydrogels (20–50%), medium swelling grade hydrogels (50–90%), high swelling grade hydrogels (90–99.5%), and superabsorbent hydrogels (>99.5%) [24, 25].

High water content hydrogels are usually more useful in the medical field because of permeability and biocompatibility [26]. Synthesis of biodegradable hydrogels with both good mechanical properties and high water content is still a major challenge, because as the amount of water increases, the mechanical properties of hydrogel become weak [27].

Because water content of hydrogels is high, these gels resemble natural living tissues more than synthetic biological materials, so the biocompatibility is higher. When the polymer network contacts the aqueous solutions, swelling occurs to achieve thermodynamic stability due to the water concentration gradient [28].

The hydrophilicity of the polymer chain and the cross-link density are parameters effective in hydrogel swelling. Hydrophilic groups of polymer networks in hydrogels are groups that like water. Therefore, they have a high affinity for water. These groups allow hydrogels to swell in water, become soft, and gain elastic properties [29]. Hydrophilic group-containing hydrogels swell at a higher rate than hydrophobic group-containing hydrogels. Because hydrophobic groups do not like water, they prevent water molecules from penetrating into the structure [30].

The structures of the high cross-linked hydrogels are more rigid, and cross-linking inhibits the movement of the polymer chain. When the amount of cross-linking agent increases, the hydrogel becomes less flexible, and the swelling ability of the hydrogel is reduced [31, 32].

2.5. Polymers used in hydrogel fabrication

Polymer materials have many uses such as catheters, heart valves, contact and intraocular lenses, dental prostheses, ophthalmic applications, urology, and gastroenterology applications [33]. In tissue engineering different polymers are used for different tissues. It is possible to classify them in four different groups.

2.5.1. Natural and synthetic polymers

Natural polymers have been used as scaffolds for tissue engineering as natural hydrogels, regarding their biological compatibility, natural biodegradability, and critical biological functions. There are four main types of natural polymers, including:

- proteins such as collagen, silk, gelatin, fibrin, lysozyme, Matrigel™, and genetically engineered proteins [34–36] like calmodulin (a calcium-binding protein), elastin-like polypeptides, and leucine [37] zipper;
- polysaccharides like hyaluronic acid (HA), dextran, agarose, and chitosan [38].
- protein/polysaccharide hybrid polymers like collagen/HA, laminin/chitosan, fibrin/cellulose, and gelatin/alginate [39]; and
- DNA [40].

Compared to natural polymers, synthetic polymers have more reproducible physical and chemical properties that are critical for the construction of scaffolding. At present, synthetic polymers have emerged as an important alternative option in the manufacture of hydrogel tissue-engineered skeletons, since they can be molecularly modified by block structures, molecular weights, mechanical strength, and biodegradability [41–43]. Synthetic polymers used in the preparation of synthetic hydrogels can be classified into three main types: non-biodegradable [44, 45], biodegradable [46], and bioactive polymers [47]. In the following section, the biodegradable synthetic polymers are disclosed because of its widespread use from among these three main types.

2.5.2. Biodegradable synthetic polymers

Biodegradability is one of the most important aspects of scaffolds for tissue engineering. It is highly desirable to ensure that the biodegradation rate coincides with new tissue regeneration at the defect site [48, 49]. Many polymers, proteins, cellulose, starch, and chitin created in the environment are limited to making hydrogel skeletons that are biodegradable but have their own biodegradability and mechanical properties. Most common biodegradable synthetic polymer groups contain poly (lactic acid) (PLA), poly(ε-caprolactone) (PCL), poly(glycolic acid) (PGA) and copolymers [50]. They can be used to modify hydrophilic polymers such as PEG to form acrylate macromers or amphiphilic polymers to produce biodegradable hydrogels by chemical or physical cross-linking [51, 52].

The mechanical stability of the gel is an important feature when a scaffold is designed. The strength of the hydrogels can be increased by incorporating cross-linking agents, comonomers, because the higher cross-linking degree provides more fragility and less elasticity.

Biodegradable synthetic hydrogels include various vinylic monomers such as 2-hydroxyethyl methacrylate (HEMA), N-isopropylacrylamide (NIPAM), 2-hydroxypropyl methacrylate (HPMA), acrylamide (AAm), acrylic acid (AAc) or macromers [53–55] N,N′-methylenebis and methoxyl poly(ethylene glycol) (PEG), monoacrylate (mPEGMA or PEGMA), and diacrylate (PEGDA) with cross-linkers such as ethylene glycol diacrylate (EGDA), and these nonbiodegradable polymers such as self-assembly of Pluronic® polymers that have poly(ethylene oxide) (PEO)-poly(propylene oxide) (PPO)-PEO structure [53–55] are modified with poly(vinyl alcohol) (PVA) via chemical cross-linking.

Synthetic polymers can be easily synthesized on a large scale and manipulated at the molecular level by polymerization, cross-linking, and functionalization. However, most synthetic hydrogels alone generally function as passive scaffolds only for cells and do not increase active cellular interactions [56]. As already mentioned, natural polymers such as proteins exhibit different tertiary structures and regulate active cellular response, biological identification, and cell-triggered remodeling. Thus, combining the properties of synthetic and natural polymers to form hybrid hydrogels has become a direct approach to creating bioactive hydrogel scaffolds for tissue engineering. These same examples of hybrid hydrogel polymers include:

- PEG-modified natural polymers [57–60], like fibrinogen, heparin, dextran, HA, and albumin;
- PNIPAm-modified natural polymers, like collagen, chitosan, and alginate [61];
- Synthetic peptide-modified proteins or polysaccharides [62]; and
- PVA and other synthetic polymer (e.g., pluronic)-modified natural polymers [63].

The hydrogel is obtained in different forms depending on the application area by selecting the appropriate polymer. It is possible to produce hydrogels which can be heat-adjusted, injectable, or in the form of film, foam, or gel. These can be controlled by different parameters such as cross-linking of structures and synthesis of polymers.

3. Application of hydrogels

The hydrogels with various features are widely used as wound dressing, drug delivery, dental equipment, pharmaceutic industry, injectable polymer systems, implants, and tissue engineering.

3.1. Applications of hydrogels in dentistry

In recent years, injectable hydrogels have emerged as a promising biomaterial for the therapeutic delivery of cells and bioactive molecules for tissue regeneration in dentistry. The controllability of deterioration and release behavior, its adaptability in the clinical setting for minimally invasive surgical procedures, and its ability to remain three-dimensional (3D) after gelling mean that these hydrogels can be used in medical practice [64]. In dental tissue engineering, injectable hydrogels are used to replace countless tissues, especially cartilage, bones, nerves, blood vessels, and soft tissues. Furthermore, it has been demonstrated that the feasibility of introducing dental pulp stem cells with hydrogels has supported the matrix and growth factors [65].

Hydrogel is biologically compatible and absorbed into the body without any side effects. The gel is completely disintegrated so no by-products or new chemicals are formed. For these reasons, they are preferred as dental fillers, as dental equipment, and even as coating for dental implants [66].

Hydrogels are administered as low-viscosity liquids that penetrate the dentin tubules when exposed to capillary action. Treatment with a gel that forms a long-term block is provided when it is swollen and wet; at this point it is also permeable to oxygen and electrolytes [67]. The use of hydrogels in dental equipment offers several advantages over other biomaterials such as (1) long-term grip despite the effects of surface abrasion, (2) its softness removes the pressure (3) brushing and abrasion only cleanse the exposed parts of the gel and leave the remainder in the tubules for a long time (4), the hardened gel is colorless and therefore maintains the normal appearance of the tooth, (5) the gel structure continues to be permeable to oxygen and electrolytes, and (6) there are no hydraulic pressures that can cause pain without fluid flow [68].

Oral hydrogels are drug-free, are completely natural as wound dressings, and are approved by the FDA (Food and Drug Administration) for all oral injuries. They can be used in tooth extractions, grafts, implants, lesions, or for wounds in the mouth [69].

Depending on the hydrostatic pressure, the sharp fluid movements in these micro-channels can cause dentin pain. The microorganisms present in the oral cavity ferment carbohydrates to form lactic acid, which causes the pH to drop. The acidic medium then causes the hydroxyapatite to dissolve in the hard walls of the micro-channels. This increases the permeability of the channels and facilitates the penetration of microorganisms. The spread of bacterial toxins causes an inflammatory reaction in the pulp, which causes symptoms such as hypersensitivity and pain. For this reason, the development of new prophylactic agents and methods to reduce tooth decay and an overly sensitive tooth neck is of great importance [70]. Ionic hydrogels, especially nanogels, have started being used in toothpastes. Micro-channels filled with a fibrous protein and a bio-hydrogel composed of a liquid phase enter the human teeth. The gel network is bound to the organic collagen matrix and the hydroxyapatite crystals on the walls of the dentin tubules [71].

3.2. Applications of hydrogels in drug delivery

Conventional drug administration can result in toxicity and side effects because of requirement of high dosages and repeated administration to stimulate a therapeutic effect, and these can reduce efficacy and patient compliance [72–74]. Researchers have focused on controlled drug delivery systems during the recent years for reducing toxicity and by enhancing efficacy of therapeutics [75]. Drug release is controlled by drug delivery systems in terms of transportation to tissues and cells in time and in space [76, 77].

Hydrogels have been considerably researched as the carrier for drug delivery systems [78]. They have low interfacial tension; thus, they exhibit minimum tendency to adsorb proteins from body fluids [79, 80]. These biomaterials have paid attention due to their peculiar properties such as temperature and pH sensitivity, swelling in aqueous medium, and sensitivity toward other stimuli [78]. Their highly porous structure can be modified through controlling the density of cross-links in the gel matrix and the affinity of the hydrogels for the aqueous environment in which they are swollen. Drug release at a rate dependent on the diffusion coefficient of the small molecule or macromolecule through the gel network and loading of drugs into the gel matrix can be provided due to their porous structure [80, 81].

Hydrogels have also been served as drug protectors particularly for proteins and peptides from in vivo environment. Furthermore, these swollen polymers are beneficial as targetable carriers for bioactive drugs with tissue specificity [78]. In general hydrogels are divided for their morphology, swelling property, and elasticity. The release mechanism of the drug from the swollen polymeric material is determined by swelling, and also elasticity affects both the stability of these drug carriers and the mechanical strength of the network [82].

The body has environmental variables, known as low pH and high temperatures. In this endeavor, it either includes physical, such as temperature, or chemical, such as in pH, ions which can be used for site-specific controlled drug delivery [80, 83]. Due to their ability to "sense" changes in environmental properties, stimuli-sensitive hydrogels, also known as "smart" hydrogels, are very different from inactive hydrogels. Their response will be to increase or decrease their degree of swelling [84, 85]. "Smart" hydrogels have a property of changing the intensity which is particularly useful in the application of drug delivery as drug release can be set off upon environmental changes.

Besides, molecules of different sizes have the ability to go into and out of hydrogels (going in is called drug loading and going out of is called drug release) which makes them permissible for dry or swollen polymeric networks as drug delivery systems for nasal, oral, ocular, buccal, parenteral, epidermal, and subcutaneous routes of administration [86].

3.2.1. Peroral drug delivery

Up to now, pharmaceutical application of hydrogels has been most commonly done orally. Taken orally, hydrogels are able to deliver drugs to four major areas: the mouth, stomach, small intestine, and colon. If their swelling properties or bioadhesive abilities are controlled, hydrogels can be useful in the release of drugs in a controlled manner at these desired areas. Additionally, they can also stick to certain specific regions in the oral pathway, leading to a locally enhanced drug concentration and, thus, promoting the drug absorption at the release site [87].

3.2.2. Drug delivery in the oral cavity

Drug delivery to the oral cavity can have many functions. Hydrogel are used ın the local treatment of diseases pertaining to the mouth (i.e., periodontal, stomatitis, viral and fungal infections) and oral cavity cancers. With this in mind, many types of bioadhesive hydrogel systems have been devised since the early 1980s. To name one, a hydrogel-based ointment can also be used for the topical treatment of certain diseases in the oral cavity. It can be used not only as a drug delivery device but also as a liposome delivery medium [88, 89].

3.2.3. Drug delivery in the gastrointestinal (GI) tract

The GI tract is undoubtedly the most popular route of drug delivery because of the easiness of administration of drugs for a manageable therapy and its large surface area for absorption to the entire body. It is, however, the most complex route, so that changeable approaches are needed to give drugs for effective therapy [90]. Like buccal delivery, hydrogel-based devices can be designed to administer drugs locally to the specific areas in the GI tract. Several hydrogels are currently being investigated as potential devices for colon-specific drug delivery. These include chemically or physically cross-linked polysaccharides, such as dextran, guar gum, azo cross-linked poly(acrylic acid), amidated pectin, and inulin [91–93]. They are designed to be highly swollen or degraded in the presence of colonic enzymes or microflora, providing colon specificity in drug delivery [94].

3.2.4. Ocular delivery

The eye has shielding mechanisms at work, like tear drainage, the power of blinking, and low permeability of the cornea. These are some of the many physiological constraints preventing a successful drug delivery to the eye [95].

The usual eye drops that contain a drug solution are most likely to be eliminated quickly from the eye; thus, medication administered will have limited absorption. This in turn leads to poor ophthalmic bioavailability (the degree to which a drug or other substances is absorbed at the site of physiological activity after being given) [96].

Researchers have been motivated to come up with a system that delivers the drug for a longer stay time in the eye. Suspension and ointments which are some of the dosage forms can remain in the eye, but they also can be uncomfortable at times due to the unpleasant feeling given to the eye from characteristics of solids and semisolids [97].

Hydrogels have elastic properties, and this is why they can also represent an ocular drainage-resistant device. Additionally, they could give a better feeling to patients, with less gritty sensations. In particular, in situ forming hydrogels are attractive as an ocular drug delivery system because of their facility in dosing as a liquid, and their long-term retention property as a gel after dosage has been given [98].

3.2.5. Transdermal delivery

Drug delivery to the skin has been traditionally conducted for a specific area use of dermatological drugs to treat skin diseases or for disinfection of the skin itself. Longer duration and a constant rate are among the benefits of transdermal drug delivery that drug delivery can be easily canceled on demand by simply removing the devices. Thus, since hydrogels are high in water content, swollen hydrogels can provide a better feeling for the skin when compared to traditional ointments and patches [96].

3.2.6. Subcutaneous delivery

Subcutaneously inserted exogenous materials may more or less evoke potentially undesirable body responses, such as inflammation, carcinogenicity, and immunogenicity. Therefore, biocompatibility is a prerequisite that makes materials implantable.

Due to their high water content, hydrogels are generally considered as biocompatible materials. Hydrogels have been used to improve cellular adhesion [99]. They also provide several promising properties: minimal mechanical irritation upon living organism implantation, due to their soft, elastic properties; prevention of protein adsorption and cell adhesion arising from the low interfacial tension between water and hydrogels; broad acceptability for individual drugs with different hydrophilicities and molecular sizes; and unique possibilities (cross-linking density and swelling) to manipulate the release of incorporated drugs. Some of these may offer an advantage for the delivery of certain delicate drugs, such as peptides and proteins [73, 80].

3.3. Applications of hydrogels in tissue engineering

Tissue engineering is an interdisciplinary field that uses the principles of engineering and life sciences that aims to improving, maintaining, and renewing the functions of tissues and organs. It was emerged by Y. C. Fung in 1985 as a technical science department by using medical and material engineering [100].

Hydrogels are frequently used in biomedical and bioengineering fields due to their controllable different properties, similarity to tissues, and their ability to form scaffolds for different tissues due to their tunable physical and mechanical properties [101, 102]. Hydrogels used in tissue engineering have low viscosity before injection and require rapid gel formation in the physiological environment of the tissue [99]. It is frequently used in bone tissue, cartilage tissue, vascular tissue, meniscus, tendon, skin, cornea, and soft tissues. Hydrogel film can be used to fabricate or repair tissue in a loss of function. In the tissue engineering studies, synthetic or natural polymers are used for hydrogel production [103–105]. In tissue studies, adhesion of cells to artificial tissues in the course of cell culture, spreading, and population is an important feature.

3.3.1. Bone tissue engineering (BTE)

The bone is a dynamic and vascularized tissue with the ability to heal naturally on injury. However, repair mechanism cannot be sufficient in large bone defects. Current approaches often impose limitations that consider autografts and various allografts. BTE is based on the use of 3D matrices that both promote cellular growth and divergence to promote bone regeneration. The hydrogel matrices may contain biological materials such as cells and growth factors [106]. One of the studies is about that to assess the effect of different natural skeletons on this coculture system, cells were encapsulated in alginate and/or collagen hydrogel scaffolds. A researcher has discovered that in addition to the cell-cell proximity between the two cell types, the natural cell-binding capabilities of hydrogels such as collagen are preferred. Researchers have discovered increased osteogenic and angiogenic potential, as evidenced by increased protein and protein expression of ALP, BMP-2, VEGF, and PECAM [107].

3.3.2. Cartilage tissue engineering

Simply, articular cartilage is a tissue composed of a single cell type (chondrocytes) embedded in an extracellular matrix (ECM). However, the structure is more complex, and, depending on depth, the three ECM contents include three depth-related layers with varying behavior of structure and chondrocyte: superficial zone, middle zone, and deep zone. In the mid-region—the largest region—the cartilaginous tissue has a low density of round chondrocytes surrounded by an ECM consisting of hard, avascular glycosaminoglycan (GAG) and type II collagen [108, 109]. Hydrogel design to improve cartilage repair has progressed in recent years. Some using hydrogels include the development of improved network cross-linking (e.g., double networks), new techniques for processing hydrogels (e.g., 3D printing), and better incorporation of biological markers (e.g., controlled release).

Hydrogel has emerged as a promising port because of its potential features in a wide range and its ability to trap cells in the material.

3.3.3. *Meniscus tissue engineering*

The meniscus plays an important role in maintaining homeostasis of the knee joint [110]. Meniscus lesions are tight, have poor healing ability, and can produce tibiofemoral osteoarthritis. Current options for reconstructive treatment focus primarily on the treatment of lesions in the peripheral vascularized area [111]. On the contrary, a small number of approaches can stimulate the repair of damaged, meniscal tissue in the middle, avascular part. Tissue engineering approaches are of great interest to repair or replace damaged meniscus tissue in this area.

Hydrogel-based biomaterials show a special interest for meniscus repair, as their cores contain relatively high rates of proteoglycans responsible for viscoelastic compaction properties and degree of hydration. Hydrogels that exhibit high water content and provide a specific three-dimensional (3D) microenvironment can be designed such that the meniscus tissue can exactly resemble this topographic composition [112].

Different polymers of both natural and synthetic origin have been manipulated to produce hydrogels that host the respective cell populations for meniscus regeneration and to provide platforms for meniscus tissue replacement. To date, these compounds have been used to design controlled delivery systems of bioactive molecules involved in the meniscus repair process or to host genetically modified cells to improve meniscus repair.

3.3.4. *Tendon tissue engineering*

Significant advances have been made in repair techniques for tendon injuries in recent years, but the treatment of finger flexor tendon injury is still one of the most difficult and important problems in the orthopedic area.

The main problem in tendon repair is adhesion between the tendon and surrounding tissue. We note that polyvinyl alcohol hydrogel (PVA-H) has a high water content with both adhesion-inhibiting function and synovial perfusion function. PVA-H has been extensively studied as a biomaterial [113, 114], and mechanical strength and biocompatibility have been excellent.

It has been reported that high water content PVA-H has selective permeability penetrating into low molecular weight nutrients such as glucose and protein, but not substances with cell-size levels such as fibroblasts and leukocytes [114, 115]. In addition, the gel prevents adhesion on the surface by preventing cell viability. Successful tendon tissue engineering requires the use of three-dimensional (3D) biomimetic scaffold molds with physical and biochemical properties of native tendon tissue. Here, the development and characterization of a new composite scaffold fabricated by co-electrospinning of poly-ε-caprolactone (PCL) and methacrylated gelatin (mGLT) are reported.

3.3.5. Skin tissue engineering

Various types of tissue engineering scaffolds have been developed and used for epidermic engineering [116]. Ideally, this scaffold should have some biologic properties (i.e., to support keratinocyte adhesion, proliferation, and differentiation) and appropriate mechanical properties [117]. The mechanical properties of the skeletons are defined as a key modulator of keratinocyte behavior with increased cell adhesion and proliferation on hard surfaces [118].

The scaffold should also be strong enough and flexible to support easy handling during surgery and to support the natural movements of the tissues [119]. In addition, these scaffolds should only ideally deteriorate after adequate recovery, and this process can take up to 8 weeks [120]. In addition, for some clinical applications, the scaffold molds need to be cross-linked quickly in place for the contour of the wound [121].

Based on these requirements, natural hydrogels are considered attractive candidates for epidermis engineering due to their unique combination of biological and physical properties, including biocompatibility, as well as on-site cross-linking capacities, mimicking extracellular matrix (ECM), adjustable mechanical, swelling, and degradation properties [122].

3.3.6. Hydrogel wound dressing

An ideal wound dressing should maintain a moist environment at the wound interface, allow gas exchange, create microorganism barrier, and remove extreme exudates. It should also be made from a readily available biomaterial that is nontoxic and nonallergic, unattached, and easily removed without trauma, requires minimal processing, has antimicrobial properties, and accelerates wound healing [123].

Hydrogels are promising materials for wound dressing and for treatment of severe burns. In these applications, the use of hydrogel dressing offers several advantages. These are light materials and contain water in significant quantities. For this reason, hydrogels imitate some of the important qualities of people's skin and texture. Hydrogels can absorb large amounts of water or biological fluid content. They also protect the three-dimensional structure. Its advantages over other wound dressings are its good mechanical properties, oxygen permeability, liquid absorption, hydration of the wound bed, shape stability, softness similar to soft peripheral tissue, and cooling effect on the wound surface. Depending on the hydration state of the tissues, the hydrogel may give or absorb water to the wound periphery [124].

Hydrogel dressings' elastic sheet, film, or gel (amorphous) can be found in form. The hydrogel in leaf form has many of the ideal wound dressing properties. The hydrogels in this form have convenient constructions and can be cut and prepared in a manner suitable for the wound. In addition, hydrogel dressings, when placed on a dry surface of the wound, provide a moist wound environment that enables the rapid healing of the wound moistening of wounds [125].

Hydrogel wound dressings can easily be shaped and leave no residue. Since the dressing does not stick to the wound surface, it is painless to remove from the wound surface and can be easily cleaned. Compared to semipermeable membranes, they have water vapor permeability. In addition, using hydrogel dressing, the drug can be applied locally to the wound, and the diffusion of the drug into the wound can be controlled by controlling the gel cross-linking grade. Hydrogels are used in dry necrotic wounds, superficial wounds, non-exudate, noninfected wounds [126, 127].

PVA has excellent and easy film-forming properties and can be mixed with synthetic and natural polymers due to its water-soluble, biodegradable, noncarcinogenic, and biocompatible character. The final properties of the mixed material depend on the properties of the materials being assembled, and the PVA properties change after mixing. Sodium alginate is one of the popular natural polymers used in wound dressing applications by combining PVA polymer as the main or additional component due to its high water swelling ability [128].

Gelatin is a protein produced by partial hydrolysis of collagen extracted from the skin (44%, usually cow and pig), bone (27%), and connective tissues or organs (28%) of animals. Gelatin has biological activities due to its natural origin, which is suitable to be used as a wound dressing material, a drug delivery carrier, and a scaffold for tissue engineering. Gelatin has the ability to form strong hydrogels and transparent films that are easily designed as insoluble hydrophilic polymers for strong regeneration and tissue implantation [129].

3.3.7. Cornea tissue engineering

Corneal disease is a leading cause of blindness all over the world [130]. Donor corneal transplantation is preferred in corneal disease. However, they are limited in supply, especially in developing countries.

For this reason, alternative treatment methods are needed to meet the demand for increased corneal replacement. Efforts have focused on the development of tissue engineering structures that use hydrogels and biopolymer materials to replace native corneas [131–135].

An ideal tissue engineering structure should imitate the native cornea closely in order to fulfill its critical functions. Ruberti and Zieske have described three design considerations that must be met for an artificial corneal formation to be functional: (i) protection, (ii) transmission, and (iii) refraction [136].

In order to meet the abovementioned design criteria, the tissue engineering structure must have the following essential characteristics [137]: (i) to stimulate the migration and proliferation of corneal epithelial cells for the formation of a functional corneal epithelium, (ii) to resist the intraocular pressure, and thus to prevent rupture, (iii) simulating the nanoscale fibrillary order of the corneal stroma to show a high degree of transparency (>90°C) and a refractive index comparable to that of the native cornea (3.8 MPa) [138] and the refractive index, (iv) transparency to support the maintenance of functional keratocytes to preserve and de novo collagen tissue synthesis, and (v) to compensate for and thereby prevent the swelling rate of the native cornea to preserve and distribute the water content [139].

3.3.8. Hydrogel prosthesis

In general, medical bionic implants include innumerable solutions that remove injuries and post-resection losses, as well as various dysfunctions resulting from the operative treatment of tumors or inflammatory processes [140]. Swelling ability of hydrogel is used for various body regions in prostheses. A simple example of this is the design of an endoprosthesis which is not removed from the site for use in the treatment of patients with biliary obstruction. Here, the hydrogel ring located in the grooves around the endoprosthesis is placed on both sides of the occlusion. The swelling of the rings prevents the device from coming out of its place, thus protecting the internal drainage [141].

Carboxymethyl cellulose (CMC) hydrogel is a biodegradable, nontoxic, and non-mutagenic viscoelastic gel. It has been in clinical use since 1984 and has been used as a monobloc breast prosthesis since 1994. In the study by Brunner et al., 122 patients with hydrogel implants were evaluated between February 2000 and February 2005. It has been found that CMC hydrogel implants have higher radiotranslucency than silicone gel and that the integrity of this material is easy to prove by clinical examination. In the event of a rupture, the implant can be replaced immediately. It can be easily attached and placed by means of a small cut due to its high elastic shell. Studies conducted in patients with CMC hydrogel-filled breast prostheses have determined that a negligible complication rate and a high satisfaction rate in patients are seen. Another advantage of the CMC hydrogel-filled implant is that it is a natural feel comparable to that of silicone gel-filled breast implants [142].

Another usage area of hydrogels is contact lenses. The use of contact lenses to treat visual impairment began in the nineteenth century. The materials used in contact lenses for visual impairment must be pure. They must possess the necessary physical and mechanical properties such as malleability, transparency, wettability, mechanical stability, and oxygen permeability. They should be able to be sterilized and must be noncarcinogenic.

Contact lenses are generally classified as "hard" or "soft" depending on their elasticity. Although hard contact lenses are more durable, users do not prefer them, since they require an adaptation period. Hard contact lenses are resistant to hydrophobic substances such as poly(methyl methacrylate) (PMMA) or poly(hexafluoroisopropyl methacrylate) (HFIM). Despite this, soft contact lenses are resistant to hydrogels [143].

FDA-approved hydrogel contact lenses are divided into four groups depending on their water contents and the reactive groups on their polymer surface. These four groups are low water content, high water content, ionize (reactive chemical groups on the polymer surface), non-ionize (those who do not carry reactive chemical group on the polymer surface). Low water content lenses should be compatible with all lens care systems. Contact lenses have some properties which are suitable for implantation. These properties are about protein accumulation, tensile strength, and color changing. Protein accumulation and color changing should be low, and lens strength has to be durable.

Sterilization of high water content lenses is not possible by heat and enzymatic way for a long time. They have high protein accumulation and oxygen permeability, and they easily lose water; thus, their tensile strength durability is low [144].

The materials of soft contact lenses that contain polymers are not natural. They can be split into two main groups. One of these groups is hydrogel materials: hydrogel materials and silicone. The second most important material used in soft contact lenses is silicone. Elastic silicone is soft in the elastomer form. In this form, soft silicone has high oxygen permeability but also has low surface wettability. For this reason, it is coated with hydrophilic materials. The most important development in contact lenses in the recent years has been on the combination of silicone and hydrogel materials. The high oxygen permeability of silicone when added to other materials lowers its disadvantages due to the high water maintenance of hydrogels [145].

Silicone hydrogel lenses contain less water than traditional hydrogel lenses. Transparent silicone hydrogel lenses with high oxygen permeability cause less dryness in the eye and are much more comfortable to use throughout the day. With the development of silicone hydrogel lenses, the contact lens industry has experienced a revolution. With these new "super permeable" transparent contact lenses, much more oxygen can reach the cornea layer. A high oxygen supply is vital for the maintenance of eye health. This development has lead to the production of "long lasting" contact lenses, which can be used 24/7. The silicone hydrogel material is so effective that it is used to make more than 50% of contact lenses that are manufactured. Compared to traditional and hydrogel contact lenses, silicone hydrogel contact lenses have been found to have about 40 times less risk of producing infective keratitis. The fact that silicone hydrogel contact lenses are adapted to the cornea physiology, have high oxygen permeability, and have advanced surface and design features should not overshadow that they too have mechanical and inflammatory effects on the cornea [146].

3.4. Future aspect

Hydrogels have a very important role in the field of biomedicine and nanotechnology. The future success of hydrogels is based on the synthesis of new polymers or on the modification of natural polymers to solve certain biological and medical difficulties. Most scientific research shows that the studies of hydrogels have a bright future. New approaches to hydrogel design are increased in the investigation of these biomaterials. Fast response, self-assembly, high and good mechanical properties, and super-porous hydrogel are just a few examples of biomaterials with an intelligent future.

Author details

Yasemin Budama-Kilinc*, Rabia Cakir-Koc, Bahar Aslan, Burcu Özkan, Hande Mutlu and Eslin Üstün

*Address all correspondence to: yaseminbudama@gmail.com

Yildiz Technical University, Faculty of Chemical and Metallurgical Engineering, Department of Bioengineering, Esenler, Istanbul, Turkey

References

[1] Hamidi M, Azadi A, Rafiei P. Hydrogel nanoparticles in drug delivery. Advanced Drug Delivery Reviews. 2008;**60**(15):1638-1649

[2] Vashist A, Ahmad S. Hydrogels: Smart materials for drug delivery. Oriental Journal of Chemistry. 2013;**29**(3):861-870

[3] Peppas NA, et al. Biomedical Applications of Hydrogels Handbook. New York: Springer Science & Business Media, 2010

[4] Tillet G, Boutevin B, Ameduri B. Chemical reactions of polymer crosslinking and post-crosslinking at room and medium temperature. Progress in Polymer Science. 2011;**36**(2): 191-217

[5] Khademhosseini A, Langer R. Microengineered hydrogels for tissue engineering. Biomaterials. 2007;**28**(34):5087-5092

[6] Shapiro JM, Oyen ML. Hydrogel composite materials for tissue engineering scaffolds. JOM. 2013;**65**(4):505-516

[7] Abou Taleb MF, Ismail SA, El-Kelesh NA. Radiation synthesis and characterization of polyvinyl alcohol/methacrylic acid–gelatin hydrogel for vitro drug delivery. Journal of Macromolecular Science, Part A. 2008;**46**(2):170-178

[8] Dengre R, Bajpai M, Bajpai S. Release of vitamin B12 from poly (N-vinyl-2-pyrrolidone)-crosslinked polyacrylamide hydrogels: A kinetic study. Journal of Applied Polymer Science. 2000;**76**(11):1706-1714

[9] Durmaz S, Okay O. Acrylamide/2-acrylamido-2-methylpropane sulfonic acid sodium salt-based hydrogels: Synthesis and characterization. Polymer. 2000;**41**(10):3693-3704

[10] Ekici S, Saraydin D. Synthesis, characterization and evaluation of IPN hydrogels for antibiotic release. Drug Delivery. 2004;**11**(6):381-388

[11] Singh S, et al. *Centella asiatica* (L.): A plant with immense medicinal potential but threatened. International Journal of Pharmaceutical Sciences Review and Research. 2010;**4**(2)

[12] Pal K, Banthia A, Majumdar D. Polymeric hydrogels: Characterization and biomedical applications. Designed Monomers and Polymers. 2009;**12**(3):197-220

[13] Ebara M, et al. Smart Biomaterials. Japan: Springer, 2014

[14] Amini AA, Nair LS. Injectable hydrogels for bone and cartilage repair. Biomedical Materials. 2012;**7**(2):024105

[15] Jin R. In-Situ Forming Biomimetic Hydrogels for Tissue Regeneration. China: INTECH Open Access Publisher; 2012, pp. 41

[16] Kuo CK, Ma PX. Ionically crosslinked alginate hydrogels as scaffolds for tissue engineering: Part 1. Structure, gelation rate and mechanical properties. Biomaterials. 2001;**22**(6): 511-521

[17] Augst AD, Kong HJ, Mooney DJ. Alginate hydrogels as biomaterials. Macromolecular Bioscience. 2006;**6**(8):623-633

[18] Sperinde JJ, Griffith LG. Synthesis and characterization of enzymatically-cross-linked poly (ethylene glycol) hydrogels. Macromolecules. 1997;**30**(18):5255-5264

[19] Wichterle O, Lim D. Hydrophilic gels for biological use. Nature. 1960;**185**(4706):117-118

[20] Langer R, Peppas NA. Present and future applications of biomaterials in controlled drug delivery systems. Biomaterials. 1981;**2**(4):201-214

[21] Giammona G, et al. New biodegradable hydrogels based on an acryloylated polyaspartamide cross-linked by gamma irradiation. Journal of Biomaterials Science, Polymer Edition. 1999;**10**(9):969-987

[22] Caliceti P, et al. Controlled release of biomolecules from temperature-sensitive hydrogels prepared by radiation polymerization. Journal of Controlled Release. 2001;**75**(1):173-181

[23] Gibas I. Janik H. Review: synthetic polymer hydrogels for biomedical applications. Chemistry and Chemical Technology. 2010;**4**(4):297-304

[24] Peppas N, et al. Hydrogels in pharmaceutical formulations. European Journal of Pharmaceutics and Biopharmaceutics. 2000;**50**(1):27-46

[25] Swami SN. Radiation synthesis of polymeric hydrogels for swelling-controlled drug release studies. Western Sydney University Thesis. 2004. http://handle.uws.edu.au:8081/1959.7/698

[26] Katime I, Novoa R, Zuluaga F. Swelling kinetics and release studies of theophylline and aminophylline from acrylic acid/n-alkyl methacrylate hydrogels. European Polymer Journal. 2001;**37**(7):1465-1471

[27] Xue S, et al. A simple and fast formation of biodegradable poly (urethane-urea) hydrogel with high water content and good mechanical property. Polymer. 2016;**99**:340-348

[28] Datta A. Characterization of Polyethylene Glycol Hydrogels for Biomedical Applications. Louisiana: Citeseer, 2007

[29] Maldonado-Codina C, Efron N. Hydrogel lenses-material and manufacture: A review. Optometry in Practice. 2003;**4**:101-115

[30] Hoffman AS. Hydrogels for biomedical applications. Advanced Drug Delivery Reviews. 2012;**64**:18-23

[31] Allen P, Bennett D, Williams D. Water in methacrylates—I. Sorption and desorption properties of poly(2-hydroxyethyl methacrylate-co-glycol dimethacrylate) networks. European Polymer Journal. 1992;**28**(4):347-352

[32] Brazel CS, Peppas NA. Mechanisms of solute and drug transport in relaxing, swellable, hydrophilic glassy polymers. Polymer. 1999;**40**(12):3383-3398

[33] Dobrzański L, et al. Biodegradable and antimicrobial polycaprolactone nanofibers with and without silver precipitates. In: XXIV International Materials Research Congress, IMRC; 2015

[34] Ehrick JD, et al. Genetically engineered protein in hydrogels tailors stimuli-responsive characteristics. Nature Materials. 2005;**4**(4):298-302

[35] Sengupta D, Heilshorn SC. Protein-engineered biomaterials: Highly tunable tissue engineering scaffolds. Tissue Engineering Part B: Reviews. 2010;**16**(3):285-293

[36] Foo CTWP, et al. Two-component protein-engineered physical hydrogels for cell encapsulation. Proceedings of the National Academy of Sciences. 2009;**106**(52):22067-22072

[37] Sakai S, et al. An injectable, in situ enzymatically gellable, gelatin derivative for drug delivery and tissue engineering. Biomaterials. 2009;**30**(20):3371-3377

[38] Liang Y, et al. An in situ formed biodegradable hydrogel for reconstruction of the corneal endothelium. Colloids and Surfaces B: Biointerfaces. 2011;**82**(1):1-7

[39] Davidenko N, et al. Collagen–hyaluronic acid scaffolds for adipose tissue engineering. Acta Biomaterialia. 2010;**6**(10):3957-3968

[40] Xing Y, et al. Self-assembled DNA hydrogels with designable thermal and enzymatic responsiveness. Advanced Materials. 2011;**23**(9):1117-1121

[41] Zhu J. Bioactive modification of poly (ethylene glycol) hydrogels for tissue engineering. Biomaterials. 2010;**31**(17):4639-4656

[42] Geckil H, et al. Engineering hydrogels as extracellular matrix mimics. Nanomedicine. 2010;**5**(3):469-484

[43] Hunt NC, Grover LM. Cell encapsulation using biopolymer gels for regenerative medicine. Biotechnology Letters. 2010;**32**(6):733-742

[44] Hejčl A, et al. HPMA-RGD hydrogels seeded with mesenchymal stem cells improve functional outcome in chronic spinal cord injury. Stem Cells and Development. 2010;**19**(10): 1535-1546

[45] Beamish JA, et al. The effects of monoacrylated poly(ethylene glycol) on the properties of poly (ethylene glycol) diacrylate hydrogels used for tissue engineering. Journal of Biomedical Materials Research Part A. 2010;**92**(2):441-450

[46] Zustiak SP, Leach JB. Hydrolytically degradable poly (ethylene glycol) hydrogel scaffolds with tunable degradation and mechanical properties. Biomacromolecules. 2010;**11**(5): 1348-1357

[47] Silva AKA, et al. Growth factor delivery approaches in hydrogels. Biomacromolecules. 2008;**10**(1):9-18

[48] Lutolf MP. Biomaterials: Spotlight on hydrogels. Nature Materials. 2009;**8**(6):451-453

[49] Lee J, Cuddihy MJ, Kotov NA. Three-dimensional cell culture matrices: State of the art. Tissue Engineering Part B: Reviews. 2008;**14**(1):61-86

[50] Varghese S, Elisseeff JH. Hydrogels for musculoskeletal tissue engineering. In: Polymers for Regenerative Medicine. Berlin: Springer, 2006, pp. 95-144

[51] Jiang Z, et al. Biodegradable and thermoreversible hydrogels of poly (ethylene glycol)-poly (ε-caprolactone-co-glycolide)-poly (ethylene glycol) aqueous solutions. Journal of Biomedical Materials Research Part A. 2008;**87**(1):45-51

[52] Deshmukh M, et al. Biodegradable poly (ethylene glycol) hydrogels based on a self-elimination degradation mechanism. Biomaterials. 2010;**31**(26):6675-6684

[53] Higuchi A, et al. Temperature-induced cell detachment on immobilized pluronic surface. Journal of Biomedical Materials Research Part A. 2006;**79**(2):380-392

[54] Schmedlen RH, Masters KS, West JL. Photocrosslinkable polyvinyl alcohol hydrogels that can be modified with cell adhesion peptides for use in tissue engineering. Biomaterials. 2002;**23**(22):4325-4332

[55] Ossipov DA, et al. Formation of the first injectable poly (vinyl alcohol) hydrogel by mixing of functional PVA precursors. Journal of Applied Polymer Science. 2007;**106**(1):60-70

[56] Brandl F, Sommer F, Goepferich A. Rational design of hydrogels for tissue engineering: Impact of physical factors on cell behavior. Biomaterials. 2007;**28**(2):134-146

[57] Jia X, Kiick KL. Hybrid multicomponent hydrogels for tissue engineering. Macromolecular Bioscience. 2009;**9**(2):140-156

[58] Hiemstra C, et al. Rapidly in situ-forming degradable hydrogels from dextran thiols through Michael addition. Biomacromolecules. 2007;**8**(5):1548-1556

[59] Zieris A, et al. FGF-2 and VEGF functionalization of starPEG–heparin hydrogels to modulate biomolecular and physical cues of angiogenesis. Biomaterials. 2010;**31**(31):7985-7994

[60] Jin R, et al. Synthesis and characterization of hyaluronic acid–poly (ethylene glycol) hydrogels via Michael addition: An injectable biomaterial for cartilage repair. Acta Biomaterialia. 2010;**6**(6):1968-1977

[61] Li F, et al. Recruitment of multiple cell lines by collagen-synthetic copolymer matrices in corneal regeneration. Biomaterials. 2005;**26**(16):3093-3104

[62] Myles JL, Burgess BT, Dickinson RB. Modification of the adhesive properties of collagen by covalent grafting with RGD peptides. Journal of Biomaterials Science, Polymer Edition. 2000;**11**(1):69-86

[63] Kimura T, et al. Preparation of poly (vinyl alcohol)/DNA hydrogels via hydrogen bonds formed on ultra-high pressurization and controlled release of DNA from the hydrogels for gene delivery. Journal of Artificial Organs. 2007;**10**(2):104-108

[64] Suzuki T, et al. Induced migration of dental pulp stem cells for in vivo pulp regeneration. Journal of Dental Research. 2011;**90**(8):1013-1018

[65] Toh W. Injectable hydrogels in dentistry: Advances and promises. Austin Journal of Dentistry. 2014;**1**(1):1001

[66] Chen YC, et al. Functional human vascular network generated in photocrosslinkable gelatin methacrylate hydrogels. Advanced Functional Materials. 2012;**22**(10):2027-2039

[67] Wang L, Stegemann JP. Thermogelling chitosan and collagen composite hydrogels initiated with β-glycerophosphate for bone tissue engineering. Biomaterials. 2010;**31**(14): 3976-3985

[68] Rosiak J M, et al. "Radiation formation of hydrogels for biomedical application." *Radiation synthesis and modification of polymers for biomedical applications* (2002): 5-47

[69] Subramani K, Ahmed W, Hartsfield JK. Nanobiomaterials in Clinical Dentistry. USA: William Andrew, 2012

[70] AppelEA,etal.Supramolecularpolymerichydrogels.ChemicalSocietyReviews.2012;**41**(18): 6195-6214

[71] Park S-J, et al. Glycol chitin–based thermoresponsive hydrogel scaffold supplemented with enamel matrix derivative promotes odontogenic differentiation of human dental pulp cells. Journal of Endodontics. 2013;**39**(8):1001-1007

[72] Langer R. Drug delivery abd targeting. Nature. 1998;**392**(6679):5-10

[73] Hoare TR, Kohane DS. Hydrogels in drug delivery: Progress and challenges. Polymer. 2008;**49**(8):1993-2007

[74] Liechty WB, et al. Polymers for drug delivery systems. Annual Review of Chemical and Biomolecular Engineering. 2010;**1**:149-173

[75] Tibbitt MW, Dahlman JE, Langer R. Emerging frontiers in drug delivery. Journal of the American Chemical Society. 2016;**138**(3):704-717

[76] Li J, Mooney DJ. Designing hydrogels for controlled drug delivery. Nature Reviews Materials. 2016;**1**:16071

[77] Tiwari G, et al. Drug delivery systems: An updated review. International Journal of Pharmaceutical Investigation. 2012;**2**(1):2

[78] Amin S, Rajabnezhad S, Kohli K. Hydrogels as potential drug delivery systems. Scientific Research and Essays. 2009;**4**(11):1175-1183

[79] Peppas NA, Colombo P. Analysis of drug release behavior from swellable polymer carriers using the dimensionality index. Journal of Controlled Release. 1997;**45**(1):35-40

[80] Sri B, Ashok V, Arkendu C. As a review on hydrogels as drug delivery in the pharmaceutical field. International Journal of Pharmaceutical and Chemical Sciences. 2012;**1**(2): 642-661

[81] Satish C, Satish K, Shivakumar H. Hydrogels as controlled drug delivery systems: Synthesis, crosslinking, water and drug transport mechanism. Indian Journal of Pharmaceutical Sciences. 2006;**68**(2):133-140

[82] Khare AR, Peppas NA. Swelling/deswelling of anionic copolymer gels. Biomaterials. 1995;**16**(7):559-567

[83] Dong L-c, Hoffman AS. A novel approach for preparation of pH-sensitive hydrogels for enteric drug delivery. Journal of Controlled Release. 1991;**15**(2):141-152

[84] Prabaharan M, Mano JF. Stimuli-responsive hydrogels based on polysaccharides incorporated with thermo-responsive polymers as novel biomaterials. Macromolecular Bioscience. 2006;**6**(12):991-1008

[85] Gupta P, Vermani K, Garg S. Hydrogels: From controlled release to pH-responsive drug delivery. Drug Discovery Today. 2002;**7**(10):569-579

[86] Vyas SP, and Khar RK. Targeted and Controlled Drug Delivery—Novel Carrier Systems, CBS, New Delhi, 2002:331-338

[87] Akiyama Y, et al. Novel peroral dosage forms with protease inhibitory activities. II. Design of fast dissolving poly (acrylate) and controlled drug-releasing capsule formulations with trypsin inhibiting properties. International Journal of Pharmaceutics. 1996;**138**(1):13-23

[88] Harris D, Robinson JR. Drug delivery via the mucous membranes of the oral cavity. Journal of Pharmaceutical Sciences. 1992;**81**(1):1-10

[89] Gandhi RB, Robinson JR. Oral cavity as a site for bioadhesive drug delivery. Advanced Drug Delivery Reviews. 1994;**13**(1-2):43-74

[90] Singh B, Sharma N, Chauhan N. Synthesis, characterization and swelling studies of pH responsive psyllium and methacrylamide based hydrogels for the use in colon specific drug delivery. Carbohydrate Polymers. 2007;**69**(4):631-643

[91] Vervoort L, et al. Inulin hydrogels as carriers for colonic drug targeting. Rheological characterization of the hydrogel formation and the hydrogel network. Journal of Pharmaceutical Sciences. 1999;**88**(2):209-214

[92] Kakoulides EP, Smart JD, Tsibouklis J. Azocross-linked poly (acrylic acid) for colonic delivery and adhesion specificity: Synthesis and characterisation. Journal of Controlled Release. 1998;**52**(3):291-300

[93] Das A, Wadhwa S, Srivastava A. Cross-linked guar gum hydrogel discs for colon-specific delivery of ibuprofen: Formulation and in vitro evaluation. Drug Delivery. 2006;**13**(2):139-142

[94] Gliko-Kabir I, et al. Low swelling, crosslinked guar and its potential use as colon-specific drug carrier. Pharmaceutical Research. 1998;**15**(7):1019-1025

[95] Eaga CM, et al. In-situ gels-a novel approach for ocular drug delivery. Der Pharmacia Lettre. 2009;**1**(1):21-33

[96] Zignani M, Tabatabay C, Gurny R. Topical semi-solid drug delivery: Kinetics and tolerance of ophthalmic hydrogels. Advanced Drug Delivery Reviews. 1995;**16**(1):51-60

[97] Xinming L, et al. Polymeric hydrogels for novel contact lens-based ophthalmic drug delivery systems: A review. Contact Lens and Anterior Eye. 2008;**31**(2):57-64

[98] Miyazaki S, et al. In situ gelling xyloglucan formulations for sustained release ocular delivery of pilocarpine hydrochloride. International Journal of Pharmaceutics. 2001;**229**(1):29-36

[99] Dobrzański L. Overview and general ideas of the development of constructions, materials, technologies and clinical applications of scaffolds engineering for regenerative medicine. Archives of Materials Science and Engineering. 2014;**69**(2):53-80

[100] Dobrzański L. The concept of biologically active microporous engineering materials and composite biological-engineering materials for regenerative medicine and dentistry. Archives of Materials Science and Engineering. 2016;**80**(2):64-85

[101] Babensee JE, et al. Host response to tissue engineered devices. Advanced Drug Delivery Reviews. 1998;**33**(1):111-139

[102] Říhová B. Immunocompatibility and biocompatibility of cell delivery systems. Advanced Drug Delivery Reviews. 2000;**42**(1):65-80

[103] Gin H, et al. Lack of responsiveness to glucose of microencapsulated islets of Langerhans after three weeks' implantation in the rat—Influence of the complement. Journal of Microencapsulation. 1990;**7**(3):341-346

[104] Matthew HW, et al. Complex coacervate microcapsules for mammalian cell culture and artificial organ development. Biotechnology Progress. 1993;**9**(5):510-519

[105] Hsu F, et al. The collagen-containing alginate/poly (L-lysine)/alginate microcapsules. Artificial Cells, Blood Substitutes, and Biotechnology. 2000;**28**(2):147-154

[106] Maisani M, et al. Design, development and validation of a new composite hydrogel for bone tissue engineering. The Bone & Joint Journal. 2017;**99**(Supp 1):106

[107] Nguyen BNB, et al. Collagen hydrogel scaffold promotes mesenchymal stem cell and endothelial cell coculture for bone tissue engineering. Journal of Biomedical Materials Research Part A. 2017;**105**(4):1123-1131

[108] Freed LE, Vunjak-Novakovic G. Culture of organized cell communities. Advanced Drug Delivery Reviews. 1998;**33**(1):15-30

[109] Wong M, Carter D. Articular cartilage functional histomorphology and mechanobiology: A research perspective. Bone. 2003;**33**(1):1-13

[110] Englund M, Lohmander L. Patellofemoral osteoarthritis coexistent with tibiofemoral osteoarthritis in a meniscectomy population. Annals of the Rheumatic Diseases. 2005;**64**(12):1721-1726

[111] Verdonk P, et al. Successful treatment of painful irreparable partial meniscal defects with a polyurethane scaffold: Two-year safety and clinical outcomes. The American Journal of Sports Medicine. 2012;**40**(4):844-853

[112] Ahmed EM. Hydrogel: Preparation, characterization, and applications: A review. Journal of Advanced Research. 2015;**6**(2):105-121

[113] Peppas NA, Merrill EW. Development of semicrystalline poly (vinyl alcohol) hydrogels for biomedical applications. Journal of Biomedical Materials Research. 1977;**11**(3):423-434

[114] Lazzeri L, et al. Physico-chemical and mechanical characterization of hydrogels of poly (vinyl alcohol) and hyaluronic acid. Journal of Materials Science: Materials in Medicine. 1994;**5**(12):862-867

[115] Young T-H, et al. Evaluation of asymmetric poly (vinyl alcohol) membranes for use in artificial islets. Biomaterials. 1996;**17**(22):2139-2145

[116] Shevchenko RV, James SL, James SE. A review of tissue-engineered skin bioconstructs available for skin reconstruction. Journal of the Royal Society Interface. 2009:0403

[117] MacNeil S. Progress and opportunities for tissue-engineered skin. Nature. 2007;**445**(7130): 874-880

[118] Trappmann B, et al. Extracellular-matrix tethering regulates stem-cell fate. Nature Materials. 2012;**11**(7):642-649

[119] Silver FH, Freeman JW, DeVore D. Viscoelastic properties of human skin and processed dermis. Skin Research and Technology. 2001;**7**(1):18-23

[120] Harding K, Morris H, Patel G. Healing chronic wounds. British Medical Journal. 2002;**324**(7330):160

[121] Keriquel V, et al. In vivo bioprinting for computer-and robotic-assisted medical intervention: Preliminary study in mice. Biofabrication. 2010;**2**(1):014101

[122] Ifkovits JL, Burdick JA. Review: Photopolymerizable and degradable biomaterials for tissue engineering applications. Tissue Engineering. 2007;**13**(10):2369-2385

[123] Jayakumar R, et al. Biomaterials based on chitin and chitosan in wound dressing applications. Biotechnology Advances. 2011;**29**(3):322-337

[124] Yu T, Ober CK. Methods for the topographical patterning and patterned surface modification of hydrogels based on hydroxyethyl methacrylate. Biomacromolecules. 2003;**4**(5):1126-1131

[125] Altay P, Başal G. Yara örtüleri. Tekstil Teknolojileri Elektronik Dergisi. 2010;**4**(1):109-121

[126] Boateng JS, et al. Wound healing dressings and drug delivery systems: A review. Journal of Pharmaceutical Sciences. 2008;**97**(8):2892-2923

[127] Amadeu TP, et al. S-nitrosoglutathione-containing hydrogel accelerates rat cutaneous wound repair. Journal of the European Academy of Dermatology and Venereology. 2007;**21**(5):629-637

[128] Corkhill PH, Hamilton CJ, Tighe BJ. Synthetic hydrogels VI. Hydrogel composites as wound dressings and implant materials. Biomaterials. 1989;**10**(1):3-10

[129] Hago E-E, Li X. Interpenetrating polymer network hydrogels based on gelatin and PVA by biocompatible approaches: Synthesis and characterization. Advances in Materials Science and Engineering. 2013;**2013**:1-9

[130] Whitcher JP, Srinivasan M, Upadhyay MP. Corneal blindness: A global perspective. Bulletin of the World Health Organization. 2001;**79**(3):214-221

[131] Carlsson DJ, et al. Bioengineered corneas: How close are we? Current Opinion in Ophthalmology. 2003;**14**(4):192-197

[132] Grolik M, et al. Hydrogel membranes based on genipin-cross-linked chitosan blends for corneal epithelium tissue engineering. Journal of Materials Science: Materials in Medicine. 2012;**23**(8):1991-2000

[133] Amano, Shiro, et al. "Decellularizing corneal stroma using N2 gas." *Molecular vision* 14 (2008):878

[134] Fagerholm P, et al. A biosynthetic alternative to human donor tissue for inducing corneal regeneration: 24-month follow-up of a phase 1 clinical study. Science Translational Medicine. 2010;**2**(46):46-61

[135] Thompson RW, et al. Long-term graft survival after penetrating keratoplasty. Ophthalmology. 2003;**110**(7):1396-1402

[136] Ruberti JW, Zieske JD. Prelude to corneal tissue engineering–gaining control of collagen organization. Progress in Retinal and Eye Research. 2008;**27**(5):549-577

[137] Shah A, et al. The development of a tissue-engineered cornea: Biomaterials and culture methods. Pediatric Research. 2008;**63**(5):535-544

[138] Ahn J-I, et al. Crosslinked collagen hydrogels as corneal implants: Effects of sterically bulky vs. non-bulky carbodiimides as crosslinkers. Acta Biomaterialia. 2013;**9**(8):7796-7805

[139] Duan D, Klenkler BJ, Sheardown H. Progress in the development of a corneal replacement: Keratoprostheses and tissue-engineered corneas. Expert Review of Medical Devices. 2006;**3**(1):59-72

[140] Dobrzański L. Applications of newly developed nanostructural and microporous materials in biomedical, tissue and mechanical engineering. Archives of Materials Science and Engineering. 2015;**76**(2):53-114

[141] Silander T, Thor K. A "nondislodgeable" endoprosthesis for nonsurgical drainage of the biliary tract. Annals of Surgery. 1985;**201**(3):323

[142] Brunner CA, Gröner RW. Carboxy-methyl-cellulose hydrogel-filled breast implants—An ideal alternative? A report of five years' experience with this device. The Canadian Journal of Plastic Surgery. 2006;**14**(3):151

[143] Kanpolat A. Contact lenses: Past, present, future. Turkiye Klinikleri Journal of Ophthalmology Special Topics. 2008;**1**(1):1

[144] Caló E, Khutoryanskiy VV. Biomedical applications of hydrogels: A review of patents and commercial products. European Polymer Journal. 2015;**65**:252-267

[145] Yildiz Y, et al. The long-term effects of silicone hydrogel contact lens wear on corneal morphology/Silikon hidrojel kontakt lens kullaniminin kornea morfolojisi uzerindeki uzun donem etkileri. Turkish Journal of Ophthalmology. 2012;**42**(2):91-97

[146] Hu X, Tan H, Hao L. Functional hydrogel contact lens for drug delivery in the application of oculopathy therapy. Journal of the Mechanical Behavior of Biomedical Materials. 2016;**64**:43-52

Chapter 13

Natural Rubber Latex Biomaterials in Bone Regenerative Medicine

Leandra E. Kerche-Silva,
Dalita G.S.M. Cavalcante and Aldo Eloizo Job

Additional information is available at the end of the chapter

http://dx.doi.org/10.5772/intechopen.69855

Abstract

Natural rubber latex (NRL) is a white and milky solution that exudes from *Hevea brasiliensis* bark when perforated, and it has been appointed as a new promising biomaterial. NRL has been proven to be a very biocompatible material, and several new biomedical applications have been proposed. NRL has been proven to stimulate angiogenesis, cellular adhesion and formation of extracellular matrix, besides promoting replacement and regeneration of tissue. NRL also can be used as an occlusive membrane for guided bone regeneration (GBR) with promising results. Therefore, the aim of this chapter is to review NRL studies and to present NRL membrane as a promising biomembrane for use in bone trauma and injury.

Keywords: natural rubber latex, biomembranes, biomaterial, bone regeneration

1. Introduction

Hevea brasiliensis, popularly known as rubber tree, is a plant species that belongs to the Euphorbiaceae family. This plant synthesizes latex by a system of laticiferous rings, organized as paracirculatory vessel systems, in the inner bark of the plant. Latex is the cytoplasm of the laticiferous cells, and its composition resembles the composition of common cells, except for having 30–45% of natural rubber [1]. The latex of *H. brasiliensis* exudes from the bark when it is perforated (**Figure 1**).

Most of the harvested latex is coagulated for the manufacture of "dry rubber" products, including automotive tires. The latex of *H. brasiliensis* can be stabilized in an uncoagulated form with the use of ammonia, which allows the latex to be used for the manufacture of other products, such as surgical gloves (**Figure 2**) [2].

Figure 1. Latex of *Hevea brasiliensis* on exuding of bark after drilling.

Figure 2. Products made by the latex of *Hevea brasiliensis*.

Natural rubber latex (NRL) from *H. brasiliensis* is a colloidal anionic system formed by rubber particles (1,4-*cis*-polyisoprene) stabilized by phospholipids and proteins molecules (**Figure 3**) [4, 5]. One-third of the weight of *H. brasiliensis* latex is made of natural rubber, but 1–2% of its weight consists of hundreds of proteins [5]. Other constituents such as lipids, Quebrachitol, ribonucleic acids and organic salts are also present [6].

In the last years, researchers have been publishing about NRL membranes (**Figure 4**) that has been proven to be an important inductor of wound healing, inductor of esophagic wall regeneration and tympanic membrane regeneration, by mechanisms involved with angiogenesis [7–9]. NRL has also been studied for reconstructing temporal muscle fascia [9] and as arterial prosthesis in animal models, healing of ocular conjunctiva and neoangiogenesis in rabbits [10].

Besides forming biomembranes, which represent a complex colloid made of rubber particles and lutoids bodies suspended in a protein-rich media, NRL can be centrifuged in high speed and separated in fractions that mainly consist of a superior phase of rubber particles, an aqueous phase called centrifuged serum (C-serum) and an inferior phase called bottom serum (B-serum) (**Figure 5**) [12–14].

The rubber particles represent 25–45% of the fresh NRL and they have a medium diameter of 50 Å to 3 μm [15, 16]. The main proteins found in the rubber particles surface are Hev b 1 and Hev b 3 [17]. Centrifuged serum is the solution composed by carbohydrates, electrolytes, proteins and amino acids. This solution is composed by the cytoplasm of the laticiferous cells and contains a great amount of proteins related to the cell metabolism [18]. This phase is implicated with most of the biological properties of NRL [19]. Bottom serum is mainly constituted of lutoids, spherical

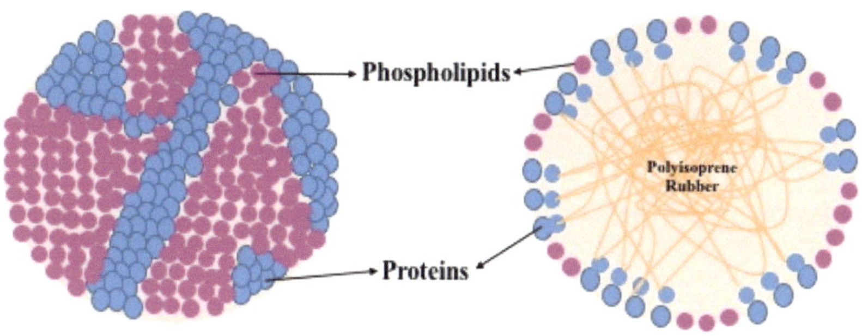

Figure 3. Rubber particles surrounded by a layer of protein-phospholipid (adapted from Ref. [3]).

Figure 4. NRL membrane.

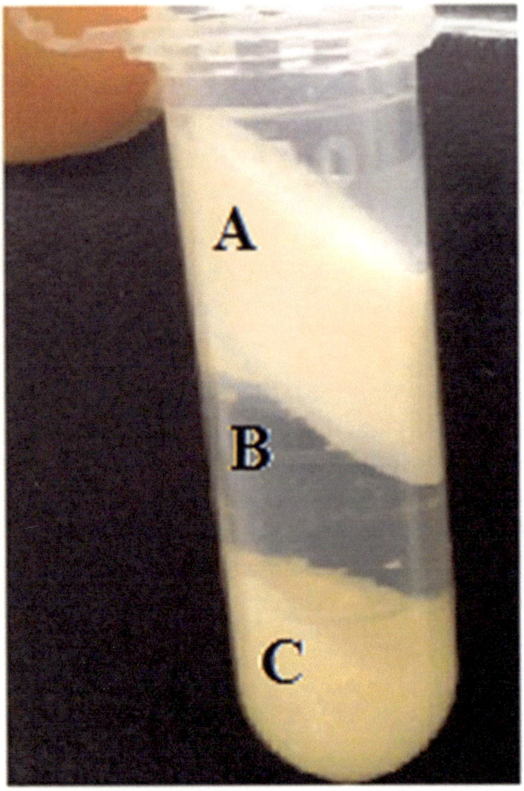

Figure 5. NRL fractions after high speed centrifugation: (A) rubber particles, (B) centrifuged serum and (C) bottom serum.

vacuoles that are osmotically sensitive and induce the latex flow to stop [20, 21]. Hevein is the main protein found in bottom serum, and it is implicated with allergenic reactions [18].

Since many biological and biomedical properties can be implicated with NRL from *H. brasiliensis*, the aim of this chapter is to review and explore the studies showing the applications of this biomaterial in bone regenerative medicine.

2. Bone remodeling

Bone is an organ capable of replacing old and disrupted tissue through a remodeling process. Mechanical changes required by skeletal functions make the remodeling process indispensable for the bones. The cells responsible for this process are osteoblasts and osteoclasts, and the first promote bone formation and the latter bone resorption. Osteoblasts are derived of the osteogenic differentiation of mesenchymal stem cells. Other important skeletal cells that have their origin in mature osteoblasts are the osteocytes, and these cells in particular are surrounded by extracellular matrix and have the ability to regulate osteoblast and osteoclast activities to maintain bone homeostasis [11].

Bones protect vital organs and provide storage for calcium and phosphate. Bone compartments designated for mechanical functions are called cortical bone, and the bone compartments designed for metabolic functions are called trabecular bone. Bones can also be distinguished by their formation. When the formation occurs in a direct way, it is called intramembranous ossification and it is characterized by the condensation of mesenchymal stem cells that become osteoblasts. This type of process occurs in the flat bones of neuro- and viscerocranium and in part of the clavicle [22]. When bone formation occurs in an indirect way, it is called endochondral ossification, and it is characterized by the differentiation of mesenchymal stem cells into cartilage first and this cartilage is later replaced by bone [23]. Endochondral ossification occurs in long bones, in vertebrae and in the skull base and the posterior part of the skull [22].

Bone modeling occurs during the growth process, and bone remodeling occurs during lifetime. These two processes take place under the control of various substances such as parathyroid hormone (PTH), calcitonin, vitamin D, growth hormone (GH), steroids, soluble cytokines and growth factors (i.e., macrophage colony-stimulating factor (M-CSF), receptor activator of nuclear κB ligand (RANKL), vascular endothelial growth factor (VEGF) and interleukin-6 (IL-6) family) [11].

Microfractures or factors related to bone microenvironment generate different stimuli that induce osteoblast to produce RANKL, which interacts with its receptor expressed by osteoclasts. This interaction activates the polarization of the osteoclasts that secrets enzymes required for bone resorption. Osteoblasts synthesize type I collagen (that represents 90% of the proteins in bone matrix) and procollagen I N-terminal peptide (PINP) that is considered a marker of bone formation [24].

Type I collagen together with other fibrillar collagens, bone proteins (osteopontin, bone sialoprotein and osteocalcin), proteoglycans, fibronectin and glycosaminoglycans compose unmineralized osteoid. The development of the osteoblast along with the stimulation of the osteogenic genes and mineral deposition are dependent on the pigment epithelium-derived factor (PEDF) [25]. Osteoblast phosphatases are responsible for the mineralization process, since they release phosphates that along with calcium form hydroxyapatite crystals [$Ca_{10}(PO_4)_6(OH)_2$] [26]. The osteoblasts that remain trapped inside the mineralized matrix acquire a stellar shape after morphologic change and form a network that produces signaling through bone tissue. These cells are called osteocytes.

2.1. Bone formation and bone remodeling molecular pathways

During osteogenesis, bone morphogenetics proteins (BMPs) and WNT signaling pathways are very important. The BMP pathways activate intracellular proteins called SMAD that control the expression of the gene RUNX2 (runt-related transcriptional factor 2), and this gene codifies a transcriptional factor that stimulates the mesenchymal stem cells to differentiate into osteogenic lineage [27]. WNT pathway is formed by proteins that are involved in many other biological processes, and it also regulates the expression of RUNX2 gene [28].

Depending on the level of differentiation of progenitor cells, WNT classic pathway induces or inhibits osteoblast formation. It also controls bone resorption when increasing the ratio of osteoprotegerin (OPG)/RANKL proteins [28]. These proteins are specifically produced by osteoblast to either inhibit (OPG) or enhance (RANKL) osteoclasts activity [29]. **Figure 6** schematically represents the complex network of molecular signaling pathways during osteogenesis.

Besides the pathways described above, systemic hormones also regulate osteogenic commitment or differentiation of mesenchymal cells. Examples of these hormones are PTH, glucocorticoids and estrogens. Local growth factor signaling, such as bone transforming growth factor-β (TGF-β1/2), insulin-like growth factor (IGF), fibroblast growth factor 2 (FGF-2), VEGF, cytokine modulators (prostaglandins) and mitogen-activated protein kinases (MAPK), also regulates the differentiation of mesenchymal stem cells [30]. Other enzyme that plays an important role in bone regulation is Peptidyl-prolyl *cis-trans* isomerase NIMA-interacting 1 (PIN1), which interacts with RUNX2, SMAD1/5 and ß-catenin proteins [31].

Osteogenesis is also regulated by epigenetic factors such as DNA methylation, microRNAs (miRNAs), histone acetylation and deacetylation, and chromatin structure modification [32, 33]. Especially short noncoding miRNAs have been proven to affect both osteoblast (bone formation) and osteoclast (bone resorption) lineage. Some miRNAs regulate osteoblastogenesis by posttranscription regulation of RUNX2 (e.g., miR-34c, miR-133a, miR135a, miR-137, miR-205, miR-217, etc.), Osterix (OSX) (e.g., miR-31, miR-93, miR-143, miR-145, etc.) and type I collagen (e.g., miR-29, miR-Let7) proteins [34].

After a bone fracture, an inflammatory response is activated and it is crucial for the process of bone regeneration, bone remodeling and healing. This inflammatory response involves the secretion of tumor necrosis factor-α (TNF-α) and interleukins (IL) IL-1, IL-6, IL-11 and IL-18

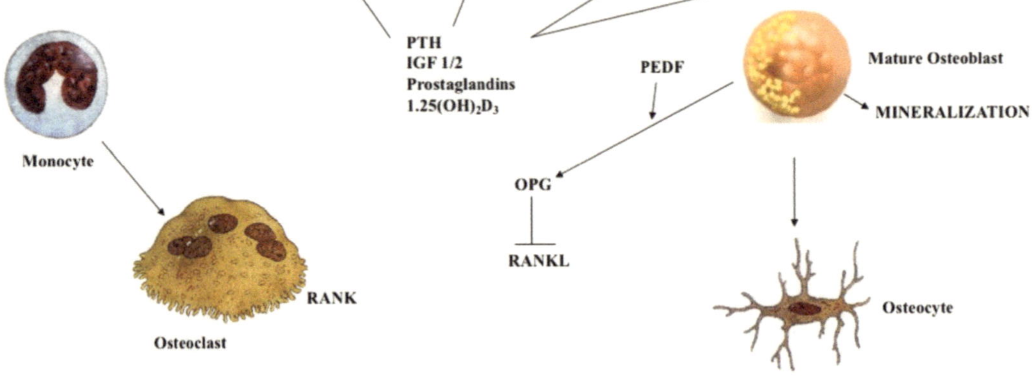

Figure 6. Schematic representation of cells and regulatory molecules involved in osteoblastogenesis and bone formation (adapted from [11]).

by macrophages, inflammatory cells and mesenchymal cells [35]. TNF-α stimulates osteoclast cells and promotes mesenchymal stem cells to start endochondral bone formation. TNF-α also stimulates apoptosis in hypertrophic chondrocytes, and a delay in the reabsorption of mineralized cartilage stops bone formation. When TNF-α is overproduced, as in diabetics, the cartilage is removed in a premature way, delaying bone formation and healing [36]. Right after the injury, during osteoclastogenesis, RANKL and OPG expressions are elevated. In bone remodeling, the expression of these proteins is diminished [37].

During bone remodeling, IL-1 and IL-6 are the most important IL. IL-1 is produced by macrophages and promotes the formation of the primary cartilaginous callus and also promotes angiogenesis in the site [38]. IL-1 also stimulates the production of IL-6 by the osteoblasts, and IL-6 also stimulates angiogenesis by the production of VEGF. It induces the differentiation of osteoblasts and osteoclasts [39].

In bone remodeling, mesenchymal, osteoblasts and chondrocytes cells produce BMPs that can work independently or in collaboration with each other and with other members of the TGF-β family to start osteoclastogenesis [40]. BMPs are structurally and functionally related, but they exhibit different patterns of expression during bone remodeling. Murine studies have shown that BMP-2 expression is more elevated in the first 24 h after the fracture, suggesting its primary function in the beginning of the bone repair. This BMP is also related to the maintenance of the normal bone mass [41, 42].

BMP-3, BMP-4, BMP-7 and BMP-8 are expressed during the healing time (14–21 days after the injury) when there is reabsorption of the calcified cartilage and osteogenesis [43]. It has been proposed that BMP-7 is the most potent inductor of differentiation in mesenchymal stem cells to osteoblasts [44].

2.2. Vasculature involved in bone formation and bone remodeling

Bone is a connective tissue that possesses high vascularization. During bone endochondral and intramembranous ossification, regeneration and remodeling, the vasculature plays an important role [45]. Ten to fifteen percent of total cardiac output goes to skeletal system [46]. The role of the blood vessels in the bone is not only to supply the bones with oxygen and nutrients but also to provide them with growth factor, hormones and neurotransmitters (e.g., brain-derived serotonin), maintaining the bone activity and cell survivor [47, 48].

Osteogenesis can occur through endochondral or intramembranous ossification; in both ways, angiogenesis is a critical stage and it is associated with the production of VEGF by hypertrophic chondrocytes or mesenchymal stem cells, respectively [49]. VEGF attracts endothelial cells acting as a chemotactic molecule that controls differentiation and function of osteoblasts and osteoclasts, participating in bone modeling and bone remodeling. A loss of the VEGF leads to incomplete bone vascularization and automatic disturbed endochondral ossification [50]. On the other hand, overexpression of VEGF can lead to osteosclerosis, highly increased bone formation that is a result of intense osteoblast differentiation, ending in altered bone morphology [51].

In endochondral ossification, cartilage vascularization begins with the formation of a primary vessel that is projected from the perichondrial vascular network to the adjacent cartilage. Then, a capillary glomerulus is formed at the leading edge and the entire vascular unit grows. Following elongation, a backward expansion of the capillary network occurs and it tightly surrounds a pair of main vessels composed by an arteriole and a venule [52].

In intramembranous ossification, capillaries of a small diameter move into thin avascular layer of mesenchyme that surrounds the mesenchymal condensation center. In this center, mesenchymal cells secrete VEGF that attracts endothelial cells. At the start of mineralization of the bone, the first blood vessels associate with an extensive external network of blood vessels [53].

During bone remodeling and bone regeneration, the blood vessels play an important role. Osteoclasts form a cutting cone that moves forward resorbing dead bone or damaged/old bone matrix. A blood vessel follows this cutting cone, delivering nutrients, hormones and growth factors to osteoblast that will produce new bone [48]. OSX-expressing osteoblast precursors are involved in the process of remodeling, accompanying the blood vessels, positioning themselves in a perivascular localization, showing the tight relationship between osteogenesis and angiogenesis [54].

3. Bone remodeling and latex

To clinically manage situations such as bone loss, injury or disease, researchers have been engaged for a long time to find a biocompatible material that is innocuous, promotes osseointegration, is manageable and has low cost for the people who need it. Natural rubber latex (NRL) extracted from *Hevea brasiliensis* has been used in the industrial manufacturing of several products, such as gloves, condoms, balloons and parts of medical and dental equipment [55].

Based on a new manufacturing process, several new biomedical applications have been proposed since NRL has been shown to be very biocompatible, stimulating cellular adhesion, the formation of extracellular matrix, and promoting the replacement and regeneration of tissue [8]. This new manufacturing process is based on the production of a biomembrane of NRL that has been used to replace vessels, esophagus, pericardium and abdominal wall [7]. In all of these experiments, the biomembrane promoted rapid tissue repair and elicited an inflammatory response that resembled the inflammatory response of normal healing process and not one of rejection process [8].

The membrane of NRL was also tested for the repair of bone defects in dental alveoli of rats [56], and the histological examination of the extraction sockets revealed a pattern of normal repair with characteristics similar to those reported for other materials [57]. Histometric evaluation of the areas close to the implant during the initial 7 days demonstrated progressive and accelerated osteogenesis by a decrease in the thickness of the fibrous capsule. At long-term, NRL implants did not induce the formation of foreign-body reaction nor persistent inflammatory reaction, and after 42 days, the NRL implants were making close contact to the bone at many sites.

The induction of angiogenesis is crucial for bone regeneration and remodeling, and a study using chick embryo chorioallantoic membrane (CAM) showed that NRL membranes possess angiogenic properties. The experiment also showed that the angiogenic capacity of the NRL membranes remains active and increases in temperatures ranging from 65 to 85°C. These results showed that the heating used to prepare NRL membrane does not affect its biological properties. The same study used centrifuged latex to demonstrate that the poly-isoprene is not the part of latex responsible for the biological properties [4].

Large fractures can mean significant reconstructive problems and sometimes require special procedures for regeneration. And this regeneration is dependent on blood clot stability, local vascularization, defect size and protection against invasion of competitive nonosteogenic tissues [58, 59]. Guided bone regeneration (GBR) was a technique developed to enhance bone repair, and in this procedure, an occlusive membrane is used to provide to the osteogenic cells better conditions to perform bone remodeling [60].

Different types of membranes have been tested for use in GBR, but resorbable membranes represent the most interesting alternative since they avoid removal surgery [61]. NRL membranes were tested as an occlusive barrier in GBR of large defects in rabbit calvaria. NRL membranes successfully enhanced bone regeneration process in the group of the treated animals, and it was shown by a statistically higher volume of mature bone in all periods of study that was up to 120 days. The NRL membranes worked as a passive barrier membrane that prevented epithelial and connective tissue migration, thus facilitating the proliferation and migration of regenerative bone cells into the wound. The NRL membranes used for GBR, as well as the ones used to treat dental alveoli defects, did not induce the formation of foreign body inflammatory reaction [62].

Another way to accelerate bone regeneration and remodeling would be incorporating BMPs to NRL membrane. As shown above, BMPs induce bone remodeling and many studies have tried to develop a BMP delivery system that could sustain gradual release of BMPs for dental and orthopedic use [63–66]. A study using bovine serum albumin (BSA) in the place of BMP (same molecular weight) and NRL membranes prepared at different polymerization temperatures showed that NRL membrane was able to release BSA for 18 days. This indicates a promising future of these membranes as active occlusive membrane in GBR and they could release BMPs for 18 days accelerating bone healing [55].

Based on previous studies, a protein was isolated from NRL (P-1) and its inductive bone repair properties were compared to recombinant human bone morphogenetic protein-2 (rhBMP-2), a commercial available human protein with good osteoinductive capabilities. To compare these two proteins, a carrier made of monoolein gel was used. P-1 is still being characterized to better understand its biological properties and its function in the laticifer cells of *Hevea brasiliensis*. However, associated with monoolein gel, P-1 was able to induce new bone formation on bone defect of rat calvaria [67].

This protein extracted from NRL (P-1) was used in combination with a fibrin sealant in the repair of rat tibial bone defects. This combination was successful, being immunoidentified by the presence of osteoblasts in the area, showing high osteogenic and osteoconductive capacity for bone healing [68].

Moura et al. [69] used different polymerized NRL membranes in rabbit calvaria with bone defect. These membranes were compared to polytetrafluoroethylene (PTFE) membranes, which are an extensively studied material considered gold standard for occlusive membranes [70]. In this study, NRL membranes performed their role as biological barriers and achieved a similar performance to the PTFE membrane. One of the polymerized NRL membranes was ammonia free, and these were the membranes that significantly improved the bone repair process producing higher bone formation, being more effective than PTFE membrane. These results were achieved since the method of preparation of these NRL ammonia-free membranes preserved the angiogenic stimulus of the membranes. These membranes also did not lead to bone tissue hypersensitization.

NRL was also coated with calcium phosphate (Ca/P) and tested for biomedical application. Biomaterials added with Ca/P present biological, chemical and mechanical properties very similar to the mineral phase of the bone besides the ability to bond to the host tissue. A hemolytic test was performed, and this material did not affect the blood cells, being so ready for animal tests [71].

These results showed above indicate NRL membranes as a promising future biomembrane that could be used to accelerate bone healing. More experiments are being done already in humans. Since NRL membranes present intense angiogenic activity and wound healing activity, NRL membranes are being commercialized in Brazil and other 60 countries around the world as a band-aid curative (BIOCURE®) for the treatment of ulcers in diabetic patients.

Author details

Leandra E. Kerche-Silva*, Dalita G.S.M. Cavalcante and Aldo Eloizo Job

*Address all correspondence to: leakerche@gmail.com

Department of Physics, Chemistry and Biology, Universidade Estadual Paulista "Júlio de Mesquita Filho", Presidente Prudente, São Paulo, Brazil

References

[1] Jacob JL, Auzac J, Prevôt JL. The composition of natural latex from *Hevea brasiliensis*. Clinical Reviews in Allergy & Immunology. 1993;**11**:325-337

[2] Arif SAM, Hamilton RG, Yusof F, Chew NP, Loke YH, Nimkar S. Isolation and characterization of the early nodule-specific protein homologue (Hev b 13), an allergenic lipolytic esterase from *Hevea brasiliensis* latex. Journal of Biological Chemistry. 2004;**279**:23933-23941

[3] Nawamawat K, Sakdapipanich JT, Ho CC, Ma Y, Vancso JG. Surface nanostructure of *Hevea brasiliensis* natural rubber latex particles. Colloids and Surfaces A. 2011;**390**:157-166

[4] Ferreira M, Mendonça RJ, Coutinho-Netto J, Mulato M. Angiogenic properties of natural rubber latex biomembranes and the serum fraction of *Hevea brasiliensis*. Brazilian Journal of Physics. 2009;**39**:564-569

[5] Rippel MM, Lee L-T, Leite CA, Galembeck F. Skim and cream natural rubber particles: Colloidal properties, coalescence and film formation. Journal of Colloid and Interface Science. 2003;**268**:330-340

[6] Bealing FJ. Quebrachitol synthesis in *Hevea brasiliensis*. Journal of Rubber Research Institute of Malaysia. 1981;**29**:111-112

[7] Mrue F. Neoformação tecidual induzida por biomembranas de latex natural com poli-lisina. Aplicabilidade na neoformação esofágica e da parede abdominal. Estudo experimental com cães [thesis (Doctorate in Medicine, Area: Surgery)]. Ribeirão Preto: Faculdade de Medicina de Ribeirão Preto, Universidade de São Paulo; 2000. p. 111

[8] Mrue F, Coutinho-Netto J, Ceneviva R, Lachat JJ, Thomazini JA, Tambelini H. Evaluation of the biocompatibility of a new biomembrane. Materials Research. 2004;**7**:277-283

[9] Oliveira JAA, Hyppolito MA, Coutinho-Netto J, Mrué F. Miringoplastia com a utilização de um novo material biossintético. Revista Brasileira de Otorrinolaringologia. 2003;**69**:649-655

[10] Pinho ECCM, Souza SJF, Schaud F, Lachat JJ, Coutinho-Netto J. Uso experimental da biomembrane de latex na reconstrução conjunctival. Arquivos Brasileiros de Oftalmologia. 2004;**67**:27-32

[11] Valenti MT, Carbonare LD, Mottes M. Osteogenic differentiation in healthy and pathological conditions. International Journal of Molecular Sciences. 2017;**18**:41-49

[12] Moir GFJ. Ultracentrifugation and staining of *Hevea* latex. Nature. 1959;**184**:1626-1628

[13] Sunderasan E, Rahman NABD, Lam KL, Yang KL, Ong MT. Proteins of dialysed C-serum supernatant sub-fractions elicit anti-proliferative activity on human cancer-origin cells. Journal of Rubber Research. 2015;**18**:49-59

[14] Yeang HY. Characterisation of rubber particle destabilization by B-serum and Bark Sap of *Hevea brasiliensis*. Journal of Natural Rubber Research. 1988;**4**:47-55

[15] Gomez JB, Moir GK. The ultracytology of latex vessels in *Hevea brasiliensis*. Monograph No. 4. Kuala Lumpur: Malaysian Rubber Research and Development Board; 1979

[16] Schoon THGF, Phoa KL. Morphology of the rubber particles in natural lattices. Arch van der Rubberc. 1956;**33**:195

[17] Rolland JM, O'Hehir RE. Latex allergy: A model for therapy. Clinical & Experimental Allergy. 2008;**38**:989-912

[18] Yeang HY. Allergenic proteins of natural rubber latex. Methods. 2002;**27**:32-45

[19] Kerche-Silva LE, Cavalcante DGSM, Danna CS, Gomes AS, Carrara IM, Cecchini AL, et al. Free-radical scavenging properties and cytotoxic activity evaluation of latex C-serum from *Hevea brasiliensis* RRIM 600. Free Radicals and Antioxidants. 2017;**7**:107-114

[20] Dickenson PB. Electron microscopical studies of latex vessel system of *Hevea brasiliensis*. Journal of Rubber Research Institute of Malaysia. 1969;**21**:543-559

[21] Southorn WA. Complex particles in *Hevea* latex. Nature. 1960;**188**:165-166

[22] Berendsen AD, Olsen BR. Bone development. Bone. 2015;**80**:14-18

[23] Wang Y, Li YP, Paulson C, Shao JZ, Zhang X, Wu M, Chen W. Wnt and Wnt signalling pathway in bone development and disease. Frontiers in Bioscience. 2014;**19**:379-407

[24] Li M, Li Y, Deng W, Zhang Z, Deng Z, Hu Y, et al. Chinese bone turnover marker study: Reference ranges for C-terminal telopeptide of type I collagen and procollagen I N-terminal peptide by age and gender. PLoS One. 2014;**9**:e103841

[25] Li F, Song N, Tombran-Tink J, Niyibizi C. Pigment epithelium-derived factor enhances differentiation and mineral deposition of human mesenchymal stem cells. Stem Cells. 2013;**31**:2714-2723

[26] Valenti MT, Carbonare LD, Mottes M. Hypophosphatasia and mesenchymal. International Journal of Stem Cell Research & Therapy. 2016;**3**:20

[27] Cao X, Chen D. The BMP signaling and *in vivo* bone formation. Gene. 2005;**357**:1-8

[28] Williams BO, Insogna KL. Where Wnts went: The exploding field of Lrp5 and Lrp6 signaling in bone. Journal of Bone and Mineral Research. 2009;**24**:171-178

[29] Honma M, Ikebuchi Y, Kariya Y, Suzuki H. Regulatory mechanisms of RANKL presentation to osteoclast precursors. Current Osteoporosis Reports. 2014;**12**:115-120

[30] Dalle Carbonare L, Innamorati G, Valenti MT. Transcription factor Runx2 and its application to bone tissue engineering. Stem Cell Reviews. 2012;**8**:891-897

[31] Islam R, Yoon WJ, Ryoo HM, Pin1, the master orchestrator of bone cell differentiation. Journal of Cellular Physiology. 2017;**232**:2339-2347

[32] Cantley MD, Zannetino ACW, Bartold PM, Fairlie DP, Haynes DR. Histone deacetylases (HDAC) in physiological and pathological bone remodelling. Bone. 2017;**95**:162-174

[33] Zaidi SK, Young DW, Montecino M, van Wijnen AJ, Stein JL, Lian JB, Stein GS. Bookmarking the genome: Maintenance of epigenetic information. Journal of Biological Chemistry. 2015;**286**:18355-18361

[34] Jing D, Hao J, Shen Y, Tang G, Li ML, Huang SH, Zhao ZH. The role of microRNAs in bone remodeling. International Journal of Oral Science. 2015;**7**:131-143

[35] Gerstenfeld LC, Cullinane DM, Barnes GL. Fracture healing as a post-natal developmental process: Molecular, spatial, and temporal aspects of its regulation. Journal of Cellular Biochemistry. 2003;**88**:873-884

[36] Kayal RA, Tsatsas D, Bauer MA, Allen B, Al-Sebaei MO, Kakar S, et al. Diminished bone formation during diabetic fracture healing is related to the premature resorption of cartilage associated with increased osteoclast activity. Journal of Bone and Mineral Research. 2007;**22**:560-568

[37] Gerstenfeld LC, Cho TJ, Kon T, Aizawa T, Tsay A, Fitch J, et al. Impaired fracture healing in the absence of TNF-alpha signaling: The role of TNF-alpha in endochondral cartilage resorption. Journal of Bone and Mineral Research. 2003;**18**:1584-1592

[38] Lee SK, Lorenzo J. Cytokines regulating osteoclast formation and function. Current Opinion in Rheumatology. 2006;**18**:411-418

[39] Yang X, Ricciardi BF, Hernandez-Soria A, Shi Y, Camacho NP, Bostrom MPG. Callus mineralization and maturation are delayed during fracture healing in interleukin-6 knockout mice. Bone. 2007;**41**:928-936

[40] Reddi AH. Bone morphogenetic proteins: From basic science to clinical applications. Journal of Bone and Joint Surgery. American Volume. 2001;**83-A**:S1–S6

[41] Tsuji K, Bandyopadhyay A, Harfe BD, Cox K, Kakar S, Gerstenfeld L, et al. BMP2 activity, although dispensable for bone formation, is required for the initiation of fracture healing. Nature Genetics. 2006;**38**:1424-1429

[42] Xiong DH, Shen H, Zhao LJ, Xiao P, Yang TL, Guo YF, et al. Robust and comprehensive analysis of 20 osteoporosis candidate genes by very high-density single nucleotide polymorphism screen among 405 white nuclear families identified significant association and gene-gene interaction. Journal of Bone and Mineral Research. 2006;**21**:1678-1695

[43] Cho TJ, Gerstenfeld LC, Einhorn TA. Differential temporal expression of members of the transforming growth factor beta superfamily during murine fracture healing. Journal of Bone and Mineral Research. 2002;**17**:513-520

[44] Bais MV, Wigner N, Young M, Toholka R, Graves DT, Morgan EF, et al. BMP2 is essential for post natal osteogenesis but nor for recruitment of osteogenic stem cells. Bone. 2009;**45**:254-266

[45] Filipowska J, Tomaszewski KA, Niedźwiedzki Ł, Walocha JA, Niedźwiedzki T. The role of vasculature in bone development, regeneration and proper systemic functioning. Angiogenesis. 2017

[46] Tomlinson RE, Silva MJ. Skeletal blood flow in bone repair and maintenance. Bone Research. 2013;**1**:314-322

[47] Brandi ML, Collin-Osdoby P. Vascular biology and the skeleton. Journal of Bone and Mineral Research. 2006;**21**:183-192

[48] Niedźwiedzki T, Filipowska J. Bone remodeling in the context of cellular and systemic regulation: The role of osteocytes and the nervous system. Journal of Molecular Endocrinology. 2015;**55**:R23-R26

[49] Liu Y, Olsen BR. Distinct VEGF functions during bone development and homeostasis. Archivum Immunologiae et Therapiae Experimentalis (Warsz). 2014;**62**:363-368

[50] Hu K, Olsen BR. Osteoblast-derived VEGF regulates osteoblast differentiation and bone formation during bone repair. Journal of Clinical Investigation. 2016;**126**:509-526

[51] Maes C, Goossens S, Bartunkova S, Drogat B, Coenegrachts L, Stockmans I, Haigh JJ. Increased skeleton VEGF enhances beta-catenin activity and results in excessively ossified bones. EMBO Journal. 2010;**29**:424-441

[52] Skawina A, Litwin JA, Gorczyca J, Miodoński A. Blood vessels in epiphyseal cartilage of human fetal femoral bone: A scanning electron microscopic study of corrosion casts. Anatomy and Embryology. 1994;**189**:457-462

[53] Thompson TJ, Owens PD, Wilson DJ. Intramembranous osteogenesis and angiogenesis in the chick embryo. Journal of Anatomy. 1989;**166**:55-65

[54] Maes C, Kobayashi T, Selig MK, Torrekens S, Sanford I, Mackem S, Kronenberg HM. Osteoblast precursors, but not mature osteoblasts, move into developing and fractured bones along with invading blood vessels. Developmental Cell. 2010;**19**:329-344

[55] Herculano RD, Pereira Silva C, Ereno C, Guimaraes SAC, Kinoshita A, Graeff CFO. Natural rubber latex used as drug delivery system in guided bone regeneration (GBR). Materials Research. 2009;**12**:253-256

[56] Balabanian CACA, Coutinho-Netto J, Lamano-Carvalho TL, Lacerda SA, Brentegani LG. Biocompatibility of natural latex implanted into dental alveolus of rats. Journal of Oral Science. 2006;**48**:201-205

[57] Brentegani LG, Bombonato KF, Lamano-Carvalho TL. Histologic evaluation of the biocompatibility of glass-ionomer cement in rat alveolus. Biomaterials. 1997;**18**:137-140

[58] Melcher AH, Dreyer CJ. Protection of the blood clot in healing circumscribed bone defects. Journal of Bone and Joint Surgery. British Volume. 1964;**44**:424-430

[59] Schenk RK. Bone regeneration biologic basis. In: Buser D, Dahlin C, Schenk RK, editors. Guided Bone Regenerations in Implant Dentistry. Chicago: Quintessence; 1994. pp. 49-100

[60] Linde A, Alberius P, Dahlin C, Bjurstam K, Sundin Y. Osteopromotion: A soft-tissue exclusion principle using a membrane for bone healing and bone neogenesis. Journal of Periodontology. 1993;**64**:1116-1128

[61] Barber D, Lignelli J, Smith BM, Bartee BK. Using a dense PTFE membrane without primary closure to achieve bone and tissue regeneration. Journal of Oral and Maxillofacial Surgery. 2007;**65**:748-752

[62] Ereno C, Guimarães SAC, Pasetto S, Herculano RD, Pereira Silva C, Graeff CFO, et al. Latex use as an occlusive membrane for guided bone regeneration. Journal of Biomedical Materials Research Part A. 2010;**95A**:932-939

[63] Jung RE, Glauser R, Schärer P, Hämmerle CHF, Sailer HF, Weber FE. Effect of rhBMP-2 on guided bone regeneration in humans: A randomized, controlled clinical and histomorphometric study. Clinical Oral Implants Research. 2003;**14**:556-568

[64] Müller F, Roher H, Vogel-Höpker A. Bone morphogenetic proteins specify the retinal pigment epithelium in the chick embryo. Development. 2007;**134**:3483-3493

[65] Oshin AO, Stewart MC. The role of bone morphogenetic proteins in articular cartilage development, homeostasis and repair. Veterinary and Comparative Orthopaedics and Traumatology. 2007;**20**:151-158

[66] Woo BH, Fink BF, Page R, Schrier JA, Woo JY, Jiang G, et al. Enhancement of bone growth by sustained delivery of recombinant human bone morphogenetic protein-2 in a polymeric matrix. Pharmaceutical Research. 2001;**18**:1747-1753

[67] Issa JPM, Defino HLA, Coutinho-Netto J, Volpon JB, Regalo SCH, Iyomasa MM, et al. Evaluation of rhBMP-2 and natural latex as potential osteogenic proteins in critical size defects by histomorphometric methods. Anatomical Record. 2010;**293**:794-801

[68] Machado EG, Issa JPM, Figueiredo FAT, Santos GR, Galdeano EA, Alves MC, et al. A new heterologous fibrin sealant as scaffold to recombinant human bone morphogenetic protein-2 (rhBMP-2) and natural latex proteins for the repair of tibial bone defects. Acta Histochemica. 2015;**117**:288-296

[69] Moura JML, Ferreira JF, Marques L, Holgado L, Graeff CFO, Kinoshita A, Comparison of the performance of natural latex membranes prepared with different procedures and PTFE membrane in guided bone regeneration (GBR) in rabbits. Journal of Mater Science: Materials in Medicine. 2014;**25**:2111-2120

[70] Lindhe J, Karring T, Lang NP. Clinical Periodontology and Implant Dentistry. 5th ed. Oxford: Wiley-Blackwell; 2008

[71] Borges FA, Almeida Filho E, Miranda MCR, Santos ML, Herculano RD, Guastaldi AC, Natural rubber latex coated with calcium phosphate for biomedical application. Journal of Biomaterials Science. Polymer edition. 2015;**26**:1256-1268

Chapter 14

Systematic Study of Ethylene-Vinyl Acetate (EVA) in the Manufacturing of Protector Devices for the Orofacial System

Reinaldo Brito e Dias, Neide Pena Coto,
Gilmar Ferreira Batalha and Larissa Driemeier

Additional information is available at the end of the chapter

http://dx.doi.org/10.5772/intechopen.69969

Abstract

Fracture of facial bones and dental elements, and laceration of soft tissue, have increased in sports over recent years. Dentist is the only professional responsible for the mouth protection design, the knowledge about suitable materials is essential. EVA is a thermoplastic material, available in the market, easy of handling and processing, and low-cost. However, it is important to understand the mechanical properties and ability to absorb and to dissipate the impact energy, when this material is submitted to different environments, such as oral cavity with saliva and different temperatures. This chapter show provides a systematic evaluation of the EVA application in orofacial protectors while focusing on sports. The research comprises two aspects: experimental tests and numerical analyses. During experimental tests, EVA was analyzed in special buccal conditions, concerning temperature and presence of saliva. Regarding the presence of saliva, more specific studies about its influence on the mechanical behavior of EVA were performed. In the numerical analyses of the EVA orofacial protector, the studies focused on its effect on the nasal bone integrity, and in the zygomatic bone protection. However, life cycle should be analyzed, since its performance deteriorates over time. Mainly due to the saliva-originated changes to the EVA mechanical characteristics, it can behave as a rigid material. For facial protection, a better performance is obtained with a combination of rigid and soft EVA material. According to the experimental and numerical results from a systematic study of EVA, its application to orofacial protection can be considered satisfactory.

Keywords: material tests, orofacial protection, trauma in sports, protection in sports

1. Introduction

According to the World Health Organization (WHO) [1], "Health is a state of complete physical, mental and social well-being and not merely the absence of disease or infirmity."

Areas of study related to life, health and disease are called human health sciences. Medicine, Biology, Biomedicine, Nursing, Speech Therapy, Pharmacy and Biochemistry, Sports Science, Physical Education, Psychology, Occupational Therapy, Nutrition, Physiotherapy, Bioengineering and Dentistry are part of this program. All these research areas focus on improving or maintaining the patient quality of life, in accordance with the conditions dictated by WHO.

In dentistry, a particularly important area is related to the endless search for materials that can more efficiently help the maintenance and/or return of the individual's well-being. Researchers in dentistry seek and study materials that may replace dental organs, may be accepted in the alveolar and dentofacial complex, or may protect the orofacial complex from injuries.

Therefore, due to the technical-scientific excellence required in its attributions, dentistry is a science that requires constant updating of materials science and applications. It is worth highlighting that, to indicate a safe and efficient clinical application for a particular material, mechanical, physical, chemical and biological properties must be known.

According to Anusavice et al. [2], four groups of materials are used and studied in dentistry: metals, ceramics, composites and polymers. These materials are separated into modalities, according to their application: preventive, restorative or auxiliary materials.

Auxiliary are materials with recognized importance and application but which do not fit into the first two modalities. It is the best option for describing the function of polymers.

Polymers are an important category of materials for dentistry. They are versatile, since they can be combined in order to improve mechanical properties, and moreover, they are reproducible and homogeneous [3].

The term polymer derives from the Greek words: poly-many and mer-unit; or, more specifically, it is a macromolecule composed of repeating units linked by a covalent bond. Its physical properties depend on the length of the molecule and its molar mass. When the polymer is formed by a single type of *mer*, it is called **homopolymer**; otherwise, it is called a **copolymer**.

According to their malleability, polymers are classified into thermoplastic and thermosetting. When the temperature is raised above its melt point, the *thermoplastic polymer* becomes softer and more fluid, allowing it to be molded. When the heat source is removed, the thermoplastic hardens in the molded shape. Since it occurs without chemical curing, it is a reversible physical transformation. In turn, with the addition of a second material and/or heat, *thermosetting polymers* soften and cure, forming cross-links that prevent the material from returning to the primary form. This process cannot be repeated.

http://dx.doi.org/10.5772/intechopen.69969

Regarding mechanical behavior, polymers are classified as *elastomers*, *plastics* and *fibers* [4].

During dentists' day-to-day operations, *resin* is the most commonly used polymer. Most of these resins are based on methacrylate, with methyl methacrylate as the main ingredient. Resins are easy to manipulate, without demanding elaborate techniques; the final resin products are esthetically acceptable and offer excellent balance when used in the oral environment, in the presence of saliva and chewing conditions, besides being low cost [5].

Ethylene-vinyl acetate (EVA), object of this study, is a thermoplastic copolymer derived from petroleum. For dental use, it is in the form of rigid or flexible flat plates, in thicknesses of 1–5 mm, without the presence of blowing agents, differently from EVA plates available in the common market. It is indicated as a shock absorber material, for producing mouth and facial protectors for sports practice, as pointed out by Coto et al. [6], as well as dental bleaching trays, orthodontic restraints and as the base for facial prostheses.

2. Ethylene-vinyl acetate

The copolymer of ethylene and vinyl acetate is a thermoplastic polymer, formed by different monomers: ethylene and vinyl acetate (**Figure 1**). Monomers merge through a polymerization process—a set of reactions among simple molecules to form a macromolecule of high molar mass.

The EVA presents semi-crystalline structure; its geometry is composed of an amorphous and a crystalline part. The damping capacity of EVA increases as the percentage of vinyl acetate decreases. As already mentioned, EVA is a macromolecule composed of repeated units linked by covalent bonds and its primary units of constitution are two monomers whose physical properties depend on their size and molecular weight. Polymeric materials generally exhibit density ranging from 0.926 to 0.950 g/cm^3, temperature resistance (glass transition temperatures

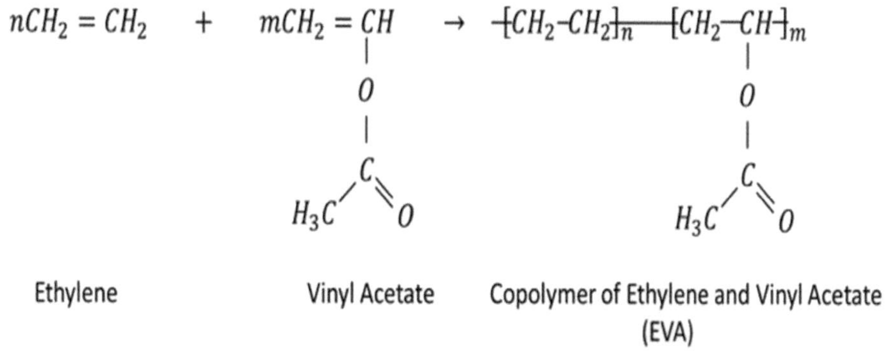

Figure 1. Polymerization reaction between ethylene and vinyl acetate, resulting in EVA.

close to 0 to −7°C). Among the main characteristics of EVA, its elastic behavior characterized by the Young Modulus ranging from 15 to 80 MPa can be highlighted. Flexible EVA, for example, behaves similarly to elastomers, and its elasticity is considerable.

In most practical situations in which EVA is applied or mechanically tested, it is possible to observe that the material's mechanical response is time dependent, that is, it is a viscoelastic material. This characteristic of viscosity is important for energy dissipation.

In the chemical industry, EVA is presented in grain form as shown in **Figure 2**.

Some mechanical properties of EVA are discussed as follows.

2.1. Stiffness

The initial stiffness of the EVA can be measured by its modulus of elasticity, that is, angle of inclination of the approximated straight line that relates stresses as a function of the strains, in elastic regime. In the elastic regime, the energy absorbed by the deformed material is totally restored by removing the stress. The higher the vinyl acetate concentration, the more flexible the EVA material is, due to the reduction in the degree of crystallization.

The degree of EVA crystallization is proportional to the latent heat of fusion (ΔH_f), and its value increases as the concentration of crystals present in EVA increases. However, EVA is not a totally crystalline polymer because, in the solid state, it contains two phases: amorphous and crystalline. In fact, the presence of a glass transition temperature (T_g) means that it contains an amorphous phase, since T_g is a thermal transition exclusive of the amorphous phase, that is, it is the temperature at which the macromolecules of the amorphous phase acquire rotational mobility. The amorphous phase of EVA is represented by a macromolecule entanglement which lacks an ordered and periodic three-dimensional structure. The crystalline phase, on the other hand, is characterized by a three-dimensional ordered and periodic structure of macromolecules folded one on the other, assuming the lamellar format. The melting temperature (T_f) is also a thermal transition, in which the crystalline phase disintegrates and the polymer becomes a viscous liquid.

Figure 2. EVA in granules.

2.2. Hardness

The hardness of a polymer is determined by the penetration of the Durometer indenter foot into a small sample (Shore Hardness). The increase in vinyl acetate content reduces the hardness of EVA, mainly due to the decrease in its degree of crystallization. Although hardness and stiffness are different properties, in some cases, it is possible to establish an empirical correlation between them for a given family of polymers. In some cases, as the degree of crystallization of EVA increases, the stiffness and hardness increase proportionally.

2.3. Transparency

The polymer crystals of EVA act as physical obstacles to the passage of light. Accordingly, as the polymer crystals concentration decreases, increasing the content of vinyl acetate, the material becomes more transparent.

2.4. Damping

It is the ability of the material to absorb the mechanical energy to mainly overcome internal friction. The damping capacity of EVA increases as the vinyl acetate content reduces. Damping capacity is sometimes unduly related to hardness. However, a hard polymer can be designed to have the same damping capacity of a soft polymer.

2.5. Viscoelasticity

In many of the practical conditions in which polymers are requested or tested, their mechanical response is found to be time-dependent, which characterizes these materials as viscoelastic, as already mentioned. This absorption may occur due to the internal friction between the macromolecules, by shape changes (rotation of the carbon–carbon bonds around its own axis) or by flow. Furthermore, in case of impact, the viscous portion is responsible, for the delay in the elastic response, which will depend on the stimulus and on the time necessary to coil and to uncoil the polymer macromolecule [4, 7–9].

3. Experimental study of EVA applied to oral protection

The study of the mechanical properties of EVA focused on mouth guards and facial protectors. Particularly for the facial protector designed in this study, a patent was applied (number BR 20 023048 9).

Several experimental tests made with the material, available in the literature [10–12], confirm this percentage, which is proportionally inverse to the EVA damping capacity. Moreover, the EVA was carefully characterized in POLITENO—Brazil (now BRASKEN) for analysis. The analyses of vinyl acetate percentage were performed by means of pyrolysis.

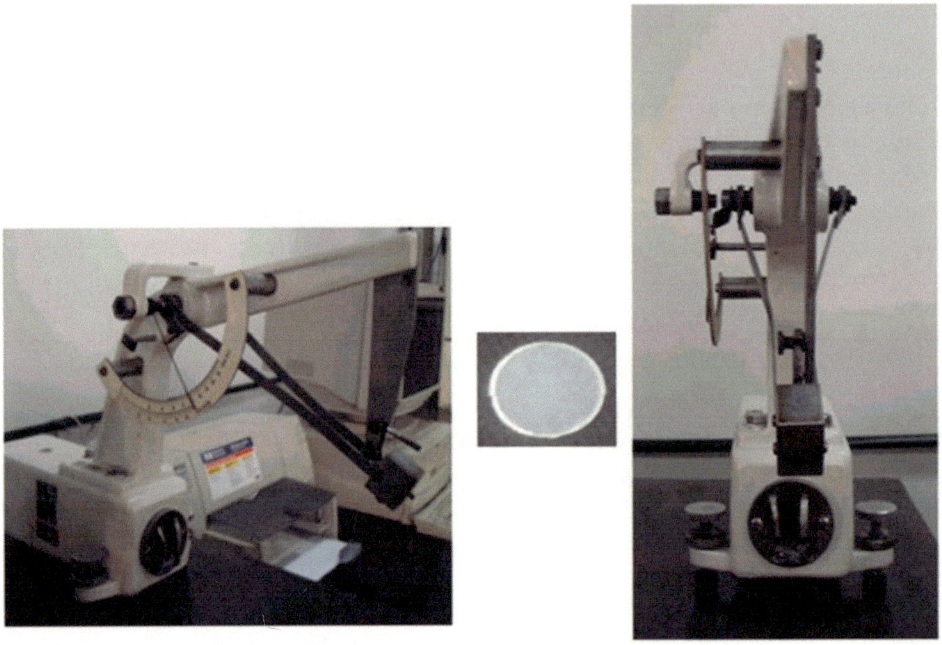

Figure 3. Compression test of the EVA specimens—ABNT NBR 8690—with Schob Pendulum.

In **Figure 3**, the Schob pendulum was used to measure the resilience of the EVA. Six experiments were performed, three used to calibrate the system and three to measure the property. The EVA was observed to have a great damping potential, since it absorbed 50% of the applied energy.

Experimental compression tests were performed to the mechanical characterization of the EVA.

Figure 4 shows the Instron® machine and the recording of the compression tests, performed by a Photron Ultima APX-RS high-speed camera (3000 frames per second). The record helped the study of the nonlinear material behavior of the EVA.

Particularly, **Figure 5** shows that EVA undergoes considerable plastic deformation before failure.

3.1. Mechanical study of the operation of a mouth guard

To reproduce conditions as close as possible to a real situation, models in epoxy resin were manufactured from a patient model (**Figure 6**).

As illustrated in **Figure 7**, the models of the upper and lower arches were fixed in a compression device that allows the lower arch to move while maintaining the upper arch fixed. The compression device was coupled to a Universal Kratos Test Machine, data acquisition system, 20 kN load cell. The aperture, initially in occlusion, was controlled by the extensometer, with a maximum opening of 18 mm. Compression tests were performed, at a velocity of 42.86 mm/min.

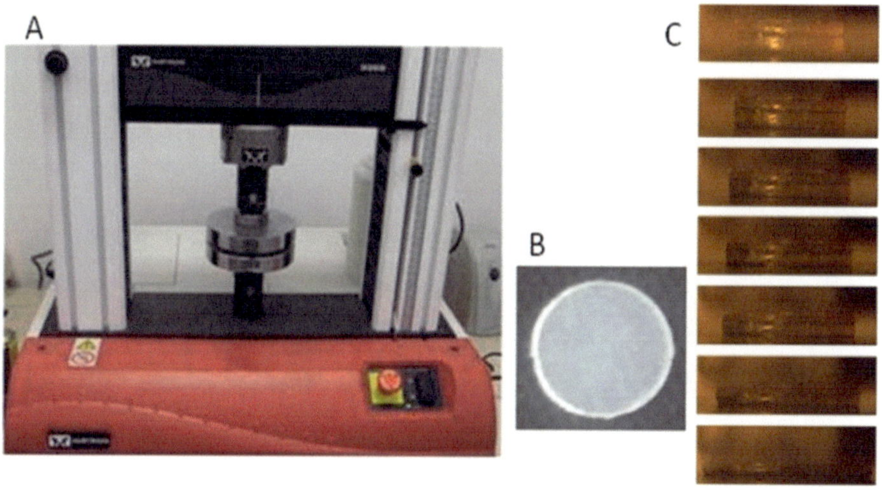

Figure 4. (A) EVA compression test in an Instron machine. (B) Detail of the geometry of the specimen: flat discs with 30 mm in diameter. (C) Compression test recording.

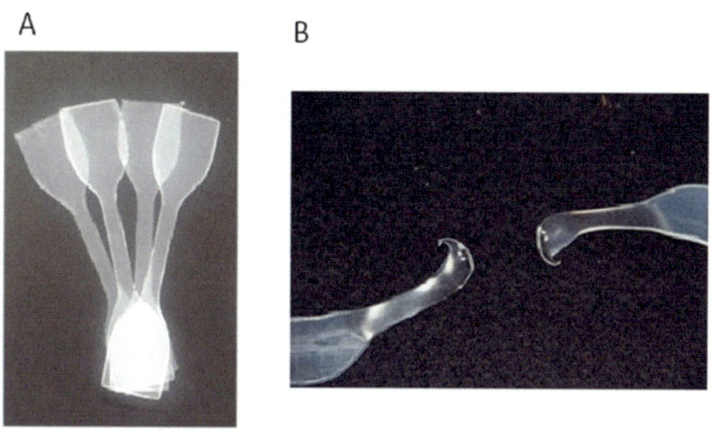

Figure 5. (A) EVA specimens for tensile test. (B) Detail of the specimen after failure.

The test was controlled by optical pyrometer, maintaining the temperature around 37–39°C, close to the mouth temperature (**Figure 8**).

Five EVA mouth guards of each thickness (3 and 4 mm) were made for each test group, using models of a superior dental arch in stone gypsum and metalvander® vacuum-form machine. The geometry respected the recommendation of American Academy for Sports Dentistry [13], that is, 2 mm below the bottom of the vestibular groove, 10 mm beyond the palatine gingival and extension up to the second upper molars.

The heating time for both thicknesses was 4 min, approximately; aspiration time was 45 s (**Figure 9**). All the protectors were immersed in cold water for 10 min.

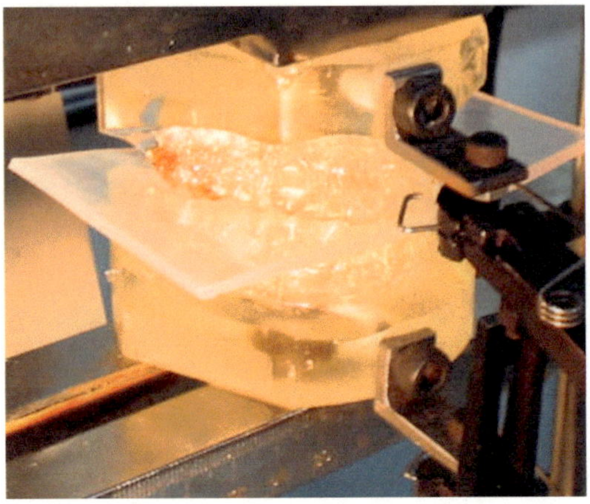

Figure 6. Model of dental arches made of epoxy resin.

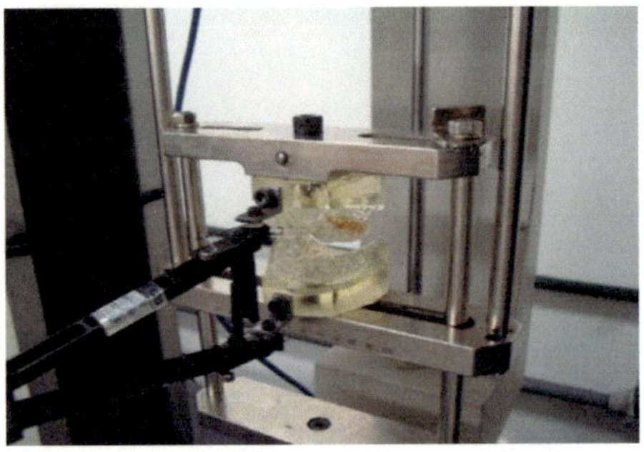

Figure 7. Test set: the Kratos® Universal Testing Machine, dental arch models and extensometer.

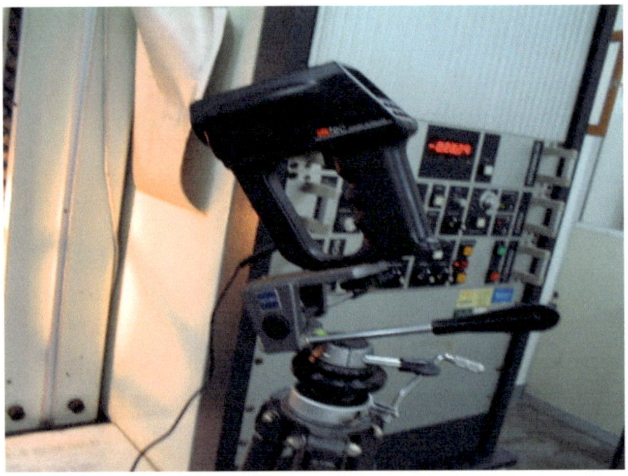

Figure 8. Maintenance of the system temperature with an optical pyrometer.

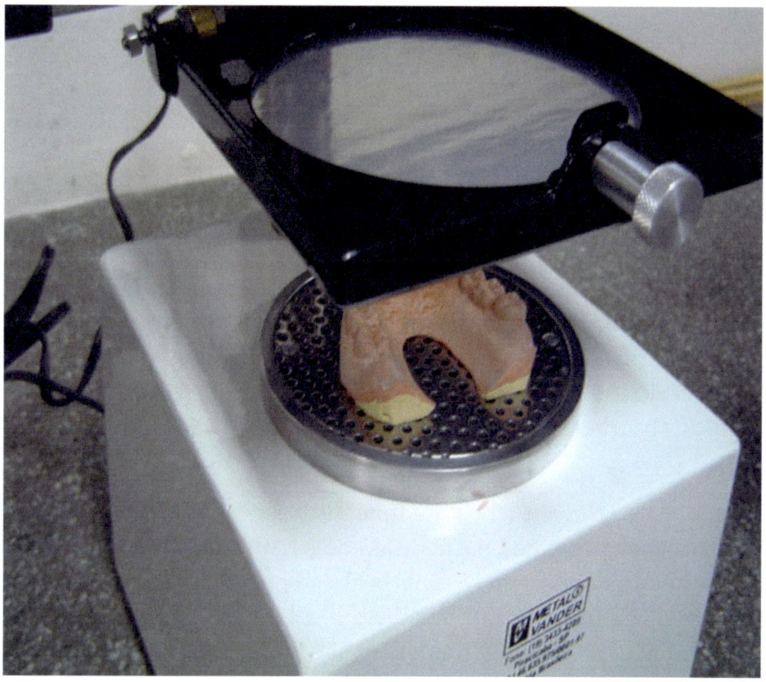

Figure 9. Manufacture of mouth guards in vacuum form Metalvander® machine.

The groups are divided as follows:

A. Mouth guard—3-mm-thick blade

- Room temperature/without saliva
- Room temperature/saturated in saliva
- Oral temperature/without saliva
- Oral temperature/saturated in saliva

B. Mouth guard—4-mm-thick blade

- Room temperature/without saliva
- Room temperature/saturated in saliva
- Oral temperature/without saliva
- Oral temperature/saturated in saliva

Table 1 shows a variation in the compression maximum load in Newtons (N) when evaluating the 3-mm-thick mouth guard (here named $\mathbf{prot_A}$) and the 4-mm-thick mouth guard (here named $\mathbf{prot_B}$).

Coefficient	Max. load (N)	Standard deviation	Significance p
$prot_A$	2046	20	0.00
$prot_B$	2219	20	0.00

Table 1. Maximum load variation (N) as a function of the thickness variable, with their respective standard deviations (SD) and significance levels ($p \leq 0.05$), for protectors A and B.

When comparing **$prot_A$** and **$prot_B$** protectors in **Table 1**, **$prot_B$** was observed to require an additional force of 173 N. It agrees with Craig and Godwin [14]: "The energy absorbed in the cyclic moment of compressive deformation should reduce the locally transmitted energy" and thus avoid the rupture of the polymer layer between the teeth. The EVA material acts as a shock absorber, guaranteeing a low energy transmission to the teeth of the dental arch [15].

These data become more relevant when the final measurements of the guard thicknesses are observed. At the end, they presented mean differences in thickness of approximately 0.55 mm, instead of the nominal 1 mm difference. This is already expected, since during the manufacture of the individualized buccal protector there is a loss of thickness between 25 and 50%, also observed in the literature [14, 16]. Analyzing **Table 1** again, one can conclude that a small difference, 0.55 mm, increased the force around 173 N.

4. Numerical analysis of EVA applied to facial protection

Studies in the biological area involving impact have become impossible to perform *in vivo* due to ethical awareness. On the other hand, engineering presented rapid technological development of tools that allow for more detailed analyses, using complex geometries and offering refined results of behavior of virtually modeled bodies [17]. Particularly, the finite element method (FEM) is a powerful tool, able to virtually mimic different complex phenomena, including the impact of an object on a human face.

However, to analyze the performance of different EVA geometries and properties (flexible and rigid forms) via FEM, it is necessary to determine the parameters and constitutive laws for the materials (tissue, bone, EVA), geometry of the studied problem (face and projectile) and boundary conditions (initial velocity of the projectile, displacement restrictions in the system).

4.1. Material parameters and constitutive laws

4.1.1. Face bones

Most of the bony framework of the face has high-level resistance, since it protects vital elements, such as the brain, the eyes and the neuromuscular structures. Yet, it is also composed of very fragile bones, such as the maxilla, nasal bones and the malar portion of the zygomatic bone [18–22].

When a facial bone is fractured, undergoing or not surgery reduction, it should not be exposed to any trauma during the bone healing process, which lasts about 30 days [23–27]. If surgical reduction is required, it should occur within the first 2–3 h after the injury occurs [28, 29].

Cases of surgical reduction may disrupt the performance of athletes. In these cases, the use of the facial protector can allow an early and safe return of the athlete to training and competitions [22, 27, 28, 30]. In general, 4–7 days are required for the face molding and for manufacturing/producing the protector.

For the present FEM analysis, the cortical bone is represented as a linear elastic, homogeneous and isotropic material. The mechanical properties—density, Young's modulus, Poisson coefficient and maximum strength—of each bone depend on its composition, as reported by Lotti et al., Handbook in 2006 [31].

Table 2 presents the maximum compressive load of each bone portion of interest for dentistry. Particularly for the cortical bone, the elastic material parameters are listed in **Table 3**.

4.1.2. Human soft tissue

The soft tissue named here is composed of the skin and the muscular portion of the studied region.

The soft tissue is a hyperelastic nonlinear material [33–37] here represented by the well-known Ogden model [35, 37].

Table 4 lists the parameters used for soft tissue in the FEA. The elastic parameters are the same as those adopted by several car manufacturers to simulate pedestrian—car impact—and Ogden parameters were obtained by Coto et al. [6], according to the definition in the finite-element software LS-Dyna.

Bone	Max. load (N)	Max compressive stress (N/mm^2)
Frontal	1000–6494	≥7.58
Zygomatic	489–2401	1.38–4.17
Mandible	668–1801	1.03–2.07
Nasal	342–450	0.13–0.34

Table 2. Face bone resistance [32].

Structure	Young Modulus (MPa)	Poisson coefficient	Density (t/mm^3)
Cortical bone	13,700	0.32	2.28

Table 3. Elastic parameters for the cortical bone.

	Elastic parameters			Ogden parameters	
	Shear Modulus (MPa)	Poisson coefficient	Density (t/mm³)	μ_1/α_1	μ_2/α_2
Tissue	0.69	0.495	1.438 E – 9	7.0/0.8	2.6/2.6

Table 4. Material model for human tissue.

4.1.3. Flexible EVA

Flexible EVA has high elasticity and low mechanical resistance. A reverse analysis method was adopted to extract the material properties from the experimental tests described. In the reverse method, material parameters are tuned such that numerical predictions match the experimental curves (**Figure 10**).

Table 5 summarizes the material parameters, used to characterize flexible EVA, according to the Ogden hyperelastic model, available in the commercial software LS-Dyna® and adopted in this study.

4.1.4. Rigid EVA

The inverse methodology was again adopted here, to characterize rigid EVA. **Figure 11** shows the similarity between the experimental and the numerical compression tests.

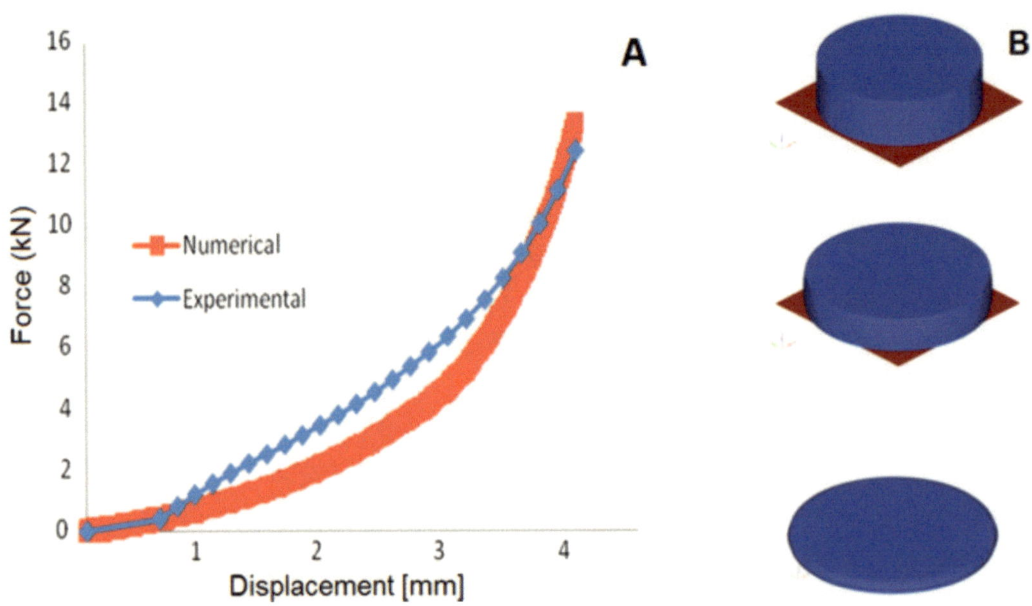

Figure 10. (A) Experimental and numerical curve for compression test. (B) Specimen configuration at different instants of the numerical analysis [6].

	μ_1/α_1	μ_2/α_2	Poisson coefficient	Shear Modulus (MPa)	Density (t/mm³)
Flexible EVA	7.0/0.8	2.6/2.6	0.48	10.0	2.0 E − 9

Table 5. The material model for flexible EVA.

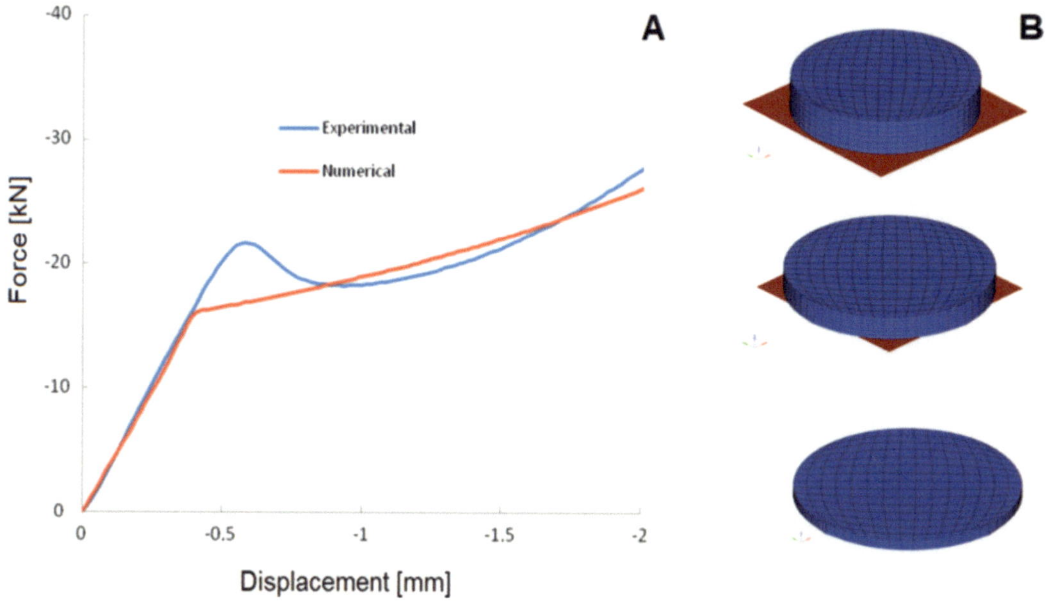

Figure 11. (A) Experimental and numerical curve for compression tests of rigid EVA. (B) Specimen configuration at different instants of the numerical analysis [37].

Table 6 shows the material parameters used for the rigid EVA, according to the Ogden model available in the software LS-Dyna®.

4.1.5. Geometry

As for numerical simulations of the human face, the geometry is a challenge, due to the great number of particularities.

To overcome this problem, a scientific partnership was established with the Renato Archer Information Technology Center (CTI Renato Archer). They provided the face images (**Figure 12**), obtained by computerized tomography (CT) and using in-house software.

	μ_1/α_1	μ_2/α_2	Poisson coefficient	Shear Modulus (MPa)	Density (t/mm³)
Rigid EVA	1.0/0.05	10.0/−4.0	0.49	175	1.2 E − 9

Table 6. Material model for rigid EVA.

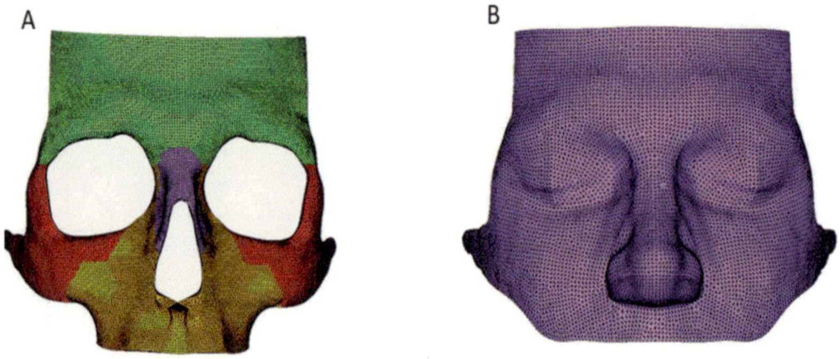

Figure 12. (A) The bone and (B) soft tissue configuration of the face, obtained from CT [37].

4.2. Numerical analyses

Using the material and geometrical parameters defined so far, it is possible to perform complex numerical analyses of the face, with different load conditions. Here, software LS-Dyna was used. The mesh generation and data analyses were performed with the pre- and post-processors HyperMesh and HyperView, respectively [38, 39].

4.2.1. *EVA as nasal protector for sport*

Coto et al. [6] studied the performance of EVA nasal protectors undergoing the impact of a rigid ball in the face with a 3D FE model (**Figure 13**). The material used was a combination of 1 mm of rigid EVA with 2 mm of flexible EVA. The author concluded that the proposed protector could absorb and dissipate the energy from the impact of a ball with mass of 0.025 kg at initial velocity of 20 m/s. The energy is high enough to fracture the nasal bone if there is no protector (**Figure 14**).

According to Coto et al. [37], rigid EVA reduced the velocity of impact and the flexible EVA increased the time interval of the impulse, thus decreasing the peak load transmitted to the bone.

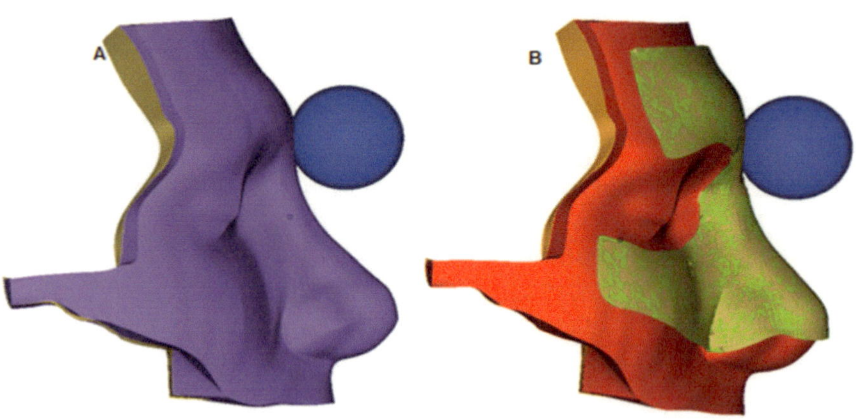

Figure 13. FE model. (A) Without the protector. (B) With the protector. Figure is extracted from Coto et al. [37].

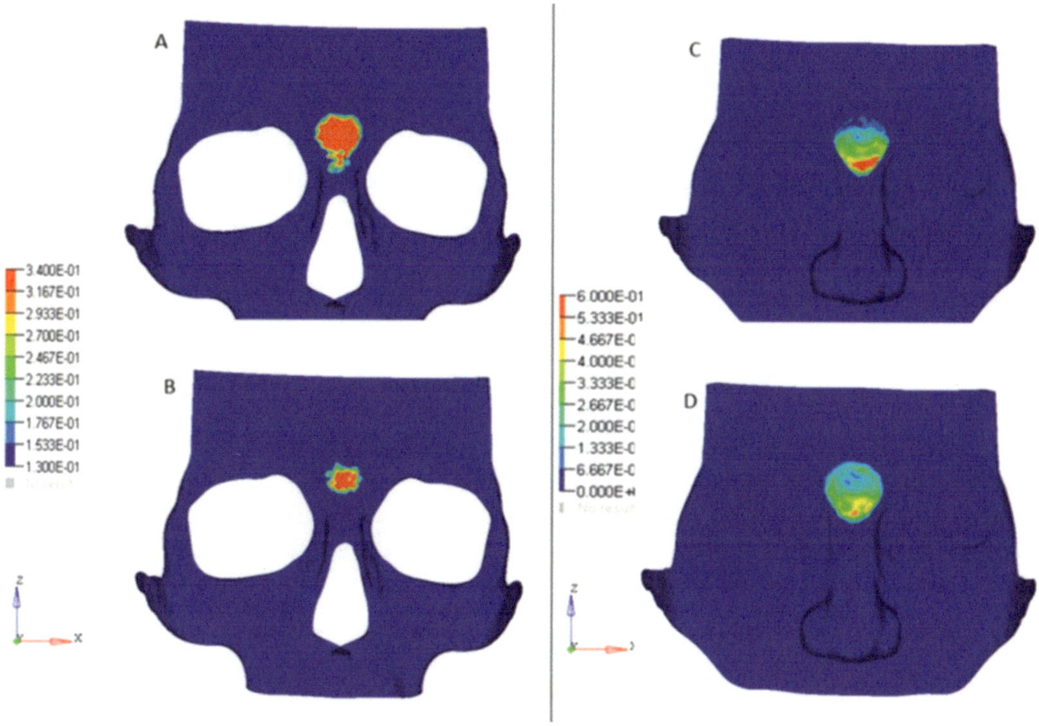

Figure 14. (A, B) Normal compressive stress in the bones of the frontal region, after impact, (A) without and (B) with nasal protector. (C, D) Normal compressive stress in the soft tissue of the frontal region, after the impact, (C) without protector and (D) with protector. Figure is extracted from Coto et al. [37].

4.3. Study of EVA to protect the zygomatic bone

The zygomatic bone forms the prominence of the cheek, part of the lateral wall and floor of the orbit. Due to its location and prominence, it presents a high risk of fracture [39–41]. The thickness is not constant in its extension. The zygomatic bone is composed of cortical and spongy bone in the thicker portion, and in the region near the frontonasal suture, it is almost exclusively formed by the cortical bone [41].

A simplified geometry of overlapping discs with a 100-mm radius was considered. The layers were composed of bone tissue (zygomatic bone portion, lower malar portion, near the nasal front suture), soft tissue and three proposed rigid and flexible EVA combinations, according to **Table 7**.

	Flexible EVA (thickness, mm)	Rigid EVA (thickness, mm)	Flexible EVA (thickness, mm)
G1	2	1	1
G2	3	1	–
G3	2	1	–

Table 7. Configurations analyzed for rigid and flexible EVA.

Figure 15 shows the geometry for G1. An extra configuration formed only by the cortical bone and soft tissue was also included in the analyses as a control group (CG) as shown in **Figure 16**.

Figure 16 also shows the projectile, here represented by a golf ball, with parameters obtained in Bartlett et al. [43] (**Table 8**). The ball had a velocity of 10 m/s at the instant of impact.

The parameters used for the cortical bone, soft tissue, flexible and rigid EVA are in **Tables 3–6**, respectively. The thicknesses of bone and soft tissue were 10.3 mm [42] and 12.3 mm [43], respectively.

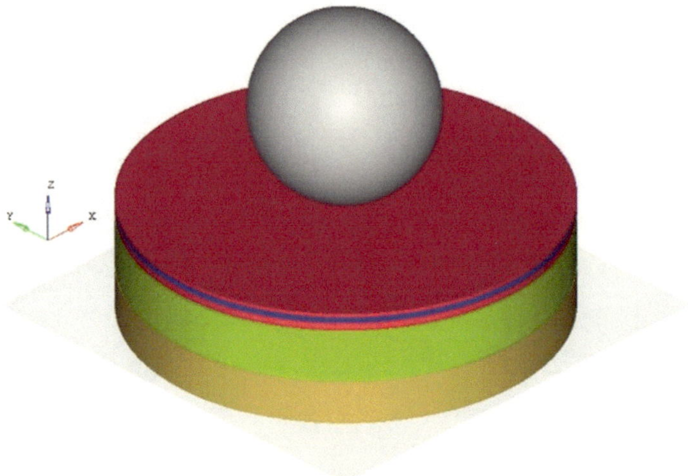

Figure 15. Simplified geometry (Group G1), soft EVA, rigid EVA, soft tissue amd bone.

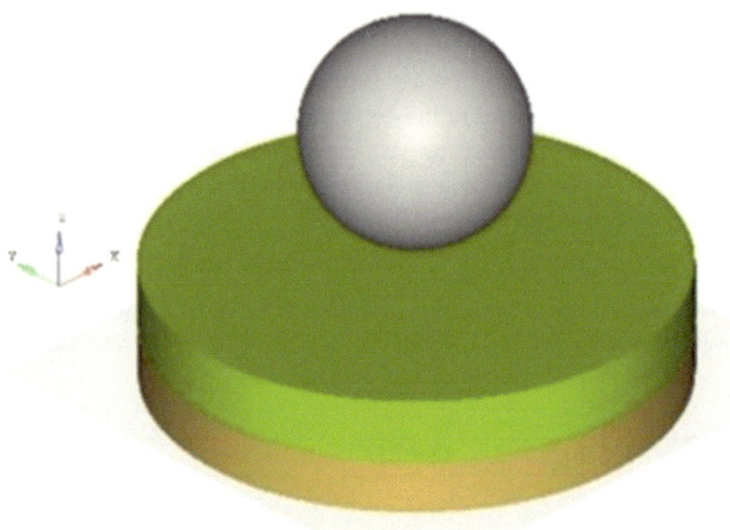

Figure 16. Control Group (CG), soft tissue and zigomatic bone.

http://dx.doi.org/10.5772/intechopen.69969

	Young Modulus (MPa)	Poisson coefficient	Density (t/mm³)	Radius (mm)	Velocity (m/s)
Golf ball	392	0.45	1.15 E − 9	21	10

Table 8. Geometric and material characteristics of the projectile.

The analyses were performed by the LS-Dyna software. The minimum compression stress was controlled. The maximum pressure allowed for the bone and the EVA (rigid or flexible) was of 2.7 MPa and 5.0 MPa, respectively. The friction value considered was 0.5 between ball and soft tissue.

5. FEA results

Figure 17 shows the pressure for the CG. The figure shows the high level of pressure at the zygomatic bone, exceeding the failure limit of 2.7 MPa.

According to the analyses, the results showed that in the three models proposed, there was the maximum performance of EVA, but the best protection to the studied bone is given by the G2 model. **Figure 18** shows the pressure profile in the EVA for G1 and G3.

Figure 19 shows the energy conversion during impact in G2.

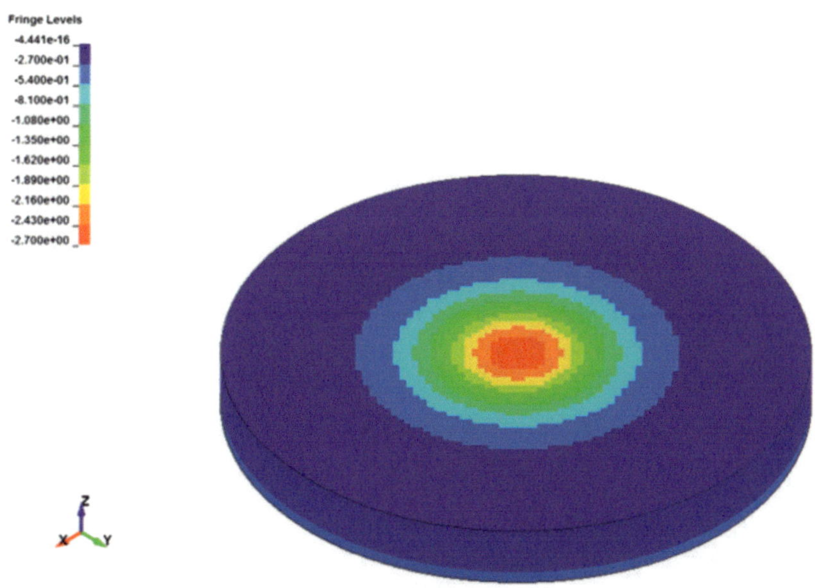

Figure 17. Pressure for the CG.

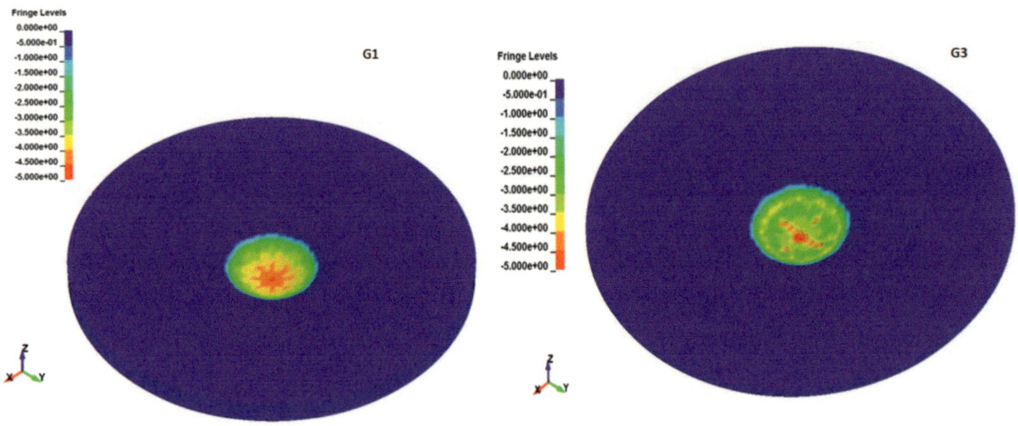

Figure 18. Results of pressure profile in the protector for G1 and G3, respectively.

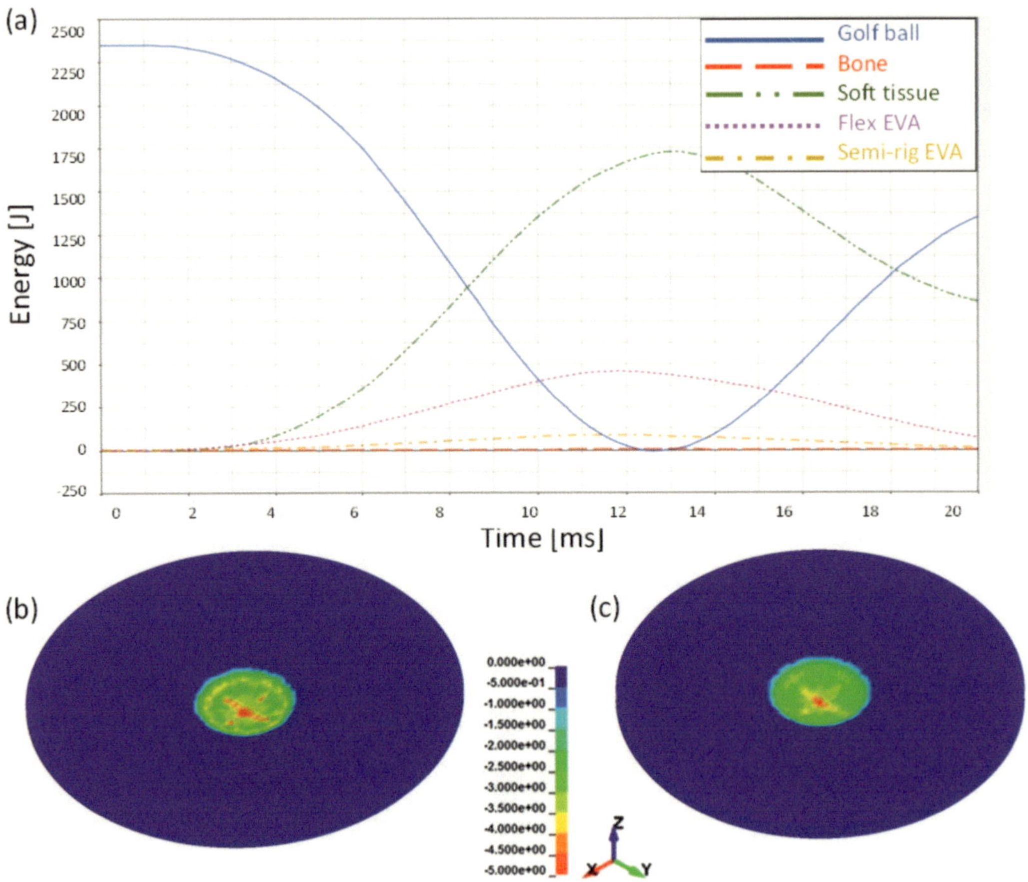

Figure 19. (a) Energy conversion during impact for G2; pressure in the (b) semirigid and (c) flexible EVA.

6. Conclusions

In human health science research, the study of materials that may replace organs is in constant evolution. Particularly in dentistry, the material should be easy to manipulate, esthetically acceptable, stable to use in the oral environment, in the presence of saliva and chewing conditions and low cost. Moreover, mechanical, physical, chemical and biological properties of any material used in the area must be known.

EVA was the object of this study. It is a thermoplastic copolymer derived from petroleum.

Initially, the material was studied in the mouth environment, and it was mechanically and chemically characterized. Finally, the material is molded and applied to facial protection.

The application is numerical, since studies in the biological area involving impact have become impossible to perform *in vivo*. FEM is a powerful tool, able to virtually mimic different complex phenomena. The quality of the results strongly depends on the correct material characterization, precise geometry of the analyzed structure and real boundary conditions (initial velocity of the projectile, displacement restrictions in the system).

The facial protector was tested during the impact of a golf ball in the nasal bone and, through a simplified model, in the zygomatic bone. The proposed protector is able to amortize the impact, and its configuration does not compromise peripheral view and does not cause discomfort to the athlete.

Author details

Reinaldo Brito e Dias[1], Neide Pena Coto[1]*, Gilmar Ferreira Batalha[2] and Larissa Driemeier[2]

*Address all correspondence to: neidecoto@gmail.com

1 School of Dentistry, University of São Paulo, Sao Paulo, Brazil

2 School of Engineering, University of São Paulo, Sao Paulo, Brazil

References

[1] WHO. Available from: http://www.direitoshumanos.usp.br/index.php/OMS-Organização-Mundial-da-Saúde/constituicao-da-organizacao-mundial-da-saude-omswho.html. 2014. [Accessed: June 10, 2014]

[2] Anusavice KJ, Shen C, Rawls R. Phillips materiais dentários. [Roberto Braga, et al. (translator)]. 12th ed. Rio de Janeiro: Elsevier; 2013

[3] Wong EW, White RC. Development of a shock absorbing biomedical elastomer for a new total elbow replacement design. Biomaterials, Medical Devices, and Artificial Organs. 1979;**7**(2):283-290

[4] Coto NP, Dias RB, Costa RA, Antoniazzi TF, de Carvalho EP. Mechanical behavior of ethylene vinyl acetate copolymer (EVA) used for fabrication of mouthguards and inter-occlusal splints. Brazilian Dental Journal. 2007;**18**(4):324-328

[5] Rawls RH. Polimeros odontológicos. In: Nausavice KJ, editor. Phillips materiais dentários. 11th ed. Rio de Janeiro: Elsevier; 2005. pp. 135-157

[6] Coto NP, Meira JBC, Dias RB, Driemeier L, Roveri GO, Noritomi PY. Assessment of nose protector for sport activities: Finite element analysis. Dental Traumatology. 2012 Apr;**28**(2):108-113

[7] Bugada DC, Rudin A. Molecular structure and melting behaviour of ethylene-vinyl acetate copolymers. European Polymer Journal. 1992;**28**(3):219-227

[8] Bhowmick AK, Stephens HL. Handbook of Elastomers—News Developments and Technology. New York: Marcel Dekker; 1988

[9] Zhang X, Zhou Q, Liu H, Liu H. UV light induced plasticization and light activated shape memory of spiropyran doped ethylene-vinyl acetate copolymers. Soft Matter. 2014 Jun 7;**10**(21):3748-3754

[10] Bishop BM, Davies EH, Von Fraunhofer JA. Materials for mouth protector. Journal of Prosthetic Dentistry. 1985;**53**(13):256-261

[11] Chen CP, Lakes RS. Design of viscoelastic impact absorbers: Optimal material property. International Journal of Solids and Structures. 1990;**26**(12):1313-1328

[12] Hoffman J, Alfter G, Rudolph NH, Göz G. Experimental comparative study of various mouth guards. Endodontics & Dental Traumatology. 1999;**15**:157-163

[13] Sports Dentistry Academy [online]. Types of Athletic Mouthguards. 2005. Available from: http.sportsdentistry.com [Accessed: June 4, 2005]

[14] Craig RG, Godwin WC. Physical properties of material for custom made mouth protectors. Journal of the Michigan State Dental Association. 1967;**47**:34-40

[15] Hans GE. An Introduction to Plastics. Weinheim, New York: VHC; 1993

[16] Park MS, Levy ML. Biomechanical aspects of sports-related head injuries. Neurologic Clinics. 2008 Feb;**26**(1):33-43

[17] Matsumoto AT, Driemeier L, Alves M. Performance of polymer reinforcements in vehicle structures submitted to frontal impact. International Journal of Crashworthiness. 2012;**17**:479-496

[18] Le Fort R. Êtude experimentale sur les fractures de la machoire superieure. Revue Chir de Paris. 1901;**23**:208-306

[19] Stanley RB Jr, Nowak GM. Midfacial fractures: Importance of angle of impact to horizontal craniofacial buttresses. Otolaryngology—Head and Neck Surgery. 1985 Apr; **93**(2):186-192

[20] Delilbasi C, Yamazawa M, Nomura K, Lida S, Kogo M. Maxillofacial fractures sustained during sports played with a ball. Oral Surgery, Oral Medicine, Oral Pathology, Oral Radiology, and Endodontology. 2004 Jan;**97**(1):23-27

[21] Higuera S, Lee EI, Cole P, Hollier LH Jr, Stal S. Nasal trauma and the deviated nose. Plastic and Reconstructive Surgery. 2007 Dec;**120**(7 Suppl 2):64S-75S

[22] Chao MT, Paletta C, Garza JR. Facial trauma, sports-related injuries. Medscape Journal of Medicine. 2008;**1**:1-14

[23] Ellis E 3rd, Kittidumkerng W. Analysis of treatment for isolated zygomatic maxillary complex fractures. Journal of Oral and Maxillofacial Surgery. 1996 Apr;**54**(4):386-400

[24] Garza JR, Baratta RV, Odinet K, Metzinger S, Bailey D, Best R, et al. Impact tolerances of the rigidly fixated maxillofacial skeleton. Annals of Plastic Surgery. 1993 Mar;**30**(3):212-216

[25] Levin L, Friedlander LD, Geiger SB. Dental and oral trauma and mouthguard use during sport activities in Israel. Dental Traumatology. 2003 Oct;**19**(5):237-242

[26] Ranalli DN, Demas PN. Orofacial injuries from sport: Preventive measures for sports medicine. Sports Medicine. 2002;**32**(7):409-418

[27] Cascone P, Petrucci B, Ramieri V, Marianetti TM. Security hi-tech individual extra-light device mask: A new protection for [soccer] players. Journal of Craniofacial Surgery. 2008 May;**19**(3):772-776

[28] Dingman RO, Natvig P. Cirurgia das fraturas faciais. São Paulo: Editora Santos; 2001

[29] Crow RW. Diagnosis and management of sports-related injuries to the face. Dental Clinics of North America. 1991 Oct;**35**(4):719-732

[30] Morita R, Shimada K, Kawakami S. Facial protection masks after fracture treatment of the nasal bone to prevent re-injury in contact sports. Journal of Craniofacial Surgery. 2007 Jan;**18**(1):143-145

[31] Handbook ES. Coefficients of Friction—Many Material Compared. 2006;**18**:942

[32] Hodgson VR. Tolerance of the facial bones to impact. American Journal of Anatomy. 1967;**120**:113-122

[33] Miller K, Chinzei K. Constitutive modelling of brain tissue: Experiment and theory. Journal of Biomechanics. 1997 Nov–Dec;**30**(11-12):1115-1121

[34] Verdejo R, Mills NJ. Heel-shoe interactions and the durability of EVA foam running-shoe midsoles. Journal of Biomechanics. 2004 Sep;**37**(9):1379-1386

[35] Gerard JM, Ohayon J, Luboz V, Perrier P, Payan Y. Non-linear elastic properties of the lingual and facial tissues assessed by indentation technique. Application to the biomechanics of speech production. Medical Engineering & Physics. 2005 Dec;**27**(10):884-892

[36] Zahouani H, Pailler-Mattei C, Sohm B, Vargiolu R, Cenizo V, Debret R. Characterization of the mechanical properties of a dermal equivalent compared with human skin in

vivo by indentation and static friction tests. Skin Research and Technology. 2009 Feb; **15**(1):68-76

[37] Coto NP, Driemeier L, Roveri GO, Meira JBC, Dias RB, Noritomi PY. Numerical study of the face bone behaviour when impacted by rigid ball. Journal of Biomechanics. 2012 Jul;**45**:1121

[38] Withnall C, Shewchenko N, Gittens R, Dvorak J. Biomechanical investigation of head impacts in football. British Journal of Sports Medicine. 2005 Aug;**39**(Suppl 1):i49-i57

[39] Verschueren P, Delye H, Depreitere B, Van Lierde C, Haex B, Berckmans D. A new test set-up for skull fracture characterisation. Journal of Biomechanics. 2007;**40**(15):3389-3396

[40] Gialain IO, Dias RB, Coto NP. Mouthguard: A new technique for the partially edentulous patient. Dental Traumatology. 2014 Oct;**30**(5):411-414

[41] Del Neri NB, Araujo-Pires AC, Andreo JC, Rubira-Bullen IR, Ferreira Júnior O. Zygomaticofacial foramen location accuracy and reliability in cone-beam computed tomography. Acta Odontologica Scandinavica. 2014 Feb;**72**(2):157-160

[42] de Almeida NH, Michel-Crosato E, de Paiva LA, Biazevic MG. Facial soft tissue thickness in the Brazilian population: New reference data and anatomical landmarks. Forensic Science International. 2013 Sep 10;**231**(1-3):404

[43] Bartlett R, Gratton C, Rolf CG, editors. Encyclopedia of International Sports Studies. Abington: Routledge; 1987

Chapter 15

Tailoring Bioengineered Scaffolds for Regenerative Medicine

Sandra Amado, Pedro Morouço,
Paula Pascoal-Faria and Nuno Alves

Additional information is available at the end of the chapter

http://dx.doi.org/10.5772/intechopen.69857

Abstract

The vision to unravel and develop biological healing mechanisms based on evolving molecular and cellular technologies has led to a worldwide scientific endeavor to establish regenerative medicine. This is a multidisciplinary field that involves basic and preclinical research and development on the repair, replacement, and regrowth or regeneration of cells, tissues, or organs in both diseases (congenital or acquired) and traumas. A total of over 63,000 patients were officially placed on organs' waiting lists on 31 December 2013 in the European Union (European Commission, 2014). Tissue engineering and regenerative medicine have emerged as promising fields to achieve proper solutions for these concerns. However, we are far from having patient-specific tissue engineering scaffolds that mimic the native tissue regarding both structure and function. The proposed chapter is a qualitative review over the biomaterials, processes, and scaffold designs for tailored bioprinting. Relevant literature on bioengineered scaffolds for regenerative medicine will be updated. It is well known that mechanical properties play significant effects on biologic behavior which highlight the importance of an extensively discussion on tailoring biomechanical properties for bioengineered scaffolds. The following topics will be discussed: scaffold design, biomaterials and scaffolds bioactivity, biofabrication processes, scaffolds biodegradability, and cell viability. Moreover, new insights will be pointed out.

Keywords: tailored scaffold, biomaterials, bioprinting, biomechanics, regenerative medicine

1. Introduction

In a society that is in constant development, the discovery of "new" scientific and technological knowledge must (i) progress at an incredibly fast pace, (ii) target a wide audience, and (iii) have a practical impact in the society. The health sciences are naturally a priority area of research, mostly because of the impact they have on the augment human life expectancy, by developing advanced and patient-specific therapies.

Only the complexity of human tissues could justify that in the 1980s tissue engineering emerged as a scientific field with an enormous potential. Targeting to regenerate the bone, cartilage, skin, or other tissues and organs, bridging the anatomy with its physiology/function is a paramount challenge to be solved. Several efforts have been made, by research groups spread worldwide, to tailor bioengineering scaffolds (sometimes denominated by tissue constructs) that could mimic native tissues. However, the achievement of three-dimensional (3D) complex organ structures is far from being tangible. Due to its nature, tissue engineering gathers scientists, engineers, and physicians in multidisciplinary teams using a variety of methods to construct biological substitutes [1]. Indeed, significant efforts are being developed worldwide in the fields of tissue engineering and regenerative medicine, but full tissue or organ regeneration remains a paramount challenge. Therefore, these multidisciplinary scientific fields apply a wide variety of methodologies, where multidisciplinary research teams can provide suitable inputs for its development [2].

One of the major goals is to produce biological substitutes to restore, maintain, or improve tissue function, using biocompatible and biodegradable support structures, i.e., scaffolds, in conjunction with human cells (**Figure 1**) [3]. Gathering tissue engineering and regenerative medicine, researchers have been interested on developing alternative approaches for restoring functionality. To do so, one of the most promising methodologies involves the use of additive manufacturing (AM) processes. AM technologies allow the production of complex 3D structures concerning mainly a high level of control, predefined geometry, size, and interconnected pores in a reproducible way. This optimized controlled organization enhances the vascularization and, thus, transports oxygen and nutrients throughout the whole structure,

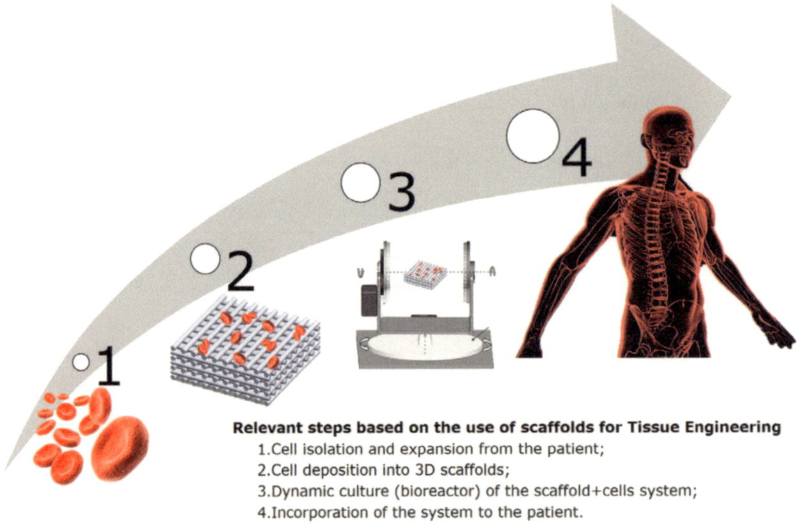

Figure 1. Relevant steps based on the use of scaffolds for tissue engineering.

providing an adequate biomechanical environment for tissue regeneration [4]. However, adapting the adequate technology with enhanced biomaterials, in order to obtain customized implants that mimic the native tissue, is nowadays a challenge with a huge potential.

This chapter intends to provide a synopses in patient-specific engineering scaffolds. A revision of the scaffold design, biomaterials, and advanced manufacturing processes will help to establish new research paradigms on tailoring bioengineered scaffolds for regenerative medicine. Recent advances will be highlighted to stimulate the readers for future insights and possibilities.

2. Scaffold design

Scaffold modeling plays a key role in tissue engineering and regenerative medicine. A well-designed 3D scaffold is a fundamental tool to guide tissue formation both in vitro and in vivo. Properties such as high surface-area-to-volume ratio, porosity, pore size, pore design, pore interconnectivity, permeability, and degradation should be taken into account when designing scaffold for different and tailored applications. These will allow a desirable biological network for cell migration, nutrient transportation, and the mechanical stiffness, and strength can be therefore obtained [5, 6]. Growth factors (GFs) and drug release (DR) should also be considered to achieve an optimized tissue growth as scaffold degraded. Moreover, some authors have shown the benefits for tissue generation of using curvature and concave surfaces compared to convex and planar ones [7].

To address and fulfill aforementioned requirements, two scaffold design approaches can be used according to the flowchart presented in **Figure 2**. The first one is based on the native tissue, whereas the second one is based on the unit digital cell model, both addressing tailored scaffold geometry. The geometry obtained can then be used on computer-aided engineering studies to optimize the performance of the tailored bioengineering scaffold. Finally, a physical optimized scaffold can be produced using 3D printing or AM technology before in vitro and/or in vivo implantation of the scaffold. Accordingly, several research works have been developed concerning tailored scaffold geometry and its fabrication. In these studies, physical scaffolds have been used directly for in vitro and/or in vivo studies. Nevertheless, the link between computer modeling and computer-aided engineering to tailor bioengineering scaffold remains a paramount challenge. When solved, it can significantly reduce animal experimental studies.

In the computer modeling based on native tissue, different noninvasive 3D scanning techniques can be considered to obtain the 3D anatomical geometric model. The most used are computed tomography (CT), μCT (micro-CT), and nCT (nano-CT), which considered different scale levels [8–10], as well as magnetic resonance image (MRI) and 3D optical techniques. All these techniques used different physical principles to obtain a series of two-dimensional (2D) images or a 3D point cloud of the sample of the native tissue studied. CT requires the exposition of the sample of the native tissue to ionizing radiation, whereas MRI uses a magnetic field and pulses of radio wave energy (avoiding radiation) both obtaining a series of 2D images. In

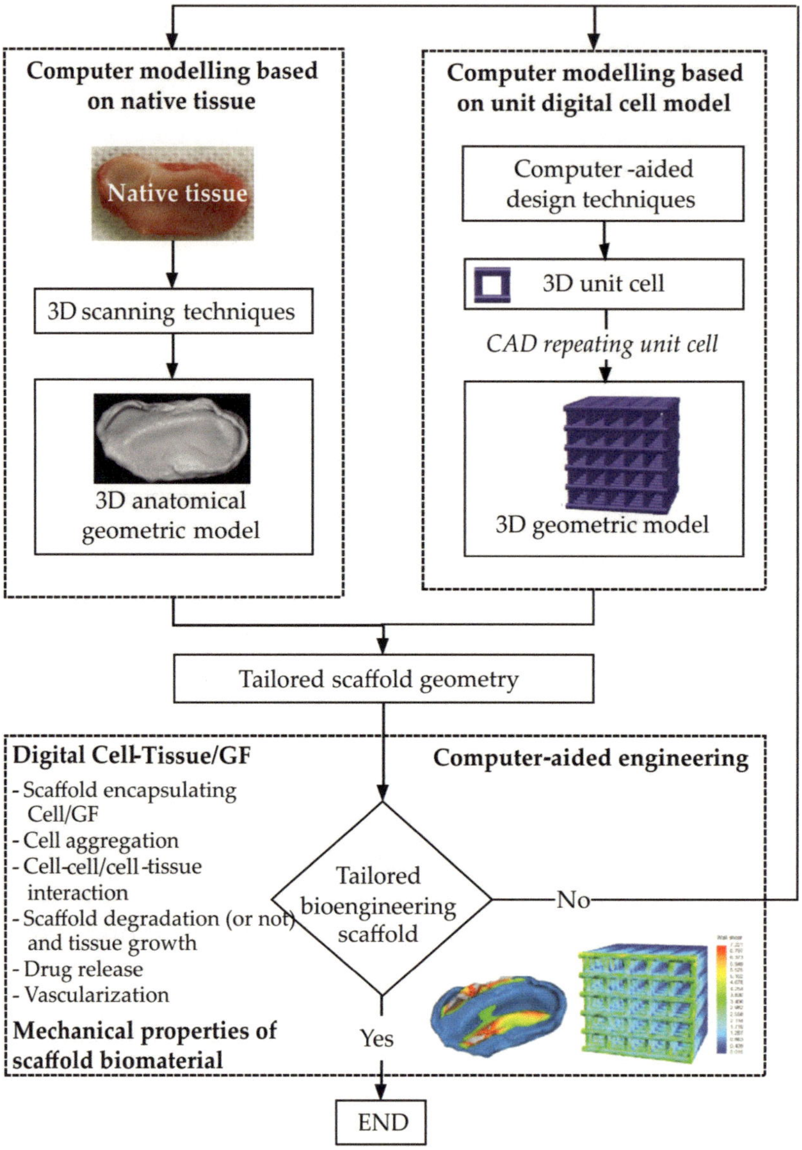

Figure 2. Computer modeling and simulation to tailor bioengineering scaffolds.

CT, these images are displayed by density, while in MRI they are compiled and segmented by its signal intensity. Additionally, both techniques can be differentiated by its resolution. The high resolution of CT allows the characterization of the micro-architecture and the mechanical properties of the tissue scaffolds [11]; however, this technique has a drawback regarding soft tissues of similar density. It is more efficient in differentiating hard tissues with sharply defined density changes, such as the interface between bone and soft tissues. To overcome this problem, contrast agents can be added [6]. Although the resolution of MRI is inferior to CT scans, with the advance of technology, it is improving, allowing the 3D representation of internal structures, such as the central nervous system, heart, and kidneys of a rat [9]. The 2D individual images obtained using the previous techniques described are then assembled and realigned, and therefore a 3D geometric anatomical model distinguishing different types of tissue is obtained.

Microscopy optical technique is also used to obtain the 3D anatomical geometric model of the native tissue. However, it can only differentiate every type of tissue down to the level of the individual cell at the cost of a huge computational effort. Other 3D optical techniques, such as 3D structured light, cannot differentiate the types of tissues presented and only allow the generation of the outer 3D geometric model of the native sample. In the case of the native tissue sample (temporomandibular joint disc (TMJ disc)) presented in the flowchart of **Figure 2**, a 3D point cloud was obtained using a white light 3D scanning system (Steinbichler—COMET 5®), and then an appropriate software was applied to replicate the 3D geometric anatomical model of the TMJ disc [12].

Hybrid modalities can also be used to construct the 3D model of the same specimen in order to take the advantage of each technique to differentiate the different types of tissues [9].

The second approach uses computer-aided design (CAD) techniques to create a 3D unit cell which is used as pattern. Then, a desired number of patterns are automatically generated and combined until a complete 3D geometric model of the scaffold is obtained with controlled architecture (**Figure 2**). Following this approach, the main scientific achievements reported are based on permanent or temporary scaffolds.

Permanent tailored engineering scaffolds have been designed mainly for bone repair of large segmental defects caused by fracture, tumor, or infection. In 2013, Wieding et al. [13] reported a numerical study which is used to determine the suitability of open porous of titanium scaffolds to act as bone scaffolds under physiological loading conditions. Uniaxial compression structural modulus of the titanium scaffolds was tailored ranging from 3.5 to 19.1 GPa as a function of the scaffold porosity from 64 to 80%. Results revealed that minimizing the amount of material of the inner core had a smaller influence than increasing the porosity when the scaffolds were under biomechanical loading. It was also noted that the scaffold design could act similarly to the intact bone. In order to tailor the mechanical properties of cellular structured scaffolds, [14] designed metal scaffolds with high porosity (62–92%) to tailor both compressive strength (4.0–113.0 MPa) and elastic modulus (0.2–6.3 GPa), respectively, were comparable to trabecular and cortical bone. Porous titanium scaffolds were also investigated by van der Stok et al. [15, 16] for grafting large bone defects. Mechanical properties were tailored, whereas high porosity of the scaffold allowed the incorporation of colloidal gelatin gels for time- and dose-controlled delivery of dual growth factors (bone morphogenetic protein-2 (BMP-2) and/or fibroblast growth factor-2 (FGF-2)), promoting a quasi-full bone regeneration. The scaffold was designed based on a decahedron pattern and composed by 120-μm-thick titanium struts with porous size ranging from 240 to 730 μm. Porous size, porosity, porous volume, compression strength, and Young's modulus were 490 μm, 88%, 55 mm^3, 14 MPa, and 0.4 GPa, respectively, allowing to achieve an optimized bone volume regeneration (~50 mm^3) for a composite scaffold with BMP-2/FGF-2. In 2012, Van Bael et al. [17] developed six distinct geometries of Ti6Al4V scaffolds in three different pore shapes (triangular, hexagonal, and rectangular) and two different pore sizes (500 and 1000 μm) aiming to understand the effect of pore geometry of Ti6Al4V bone scaffolds on the in vitro biological behavior of human periosteum-derived cells. The main result showed that a functional Ti6Al4V-graded scaffold, with specific morphological and mechanical properties, will contribute to enhance cell seeding and at the same time can maintain nutrient transport throughout the whole scaffold during in vitro culturing by avoiding pore occlusion.

Temporary (or biodegradable) tailored engineering scaffolds have been designed as a tissue engineering approach that uses degradable porous biomaterial incorporating biological cells and/or molecules to regenerate tissues such as the bone, cartilage, skeletal muscle, nerve, and blood vessels. Scaffold design must be able to create hierarchical porous structures to fulfill all mechanical and biological requirements. In 2005, Hollister [18] introduced the concept of hierarchical scaffold design as geometric features at scales from the nanometer to millimeter level that will determine how well the scaffold meets conflicting mechanical function and mass transport needs. In 2011, Khoda et al. [19] developed a functionally gradient variational porosity architecture (hierarchical design) for hollowed scaffolds. In 2014, Giannitelli et al. [20] reviewed tailored scaffold architecture with microstructural features. Authors highlighted the growing interest in the development of innovative scaffold designs to overcome often conflict requirements (such as biological and mechanical ones). Considering different pore size gradients, Sobral et al. [21] designed and manufactured. The goal was to enhance cell seeding efficiency and control the spatial organization of cells within the scaffold. Some authors [22] also emphasized the importance of scaffold pore size gradients in osteogenic differentiation of human mesenchymal stromal cells. In 2010, Puppi et al. [23] in deep reviewed the design of biodegradable and bioactive polymeric scaffolds, with properly suited architecture and tailored properties for bone and cartilage tissue regeneration. According to the authors, a good scaffold design must account that macro- and microstructural properties affect cells survival, signaling, growth, propagation, and reorganization and play also a major role in modeling cell shape and gene expressions, both related to cell growth and preservation of native phenotypes [24, 25]. In addition, several scaffold designs were developed and then manufactured using different AM processes. For example, Fierz et al. [26] designed three labeled anisotropic 3D hydroxyapatite scaffolds (pixel-wise and labeled layer-wise) with tailored pores ranging from the nanometer to millimeter scale for the reconstruction of centimeter-sized osseous defects. Seventy percent micrometer-wide pores were successfully interconnected, and virtual spheres (diameter of up to 350 ± 35 μm) were used to simulate cell migration along the pores linked with central channel. Melchels et al. [27] designed poly-DL-lactic acid (PDLLA) porous scaffolds with a gyroid architecture. This architecture was mathematically defined, allowing a precise control of porosity and pore size of a fully interconnected pore network. As noted by the authors, cell seeding of porous structures prepared from hydrophobic polymers, such as PDLLA, was difficult. Moreover, the penetration of a cell suspension was further hindered by the high tortuosity and poor interconnectivity of pore networks when manufactured by salt-leaching or freeze-drying conventional methods. Therefore, very open scaffold structure of the gyroid architecture that facilitates the penetration of water into PDLLA scaffold was manufactured by stereolithography. It was highlighted that the cells were well attached and homogeneously distributed throughout the porous scaffold. Good mechanical properties can be tailored in predesigned (porous) architectures from PDLLA based on gyroid architecture.

In 2012, Melchels et al. [28] reviewed additive manufacturing of tissues and organs. Authors also addressed tailored engineering scaffolds for breast reconstruction, focusing pore size and porosity for the generation of three scaffold models. Cipitria et al. [29] developed a poly (ε-caprolactone) (PCL) scaffold incorporating recombinant human bone morphogenetic protein 7 (rhBMP-7) for the regeneration of critical-sized defects in sheep tibiae. PCL scaffold with b-tricalcium phosphate (mPCL-TCP) to promote bone regeneration was designed based on a honeycomb structure with large interconnected pores to facilitate cellular bridging, ingrowth of bone tissue, and efficient mass transport and vascular infiltration. Moreover, Domingos et al. [30] developed PCL scaffolds for tissue engineering purposes. Authors addressed internal/external scaffold geometry, different material deposition strategies, and the biocompatibility of the material used. 3D PCL porous scaffolds (rectangular porous prisms) were designed with an average porosity of ~76% using commercial computer-aided design software. These structures were then produced via bioextrusion in a 0/90 lay-down pattern trying to reproduce a honeycomb-like pattern of fully interconnected square pores. Similar bioextruded scaffolds were designed (regular dimensions of 600 × 600 mm) to have a well-defined internal geometry with square interconnected pores and uniform distribution. The overall porosity of the structures was found to be ~76%. In vitro degradation of the scaffold was studied as a function of the degradation environment, pore size, and geometry [30, 31]. Scaffold degradation plays a key role when tailoring scaffold properties. In 2016, Morouço et al. [32] developed three types of PCL scaffolds reinforced with cellulose nanofibers (CNF), with and without the addition of hydroxyapatite nanoparticles (HANP), aiming to tailor scaffold properties for tissue engineering applications. The authors studied scaffold porosity, mechanical properties, and biocompatibility as a function of three material combinations. PCL, PCL/CNF, and PCL/CNF/HANP scaffolds were described with porous fully interconnected and porosity (%) of 49.0, 49.5, and 50.0; compressive modulus (MPa) of 54.42, 64.58, and 70.88; and maximum compressive stress (MPa) of 10.96, 11.35, and 12.12, respectively. These structures were then produced via bioextrusion in a 0/90 lay-down pattern. Some authors [33] studied hybrid hierarchical 3D scaffolds with well-controlled architecture for both macro- and microscale. Hybrid and hierarchical 3D structures include thick filaments with the diameter of hundreds of microns, and thin filaments with sub-10 μm dimensions were developed. The microscale features can help in cell seeding, alignment, and guidance. Trying to mimic morphological and mechanical behavior of a blood vessel, Vaz et al. [34] proposed a tailored tissue engineering scaffold. Design parameters such as bilayered tubular scaffold, stiff and oriented outside fibrous layer, and a pliable and randomly oriented fibrous inner layer were considered, combining two biomaterials (PLA/PCL). Structural and mechanical properties of the scaffolds were examined using scanning electron microscopy (SEM) and tensile testing. Cell viability was investigated using 3T3 mouse fibroblasts and the tubular scaffold in an appropriated in vitro environment. The proposed scaffold presented appropriate characteristics to be considered a candidate for blood vessel tissue engineering. Other authors also proposed to tailor tissue engineering scaffolds trying to mimic extracellular matrix morphology of natural tissue for blood vessel applications [35, 36].

In 2012, Chantarapanich et al. [37] developed a computer-aided design library based on polyhedrons for tissue engineering applications. Close-cellular scaffold included truncated octahedron, rhombicuboctahedron, and rhombitruncated cuboctahedron, while open-cellular scaffold included hexahedron, truncated octahedron, truncated hexahedron, cuboctahedron, rhombicuboctahedron, and rhombitruncated cuboctahedron. Both relationship between pore size and porosity of close-cellular scaffolds and relationship between pore size/beam thickness and porosity of open-cellular scaffolds were studied. The study concluded that some design combinations were not good for making the open-cellular scaffold, generating enclosed pores inside the scaffold, and, therefore, they were excluded from the digital library. Compressive stresses were computed as a function of polyhedron-based geometries which can also be helpful for tailoring mechanical properties of the scaffolds.

In the computer-aided engineering based on tailored scaffold geometry, several digital features should be taken into account to obtain computer-tailored bioengineering scaffolds. Such features encompass cell and growth factor encapsulating, cell aggregation, cell-cell and cell-tissue interaction, vascularization, scaffold degradation (or not if permanent) and tissue growth, drug release, and scaffold mechanical behavior (**Figure 2**). To help digital prediction of cell/tissue phenomena, several automated methods exist, namely, cell counting, cell geometry determination, chromosomal counting, correlation of DNA expression determined through microarrays, interpreting fluorescence data, determining cell's lineage, and cross correlating gene expression with predicted in vivo pathology. All of these features have predictive value for determination of tissue viability and the differentiative rate of cells seeded with the goal of tissue culture. A detailed description about both accumulation of the expression data and large-scale computer cross correlation (between this expression and expressions commonly used in pathology) is provided in Ref. [9], as well as a number of specific tools for tissue analysis/identification.

Despite of the aforementioned research works, new scaffold designs integrating cell/GF/tissue phenomena and scaffold mechanical behavior (geometric characteristics and materials) are needed for regenerative medicine. These complex hierarchical 3D structures must be designed according to the structural heterogeneity of the host tissue and/or scaffold environment.

3. Biomaterials and scaffold bioactivity

Tissues possess different structures and properties that a tissue engineering scaffold should be tailored to. A general requirement for all biomaterial scaffolds is to reproduce an extracellular matrix (ECM) environment for supporting cell growth outside of the body. Moreover, scaffold should host cell adhesion, proliferation, and ECM production. Hence, the scaffold should surrogate the missing ECM. Tissue engineering products can be designed to conduct, induct, or block tissue responses and architectures [38]. Besides providing the three-dimensional growth of cells in an organized way, an ideal scaffold should be characterized by biocompatibility, biodegradability, appropriate mechanical properties, interconnectivity of pores with appropriate size to retain cells, and low exchanges of nutrients and waste products [39]. Tailoring biomaterials for enhanced biofunctionality can be achieved using a variety of approaches that involve the introduction of chemical, topographical, or mechanical cues via top-down or bottom-up approaches [40]. Therefore, the selection of the starting materials and of the fabrication techniques is of paramount importance. Numerous natural and synthetic materials can be used for the fabrication of scaffolds including polymers, ceramics, bioactive glass, calcium phosphates, and biometals. For example, scaffolds fabricated from bioactive ceramic materials such as hydroxyapatite and tricalcium phosphate show promise because of their biological ability to support bone tissue regeneration. However, the use of ceramics as scaffold materials is limited because of their inherent brittleness and difficult processability [39]. In 2006, Rezwan et al.'s [41] review showed that conventional material processing methods have been adapted and extended for incorporation of inorganic bioactive phases into porous and interconnected 3D polymer networks. The biomaterials were extended from purely synthetic materials to material/biologic hybrids, engineering at the same time bioactivity and biodegradability [41]. Addressing this issues, in 2015, Fiedler et al. [42] focused on the mechanical characterization of PCL-bioglass composites and concluded that the addition of bioglass was found to decrease the elastic gradient and yield stress if two scaffolds of the same density are compared and the highest bioglass content (35%) seems beneficial as it (i) does not significantly deteriorate the scaffold mechanical properties and (ii) promotes bioactivity.

The next generation of synthetic biodegradable, bioactive, living composite biomaterials that feature high adaptiveness to the biological environment [41] considers the incorporation of biomolecules as promising and is currently under extensive research. Incorporating biomolecules such as growth factors during scaffold processing with the aim to accelerate local tissue healing however are not simple as biomolecules are sensitive to elevated temperatures and extreme chemical conditions. A promising strategy is the immobilization of proteins and growth factors in the post-processing phase via surface functionalization of the scaffold [43].

"Soft" material routes like sol-gel processing might be a strategy to incorporate biomolecules during scaffold fabrication. To the authors' knowledge, however, sol-gel-derived bioactive organic/inorganic hybrids have not yet been formed into highly interconnected porous structures, which would be essential for application of these composites as scaffolds. Another related challenge was the elucidation of the local impact of growth factors on the cell and tissue systems, including long-term effects [41]. As pointed out in Section 2, mechanical property is one of the most critical parameters that determine the performance of a designed implant. It mainly depends on the process and structural properties of the biomaterials. Therefore, it is possible to achieve desired mechanical properties through modifying the structural characteristics of a biomaterial. Biological behavior of cell assessment after surface modifications is required to check its biocompatibility and bioactivity [38]. The study of the interactions of biochemical and geometrical cues on stem cell differentiation and alignment should be also considered. The capability to spatially control stem cell orientation and differentiation toward multiple phenotypes simultaneously, i.e., myocyte, tenocyte, and osteoblast, allows cells grown in vitro to more closely mimic aspects of native tissue organization and structure [44]. Although the precise mechanism behind geometry-induced cell alignment is presently unknown, it is likely that the alignment of cells observed on fibers may be attributed to a combination of factors including physical space constraint and relative stiffness of the underlying substrate (fiber); ultimately affecting changes in both cell spreading and cell stiffness, cells may be predisposed toward a specific orientation through the modulation of mechanotransduction pathways via cytoskeletal rearrangements [9].

Multilayer scaffolds and combinations of several biomaterials are a better option to create graded structures that resemble the biological interface. The development of multilayer scaffolds and the controlled release of bioactive molecules to promote in situ regeneration of biological tissues are some of the latest technologies that intended to improve on the available traditional treatments. To confirm the potential of these novel approaches, long-term evaluation is necessary with special focus on studying the biological and mechanical properties of the synthesized tissues [45].

Scaffolds should be designed more as a bioactive system rather than just passive cell carriers. Thus, integration of fabrication techniques with surface modification may also act as route to obtain nanofibrous scaffolds with better understanding of cell scaffolds both in vivo and in vitro. Similarly and significantly, the biomaterial as design strategy can be used in a better way to relate science and engineering, and use this advanced knowledge to engineer more advanced tissue scaffolds [46, 47].

4. Biofabrication processes

Aiming to tailor bioengineering scaffolds that closely mimic the native tissues, AM technologies are suitable to dispense biomaterials (with live cells or cell aggregates) at specific, and desired, locations [48]. The usage of these technologies has been commonly divided in three categories: (i) the jet-based techniques, (ii) robotic dispensing techniques, and (iii) laser-induced forward transfer [49, 50]. Each of these techniques has advantages and drawbacks (**Table 1**). Thus, understanding its limitations and potentials is a must-do to choose the right approach for the specific tissue that is aimed to regenerate. Furthermore, some advancements have been recently achieved with integrated/hybrid systems. These systems combine different techniques within the same equipment aiming to generate a multifunctional graded construct with tailored properties similar to the native tissue.

In the available literature, it is possible to find investigations using different approaches, for the same type of tissue. In fact, there seems to be a trend to some research groups get specialized in some type of technology and use it for various goals.

	Jet based	Robotic dispensing	Laser-induced forward transfer
Resolution	+	+/–	++
Fabrication speed	+/–	++	–
Hydrogel viscosity	–	+	+/–
Gelation speed	++	+/–	++
Cell density	–	+	+/–

Adapted from Ref. [51].

Table 1. Comparison of the three AM approaches for tissue engineering.

Regarding the jet-based techniques, with a common resolution of 10–50 μm, it is difficult to obtain an adequate structural support. It consists of dispensing a jet of small droplets of liquid material, also called as bioink, in a spatially controlled manner. There are two different approaches, thermal inkjet printing and piezoelectric-actuated inkjet printing, having the former lower suitability for 3D bioengineering scaffolds. Using a piezoelectric actuator, research has been able to suppress some of the thermal constraints [51]. For instance, good viability of printed cell populations was obtained for human fibroblast cell line [52], and recently a silk-based ink eliminated the usage of any cytotoxic organic or inorganic solvents [53]. Even though jet-based techniques are the pioneer techniques used for tissue engineering, translating it to the construct of large 3D structures is a challenge to overcome, mostly because of the low-viscous solutions that do not provide strong and complex 3D structures.

The most successful attempts to engineer cell-containing bioengineering scaffolds have been achieved through robotic dispensing systems. These technologies are based on a controlled extrusion of a material in a continuous fashion, instead of liquid droplets (**Figure 3**) and are developed at the Center for Rapid and Sustainable Development (CDRSP) of the Polytechnic Institute of Leiria, Portugal. Therefore, this approach enables the printing of hydrogels encapsulating cells in a very controlled architecture [28]. The most common methods are the pneumatic [54] or mechanical [55] dispensing systems, comprising (i) a dispensing system and a stage with the capability of moving along the *x*, *y*, and *z* axes; (ii) a light source to illuminate the working area and/or for photoinitiator activation; and (iii) a piezoelectric humidifier [56], with some of them using multiple printing heads to permit the dispensing of various materials without retooling [57]. However, researcher should bear in mind that optimal balance should be aimed between pressure and nozzle size (to obtain higher cell viability).

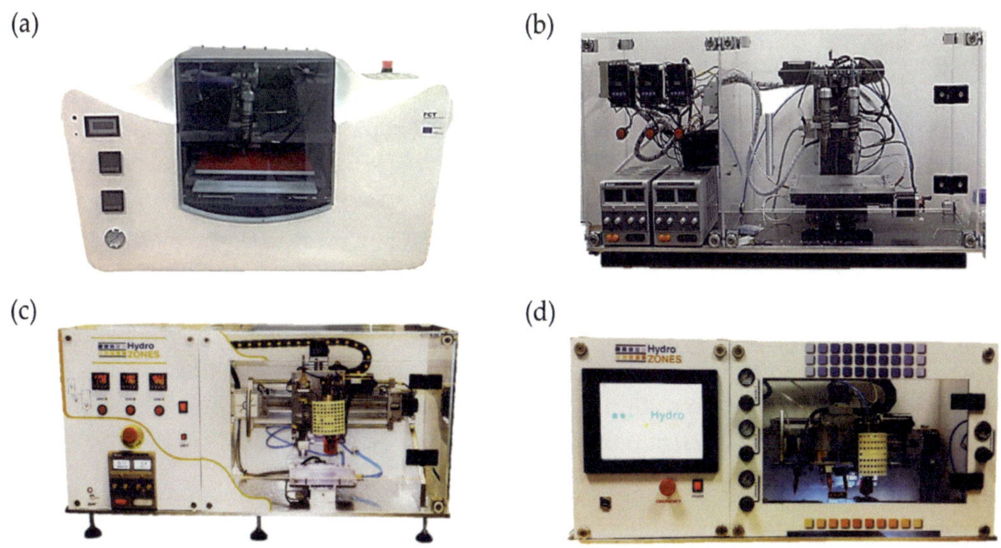

Figure 3. Equipments developed at the CDRSP, Portugal: (a) one-head extrusion system, (b) dual-head extrusion system, (c) system combining an extrusion head and a syringe for hydrogel deposition, and (d) hybrid system combining extrusion with up to three hydrogels.

Recently, a US research group presented an integrated tissue-organ printer (ITOP) for the production of human-scale tissue bioengineering scaffolds of any shape, in a single structure [50]. Combining different procedures, it was possible to successfully engineer (i) a mandible bone, (ii) an ear-shaped cartilage, and (iii) a skeletal muscle. Furthermore, some institutions are now combining different technologies, for instance, merging a robotic dispensing system with a jet-based printing for muscle-tendon unit repair [58].

Lastly, the laser-induced forward transfer (LIFT) is a not-so-common technology for tissue engineering, but is gaining significant importance in this domain [56]. It is based on using three layers of different components: the first layer based on a donor slide, covered by a laser energy-absorbing layer and completed with a cell-bioink component [51]. There are three main advantages for this technology: it is suitable for using (i) a wide range of materials, (ii) a very high precise deposition (but in small 3D structures), and (iii) a clog-free process without the use of nozzles. However, it requires a rapid gelation process, and researchers should bear in mind that several factors should be considered (e.g., laser wavelength, bioink viscosity; for more info read Ref. [59]). Apart from these constraints, successful cell viability (>90%) has been reported for printing skin cell lines and human mesenchymal stem cells and to prepare a cardiac patch [60].

5. Scaffold biodegradability and cell viability

The main objective of tissue engineering is to allow the cells of the body to replace the implanted scaffold over a period. Because bioengineering scaffolds are not intended as permanent implants (besides some of them have shown good results mainly in bone regeneration), they must therefore be biodegradable, so that the need of surgical removal can be avoided. Furthermore, the degradation products should be nontoxic and should be able to swiftly exit from the body without interference with other organs. In addition to this, the intermediate product, the timing of the degradation process, and the route and mechanism of degradation are equally important aspects that need to be taken care [47]. Scaffold materials should fulfill several requirements. A scaffold is not just a passive support for cell growth, but a device whose properties affects the regeneration cascade. Mechanical properties, surface properties, and morphology are in turn relevant to the specific application. Degradation kinetics and the rate at which scaffold properties change with degradation should always be predictable. In particular, the degradation behavior of biomaterials can follow several mechanisms and is controlled by different factors. Understanding the degradation kinetics and mechanism of biomaterials is necessary to optimize their possible usage. The rate of degradation is also strictly connected to the degree of porosity [38].

One of the general variables that need to be thoroughly considered to successfully bioprint viable and functional tissue bioengineering scaffolds is the inclusion of supportive biomaterials, generally in the form of proteins and polymers, which (1) facilitate the deposition method by mechanical means and (2) provide support and protection to the cells during and after the tissue construct fabrication process. These biomaterials can encompass the physical environment inside of which the cells will reside, as well as the biochemical signal cells need to function as they would in the body [46].

Scaffolds represent the space available for the tissue to develop and the physical support for cell growth. Scaffold mechanical properties should allow shape maintenance during tissue regeneration and enable stress transfer and load bearing. Moreover, during the first stage of tissue reconstruction, wound contraction forces act against the process, and enough mechanical strength and stiffness of the scaffold is required. Scaffold porosity is a fundamental characteristic for providing available space for cells to migrate and for vascularization of the tissue. Furthermore, the larger the surface available, the more cell interactions will arise. In general, the biological activity of a scaffold is determined by ligand density. Scaffold composition and porous fraction, that is, the total surface of the structure exposed to cells, determine the ligand density. Highly specific surface areas allow for cell attachment and anchorage, and a high pore volume fraction enables cell growth, migration, and effective transportation of fluids and nutrients. In particular, microporosity is important for capillary ingrowth and interactions between cells and matrix, while macroporosity is relevant to nutrient supply and waste removal of cell metabolism. The rate of degradation is also strictly connected to the degree of porosity [38].

As in the development of the tissue-engineered organs, regeneration of functional tissue requires maintenance of cell viability and differentiated function, encouragement of cell proliferation, modulation of the direction and speed of cell migration, and regulation of cellular adhesion [61]. Cell viability may be judged by morphological changes or by changes in membrane permeability and/or physiological state inferred from the exclusion of certain dyes or the uptake and retention of others. Cultured cells are seeded onto a three-dimensional biocompatible scaffold that will slowly degrade and resorb as the soft and hard structures grow and assimilate in vitro and/or in vivo [2]. Cell viability during 3D bioprinting is dependent on the shear stress experienced during extrusion, which in turn is dependent on the viscosity of the solution, the applied pressure, and the needle diameter. In addition, any post-printing bioink cross-linking may also impact on cell viability [62]. Cell viability can be measured with Live/Dead Viability/Cytotoxicity assay after printing [63] and could vary with dispensing pressure and nozzle diameter. It decreases as the pressure increases and the nozzle diameter decreases, and it is seen that the effect of pressure is significantly larger than the effect of the nozzle diameter. At higher pressures, there is an increase in the number of apoptotic cells as well as necrotic cells [64].

Tissue bioengineered scaffolds targeted for in vivo applications are typically restricted to a thickness of only a few hundreds of microns, owing to the diffusion limitations of oxygen and nutrients [43]. One of the major challenges in tissue engineering for translation in clinical applications is the vascularization of bioengineering scaffolds of clinically relevant size. Insufficient vascularization inhibits nutrient and host cell delivery or migrations and leads to improper cell integration or cell death. While vascularization remains a challenge to maintain viability of large biofabricated tissue bioengineering scaffolds, recent advances in the field demonstrate that novel biofabrication techniques may resolve this problem [65].

6. New insights: 3D to 4D

Doing a survey on the Web of Science®, it is noticeable that the number of original articles on tissue engineering and regenerative medicine has experienced a tremendous increase over the past 10 years (**Figure 4**; review papers and proceedings not included). Likewise, bioprinting is attracting a lot of researchers presenting an exponential increase in the last 3 years. Meanwhile, 3D bioprinting market was valued at $98.6 million in 2015, and an annual growth of 36% for the next 6 years is expected [66].

Nevertheless, 3D bioprinting has been focused on the development of bioengineering scaffolds that lack a crucial element for mimicking native live tissues: its ability to acutely change according to its function. That is why leading research groups have recently proposed the four-dimensional (4D) bioprinting (time is integrated with 3D bioprinting) as an enhanced approach for tissue engineering and regenerative medicine: the development of stimuli-responsive biomaterials that can be printed and dynamic to intended stimulation. However, several challenges arise, namely, (i) bioinks have to be optimized to achieve successful bioprinting; (ii) processes must be mechanically designed to obtain robust shape-changing capability of the bioengineering scaffolds [67]; (iii) specific bioreactors for complex tissue function maturation need to be invented; and (iv) evaluation procedures should be defined to examine the functionality response.

Therefore, the most promising approach is to optimize the cell-bioengineering scaffold interactions, becoming feasible to explore the usage of computer modeling to examine the further responses. Developing "smart" biomaterials (also referred as "intelligent," "stimuli responsive," "stimuli sensitive," or "environmentally sensitive") to allow the dynamic changes of the structure, upgrade of the printing processes into defined architecture for targeting tissues,

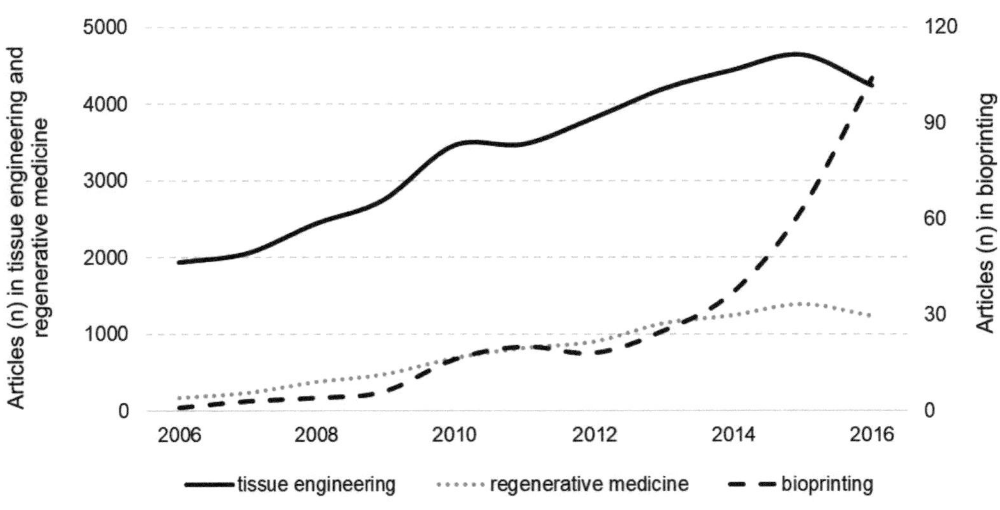

Figure 4. Number of original articles on tissue engineering and regenerative medicine 2006–2016.

automation of stimulus, and standardizing the assessment procedures to evaluate the result is crucial for enhanced regenerative medicine approaches [68].

Acknowledgements

This publication is supported by the Portuguese Foundation for Science and Technology (FCT) through the following projects: UID/Multi/04044/2013 and PTDC/EMS-SIS/7032/2014.

Author details

Sandra Amado[1,2,3]*, Pedro Morouço[1], Paula Pascoal-Faria[4] and Nuno Alves[1,5]

*Address all correspondence to: sandra.amado@ipleiria.pt

1 Centre for Rapid and Sustainable Product Development, Polytechnic Institute of Leiria, Leiria, Portugal

2 CIPER-FMH, Centro Interdisciplinar de Estudo de Performance Humana, Faculdade de Motricidade Humana (FMH), Universidade de Lisboa (UL), Lisboa, Portugal

3 School of Health Sciences at the Polytechnic Institute of Leiria, Leiria, Portugal

4 Centre for Rapid and Sustainable Product Development & Mathematics Department, School of Technology and Management, Polytechnic Institute of Leiria, Leiria, Portugal

5 Mechanical Engineering Department, School of Technology and Management, Polytechnic Institute of Leiria, Leiria, Portugal

References

[1] Morouço P, Alves N, Amado S. The role of biomechanics in tissue engineering. Austin Journal of Biomedical Engineering. 2016;**3**(1):1-2

[2] Hutmacher DW. Scaffold design and fabrication technologies for engineering tissues—state of the art and future perspectives. Journal of Biomaterials Science Polymer Edition. 2001;**12**(1):107-124

[3] Malda J, Groll J. A step towards clinical translation of biofabrication. Trends in Biotechnology. 2016;**34**(5):356-357

[4] Zadpoor AA, Malda J. Additive manufacturing of biomaterials, tissues, and organs. Annals of Biomedical Engineering. 2017;**45**(1):1-11

[5] Parker GJM, Wheeler-Kingshott CAM, Barker GJ. Diffusion tensor imaging. IEEE Transactions on Medical Imaging. 2002;**21**(5):505-512

[6] Fang Z, Starly B, Sun W. Computer-aided characterization for effective mechanical properties of porous tissue scaffolds. CAD Computer-Aided Design. 2005;**37**(1):65-72

[7] Zadpoor A.A.. Bone tissue regeneration: The role of scaffold geometry. Biomaterials Science. 2015;**3**(2):231-245

[8] Folch A, Mezzour S, Düring M, Hurtado O, Toner M, Müller R. Stacks of microfabricated structures as scaffolds for cell culture and tissue engineering. Biomedical Microdevices. 2000;**2**(3):207-214

[9] Sun W, Darling A, Starly B, Nam J. Computer-aided tissue engineering: Overview, scope and challenges. Biotechnology and Applied Biochemistry. 2004;**39**(Pt 1):29-47

[10] Landers R, Hubner U, Schmelzeisen R, Mulhaupt R. Rapid prototyping of scaffolds derived from thermoreversible hydrogels and tailored for applications in tissue engineering. Biomaterials. 2002;**23**(23):4437-4447

[11] Lin A, Barrows T, Cartmell S, Guldberg R. Microarchitectural and mechanical characterization of oriented porous polymer scaffolds. Biomaterials. 2003;**24**(3):481-489

[12] Angelo DF, Morouço P, Alves N, Viana T, Santos F, González R, et al. Choosing sheep (*Ovis aries*) as animal model for temporomandibular joint research: Morphological, histological and biomechanical characterization of the joint disc. Morphologie. 2016; **100**(331):223-233

[13] Wieding J, Souffrant R, Mittelmeier W, Bader R. Finite element analysis on the biomechanical stability of open porous titanium scaffolds for large segmental bone defects under physiological load conditions. Medical Engineering and Physics. 2013;**35**(4):422-432

[14] Cheng XY, Li SJ, Murr LE, Zhang ZB, Hao YL, Yang R, et al. Compression deformation behavior of Ti-6Al-4V alloy with cellular structures fabricated by electron beam melting. Journal of the Mechanical Behavior of Biomedical Materials. 2012;**16**(1):153-162

[15] Van Der Stok J, Koolen MKE, De Maat MPM, Amin Yavari S, Alblas J, Patka P, et al. Full regeneration of segmental bone defects using porous titanium implants loaded with BMP-2 containing fibrin gels. European Cells & Materials. 2015;**29**:141-154

[16] van der Stok J, Wang H, Amin Yavari S, Siebelt M, Sandker M, Waarsing JH, et al. Enhanced bone regeneration of cortical segmental bone defects using porous titanium scaffolds incorporated with colloidal gelatin gels for time- and dose-controlled delivery of dual growth factors. Tissue Engineering Part A. 2013;**19**(23-24):2605-2614

[17] Van Bael S, Chai YC, Truscello S, Moesen M, Kerckhofs G, Van Oosterwyck H, et al. The effect of pore geometry on the in vitro biological behavior of human periosteum-derived cells seeded on selective laser-melted Ti6Al4V bone scaffolds. Acta Biomaterialia. 2012;**8**(7):2824-2834

[18] Hollister SJ. Porous scaffold design for tissue engineering. Nature Materials. 2005 Jul;**4**(7):518-524

[19] Khoda KM, Ozbolat IT, Koc B. A functionally gradient variational porosity architecture for hollowed scaffolds fabrication. Biofabrication. 2011;**3**(3):34106

[20] Giannitelli S, Mozetic P, Trombetta M, Rainer A. Combined additive manufacturing approaches in tissue engineering. Acta Biomaterialia. 2015;**24**:1-11

[21] Sobral JM, Caridade SG, Sousa RA, Mano JF, Reis RL. Three-dimensional plotted scaffolds with controlled pore size gradients: Effect of scaffold geometry on mechanical performance and cell seeding efficiency. Acta Biomaterialia. 2011;**7**(3):1009-1018

[22] Di Luca A, Szlazak K, Lorenzo-Moldero I, Ghebes CA, Lepedda A, Swieszkowski W, et al. Influencing chondrogenic differentiation of human mesenchymal stromal cells in scaffolds displaying a structural gradient in pore size. Acta Biomaterialia. 2016;**36**(March):210-219

[23] Puppi D, Chiellini F, Piras MM, Chiellini E. Polymeric materials for bone and cartilage repair. Progress in Polymer Science. 2010;**35**(4):403-440

[24] Chua C, Leong K, Lim C. Classification of rapid prototyping systems. In: Chua C, Leong K, Lim C, editors. Rapid Prototyping: Principles and Applications. Singapore: World Scientific Publishing; 2003. pp. 19-23

[25] Karageorgiou V, Kaplan D. Porosity of 3D biomaterial scaffolds and osteogenesis. Biomaterials. 2005;**26**:5474-5491

[26] Fierz FC, Beckmann F, Huser M, Irsen SH, Leukers B, Witte F, et al. The morphology of anisotropic 3D-printed hydroxyapatite scaffolds. Biomaterials. 2008;**29**(28):3799-3806

[27] Melchels FPW, Feijen J, Grijpma DW. A poly(d,l-lactide) resin for the preparation of tissue engineering scaffolds by stereolithography. Biomaterials. 2009;**30**(23-24):3801-3809

[28] Melchels FPW, Domingos MAN, Klein TJ, Malda J, Bartolo PJ, Hutmacher DW. Additive manufacturing of tissues and organs. Progress in Polymer Science. 2012;**37**(8):1079-1104

[29] Cipitria A, Reichert JC, Epari DR, Saifzadeh S, Berner A, Schell H, et al. Polycaprolactone scaffold and reduced rhBMP-7 dose for the regeneration of critical-sized defects in sheep tibiae. Biomaterials. 2013;**34**(38):9960-9968

[30] Domingos M, Dinucci D, Cometa S, Alderighi M, Bártolo PJ, Chiellini F. Polycaprolactone scaffolds fabricated via bioextrusion for tissue engineering applications. International Journal of Biomaterials. 2009;**2009**:239643

[31] Domingos M, Chiellini F, Cometa S, De Giglio E, Grillo-Fernandes E, Bártolo P, et al. Evaluation of in vitro degradation of PCL scaffolds fabricated via BioExtrusion. Part 1: Influence of the degradation environment. Virtual and Physical Prototyping. 2010;**5**(2):65-73. Available from: http://www.tandfonline.com/doi/abs/10.1080/17452751003769440

[32] Morouço P, Biscaia S, Viana T, Franco M, Malça C, Mateus A, et al. Fabrication of poly(ε-caprolactone) scaffolds reinforced with cellulose nanofibers, with and without the addition of hydroxyapatite nanoparticles. BioMed Research International. 2016;**2016**:1-10

[33] Wei C, Dong J. Hybrid hierarchical fabrication of three-dimensional scaffolds. Journal of Manufacturing Processes. 2014;**16**(2):257-263

[34] Vaz CM, van Tuijl S, Bouten CVC, Baaijens FPT. Design of scaffolds for blood vessel tissue engineering using a multi-layering electrospinning technique. Acta Biomaterialia. 2005;**1**(5):575-582

[35] Nemeno-Guanzon JG, Lee S, Berg JR, Jo YH, Yeo JE, Nam BM, et al. Trends in tissue engineering for blood vessels. Journal of Biomedicine and Biotechnology. 2012;**2012**

[36] Hoenig MR, Campbell GR, Rolfe BE, Campbell JH. Tissue-engineered blood vessels: Alternative to autologous grafts? Arteriosclerosis, Thrombosis, and Vascular Biology. 2005;**25**(6):1128-1134

[37] Chantarapanich N, Puttawibul P, Sucharitpwatskul S, Jeamwatthanachai P, Inglam S, Sitthiseripratip K. Scaffold library for tissue engineering: A geometric evaluation. Computational and Mathematical Methods in Medicine. 2012;**2012**

[38] Haycock J. 3D Cell Culture [Internet]. Vol. 695, Image (Rochester, N.Y.). 2011. pp. 261-280

[39] Seol YJ, Park DY, Park JY, Kim SW, Park SJ, Cho DW. A new method of fabricating robust freeform 3D ceramic scaffolds for bone tissue regeneration. Biotechnology and Bioengineering. 2013;**110**(5):1444-1455

[40] Camarero-Espinosa S, Cooper-White J. Tailoring biomaterial scaffolds for osteochondral repair. International Journal of Pharmaceutics. 2016;**523**(2): 476-489

[41] Rezwan K, Chen QZ, Blaker JJ, Boccaccini AR. Biodegradable and bioactive porous polymer/inorganic composite scaffolds for bone tissue engineering. Biomaterials. 2006;**27**(18):3413-3431

[42] Fiedler T, Videira AC, Bártolo P, Strauch M, Murch GE, Ferreira JMF. On the mechanical properties of PLC-bioactive glass scaffolds fabricated via BioExtrusion. Materials Science and Engineering C. 2015;**57**:288-293

[43] Cosson S, Otte EA, Hezaveh H, Cooper-White JJ. Concise review: Tailoring bioengineered scaffolds for stem cell applications in tissue engineering and regenerative medicine. Stem Cells Translational Medicine. 2015;**4**(2):156-164

[44] Ker EDF, Nain AS, Weiss LE, Wang J, Suhan J, Amon CH, et al. Bioprinting of growth factors onto aligned sub-micron fibrous scaffolds for simultaneous control of cell differentiation and alignment. Biomaterials. 2011;**32**(32):8097-8107

[45] López-Ruiz E, Jiménez G, García MÁ, Antich C, Boulaiz H, Marchal JA, et al. Polymers, scaffolds and bioactive molecules with therapeutic properties in osteochondral pathologies: What's new? Expert Opinion on Therapeutic Patents. 2016;**26**(8):877-890

[46] Cheung DYC, Duan B, Butcher JT. Essentials of 3D Biofabrication and Translation [Internet]. Essentials of 3D Biofabrication and Translation. 2015. pp. 351-370

[47] Kariduraganavar MY. Advances in polymers and tissue engineering scaffolds. In: Inamuddin, editor. Green Polymer Composites Technology Properties and Applications. Boca Raton: Taylor & Francis Group; 2016;343-354

[48] Gao B, Yang Q, Zhao X, Jin G, Ma Y, Xu F. 4D bioprinting for biomedical applications. Trends in Biotechnology. 2016;**34**(9):746-756

[49] Melchels F, Malda J, Fedorovich N, Alblas J, Woodfield T. Organ Printing; Comprehension Biomaterials; 2011;**5**:587-606

[50] Kang H-W, Lee SJ, Ko IK, Kengla C, Yoo JJ, Atala A. A 3D bioprinting system to produce human-scale tissue constructs with structural integrity. Nature Biotechnology. 2016;**34**(3):312-319

[51] Malda J, Visser J, Melchels F, JÜngst T, Hennink W, Dhert W, et al. 25th anniversary article: Engineering hydrogels for biofabrication. Advanced Materials. 2013;**25**:5011-5028

[52] Saunders R, Gough J, Derby B. Delivery of human fibroblast cells by piezoelectric drop-on-demand inkjet printing. Biomaterials. 2008;**29**:193-203

[53] Tao H, Marelli B, Yang M, An B, Onses M, Rogers J, et al. Inkjet printing of regenerated silk fibroin: From printable forms to printable functions. Advanced Materials. 2015;**27**:4273-4279

[54] Chang C, Boland E, Williams S, Hoying J. Direct- write bioprinting three-dimensional biohybrid systems for future regenerative therapies. Journal of Biomedical Materials Research. Part B, Applied Biomaterials. 2011;**98**:160-170

[55] Visser J, Peters B, Burger T, Boomstra J, Dhert W, Melchels F, et al. Biofabrication of multi-material anatomically shaped tissue constructs. Biofabrication. 2013;**5**(3):35007

[56] Murphy S, Atala A. 3D bioprinting of tissues and organs. Nature Biotechnology. 2014; Aug;**32**(8):773-785

[57] Mironov V, Visconti R, Kasyanov V, Forgacs G, Drake C, Markwald R. Organ printing: Tissue spheroids as building blocks. Biomaterials 2009;**30**:2164-2174

[58] Merceron T, Burt M, Seol Y, Kang H, Lee S, Yoo J, et al. A 3D bioprinted complex structure for engineering the muscle-tendon unit. Biofabrication. 2015;**7**:35003

[59] Guillemot F, Souquet A, Catros S, Guillotin B. Laser- assisted cell printing: Principle, physical parameters versus cell fate and perspectives in tissue engineering. Nanomedicine. 2010;**5**:507-515

[60] Gaebel R, Ma N, Liu J, Guan J, Koch L, Klopsch C, et al. Patterning human stem cells and endothelial cells with laser printing for cardiac regeneration. Biomaterials. 2011;**32**:9218-9230

[61] Khang G. Handbook of Intelligent Scaffolds for Tissue Engineering and Regenerative Medicine; Singapore: Taylor & Francis Group; 2012. 1-972 pp

[62] Daly AC, Critchley SE, Rencsok EM, Kelly DJ. A comparison of different bioinks for 3D bioprinting of fibrocartilage and hyaline cartilage. Biofabrication. 2016;**8**(4):45002

[63] Cui X, Breitenkamp K, Finn MG, Lotz M, D'lima DD. Direct human cartilage repair using three-dimensional bioprinting technology. Tissue Engineering Part A. 2012;**18**:1304-1312

[64] Nair K, Gandhi M, Khalil S, Yan KC, Marcolongo M, Barbee K, et al. Characterization of cell viability during bioprinting processes. Biotechnology Journal. 2009;**4**(8):1168-1177

[65] Lim K, Morouço P, Levato R, Melchels F, Malda J. Organ Biofabrication. Comprehension Biomaterials II; 2017;**5**:236-266

[66] Market Research P&S. Global 3D Bioprinting Market Size, Share, Development, Growth and Demand Forecast to 2022, 2016. Available from (https://www.psmarketresearch.com/press-release/3d-bioprinting-market)

[67] Li Y-C, Zhang YS, Akpek A, Shin SR, Khademhosseini A. 4D bioprinting: The next-generation technology for biofabrication enabled by stimuli-responsive materials. Biofabrication. 2016 2;**9**(1):012001

[68] Furth ME, Atala A, Van Dyke ME. Smart biomaterials design for tissue engineering and regenerative medicine. Biomaterials. 2007;**28**(34):5068-5073

Chapter 16

Biomaterials and Stem Cells: Promising Tools in Tissue Engineering and Biomedical Applications

Małgorzata Sekuła and Ewa K. Zuba-Surma

Additional information is available at the end of the chapter

http://dx.doi.org/10.5772/intechopen.70122

Abstract

Biomaterial sciences and tissue engineering approaches are currently fundamental strategies for the development of regenerative medicine. Stem cells (SCs) are a unique cell type capable of self-renewal and reconstructing damaged tissues. At the present time, adult SCs isolated from postnatal tissues are widely used in clinical applications. Their characteristics such as a multipotent differentiation capacity and immunomodulatory activity make them a promising tool to use in patients. Modern material technologies allow for the development of innovative biomaterials that closely correspond to requirements of the current biomedical application. Biomaterials, such as ceramics and metals, are already used as implants to replace or improve the functionality of the damaged tissue or organ. However, the continuous development of modern technology opens new insights of polymeric and smart material applications. Moreover, biomaterials may enhance the SCs biological activity and their implementation by establishing a specific microenvironment mimicking natural cell niche. Thus, the synergistic advancement in the fields of biomaterial and medical sciences constitutes a challenge for the development of effective therapies in humans including combined applications of novel biomaterials and SCs populations.

Keywords: adult stem cells, biomaterials, regenerative medicine, tissue engineering

1. Introduction

Regenerative medicine represents a new interdisciplinary field of clinical science focused on the development and implementation of novel strategies to enhance the process of regeneration of impaired cells, tissues and organs as well as replacing damaged cells with new, fully functional cells of the required phenotype [1, 2].

To improve the effectiveness of such regeneration processes, one of the potential approaches is application of stem cell (SC)-based therapy. In addition, the combination of stem cells with biocompatible materials that may constitute a scaffold for the seeded cells may lead to enforcement of biological activity of stem cells and as such accelerate the process of regeneration or restoration of impaired tissue [3, 4].

2. Therapeutic applications of stem cells

2.1. Classification and characteristics of stem cells (SCs)

Stem cells (SCs) are a unique type of cells characterized by the ability to (i) self-renewal through unlimited cell divisions and (ii) differentiate into other types of specialized cells, including epithelial, muscle, neuronal cells and others [5, 6].

Based on the origin and source of isolation, SCs may be included in two main groups: (i) embryonic SCs (ESCs)—derived from embryos at different stages of development and (ii) adult SCs (ASCs)—isolated from several postnatal and adult tissue sources, including the umbilical cord, cord blood, bone marrow, adipose tissue, central nervous system, retina, skeletal muscle and other mature tissues [7, 8]. The differentiation capacity of embryonic SCs allows them to form any individual organs and fully differentiated cells of the whole body, which corresponds to pluripotency of these SCs. In opposite, most of adult SCs are multipotent and lineage-restricted (monopotent) and generally give rise to certain cell types of one germ layer or cell lineage. They residue in several niches, including bone marrow, liver, muscle, brain and others, where they may be activate towards tissue- or organ-specific cells under certain physiologic or experimental conditions. Moreover, it has been shown that adult SCs may provide efficient regeneration of impaired organs in both preclinical and clinical conditions [7].

Based on the differentiation capacity, SCs populations belong to the following types [5, 9–11]:

- Totipotent SCs (TSCs)—The most developmentally primitive and potent SCs are capable to differentiate into any cell type from three germ layers (mesoderm, ectoderm and endoderm) forming whole organism as well as into extra-embryonic tissues such as placenta; the best examples of TSCs are zygote and first blastomeres [10–13].
- Pluripotent SCs (PSCs)—The cells sustaining the capacity to differentiate into all cell types from three germ layers, but they are not able to give rise to placenta; PSCs are naturally present in developing embryo in stage of morula, in inner cell mass (ICM) of developing blastocyst and in the epiblast of gastrula and in limited number may also be found in adult tissues as remnants from embryonic development [9, 14]; PSCs may also be de novo created via genetic reprogramming of somatic cells and are called 'induced PSCs' (iPS cells) [11, 15].
- Multipotent SCs—The SCs typically capable to give rise to all cell types within one germ layer; the best described examples are mesenchymal SCs (MSCs) isolated from several adult and postnatal tissues [10, 16–18];

- Unipotent SCs (progenitors)—The cells capable to differentiate into one or two particular types of specialized cell present in particular tissue type; this group includes several population of tissue-committed progenitors such as endothelial progenitor cells, cardiac SCs, satellite cells of skeletal muscles, neural progenitors and others [5, 9, 10].

2.2. Types of SCs with potential clinical application

Recently, more attention has been directed to potential utilization of SCs in clinical applications in patients. Due to the legal and ethical restrictions, the employment of embryonic SCs, which possess the largest spectrum of differentiation capacity, is controversial and prohibited in many countries. Moreover, SCs with pluripotent characteristics may lead to adverse side effects following their injection including teratoma formation [19, 20]. Therefore, the therapies employing adult SCs play a major role in human treatment as safe and effective approaches. Transplantations of autologous SCs isolated from mobilized peripheral blood or bone marrow are currently widely used in haematological patients with malignancies such as leukaemia and lymphoma [21]. Moreover, haematopoietic stem cells (HSCs) residing predominantly in adult bone marrow are widely used for bone marrow reconstitution in patients suffering from several genetic and autoimmune diseases, blood cancers and haematopoietic defects [5, 7].

Current growing expectations for further advancements in regenerative medicine are highly focused on mesenchymal stem/stromal cells (MSCs) belonging to adult SCs isolated from several tissue types [16, 18, 22, 23]. MSCs are multipotent, non-haematopoietic cells, which can be isolated from various sources including bone marrow, adipose tissue, cord blood, umbilical cord, Wharton's jelly and other tissues of the adult organism [22, 24]. Isolation of this type of cells does not raise any ethical concerns and is a relatively easy procedure. Moreover, MSCs are characterized by low immunogenicity with simultaneous immunomodulatory effect, and after transplantation, teratoma formation does not occur in the recipient organism [16, 18, 22, 23, 25]. Furthermore, potential regenerative applicability of MSCs is also enhanced by their paracrine activity related to several molecules released to their environment that may impact on other neighbouring cells affecting their functions [22, 25]. MSCs produce and release bioactive molecules, including multitude growth factors (e.g. TGF-β1, bFGF, BMP-4), anti-inflammatory factors (e.g. IL-10, PGE2, HGF) and cytoprotective agents (e.g. IL-6, MCP-1, IGF-1), which promote resident cells to divide and remodel the damaged tissue [26]. All these listed features make MSCs as promising tool for biomedical research.

MSCs possess a robust proliferation capacity as well as a potential to differentiate into several lineages of mesodermal origin, including bone, cartilage and adipose tissue [23]. Moreover, they have been also shown to give rise to other cell types, such as endothelial, cardiac or liver cells, which may also be utilized in tissue regeneration [16, 27]. These unique biological values may be utilized for the development of personalized treatment strategy for several diseases and provide the progress in establishing modern cell-based therapy. Cell-based therapy represents a promising perspective of treatment directed at the regeneration of damaged tissues or organs using stem cells or progenitor cells both in the autologous and allogeneic system [16, 28–30].

According to the current U.S. National Institutes of Health database including clinical trials conducted worldwide, there are currently more than 240 clinical trials being conducted in the world employing MSCs in patients [31]. Examples of the application of MSCs isolated from different sources in the treatment of selected diseases are shown in **Table 1**.

One of the great new opportunities in medical science is the possibility of obtaining the induced pluripotent SCs (iPS cells) by genetic reprogramming of mature cells into the stage of pluripotency [15]. Due to the discovery of this phenomenon, professor Shinya Yamanaka was honoured with the Nobel Prize in medicine and physiology in 2012. Since then, iPS cells constitute an excellent model for *in vitro* studies of molecular mechanisms associated with the development and progression of several diseases, including Parkinson's disease [32], Huntington disease [33], Down syndrome [34] and others. Moreover, dozens of laboratories are questing for optimal utilization of these cells in tissue regeneration. However, due to the possibility of teratoma formation after iPS transplantation, their applications in medicine are still limited.

Thus, despite the fact that several new rising SCs types are being examined and optimized for future applications, the most commonly applicable SCs in cell therapies of distinct human diseases are adult stem cells including predominantly MSCs derived from bone marrow, adipose tissue and umbilical cord as well as HSCs harvested from bone marrow, mobilized peripheral blood and cord blood [16, 18, 28–30].

Type of MSCs	Condition	ClinicalTrials.gov identifier
Umbilical cord-derived MSCs	Hepatic cirrhosis	NCT02652351
	Aplastic anaemia	NCT03055078
	Stroke	NCT02580019
	Pneumoconiosis	NCT02668068
	Rheumatoid arthritis	NCT02643823
	Sweat gland diseases	NCT02304562
Bone marrow-derived MSCs	Acute myocardial infarction	NCT01652209
	Chronic myocardial ischaemia	NCT02460770
	Acute respiratory distress syndrome	NCT02097641
	Middle cerebral artery infarction	NCT01461720
	Prostate cancer	NCT01983709
	Stroke	NCT02564328
Adipose-derived MSCs	Infantile spinal muscular atrophy	NCT02855112
	Multiple sclerosis	NCT02326935
	Hair restoration	NCT02865421

Source: Ref. [31].

Table 1. The application of MSCs in selected clinical trials.

2.3. Utilization of SC derivatives as a potential alternative to cell-based therapy

Immunomodulatory properties of SCs are important features involved in tissue repair, which are directly related to their paracrine activity. Despite the directly released molecules, mammalian cells, including SCs, are able to produce extracellular vesicles (EVs) carrying bioactive factors, which may additionally be involved in the modulation of the repair process of damaged tissues [35]. EVs represent heterogeneous population of small, circular structures surrounded with the protein-lipid membrane that are released by cells including SCs. Importantly, the size and molecular composition of EVs are different and unique depending on the cell type of origin and the mechanism of their biogenesis. Depending on the size of EVs, they may be distinguished in apoptotic bodies (1–5 μm), microparticles (100 nm–1 μm) and exosomes (30–100 nm) fractions [36, 37].

Several recent scientific reports indicate that EVs express surface markers characterizing the cells from which they are released, along with EV-specific antigens including tetraspanins (CD9, CD63 and CD81), endosome or membrane-binding proteins (TG101), signal transduction or scaffolding proteins (syntenin) [36, 37]. Importantly, EVs may also include various types of bioactive components (e.g. mRNA, miRNA and enzymes), as well as receptors, adhesion or signalling proteins [38, 39]. Importantly, the contents of EVs can be effectively transferred to the target cells, change their function and impact in the regeneration of impaired tissues. Moreover, the presence of protein-lipid membrane on the surface of EVs can protect their bioactive content from extracellular enzymes and therefore the cargo may be delivered in a fully functional form into targeted cells [38, 40]. Thus, EVs are recognized as mediators of intercellular communication and constitute an alternative or reinforcement of a standard cell-based therapy.

The biological relevance of EVs has been established in different experimental settings. Depending on the origin and content of EVs, they may enhance immune system, endorse anti-tumour responses and thus may provide important tools for novel anti-tumour therapies, such as melanoma treatment [41]. EVs may also be utilized as drug delivery vehicles [42], in regenerative medicine [43] and immune therapy [44]. Recently, our study also indicated that SC-derived EVs may be utilized as a novel tool for regenerative therapies of ischemic tissue including in heart repair [38, 40].

However, further studies are required for comprehensive analysis of the mechanisms of EVs action and potential clinical applications of these promising SC derivatives.

3. Medical application of selected natural and synthetic biomaterials

3.1. Material requirements for biomaterials

Biomaterial by definition is a 'substance (other than a drug), synthetic or natural, that can be used as a system or part of a system that treats, augments, or replaces any tissue, organ, or function of the body' [45]. Thus, according to the definition biomaterials are progressively used in tissue engineering. They may be utilized for the construction of implants to replace lost or damaged organs or tissues and may also constitute a scaffold for enhanced stem cells to reconstruct not fully functional tissue [45, 46].

Due to the wide range of potential applications of biomaterials in regenerative medicine, their physical and chemical properties may be different [45, 47]. However, in order to use a biomaterial in medical application, it should follow relevant requirements such as biocompatibility and biofunctionality [45, 47]:

- Biocompatibility is the ability to integrate with the recipient's cells in a safe manner and without adverse side effects.
- Biofunctionality is the ability to perform a specific biological function, based on the relevant parameters of the physical and mechanical properties.

Other important properties of biomaterials, which are affecting the potential application in medicine, include [45, 48, 49] the following:

- Biodegradation—Decomposition of the material in a natural way, when degradation products remain in the human body but without adverse side effects.
- Bioresorbability—Decomposition of the material in a natural way at a certain period of time after implantation. Non-toxic-degraded products are removed from the body *via* metabolic pathways (hydrolytic or enzymatic degradation).
- Non-toxicity—From the surface or porous of the material does not elute any toxic components, such as surfactants, stabilizers, catalysts, pigments and UV absorbents, which were used during production and that are incompatible with living organisms.
- Mechanical properties—Biomaterial should possess particular mechanical properties consistent with the anatomical site into which it will be implanted.

3.2. Applications of biomaterials

Several biomaterials useful for distinct applications in medical sciences, including in tissue repair and organ reconstruction, have already been developed over the last few decades [45, 47]. The biomaterial sciences are currently one of the highly advancing fields, which also closely cooperate with biotechnological and medical studies. Recent advancement in regenerative medicine strongly requires such strong support from biomaterial sciences, which may provide novel solutions for tissue repair [4, 49].

Among the biomaterials recognized and developed for potential medical purposes, here are multitude materials commonly present in natural sources or *de novo* designed and created for such purposes.

3.2.1. Naturally derived biomaterials

Natural materials commonly present in nature such as agarose, collagen, alginate, chitosan, hyaluronate or fibrin fully cooperate with living tissues of the recipient and possess low cytotoxicity [47, 48]. Moreover, they may exhibit specific protein-binding sites that improve integration with cells after transplantation [48]. Thus, they are considered predominantly interesting for tissue engineering applications.

One of the most common natural biomaterials is collagen—an important component of connective tissue, including bones, tendons, ligaments and skin [46, 50]. Collagen is simply absorbed into the body, is non-toxic and exhibits a low immune response and as such is a perfect biocompatible material with an adequate mechanical strength and flexibility for several applications. Moreover, collagen enhances cell adhesion to such surface, stimulates also biological interactions between cells and facilitates restoration of the natural microenvironment of cell niche and thereby may support the reconstruction of several damaged tissues [46, 48, 50].

Collagen may be employed for tissue engineering in the form of sponges, gels, hydrogels and sheets. It may also be chemically crosslinked in order to enhance or alter the rate of degradation of the fibres [51]. Currently, collagen preparations are used predominantly in wound healing and cartilage regeneration. Injectable form of collagen is used for cosmetic and aesthetic medicine as a tissue filler. In addition, collagen-based membranes are used in the periodontal treatment as a barrier preventing the migration of epithelial cells. It also forms a favourable microenvironment for stem cells to facilitate reconstruction of the damaged area [50, 51].

3.2.2. Synthetic biomaterials

Synthetic materials are considered as an alternative to natural materials. Due to their defined chemical composition and the ability to control the mechanical and physical properties, they are extensively used in therapeutic applications and basic biological studies [48, 52–55].

Due to distinct variants of polymerization reaction and formation of co-polymers, multiple synthetic polymers with wide range of physical and chemical properties may be achieved in chemical laboratories. Moreover, novel technologies in the synthesis and formation of more complex structures allow for the production of advanced composites [54]. Synthetic polymers, such as poly(ethylene) (PE), polyurethanes (PUR), polylactides (PLA) and poly(glycolide) (PGA), are widely employed as implants and components of medical devices [56]. Moreover, polymers may constitute suitable scaffold for cell propagation and enhance their biological activity, including neural stem cells, retinal progenitor cells or smooth muscle cells [55, 57, 58]. Thus, this group of biomaterials is currently in a special focus of scientists working on combined approaches using biocompatible scaffolds and stem cells for tissue repair [55, 57, 58].

Biodegradable polymers, including polyhydroxycarboxylic acids, such as PGA, PLA, poly(3-hydroxybutyrate), poly(4-hydroxybutyrate) and poly(∈-caprolactone) (PCL) are of wide interest in the development of novel technologies [56]. One of their potential applications is utilization in the treatment of cardiovascular diseases. Our recent studies have shown the positive impact of both PCL and PLA scaffolds on proliferation, migration and proangiogenic potential of mesenchymal SCs derived from umbilical cord tissue *in vitro*, suggesting the possible applications of these materials in cardiovascular repair *in vivo* (unpublished data) [59].

Synthetic polymers may also be used in biodegradable stents implanted after a heart attack and greatly contribute to patient recovery [56]. Importantly, the material should have suitable decomposition kinetics. Too long decomposition time (i.e. in the case of PLA or PGA) may lead to late stent thrombosis or blockages [56, 60]. One of a possible solution of this problem is to use rapidly biodegradable polymer stents coated with SCs to help rebuild damaged tissue and additionally stimulate resident cells to grow.

Other types of common synthetic materials useful for biomedical applications are ceramics. It has been well described that ceramic scaffolds, such as, for example, hydroxyapatite (HA) and tri-calcium phosphate (TCP), are characterized by biocompatibility, high mechanical stiffness (Young's modulus), very low elasticity and a hard brittle surface [49]. Due to their chemical and structural similarity to the mineral phase of native bone, these materials may enhance osteoblast proliferation and therefore they are widely utilized in bone regeneration [61, 62]. Moreover, ceramics may be exploited in dental and orthopaedic procedures to fill bone defects or as a bioactive coating material for implants to increase their integration after transplantation [63, 64]. However, their clinical applications are still limited due to the difficulties with the ability to change the shape of the material dedicated for transplantation and controlling time of their degradation rate [49, 65].

Similarly, titanium (Ti)-based metallic materials have been widely optimized for bone repair due to their mechanical properties and resistance to corrosion following the transplantation [66–68]. It has been shown that titanium scaffolds are effectively colonized by osteoblasts responsible for bone formation and this process may be enhanced *via* additional modifications of the scaffold surface by its roughening, coating with HA or graphene oxide (GO), as well as its biofunctionalization with bioactive molecules such as heparin and bone morphogenetic protein 2 (BMP-2) [69–72].

Importantly, graphene in its different forms is currently being considered as a potential new promising material for biomedical applications including tissue repair [73, 74]. This 2D carbon biocompatible material exhibits great electrical, conductive and physical properties, which make it interesting for potential applications for drug delivery and scaffold coating in regenerative therapies [74, 75]. It has been shown that graphene may enhance osteogenic differentiation of SCs [72, 73]. Moreover, our recent data also suggest the beneficial impact of graphene oxide (GO) on proliferative capacity, viability and differentiation potential of umbilical cord tissue-derived MSCs, which confirms the possibility of future graphene employment in tissue repair [76].

3.2.3. Hydrogels

Hydrogels are frequently used biomaterials in the biomedical applications and represent systems consisting of two or more compartments comprising a three-dimensional (3D) network of polymer chains and water that fills the spaces between the macromolecules [77, 78]. The main characteristics of hydrogels include the biocompatibility and ability to swell in solution until they reach a state of equilibrium. These allow them to be injected into the body in a non-invasive manner [77, 78].

Hydrogels demonstrate transparency and bioadhesive properties and they are widely used in the pharmaceutical and dermatological industries by local administration or filling the defects caused by injury [77]. They may also be utilized as an injectable material for bone and cartilage tissue engineering, which may be combined with appropriate cell injection [53, 78, 79]. It has been shown that *in situ* implementation of hydrogels promotes osteoblast differentiation [53, 79]. Therefore, injectable therapy constitutes a promising approach for non-invasive technique of transplantation, where also cell-based component may be added to enhance tissue repair.

3.2.4. Smart materials

Smart materials represent a new generation of biomaterials, exceeding the functionality of the currently widely used construction materials. Smart materials are characterized by the ability to alter their physical characteristics in a controlled manner including changing the shape, colour, stiffness or stickiness in response to several external stimuli, such as temperature, hydrostatic pressure, electric and magnetic field or radiation [80]. These changes are related to the revealing or eliciting the new functionality of the material and may be utilized in biomedical applications. Through the common connection between the internal sensor, the activator and a specific control mechanism, smart materials are able to respond to external stimuli. Importantly, these mechanisms are also responsible for the return to the original state, when a stimulant disappeared [80, 81].

Smart materials include several types such as listed below [52, 80, 82–84]:

- **Colour changing materials**—Materials that change colour in a reversible manner, depending on electrical, optical or thermal changes. These types of materials are exploited, for example, in optoelectronic components, lenses, lithium batteries, ferroelectric memory, temperature sensors or as the indicators of battery consumption [80, 81].
- **Light-emitting materials**—Materials emitting visible or invisible light, as a result of external stimuli such as short wavelength radiation (e.g. X-rays, ultraviolet light), temperature and electric voltage. They are utilized in electronics, filters for glasses, devices that detect UV rays, in criminology and in geology to identify minerals and rocks. They may also be exploited as a component of protective clothing, safety elements and warning materials [80, 81].
- **Shape memory materials**—Metal alloys that change shape as a result of temperature increase or decrease, respectively, to the set value. The reversibility of the process is to return to its original shape by changing the temperature or under the influence of the applied motion (the effect of pseudoelasticity). These materials are used in temperature sensors, electronics, robotics, telecommunications and production of medical devices (micro-pump, surgical clamps, orthodontic wire, long- and short-term implants, suture tightening on a stiffen wound, orthopaedic devices, bone nails, clamps, surgical instruments and others) [83, 84].
- **Self-assembling materials**—Materials that exhibit the intrinsic ability to spontaneously connect individual elements into an ordered 2D or 3D structure. In addition, they can also

have the ability to bind metal atoms, ions, molecules or semiconductors. They are widely used in biological research and nanotechnology, that is, in the tissue regeneration, as components for the storage of drugs, crystal engineering, as artificial proteins with pH-sensitive structure, as semi-permeable membrane as well as for the production of electronic processors and displays [52, 82].

- **Self-repairing materials**—Structural damage of this type of material is automatically and autonomously recovered by inducing a change in the shape or the self-assembly of the molecules. This process is not a method of complete repair of the impaired material; however, it may be used in the military, automotive, aviation and electronics industries [52, 82].

4. Novel aspects of the application of stem cells and biomaterials in tissue engineering and regenerative medicine

4.1. Biomaterials approaches for enhancement of SCs-based therapy

Modern approaches in current regenerative medicine include developing biocompatible scaffolds and combining them with living cell of selected type and bioactive molecules, in order to enhance the regeneration process of damaged tissues and organs [47].

Growing evidence indicate different populations of stem cells as a promising tool that may be utilized in tissue engineering and repair. Importantly, despite the regenerative properties of SCs, the restoration processes in damaged tissue are long and may not often be fully effective for functional recovery of damaged tissue. On the other hand, appropriate stimulation of reparative capacity of SCs may be achieved by modulation of chemical and physical properties of optimized biomaterials [47, 70, 77]. Therefore, simultaneous application of optimized and well-combined SCs and biomaterials may open new perspectives for the synergistic effective cooperation of both such components to improve the efficiency of the regeneration process [77]. Biomaterials may enhance the biological activity of SCs by establishing a specific niche related to their native microenvironment. This type of cell-biomaterial interactions leads to stimulation of cell adhesion, proliferation and directed differentiation of the cells implemented at the injured site [47, 70, 77]. Therefore, therapy based on biomaterials and SCs opens new possibilities for the development of innovative medicine [47, 77].

Currently, growing evidence is focused on encapsulation of native SCs prior to their transplantation [47, 85]. Cells encapsulation technique is based on the immobilization of cells in a semi-permeable membrane, which protects cells against mechanical damage and immune system response. Notably, the construction of the microcapsules allows bidirectional diffusion of nutrients, oxygen and wastes and therefore provides appropriate conditions for cell development [47, 85].

Encapsulated cells may be subjected to transplantation and directed differentiation. The material used to construct the microcapsules should possess particular physical properties,

http://dx.doi.org/10.5772/intechopen.70122

such as biocompatibility, mechanical stability, permeability, appropriate size, strength and durability [47]. One of the most common encapsulation materials is alginate. Due to the fact that the procedure for cell encapsulation using alginate can be performed under physiological conditions (physiological temperature and pH) and using isotonic solutions, it is widely distributed through clinical and industrial applications. Moreover, this natural biodegradable polymer that mimics the extracellular matrix and promotes cell functions and metabolism has been established in cartilage regenerative approaches [86, 87]. Microencapsulation technology represents a novel cell culture system that allows maintaining cell viability and differentiation of interested cell lines. It also may support the extracellular matrix production and cell organization in reconstructed tissue [86].

5. Conclusions

Significant advancement of regenerative medicine, nanomedicine and biomaterials engineering offers extended possibilities to obtain novel, effective achievements, which may be utilized in biomedical applications. The effect of interdisciplinary activity resulted in the development of bioactive scaffolds that promote cell propagation and enhance their biological activity. However, some difficulties in biomaterial- and cell-based therapy are still unclear and need to be addressed for widespread investigations. Nevertheless, integrative research in biomaterials and medicine fields is a challenge to develop effective therapies for cancer, civilization diseases and provide further development of tissue engineering.

Acknowledgements

This work is supported by grants from the National Science Centre (NCN): SONATA BIS-3 (UMO-2013/10/E/NZ3/007500), SYMFONIA 3 (UMO-2015/16/W/NZ4/00071) and the National Centre for Research and Development (NCBR): STRATEGMED III (BioMiStem project; ID 303570) to EZS. The Faculty of Biochemistry, Biophysics and Biotechnology at the Jagiellonian University, Krakow, Poland, is a partner of the Leading National Research Center (KNOW) supported by the Ministry of Science and Higher Education.

Author details

Małgorzata Sekuła[1] and Ewa K. Zuba-Surma[2]*

*Address all correspondence to: ewa.zuba-surma@uj.edu.pl

1 Malopolska Centre of Biotechnology, Jagiellonian University, Krakow, Poland

2 Department of Cell Biology, Faculty of Biochemistry, Biophysics and Biotechnology, Jagiellonian University, Krakow, Poland

References

[1] Galliot B, Crescenzi M, Jacinto A, Tajbakhsh S. Trends in tissue repair and regeneration. Development. 2017;**144**:357-364. DOI: 10.1242/dev.144279

[2] Forbes SJ, Rosenthal N. Preparing the ground for tissue regeneration: From mechanism to therapy. Nature Medicine. 2014;**20**:857-869. DOI: 10.1038/nm.3653

[3] Sahito RG, Sureshkumar P, Sotiriadou I, Srinivasan SP, Sabour D, Hescheler J, et al. The potential application of biomaterials in cardiac stem cell therapy. Current Medicinal Chemistry. 2016;**23**:589-602. DOI: 10.2174/0929867323066160303151041

[4] Shafiq M, Jung Y, Kim SH. Insight on stem cell preconditioning and instructive biomaterials to enhance cell adhesion, retention, and engraftment for tissue repair. Biomaterials. 2016;**90**:85-115. DOI: 10.1016/j.biomaterials.2016.03.020

[5] Hima Bindu A, Srilatha B. Potency of various types of stem cells and their transplantation. Journal of Stem Cell Research & Therapy. 2011;**1**:1-6. DOI: 10.4172/2157-7633.1000115

[6] Ghodsizad A, Voelkel T, Moebius J, Gregoric I, Bordel V, Straach E, et al. Biological similarities between mesenchymal stem cells (MSCs) and fibroblasts. Journal of Cytology & Histology. 2010;**1**:1-6. DOI: 10.4172/2157-7099.1000101

[7] Novik AA, Kuznetsov A, Melnichenko VY, Fedorenko DA, Ionova TI, Gorodokin GV. Non-myeloablative autologous haematopoietic stem cell transplantation with consolidation therapy using mitoxantrone as a treatment option in multiple sclerosis patients. Stem Cell Research & Therapy. 2011;**1**:1-5. DOI: 10.4172/2157-7633.1000102

[8] Toma JG, Akhavan M, Fernandes KJ, Barnabé-Heider F, Sadikot A, Kaplan DR, et al. Isolation of multipotent adult stem cells from the dermis of mammalian skin. Nature Cell Biology. 2001;**3**:778-784. DOI: 10.1038/ncb0901-778

[9] Ratajczak MZ, Ratajczak J, Suszynska M, Miller DM, Kucia M, Shin DM. A novel view of the adult stem cell compartment from the perspective of a quiescent population of very small embryonic-like stem cells. Circulation Research. 2017;**120**:166-178. DOI: 10.1161/CIRCRESAHA.116.309362

[10] Daley GQ. Stem cells and the evolving notion of cellular identity. Philosophical Transactions of the Royal Society of London Series B, Biological Sciences. 2015;**370**:20140376. DOI: 10.1098/rstb.2014.0376

[11] Cahan P, Daley GQ. Origins and implications of pluripotent stem cell variability and heterogeneity. Nature Reviews Molecular Cell Biology. 2013;**14**:357-368. DOI: 10.1038/nrm3584

[12] Thomson JA, Itskovitz-Eldor J, Shapiro SS, Waknitz MA, Swiergiel JJ, Marshall VS, et al. Embryonic stem cell lines derived from human blastocysts. Science. 1998;**282**:1145-1147. DOI: 10.1126/science.282.5391.1145

[13] Suwinska A, Czolowska R, Ozdzenski W, Tarkowski AK. Blastomeres of the mouse embryo lose totipotency after the fifth cleavage division: Expression of Cdx2 and Oct4 and developmental potential of inner and outer blastomeres of 16- and 32-cell embryos. Developmental Biology. 2008;**322**:133-144. DOI: 10.1016/j.ydbio.2008.07.019

[14] Kucia M, Reca R, Campbell FR, Zuba-Surma E, Majka M, Ratajczak J, et al. A population of very small embryonic-like (VSEL) CXCR4(+)SSEA-1(+)Oct-4+ stem cells identified in adult bone marrow. Leukemia. 2006;**20**:857-869. DOI: 10.1038/sj.leu.2404171

[15] Takahashi K, Yamanaka S. Induction of pluripotent stem cells from mouse embryonic and adult fibroblast cultures by defined factors. Cell. 2006;**126**:663-676. DOI: 10.1016/j.cell.2006.07.024

[16] Chugh AR, Zuba-Surma EK, Dawn B. Bone marrow-derived mesenchymal stems cells and cardiac repair. Minerva Cardioangiologica. 2009;**57**:185-202

[17] Kobolak J, Dinnyes A, Memic A, Khademhosseini A, Mobasheri A. Mesenchymal stem cells: Identification, phenotypic characterization, biological properties and potential for regenerative medicine through biomaterial micro-engineering of their niche. Methods. 2016;**99**:62-68. DOI: 10.1016/j.ymeth.2015.09.016

[18] Malek A, Bersinger NA. Human placental stem cells: Biomedical potential and clinical relevance. Journal of Stem Cells. 2011;**6**:75-92

[19] Kamada M, Mitsui Y, Matsuo T, Takahashi T. Reversible transformation and de-differentiation of human cells derived from induced pluripotent stem cell teratomas. Human Cell. 2016;**29**:1-9. DOI: 10.1007/s13577-015-0119-1

[20] Masuda S, Yokoo T, Sugimoto N, Doi M, Fujishiro SH, Takeuchi K, et al. A simplified in vitro teratoma assay for pluripotent stem cells injected into rodent fetal organs. Cell Medicine. 2012;**3**:103-112. DOI: 10.3727/215517912X639351

[21] Wang B, Ren C, Zhang W, Ma X, Xia B, Sheng Z. Intensified therapy followed by autologous stem-cell transplantation (ASCT) versus conventional therapy as first-line treatment of follicular lymphoma: A meta-analysis. Journal of Hematology & Oncology. 2013;**31**:29-33. DOI: 10.1002/hon.2015

[22] Hass R, Kasper C, Böhm S, Jacobs R. Different populations and sources of human mesenchymal stem cells (MSC): A comparison of adult and neonatal tissue-derived MSC. Cell Communication and Signaling. 2011;**9**:1-59. DOI: 10.1186/1478-811X-9-12

[23] Jin HJ, Bae YK, Kim M, Kwon SJ, Jeon HB, Choi SJ, et al. Comparative analysis of human mesenchymal stem cells from bone marrow, adipose tissue, and umbilical cord blood as sources of cell therapy. International Journal of Molecular Sciences. 2013;**14**:17986-18001. DOI: 10.3390/ijms140917986

[24] Wei X, Yang X, Han ZP, Qu FF, Shao L, Shi YF. Mesenchymal stem cells: A new trend for cell therapy. Acta Pharmacologica Sinica. 2013;**34**:747-754. DOI: 10.1038/aps.2013.50

[25] Gnecchi M, Zhang Z, Ni A, Dzau VJ. Paracrine mechanisms in adult stem sell signaling and therapy. Circulation Research. 2008;**103**:1204-1219. DOI: 10.1161/CIRCRESAHA.108.176826

[26] Mirotsou M, Jayawardena TM, Schmeckpeper J, Gnecchi M, Dzau VJ. Paracrine mechanisms of stem cell reparative and regenerative actions in the heart. Journal of Molecular and Cellular Cardiology. 2011;**50**:280-289. DOI: 10.1016/j.yjmcc.2010.08.005

[27] Labedz-Maslowska A, Lipert B, Berdecka D, Kedracka-Krok S, Jankowska U, Kamycka E, et al. Monocyte chemoattractant protein-induced protein 1 (MCPIP1) enhances angiogenic and cardiomyogenic potential of murine bone marrow-derived mesenchymal stem cells. PloS One. 2015;**10**:e0133746. DOI: 10.1371/journal.pone.0133746

[28] Mobasheria A, Kalamegame G, Musumecif G, Batt ME. Chondrocyte and mesenchymal stem cell-based therapies for cartilage repair in osteoarthritis and related orthopaedic conditions. Maturitas. 2014;**78**:188-198. DOI: 10.1016/j.maturitas.2014.04.017

[29] Samanta A, Kaja AK, Afzal MR, Zuba-Surma EK, Dawn B. Bone marrow cells for heart repair: Clinical evidence and perspectives. Minerva Cardioangiologica. 2017;**65**:299-313

[30] Afzal MR, Samanta A, Shah ZI, Jeevanantham V, Abdel-Latif A, Zuba-Surma EK, et al. Adult bone marrow cell therapy for ischemic heart disease: Evidence and insights from randomized controlled trials. Circulation Research. 2015;**117**:558-575. DOI: 10.1161/CIRCRESAHA.114.304792

[31] Health AsotUNIo. 2017. Available from: https://clinicaltrials.gov/ [cited: 18 February 2017]

[32] Byers B, Lee HL, Reijo Pera R. Modeling Parkinson's disease using induced pluripotent stem cells. Current Neurology and Neuroscience Reports. 2012;**12**:237-242. DOI: 10.1007/s11910-012-0270-y

[33] Tousley A, Kegel-Gleason KB. Induced pluripotent stem cells in Huntington's disease research: Progress and opportunity. Journal of Huntington's Disease. 2016;**5**:99-131. DOI: 10.3233/JHD-160199

[34] Brigida AL, Siniscalco D. Induced pluripotent stem cells as a cellular model for studying Down syndrome. Journal of Stem Cells & Regenerative Medicine. 2012;**12**:54-60

[35] Yanez-Mo M, Siljander PR, Andreu Z, Zavec AB, Borras FE, Buzas EI, et al. Biological properties of extracellular vesicles and their physiological functions. Journal of Extracellular Vesicles. 2015;**4**:1-60. DOI: 10.3402/jev.v4.27066

[36] György B, Szabó TG, Pásztói M, Pál Z, Misják P, Aradi B, et al. Membrane vesicles, current state-of-the-art: Emerging role of extracellular vesicles. Cellular and Molecular Life Sciences. 2011;**68**:2667-2688. DOI: 10.1007/s00018-011-0689-3

[37] Lötvall J, Hill AF, Hochberg F, Buzás EI, Di Vizio D, Gardiner C, et al. Minimal experimental requirements for definition of extracellular vesicles and their functions: A position statement from the International Society for Extracellular Vesicles. Journal of Extracellular Vesicles. 2014;**3**:1-21. DOI: 10.3402/jev.v3.26913

[38] Bobis-Wozowicz S, Kmiotek K, Sekula M, Kedracka-Krok S, Kamycka E, Adamiak M, et al. Human induced pluripotent stem cell-derived microvesicles transmit RNAs and proteins to recipient mature heart cells modulating cell fate and behavior. Stem Cells. 2015;**33**:2748-2761. DOI: 10.1002/stem.2078

[39] Camussi G, Deregibus MC, Bruno S, Grange C, Fonsato V, Tetta C. Exosome/microvesicle-mediated epigenetic reprogramming of cells. American Journal of Cancer Research. 2011;**1**:98-110

[40] Bobis-Wozowicz S, Kmiotek K, Kania K, Karnas E, Labedz-Maslowska A, Sekula M, et al. Diverse impact of xeno-free conditions on biological and regenerative properties of hUC-MSCs and their extracellular vesicles. Journal of Molecular Medicine. 2017;**95**:205-220. DOI: 10.1007/s00109-016-1471-7

[41] Escudier B, Dorval T, Chaput N, André F, Caby M, Novault S, et al. Vaccination of metastatic melanoma patients with autologous dendritic cell (DC) derived-exosomes: Results of the first phase I clinical trial. Journal of Translational Medicine. 2005;**3**:1-13. DOI: 10.1186/1479-5876-3-10

[42] Pascucci L, Coccè V, Bonomi A, Ami D, Ceccarelli P, Ciusani E, et al. Paclitaxel is incorporated by mesenchymal stromal cells and released in exosomes that inhibit in vitro tumor growth: A new approach for drug delivery. Journal of Controlled Release. 2014;**192**:262-270. DOI: 10.1016/j.jconrel.2014.07.042

[43] Xin H, Li Y, Liu Z, Wang X, Shang X, Cui Y, et al. MiR-133b promotes neural plasticity and functional recovery after treatment of stroke with multipotent mesenchymal stromal cells in rats via transfer of exosome-enriched extracellular particles. Stem Cells. 2016;**31**:2733-2746. DOI: 10.1002/stem.1409

[44] Kordelas L, Rebmann V, Ludwig A, Radtke S, Ruesing J, Doeppner T, et al. MSC-derived exosomes: A novel tool to treat therapy-refractory graft-versus-host disease. Leukemia. 2014;**28**:970-973. DOI: 10.1038/leu.2014.41

[45] Williams DF. The Williams Dictionary of Biomaterials. Liverpool University Press. 1999. ISBN: 0853237344

[46] Jiang J, Papoutsakis ET. Stem-cell niche based comparative analysis of chemical and nano-mechanical material properties impacting ex vivo expansion and differentiation of hematopoietic and mesenchymal stem cells. Advanced Healthcare Materials. 2013;**2**:25-42. DOI: 10.1002/adhm.201200169

[47] Perán M, García MA, López-Ruiz E, Bustamante M, Jiménez G, Madeddu R, et al. Functionalized nanostructures with application in regenerative medicine. International Journal of Molecular Sciences. 2012;**13**:3847-3886. DOI: 10.3390/ijms13033847

[48] Reis R, Cohn D. Polymer Based Systems on Tissue Engineering, Replacement and Regeneration. Springer Science & Business Media; Nato Science Series II 2002. DOI: 10.1007/978-94-010-0305-6. ISBN: 9781402010002

[49] O'Brien F. Biomaterials & scaffolds for tissue engineering. Materials Today. 2011;**14**:88-95. DOI: 10.1016/S1369-7021(11)70058-X

[50] Khan R, Khan MH. Use of collagen as a biomaterial: An update. Journal of Indian Society of Periodontology. 2013;**17**:539-542. DOI: 10.4103/0972-124X.118333

[51] Patino MG, Neiders ME, Andreana S, Noble B, Cohen RE. Collagen as an implantable material in medicine and dentistry. Journal of Oral Implantology. 2002;**28**:220-225. DOI: 10.1563/AAID-JOI-D-13-00063

[52] Chen JK, Chang CJ. Fabrications and applications of stimulus-responsive polymer films and patterns on surfaces: A review. Materials. 2014;**7**:805-875. DOI: 10.3390/ma7020805

[53] Kondiah PJ, Choonara YE, Kondiah PPD, Marimuthu T, Kumar P, du Toit LC, et al. A review of injectable polymeric hydrogel systems for application in bone tissue engineering. Molecules. 2016;**21**:1-31. DOI: 10.3390/molecules21111580

[54] Maitz M. Applications of synthetic polymers in clinical medicine. Biosurface and Biotribology. 2015;**1**:161-176. DOI: 10.1016/j.bsbt.2015.08.002

[55] Tomita M, Lavik E, Klassen H, Zahir T, Langer R, Young MJ. Biodegradable polymer composite grafts promote the survival and differentiation of retinal progenitor cells. Stem Cells. 2005;**23**:1579-1588. DOI: 10.1634/stemcells.2005-0111

[56] Strohbach A, Busch R. Polymers for cardiovascular stent coatings. International Journalof Polymer Science. 2015;**2015**:1-11. DOI:10.1155/2015/782653

[57] Bhang S, Lim J, Choi C, Kwon Y, Kim B. The behavior of neural stem cells on biodegradable synthetic polymers. Journal of Biomaterials Science, Polymer Edition. 2007;**18**:223-239. DOI: 10.1163/156856207779116711

[58] Yim EK, Reano RM, Pang SW, Yee AF, Chen CS, Leong KW. Nanopattern-induced changes in morphology and motility of smooth muscle cells. Biomaterials. 2005;**26**:5405-5413. DOI: 10.1016/j.biomaterials.2005.01.058

[59] Sekula M, Domalik-Pyzik P, Morawska-ChochóŁ A, Czuchnowski J, Madeja Z, Zuba-Surma E, et al., editors. Utilization of biocompatible and biodegradable polymers in stem cell research and biomedical applications. In: 27th European Conference on Biomaterials (ESB 2015); Krakow, Poland; 2015

[60] Sengel CT. Delivery of nanoparticles for the treatment of cardiovascular diseases. Global Journal of Obesity, Diabetes and Metabolic Syndrome. 2015;**2**:18-21. DOI: 10.17352/2455-8583.000010

[61] Ambrosio AM, Sahota JS, Khan Y, Laurencin CT. A novel amorphous calcium phosphate polymer ceramic for bone repair: I. Synthesis and characterization. Journal of Biomedical Materials Research. 2001;**58**:295-301. DOI: 10.1002/1097-4636(2001)58:3<295 ::AID-JBM1020>3.0.CO;2-8

[62] Smith IO, McCabe LR, Baumann MJ. MC3T3-E1 osteoblast attachment and proliferation on porous hydroxyapatite scaffolds fabricated with nanophase powder. International Journal of Nanomedicine. 2006;**1**:189-194. DOI: 10.2147/nano.2006.1.2.189

[63] Al-Sanabani J, Madfa A, Al-Sanabani F. Application of calcium phosphate materials in dentistry. International Journal of Biomaterials. 2013;**2013**:1-12. DOI: 10.1155/2013/876132

[64] McEntirea BJ, Bala BS, Rahamanc MN, Chevalierd J, Pezzottie G. Ceramics and ceramic coatings in orthopaedics. Journal of the European Ceramic Society. 2015;**23**:4327-4369. DOI: 10.1016/j.jeurceramsoc.2015.07.034

[65] Wang M. Developing bioactive composite materials for tissue replacement. Biomaterials. 2003;**24**:2133-2151. DOI: 10.1016/S0142-9612(03)00037-1

[66] Nair M, Elizabeth E. Applications of titania nanotubes in bone biology. Journal of Nanoscience and Nanotechnology. 2015;**15**:939-955. DOI: 10.1166/jnn.2015.9771

[67] Vanderleyden E, Mullens S, Luyten J, Dubruel P. Implantable (bio)polymer coated titanium scaffolds: A review. Current Pharmaceutical Design. 2012;**18**:2576-2590. DOI: 10.2174/138161212800492903

[68] Oliveira NT, Guastaldi AC. Electrochemical stability and corrosion resistance of Ti-Mo alloys for biomedical applications. Acta Biomaterialia. 2009;**5**:399-405. DOI: 10.1016/j.actbio.2008.07.010

[69] Gao Y, Zou S, Liu X, Bao C, Hu J. The effect of surface immobilized bisphosphonates on the fixation of hydroxyapatite-coated titanium implants in ovariectomized rats. Biomaterials. 2009;**30**:1790-1796. DOI: 10.1016/j.biomaterials.2008.12.025

[70] Elias CN, Oshida Y, Lima JH, Muller CA. Relationship between surface properties (roughness, wettability and morphology) of titanium and dental implant removal torque. Journal of the Mechanical Behavior of Biomedical Materials. 2008;**1**:234-242. DOI: 10.1016/j.jmbbm.2007.12.002

[71] Kim SE, Song SH, Yun YP, Choi BJ, Kwon IK, Bae MS, et al. The effect of immobilization of heparin and bone morphogenic protein-2 (BMP-2) to titanium surfaces on inflammation and osteoblast function. Biomaterials. 2011;**32**:366-373. DOI: 10.1016/j.biomaterials.2010.09.008

[72] Zancanela DC, Simao AM, Francisco CG, de Faria AN, Ramos AP, Goncalves RR, et al. Graphene oxide and titanium: Synergistic effects on the biomineralization ability of osteoblast cultures. Journal of Materials Science: Materials in Medicine. 2016;**27**:71. DOI: 10.1007/s10856-016-5680-y

[73] Xie C, Sun H, Wang K, Zheng W, Lu X, Ren F. Graphene oxide nanolayers as nanoparticle anchors on biomaterial surfaces with nanostructures and charge balance for bone regeneration. Journal of Biomedical Materials Research Part A. 2017;**5**:1311-1323. DOI: 10.1002/jbm.a.36010

[74] Kumar S, Chatterjee K. Comprehensive review on the use of graphene-based substrates for regenerative medicine and biomedical devices. ACS Applied Materials & Interfaces. 2016;**8**:26431-26457. DOI: 10.1021/acsami.6b09801

[75] Bikhof Torbati M, Ebrahimian M, Yousefi M, Shaabanzadeh M. GO-PEG as a drug nanocarrier and its antiproliferative effect on human cervical cancer cell line. Artificial Cells, Nanomedicine, and Biotechnology. 2017;**45**:568-573. DOI: 10.3109/21691401.2016.1161641

[76] Liu Y, Chen T, Du F, Gu M, Zhang P, Zhang X, et al. Single-layer graphene enhances the osteogenic differentiation of human mesenchymal stem cells in vitro and in vivo. Journal of Biomedical Nanotechnology. 2016;**12**:1270-1284. DOI: 10.1166/jbn.2016.2254

[77] Assunção-Silva RC, Gomes ED, Sousa N, Silva NA, Salgado AJ. Hydrogels and cell based therapies in spinal cord injury regeneration. Stem Cells International. 2015;**2015**:1-24. DOI: 10.1155/2015/948040

[78] Wang T, Lai JH, Yang F. Effects of hydrogel stiffness and extracellular compositions on modulating cartilage regeneration by mixed populations of stem cells and chondrocytes in vivo. Tissue Engineering Part A. 2016;**22**:1348-1356. DOI: 10.1089/ten.TEA.2016.0306

[79] Niranjan R, Koushik C, Saravanan S, Moorthi A, Vairamani M, Selvamurugan N. A novel injectable temperature-sensitive zinc doped chitosan/β-glycerophosphate hydrogel for bone tissue engineering. International Journal of Biological Macromolecules. 2013;**54**:24-29. DOI: 10.1016/j.ijbiomac.2012.11.026

[80] Susmita K. Introduction, classification and applications of smart materials: An overview. American Journal of Applied Sciences. 2013;**10**:876-880. DOI: 10.3844/ajassp.2013.876.880

[81] Wang ZL, Kang ZC. Functional and Smart Materials Structural Evolution and Structure Analysis. Plenum Press New York; 1998. DOI: 10.1007/978-1-4615-5367-0. ISBN-13: 978-1-4613-7449-7

[82] Bekas DG, Tsirka K, Baltzis D, Paipetis AS. Self-healing materials: A review of advances in materials, evaluation, characterization and monitoring techniques. Composites Part B: Engineering. 2016;**87**:92-119. DOI: 10.1016/j.compositesb.2015.09.057

[83] El Feninat F, Laroche G, Fiset M, Mantovani D. Shape memory materials for biomedical applications. Advanced Engineering Materials. 2002;**4**:91-104. DOI:10.1002/1527-2648(200203)4:3<91::AID-ADEM91>3.0.CO;2-B

[84] Chan BQY, Kenny Low ZW, Jun Wen Heng S, Chan SY, Owh C, Jun Loh X. Recent advances in shape memory soft materials for biomedical applications. Applied Materials & Interfaces. 2016;**8**:10070-10087. DOI: 10.1021/acsami.6b01295

[85] Murua A, Portero A, Orive G, Hernández RM, de Castro M, Pedraz JL. Cell microencapsulation technology: Towards clinical application. Journal of Controlled Release. 2008;**132**:76-83. DOI: 10.1016/j.jconrel.2008.08.010

[86] Ghidoni I, Chlapanidas T, Bucco M, Crovato F, Marazzi M, Vigo D, et al. Alginate cell encapsulation: New advances in reproduction and cartilage regenerative medicine. Cytotechnology. 2008;**58**:49-56. DOI: 10.1007/s10616-008-9161-0

[87] Hunt NC, Grover LM. Cell encapsulation using biopolymer gels for regenerative medicine. Biotechnology Letters. 2010;**32**:733-742. DOI: 10.1007/s10529-010-0221-0

Chapter 17

Identification of Fe_3O_4 Nanoparticles Biomedical Purpose by Magnetometric Methods

Zoia Duriagina, Roman Holyaka, Tetiana Tepla,
Volodymyr Kulyk, Peter Arras and Elena Eyngorn

Additional information is available at the end of the chapter

http://dx.doi.org/10.5772/intechopen.69717

Abstract

The application of magnetic nanoparticles for biomedical research is an interdisciplinary problem. The use of nano- and microsized powder materials as developed technology for obtaining bionanomaterials with magnetocatalytic properties has been investigated. Control over immobilization can be carried by means of magnetic properties. Synthesis of superparamagnetic nanoparticles is developed not only for the benefit of fundamental science, but also for many technologies, such as technologies of magnetic storage media, magnetic ink for printers, but mainly for biosensors and medical applications. All the biomedical applications require that the nanoparticles have high enough levels of saturation of magnetization; their size should be less than 100 nm with a small deviation in size. Appropriate coating of the surface of magnetic nanoparticles should be nontoxic, biocompatible with the target of bioorganic compound. The techniques of measurement of magnetic nanoparticle properties by means of vibrational magnetometers, as well as by means of a set of smart sensor devices in accordance with new concept of Internet of Things (IoTh), were described. The first method is based on vibrating sample magnetometer technique. The second method is based on direct measurement of three dimensions (3D) of nanoparticles' magnetic field components.

Keywords: magnetic nanoparticles, drug transportation, magnetic field sensor

1. Introduction

Use of nanomaterials has become one of the innovation research directions in materials science, biochemistry and medicine. Small sizes of nanoparticles (NPs) lead to the appearance of new unique functional properties. The methods of such material preparation are improving every year and become more accessible. The technology of material structure "design" by combining different types of materials, starting from metals, their compounds (oxides, nitrides, borides and hydrides) and ending with organic and inorganic polymers [1], has recently been widely used. Nanomaterials based on metals operate in power generating, optical industry and biomedicine. The unique superparamagnetic, optical and electrochemical properties are accentuated first of all. Use of magnetic nanoparticles (MNPs) for biomedical researches is an interdisciplinary problem and to solve it experts in different areas are required—medical workers, biologists, specialists in the field of materials science and electrodynamics. Application of nanomaterials with coating for biomedical purposes foresees such functional properties of the given system as adsorption, adhesion strength, biocompatibility, and certain magnetic and optoelectronic properties that will identify the stability of the obtained structure-energy state. The task of the materials science researchers is to develop complicated nanoobjects with optimal properties and morphological structure and to create the system of monitoring (Internet of Things (IoTh)) by their life activity cycle. The perfect biomedical systems can be considered such that the stable functional properties are possessed.

In particular, magnetism of nanoparticles is an important information carrier and what is especially valuable can be implemented through industrial means. This action at a distance in combination with a typical magnetic field penetration into human body tissues opens many new applications, including transport and purposeful delivery of biomagnets to a corresponding biological object [2]. Magnetism can be exhibited in a greater number of new nanomaterials. It is known that, under transition to nanosizes in metals and their compounds, new specific characteristics appear. Thus, magnetic properties of bulk gold and platinum are nonmagnetic, but at the nanosize they are magnetic. Surface atoms are not only different to bulk atoms, but they can also be modified by interaction with other chemical species, that is, by capping the nanoparticles. This phenomenon opens the possibility to modify the physical properties of the nanoparticles by capping them with appropriate molecules. Actually, it should be possible that nonferromagnetic bulk materials exhibit ferromagnetic-like behavior when prepared in a nanorange. One can obtain magnetic nanoparticles of Pd, Pt and the surprising case of Au (that is diamagnetic in bulk) from nonmagnetic bulk materials. In the case of Pt and Pd we can obtain the ferromagnetism. However, gold nanoparticles become ferromagnetic when they are capped with appropriate molecules: the charge localized at the particle surface gives rise to ferromagnetic-like behavior. This observation suggested that modification of the d band structure by chemical bonding can induce ferromagnetic-like character in metallic clusters [3].

2. Use of magnetic nanoparticles in medicine

Magnetic biomaterials are widely used in medicine, especially in cardiology, neurosurgery, oncology, radiology and cellular biology for cell separation, to perform immunological analysis, magnetic resonance spectroscopy, information preservation and so on [1–8]. They are also used as the X-ray contrast means and magnetosensitive composites for drugs and genes delivery, in radionuclide therapy and hyperthermia (**Figure 1**). Such applications are

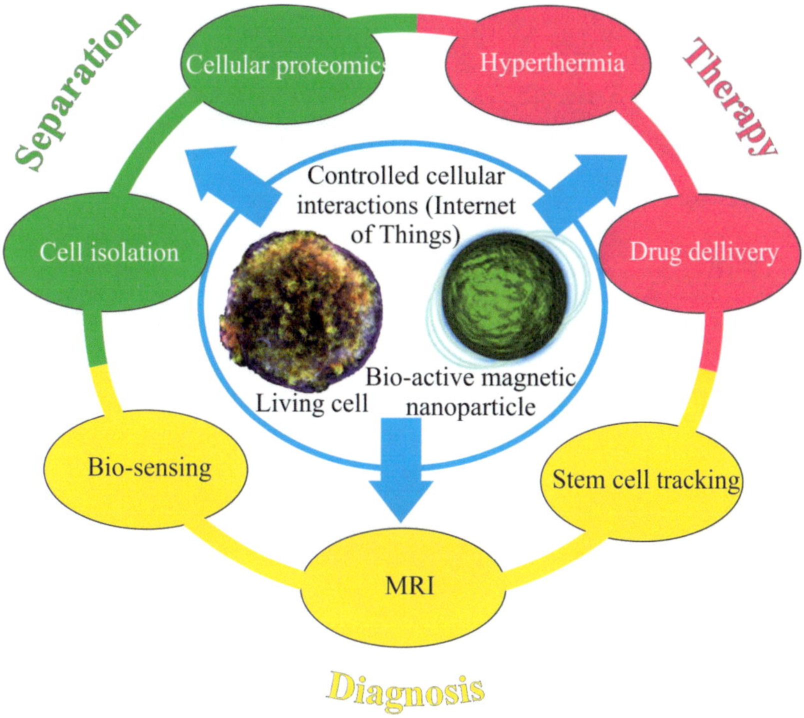

Figure 1. Use of magnetic nanoparticles in medicine.

perspective in case the clearly defined and checked interaction between magnetic nanoparticles and living cells is provided.

It is known that living organisms are built of cells with usual sizes of about 10 μm. The objects from which a cell consists of are much less than 1 μm. It should be considered that the synthesized nanoparticles are of sizes 2–450 nm, commensurable with the sizes of intercellular biological objects, viruses, proteins and genes. Thus, the nanoparticles by the sizes and mass occupy the intermediate place between single molecule and living cells. Still the main advantage of these materials is their ability to perform the preset functions under the effect of external magnetic field (**Figure 2**).

This occurs due to the formation of potential barriers at their boundary that limit the charge carrier movement in different directions and give the electron processes a mainly quantum character with dominating role of the interface. It should be noted that in this case a nonlinear dependence of equilibrium concentration of defects at the interface is observed that increases the dependence of properties on nanostructure sizes [10]. This is especially important when delivering drugs and means of diagnostics. Depending on the sizes, other properties of nanoparticles as toxicity, adsorption ability and magnetism also change. In particular, when the size of nanoparticles is less than 10 nm, they pass into a superparamagnetic state that, when adsorbing the energy of the external high-frequency electromagnetic field, promotes conversion of the energy state into the thermal one. This enables a tumor heating to a temperature of 43°C and thus its destruction [11, 12].

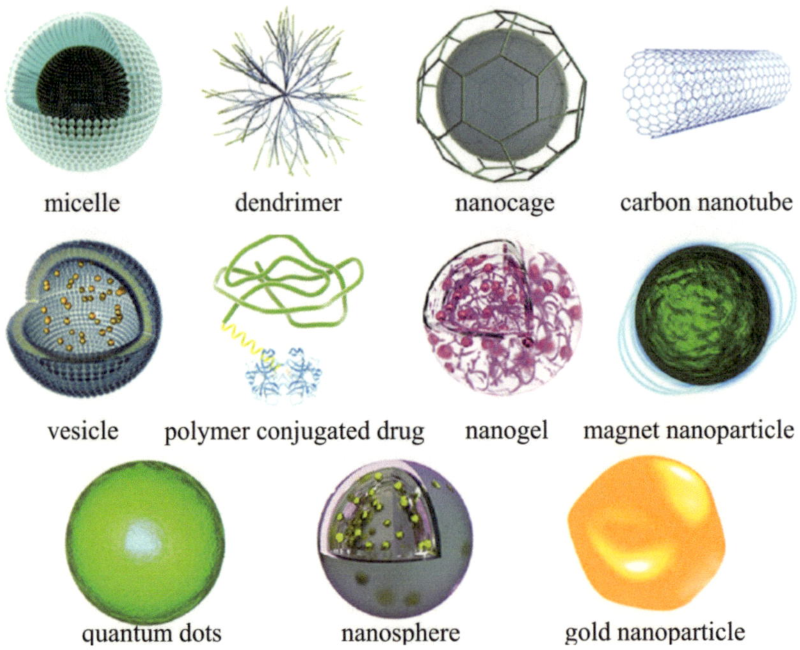

Figure 2. Several classes of NP used for the design of smart micro/nanocarriers, including micelles, dendrimers, nanocages, CNT, polymeric conjugates, nanogels, magnetic NP, quantum dots (QD), nanospheres and gold (Au) NP [9].

It has recently been proven that the effects of heat are more cytotoxic for cancer cells than they are for surrounding normal cells. Because of this, researchers have focused on hyperthermia as a method to selectively treat cancer cells. In Gilchrist's invaluable study, the cancer cells were heated locally using magnetic NP with a 1.2-MHz magnetic field. Since then, other observations have demonstrated that magnetic-induced hyperthermia in animals that have been injected directly with MNP can produce tumor regression through the application of magnetic fields to the solid tumors [9].

For therapeutic aims, magnetic nanoparticles are rarely used in a pure form. Usually, they are encapsulated or situated in bioinert matrices made of different organic compounds (**Figure 3**). This allows decreasing of the possible toxic effect of magnetic phase and simultaneous increasing of its stability due to immobilization on the surface of such capsules or matrices of medical aids. Encapsulation is usually carried out in suspensions of ultradispersive ferromagnetic or superparamagnetic particles containing stabilizing reagents—so called magnetic liquids [2, 5].

The structure of a multifunctional/multimodality MNP with a magnetic core, a polymeric coating and targeting ligands extended from the surface of MNP with the aid of polymeric spacers. Therapeutic payloads (drugs and genes) and imaging reporters (fluorophores and radionuclides) can be either embedded in the coating, or conjugated on the surface [5].

Magnetic contrast agents using which an image of a human body is obtained with a nuclear magnetic resonance tomography are widely known. Particles of these agents consist of a core

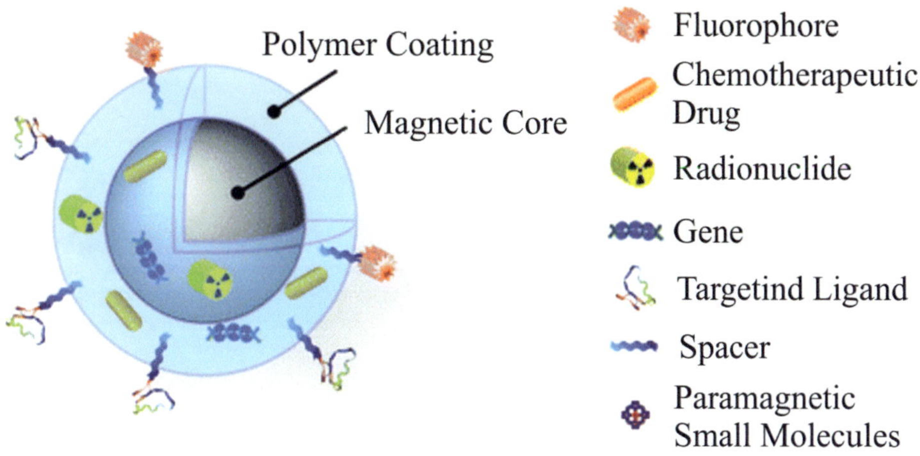

Figure 3. Graphic illustration of the structure of a multifunctional/multimodality MNP and different types of magnetic cores.

form magnetite or maghemite and a shell from dextran or silicon. Colloid solutions based on biomolecule-sorbed magnetic iron oxides, surrounded by a polymer layer, can be used for interaction with biological targets and direct them into a tumor with the aim of treatment or diagnostics. In this case, the controlled movement of magnetic particles in a body occurs because of the use of magnetic field gradient, regulated such that particles penetrate directly into the affected area [4].

Magnetic nanoparticle-based drugs can be divided into two groups: for external use (ointments, applicators, etc.) and for internal use (liquids, suspensions and magnetic containers). To improve physicochemical and medical biological properties when delivering drugs, the sizes of magnetic material particles in the blood stream should not exceed 100 nm. Under such conditions, they can be also used for hypoferritic anemia treatment. Today, the drugs obtained by adding precious metals to iron are paid more and more attention to. They are used as anti-inflammatory, antibacterial, anticancer and anesthetic compounds of the ointments for external use.

Magnetic carriers of medical purpose should be biocompatible with a human organism, nontoxic and nonallergenic. To improve biocompatibility, magnetic carriers are coated with chitosan, dextran, starch, carbon, gelatin, polymer starch coatings and so on (**Figure 4**). It should be noted that silicon oxide also increases biocompatibility of nanoparticles, and in this case, iron is localized inside the SiO_2 particles as in microcapsule. Release rate of drugs in these conditions can be regulated by changing the size and porosity of the obtained systems [4].

Recently, nano- and microsize powder materials are based on Fe_3O_4, especially in combination with bioselective elements—ferments have been widely used. This is important for developing technologies of manufacturing bionanomaterials with catalytic properties. High chemical activity of Fe_3O_4particles is caused by their higher ability to ion or atomic exchange, adsorption and formation of surface ligaments with other adsorbed particles. This guarantees

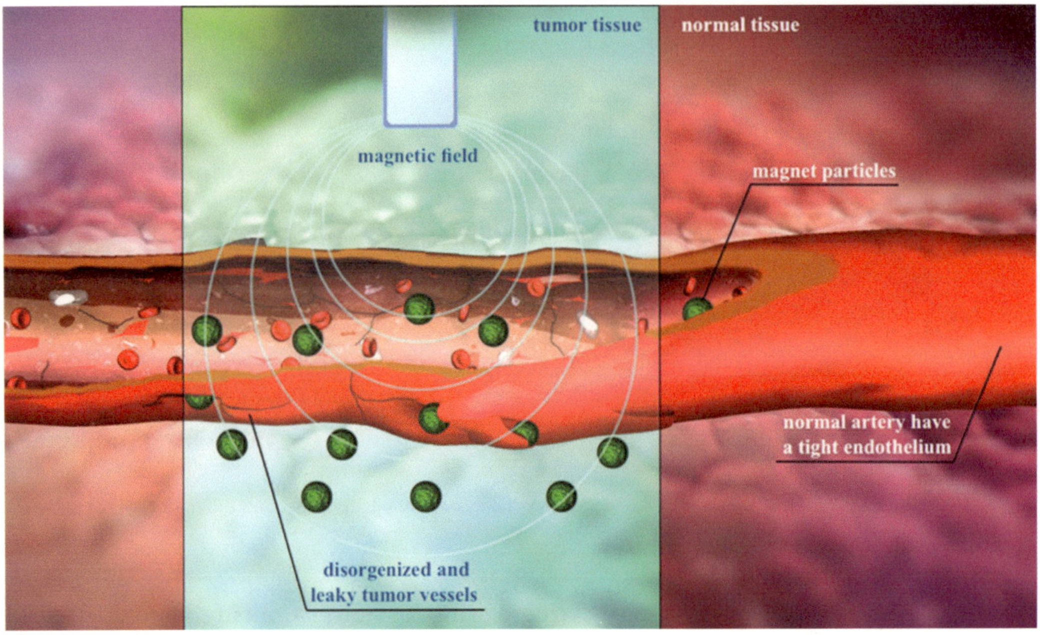

Figure 4. Schematic representation of the proposed magnetic sensitive carrier and triggered drug release mechanism [9].

the creation of bioparticles (immobilized ferments on the particle surface) with their further use in biosensorics and in enzymatic reactions. It is known that powder particles of Fe_3O_4 retain magnetization even if no external magnetic field is present, possessing own magnetic moment. However, remaining magnetization has a negative effect—a tendency of nanoparticles to agglomeration. Therefore, for the effective use of such magnetic particles, their chemical stability should be ensured by applying the corresponding coating on their surface.

2.1. Methods of obtaining magnetic nanoparticles

Physical and chemical properties of magnetic nanoparticles are determined by their structure and method of preparation. In most cases, these are particles of size from 1 to 100 nm with clearly expressed superparamagnetism. The most widely used methods of magnetic nanoparticle synthesis are as follows:

- Deposition—synthesis of iron oxides from water solutions of their salts in inert gas atmosphere at ambient or elevated temperature.
- Thermal decomposition—synthesis of Fe_3O_4 nanoparticles during decomposition of organic metal compounds in boiling organic solvents that contain stabilizing surface active agents (SAA).
- Microemulsion technology.

To synthesize magnetite-based nanocomposites, the following methodical approaches are mainly used [13–16].

- Precipitation in the conditions of oxidizing hydrolysis of iron sulfate Fe^{2+} in acid environment.
- Precipitation in the conditions of alkali hydrolysis of iron chlorides Fe^{2+} and Fe^{3+}.
- Precipitate deposition from the solutions of iron chlorides Fe^{2+} and Fe^{3+} under alkali hydrolysis of urea.

When modifying the conditions of magnetite core synthesis and forming the core shell, it is possible to obtain magnetic nanocomposites that differ not only by size but also by the ratio of "core–shell" sizes, magnetic susceptibility and complexity of surface microrelief that will specify the absorption properties.

However, nanoparticle synthesis has remained until now a complicated task. This is related with the difficulty of formation of homodispersive population of checked-size magnetic particles. It should be noted that there is no still a complete understanding of the mechanisms of core formation and growth. Besides the technology of obtaining crystalline nanoparticles with high saturation, magnetization needs further improvement. Moreover, nanoparticles lose their stability with time. This occurs due to the decrease of their free surface energy as a result of agglomeration. It is necessary to develop strategies of the surface coating preparation to simplify the process and effectively prevent agglomeration and segmentation of superparamagnetic particles. As a result, a stable solution for injections or frozen-dried (lyophilized) powder, that is easily dissolved, can be obtained [17].

Investigation of superparamagnetic nanoparticles, especially Fe_3O_4 nanoparticles, should be aimed at establishing the interrelations between their structure and pharmacokinetics. Nature of the coating on the iron oxide surface will determine not only the size of colloids but also will affect the pharmacokinetics and metabolic properties. This will enable modeling of their capture by reticuloendothelial system (RES) and facilitate diffusion to the tumor tissue. Processes of modification of magnetic nanoparticle surface, which are used for connection of biovectors, must be also improved. This is decisive in optimization of superparamagnetic nanoparticles likeness with biological objects [18].

2.2. Properties of Fe_3O_4 magnetite

Iron oxides (in particular magnetite) are among the most investigated materials in human history. Synthesis of superparamagnetic nanoparticles of iron oxide Fe_3O_4 is developing for the sake of fundamental science and technological applications, as for example magnetic data carriers, for biosensor, medical applications and magnetic inks [19–21]. Superparamagnetic nanoparticles of iron oxide with corresponding surface microgeometry can be used for increasing the image contrast, restoration of tissues, detoxification of biological fluids, hyperthermia, directed delivery of drugs and cell separation.

Detailed characterization of magnetite NPs is therefore necessary in order to obtain an accurate relationship between their electronic, magnetic and structural properties. Nominally claimed magnetite NPs are often (and to various extents) composed of nonstoichiometric oxide phases, and their instability in air ultimately causes oxidation to maghemite. The oxidization rate of magnetite NPs in ambient conditions is size-dependent and can range from several months for NPs <10 nm, up to several years for NPs ~100 nm. However, there is no general consensus regarding the complex oxidation mechanism, which presumably proceeds through a continual set of intermediate phases accompanied by cations and vacancies (C–V) reordering. Magnetite and maghemite both crystallize in the face-centered cubic (fcc) spinel structure, whose unit cell is composed of 32 O^{2-} ions placed at the 32e crystallographic position, and 24 Fe ions distributed over the 64 tetrahedral 8a (A) and 32 octahedral 16d (B) crystallographic positions (**Figure 5**). Magnetite and maghemite can be represented with a single formula: $(Fe^{3+})_A[(Fe^{2+})_{1-3\delta}(Fe^{3+})_{1+2\delta\ \delta}]_B O^4$, where stands for vacancies and $0 \leq \delta \leq 1/3$. In pure magnetite ($\delta = 0$), all A sites are occupied by Fe^{3+} ions, while B sites are equally occupied with Fe^{2+} and Fe^{3+}. In pure maghemite ($\delta = 1/3$), all Fe ions are in 3+ state, with a tendency to regular arrangement

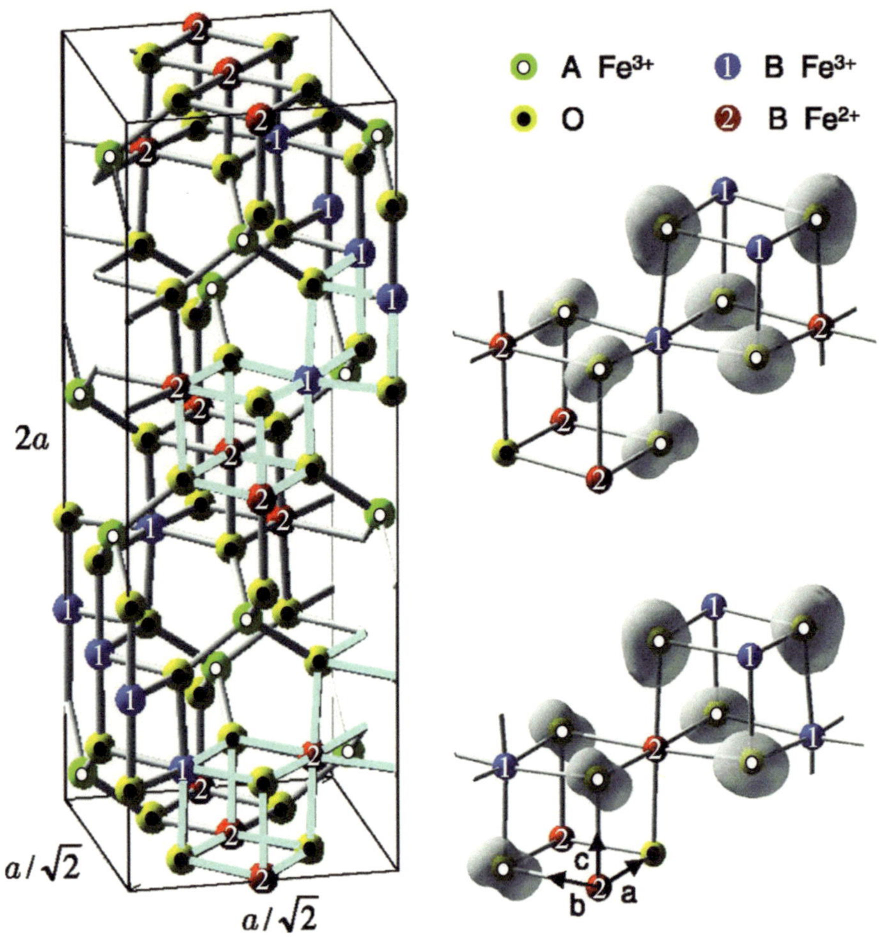

Figure 5. Left: monoclinic crystal structure of Fe_3O_4 in the low-temperature phase corresponding to a subcell of a2×a2×2a with P2/c symmetry. Right: 3D isosurfaces of unoccupied density of states of the O 2p integrated between the fermi level and 1 eV above within the corresponding Fe_3O_4 cubes highlighted in the crystal structure. For simplicity, we denote the octahedral Fe site with nominal 2+ and 3+ valences as Fe^{2+} and Fe^{3+}, respectively [22].

at B sites (two occupied followed by one vacant site). Vacancies occur preferentially at octahedral sites, but they can also mix over octahedral and tetrahedral sites. The degree of vacancy ordering in maghemite decreases with a particle size, and it is believed that for NPs smaller than 20 nm vacancy ordering vanishes. Magnetite and maghemite are both ferrimagnetic (the two uneven ferromagnetic sublattices Fe_A and Fe_B are antiferromagnetically aligned), with comparable saturation magnetizations (MS = 90 emu/g for magnetite; MS = 83.5 emu/g for well-ordered crystalline maghemite samples) and extraordinary high Curie temperatures (TC = 858 K magnetite; TC = 790–893 K maghemite, depending on the degree of C–V ordering).

The important difference between magnetite and maghemite is that maghemite is an insulator with energy gap Eg ≈ 2 eV, while magnetite is half-metal with much narrower Eg = 0.1–0.5 eV (depending on the sample quality). Furthermore, bulk magnetite undergoes so-called Verwey order-disorder phase transition to the insulating state at temperatures 120–125 K, accompanied with structural change from cubic to monoclinic lattice symmetry and various anomalies in the physical properties. The decisive influence on this complex transformation is ascribed to charge and orbital ordering involved in the three-site distortions. The exact structural parameters of the low temperature (LT) crystal structure, which is thought to have at least four inequivalent octahedral Fe sites, are extremely difficult to determine.

Things are even more elusive when the sample size is in the nanometer range. According to some recent reports for NPs with the mean size ~ 50 nm, the Verwey temperature (TV) shifts down to 20 K and it cannot be observed for smaller particles. The Verwey transition is weakly size-dependent in magnetite NPs larger than 20 nm, slightly suppressed in NPs smaller than 20 nm, and completely vanishes for NPs smaller than 6 nm. These inconsistencies are often ascribed to the fact that final properties of magnetite NPs strongly depend also on structural order [23].

2.3. Investigation of the structure and properties of Fe_3O_4 nanoparticles

In this research, the Fe_3O_4–NP characterization was performed by infrared Fourier spectroscopy using spectrophotometer Bruker Vertex 70 with attachment attenuated total reflectance (ATR). Powder samples were dried on the microscope slide surface and irradiated by a laser beam.

The intensive peak from Fe_3O_4-NP at 600 cm^{-1}, similarly to micro X-ray spectral analysis, indicated the presence of Fe-O bonds in Fe_3O_4–NP (**Figure 6**, **Table 1**). Peak intensity increases from the value of 1200 cm^{-1}, and in the range of 600–1200 cm^{-1} a significant background of the wave number area of the substances of glass on which the samples were dried is observed. In this range, only a signal from Si-O (for a wave number 1053 cm^{-1}) is clearly seen. The blind area should contain signals from C-O (for a wave number of about 1100 cm^{-1}) and signals of group Si–O–C (for a wave number of about 1108 cm^{-1}).

Obtained data can be used as an introductory information for processing monitoring of dynamics in situ of nanoparticle modification.

Size of Fe_3O_4 particles strongly affects its microstructure and properties. **Figure 7** presents a difference in microstructure of Fe_3O_4 micro- and nanoparticles.

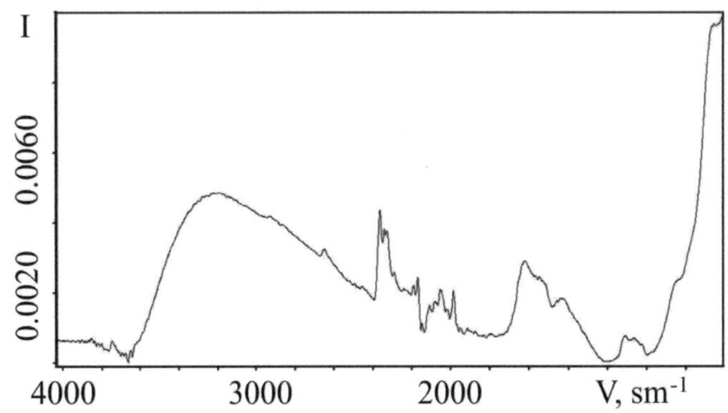

Figure 6. IR-Fourier spectra of Fe_3O_4 nanoparticles.

To estimate the sizes of Fe_3O_4-NP atomic-force microscopy was used, which software products allowed the establishment of the scanned particle structure (**Figure 8**). Investigation data prove the results of microscopic investigations on the spherical nanoparticle accumulation. Two types of structural inhomogeneity can be distinguished on the investigated sample surface: conglomerates of conical-like nanoparticles and granular texture of substrate. Surface topography is characterized by a rough relief with morphological regions of blocked structure. Blocks are characterized by nonisometric round form with no surface faceting. Their height above the substrate surface is in the range from 5 to 10 nm and base diameter from 30 to 50 nm. In this case, the dominant size of nanoparticles is 8 nm.

It is established at the same time that the size of Fe_3O_4microparticles is about 1μm (**Figure 9(b)**).

Use of Fe_3O_4nanoparticles with the aim of their functionalization (application of shells, medical aids and markers on them) or introduction in a living organism for hyperthermia foresees the application of surface coating. In this case, the analysis of the dimensions and properties of nanoparticles using a simple method becomes more complex [24]. In such cases, magnetic methods become one of the methods of particle categorization.

2.4. Methods of magnetic properties investigation

Among the magnetic research methods in materials science, the magnetic phase analysis is especially widely used. Its possibilities and effectiveness to a great extent are determined

Wave number (cm^{-1})	Chemical bond
3230	-OH valence vibrations
1617	-OH deformation ("wagging")
1421	Not described
600	Fe-O

Table 1. Correspondence of IR-Fourier spectra peaks of Fe_3O_4–NP to chemical bonds [13–16].

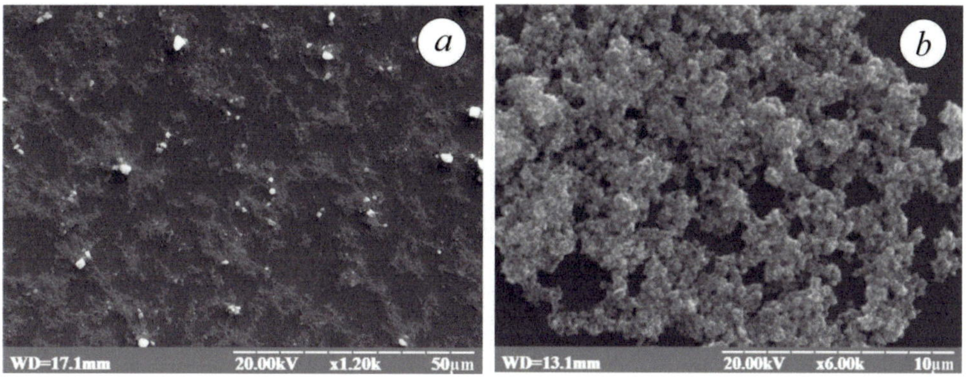

Figure 7. Microstructure of synthesized Fe_3O_4: (a) Fe_3O_4 nanoparticles; (b) Fe_3O_4 microparticles. Microstructures were obtained according to the scanning microscopy data.

by technical characteristics of the equipment. The quantitative phase magnetic analysis uses properties of ferromagnetics, which they acquire in strong magnetic fields—in the state of technical saturation. Primary magnetic properties that are structurally insensitive are obtained from the curve of temperature dependence of saturation magnetization. Such characteristics are magnetization and Curie point. These values give the information about phase composition of material and its changes in the process of certain thermal operations and also in the process of deformation. Curie point and saturation magnetization are called primary magnetic properties because their values are determined by the nature of the ferromagnetic phase (crystal lattice, electron structure of atoms and chemical phase composition) [25].

Changes in magnetization and Curie point of separate phases observed in the process of investigation of a certain material are not caused by their particle growth but are conditioned by the change of chemical composition of phases and their crystalline composition only. Based on these data, it is possible to study the kinetics of phase formation at the very early stages of the process (for particles sizes up to 10^{-6} cm) by the growth of magnetization and Curie point to their normal values. Thus, a weak dependence of saturation magnetization and Curie point

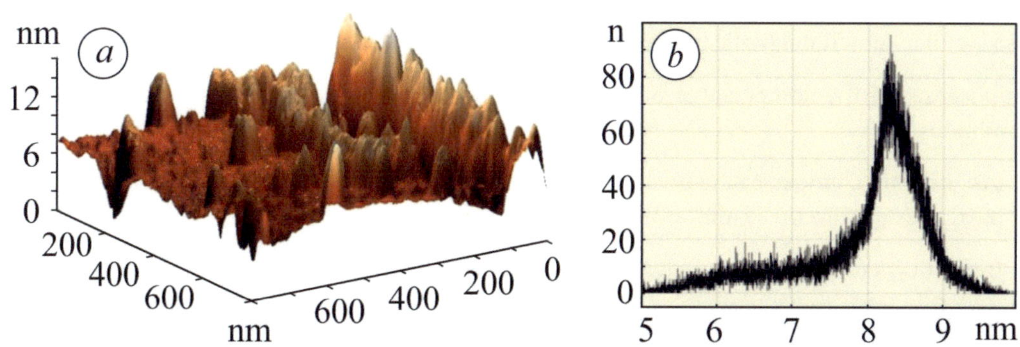

Figure 8. Character of distribution of synthesized Fe_3O_4-NP on the surface (a) and histogram of their distribution by size (b). Lateral dimension of particles in nanometers is given on X-axis, and a number of scanned particles on Y-axis.

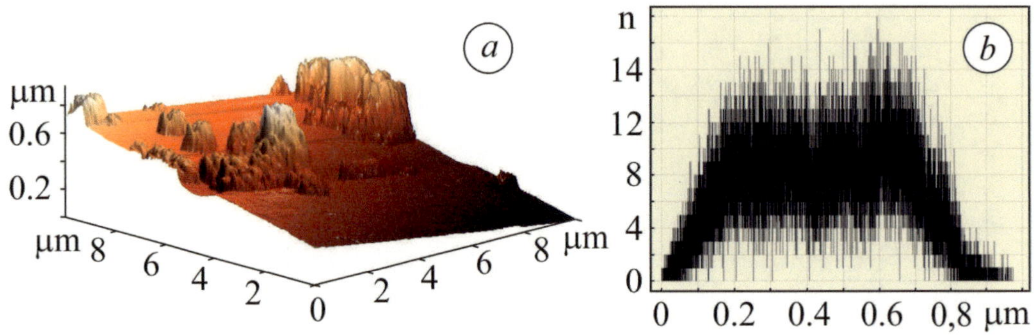

Figure 9. Character of distribution of synthesized microparticles of Fe_3O_4(a) and histogram of their distribution by size (b). Transverse dimension of particles in nanometers is given on X-axis, and a number of scanned particles on Y-axis (b).

on the stress state, form and degree of ferromagnetic dispersion allows us to choose these physical parameters as the quality characteristics of phases. Just these factors that strongly affect the results of the quantitative X-ray analysis, in the magnetic phase analysis, practically have no effect on the investigation results.

To carry out magnetic phase analysis, it is necessary to build experimentally a temperature dependence of saturation magnetization of the tested material sample. Equipment for the analysis should provide measuring of magnetic properties in a wide range of temperatures and heating and cooling rates. Peculiarity of the determination of the transition temperature of a number of dispersive magnetic systems to the paramagnetic state is that their magnetic phase analysis must be performed at high heating rates. Since in a number of cases the mass of the investigated samples can be very small (nanomaterials, fine-dispersive powders, amorphous films, surface layers after different types of heat treatment) and as a result the value of the magnetic moment is relatively small (from 10^{-2} to 10^{-3} A·м^2), the equipment must possess a very high sensitivity.

For such investigations, it is most reasonable to use the method of vibrating specimen that is realized through vibration magnetometers [26].

Measurement accuracy with vibration magnetometer depends on the calibration accuracy. We have proposed the method of comparison when magnetometer is calibrated using a standard with a known magnetic moment. As a standard, carbonyl iron of certain batches is used.

Application of the comparative method for calibration requires ensuring of the following conditions:

- insignificant changes in the position and deviation from the perfect sphere-like symmetry have no influence on a calibration constant;
- frequency and amplitude of vibrations remain constant;
- sample holder (container) introduces only an insignificant value to the signal value and can be subtracted from it; and
- signal voltage—linear function of the magnetic moment of the sample.

Two measuring methods are used in vibration magnetometers—direct or compensation. The last one, in its turn, is of two types: the method of current shell and differential method. Compensation method allows avoiding the dependence of measuring results on the values of amplitude and frequency of vibrations. It is optimal to perform high-temperature measuring preliminary assuming the measures of stabilization of sample mechanical vibration. Constant amplitude and frequency of sample vibrations is a required condition of providing an adequate accuracy of measuring by the direct method.

The main cause of vibration amplitude instability is the changes in the moving parts of the magnetometer. Stability of measuring parameters is determined, in addition, by the stability of the alternating current generator which feeds the vibrator. Vibration amplitude stabilization was realized by providing a negative feedback of the generator.

Such a device for investigation of the magnetic properties possesses a number of unique characteristics. At room temperature, it is possible to build a hysteresis loop, partial hysteresis loops, initial curves of magnetization and demagnetization, and dependence of a magnetic moment on the sample orientation. At elevated temperatures, thermomagnetic measurements, high-temperature hysteresis measurements, and time dependences of magnetic moment at different temperatures can be carried out. Temperature measuring interval ranges from 80 to 1100 K at power 3.5 кWt. Magnetic measurement in this work was performed using a vibration magnetometer [15]. Curves of the investigated samples over magnetization were recorded in magnetic fields from −200 кA/m to +200 кA/m by measuring the dependence of the given magnetization I/I_{200} on the magnetic field strength H, I_{200}—magnetization at magnetic field strength 200 кA/m.

As one can see, curves of Fe_3O_4-NP sample remagnetization have a nonhysteresis form with a zero coercive force Hc and residual magnetization Ir (**Figure 10**(a)). Probably particles of Fe_3O_4 due to a high degree of dispersivity are in a superparamagnetic state—state that is typical of microscopic and nanoscopic particles of ferromagnetic materials. In this case, a magnetic moment of the like particle changes its orientation spontaneously and randomly or due to thermal fluctuations. When external magnetic field is absent, superparamagnetics

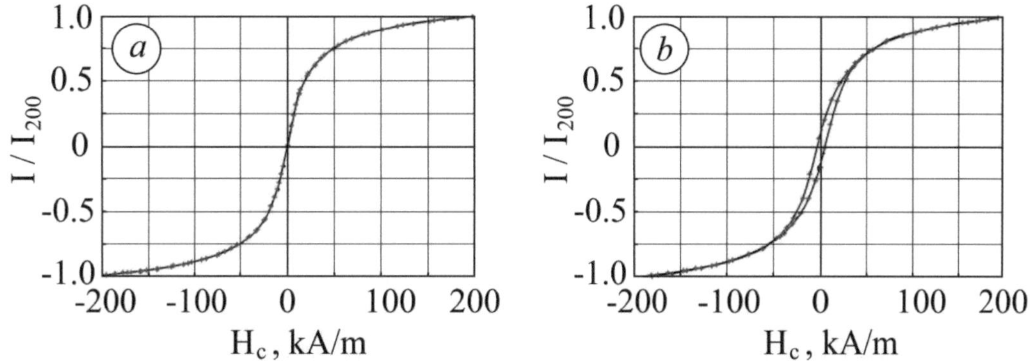

Figure 10. Curves of remagnetization of Fe_3O_4-NP (a) and Fe_3O_4-MP (b) samples.

have on average a zero magnetic moment, i.e., they behave like paramagnetics with a high magnetic susceptibility. As known, superparamagnetic properties of Fe_3O_4 particles at room temperature are exhibited, when reaching an average diameter D < 25 nm. In our case, the particles size is 5–10 nm.

With increasing sizes of Fe_3O_4 particles to microsizes, the magnetic properties do not change. These particles have already a hysteresis loop (**Figure 10(b)**). Samples' coercive force is H_c =4 кA/m, and ratio is I/I_{200}=0.1.

2.5. Use of the Internet of Things to measure of magnetic properties

For a remote study of magnetic properties of nanoparticles, it is proposed in this work to use the approaches of the Internet of Things—a network that consists of the interrelated physical objects or devices that contain mounted gauges. Their software allows one to pass and exchange data between physical world and computer systems using the conventional procedure of communication. In addition to gauges, the network contains executing devices, built-in physical objects interrelated by wire or wireless nets. These interrelated objects possess function of reading, putting in operation, programming and identification of objects and also allow one to exclude the necessity of person participation due to application of the intellectual interfaces.

Technology of magnetic particle detection and characterization is based on many approaches of magnetic sensing techniques [27, 28].

Here we present novel 3D magnetic sensors which offer three-dimensional sensing of orthogonal B_X, B_Y, and B_Z projections of magnetic field vector. Nowadays, many types of techniques and magnetic sensors provide such possibility [29, 30]. Among them, three types, namely magnetoresistors, Hall sensors and magnetotransistors, are preferable.

The first type, magnetoresistors, is based on the effect of magnetoresistance which is a tendency of material to change the value of its electrical resistance in an externally applied magnetic field. There are a variety of effects that can be called magnetoresistance: some occur in bulk nonmagnetic metals and semiconductors, such as geometrical magnetoresistance, Shubnikov de Haas oscillations, or the common positive magnetoresistance in metals. Other effects occur in magnetic metals, such as negative magnetoresistance in ferromagnets or anisotropic magnetoresistance (AMR). Finally, in multicomponent or multilayer systems (e.g., magnetic tunnel junctions), giant magnetoresistance, tunnel magnetoresistance and extraordinary magnetoresistance can be observed [31].

As an example, Honeywell International Inc. produced a series of magnetic sensors based on anisotropic magnetoresistance effect. AMR is a property of material in which the dependence of electrical resistance on the angle between the direction of electric current and the direction of magnetization is observed. AMR arises from the simultaneous action of magnetization and spin-orbit interaction, and its mechanism depends on the material. It can be for example due to a larger probability of s-d scattering of electrons in the direction of magnetization, which is controlled by the applied magnetic field.

AMR technology provides advantages over other magnetic sensor technologies. These anisotropic, directional sensors feature precision in-axis sensitivity, linearity, and low cross-axis sensitivity. For example, the HMC5883L is a small-size surface-mount, multi-chip module designed for low-field 3D magnetic sensing with a digital interface for applications such as low-cost compassing and magnetometry. The HMC5883L includes the state-of-the-art, high-resolution HMC118X series magnetoresistive sensors plus an application-specific integrated circuit, containing amplification, automatic degaussing strap drivers, offset cancellation and a 12-bit analog-to-digit converter.

The second type, Hall sensors, is based on Hall effect which is due to the nature of the current in a conductor. Current consists of the movement of many small charge carriers, typically electrons, holes or ions. When magnetic field is present, these charges experience a force, called the Lorentz force. When such a magnetic field is absent, the charges follow approximately straight "line of sight" paths between collisions with impurities, phonons and so on. However, when a magnetic field with a perpendicular component is applied, their paths between collisions are curved so that moving charges accumulate on one face of the material. This leaves equal and opposite charges exposed on the other face where there is a scarcity of mobile charges. The result is an asymmetric distribution of charge density across the Hall element, arising from the force that is perpendicular to both the "line of sight" path and the applied magnetic field. The separation of charge establishes an electric field that opposes the migration of further charge, so a steady electrical potential is established for as long as the charge is flowing.

We have proposed new type of thin-film magnetic field sensors for three-dimensional sensing of orthogonal B_X, B_Y and B_Z projection of magnetic field vector [32–37], new method for measuring magnetic field [38] as well as their modeling and signal processing [39–44].

The general view of such thin-film 3D sensors, its active area, structure and photographs are presented in **Figures 11–13**, correspondingly. The sensor (**Figure 12(a)**) active area is the thin sensitive semiconductor film (2) of corresponding configuration formed at the semi-insulating GaAs substrate (1). Sensor contacts are formed by the metallization layer (3), which typically is the gold film together with other metals, for example, with titanium sublayer. For manufacturing of the structures, three photolithographies are used. The first one is intended

a *b*

Figure 11. General view of thin film 3D sensors (a) and its active area (b).

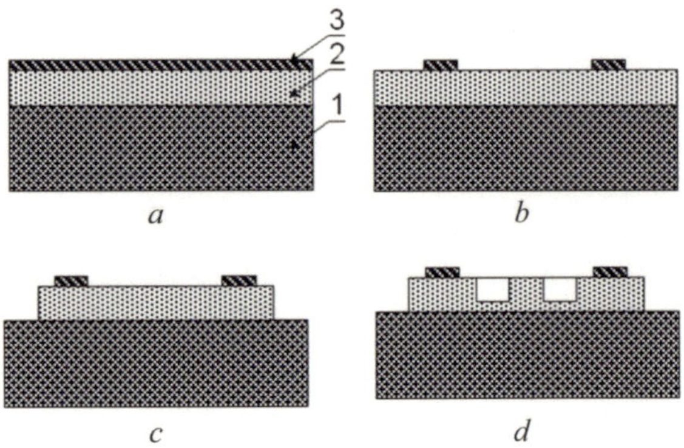

Figure 12. Structure and its formation stages (a), (b), (c), (d) of 3D thin film sensor: 1—substrate, 2—semiconductor active area; 3—contact metallization layer.

for creation of the contact system (**Figure 12(b)**); the second one is for etching a semiconductor film into the whole depth (**Figure 12(c)**); and the third one is for etching the semiconductor film to approximately 10% of their thickness (**Figure 12(d)**).

To show how a 3D thin-film sensor works, let us consider its main component—transducer #1 (**Figure 14**). The magnetic field vector projections B_X, B_Y are located in the transducer plane, and B_Z projection is perpendicular to this plane.

The operation principle of transducer of type #1 is the following: transducer is connected to the power source, typically direct current source; for this, the central current contact 3 is connected to the first power supply output, and side current outputs 4, 5 are connected together to the second power supply output. Thus, the current in the active area of the transducer is divided equally and run in mutually antithetical directions in relation to the current output.

Neglecting the current in the 6 and 7 voltage contacts' circuit, one may consider that the voltages at those outputs are equal to the corresponding voltages of the intermediate areas 10 and

Figure 13. Photos of 3D thin-film sensors: wafer (a), chip (b) and active area (c).

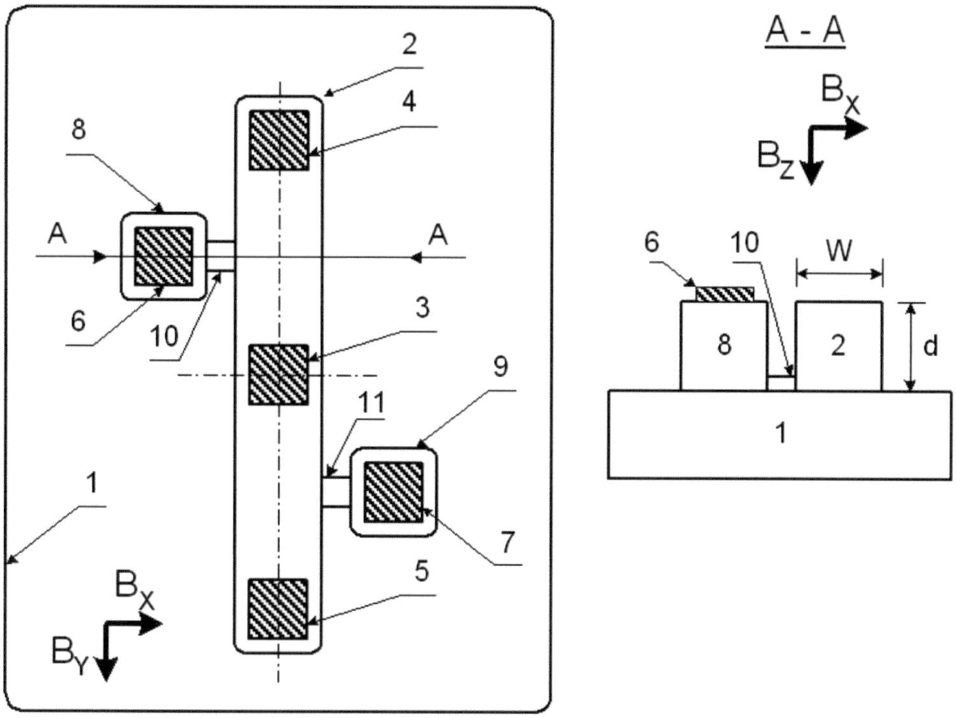

Figure 14. Structure of type #1 transducer: 1—substrate, 2—semiconductor active area; 3—central current contact; 4, 5—side current contacts; 6, 7—voltage contacts; 8, 9—auxiliary areas; 10, 11—intermediate areas.

11. So the construction of the type #1 transducer provides the possibility of forming the output voltage at the 6 and 7 voltage contacts, which is equal to the voltage difference between the areas of lower surface of the active area in the neighborhood for the contacting to the intermediate areas 10 and 11. The voltage measurement at the areas of lower active area surface is obvious that the thickness of intermediate areas 10, 11 was considerably thinner (at least 10 times) than the thickness of active area 2.

The principal difference of the type #1 transducer from analogs, for example Refs. [29, 30], is that in the first one the voltage difference is measured between the areas of lower active area surface, and in the analogs—between the areas of the upper active area surface. In its turn, the absence of contacts with the upper active area surface makes it possible not to perform the active area surface insulation.

In general case, the voltages at the voltage outputs 6, 7 are composed of three components. The first component V_R is caused by the voltage drop at the semiconductor active area 2. Taking into consideration the symmetry of the active area in relation to the first current contact 3, the first voltage component at both voltage contacts 6, 7 is equal $V_R(6) = V_R(7)$. The second component V_Z is caused by the influence of B_Z magnetic field vector projection, which is perpendicular to the transducer plane. Taking into consideration the transducer construction and current flow directions in it, the second voltage component at the voltage contacts 6 and 7 is also equal $V_Z(6) = V_Z(7)$. The third component V_X is caused by the influence of B_X projection of induction vector. Contrary to the two stated above, this component at the voltage contacts 6 and 7 has the opposite signs $V_X(6) = -V_X(7)$. In particular, in the upper part of active area, two charge carriers under the influence of electromotive force deviated to the upper semiconductor layer surface, and then, in the lower part—to the lower surface (in the direction of substrate). The voltage difference caused by this carrier deviation is transferred to the voltage contacts 6 and 7 through the intermediate areas 10 and 11. It is to be noted that the B_Y magnetic field induction vector projection, parallel to the direction of current flow through the active area, does not cause the carrier deviation, and therefore, it may be neglected: $V_Y(6) = V_Y(7) = 0$.

Thus, the voltage difference between the voltage outputs 6 and 7 (output voltage of type#1 transducer) is determined only by B_X magnetic field induction vector projection, and in first approximation, it does not depend on other vector projections:

$$V_{OUT} = \left[V_R(6) + V_X(6) + V_Y(6) + V_Z(6)\right] - \left[V_R(7) + V_X(7) + V_Y(7) + V_Z(7)\right] = V_X, \quad (1)$$

where $V_X = V_X(6)+V_X(7)$.

The especially high efficiency of transducer of type#1 application is realized in case of their application in 3D sensor for simultaneous measurement of three projections B_X, B_Y and B_Z of the magnetic field induction vector. Such 3D sensor contains three coupled transducers at a single substrate. The construction of two transducers of #1 type is similar to **Figure 14**. They are mutually orthogonally rotated and provide the sensitivity to the BX, BY vector projections of the field induction vector. The third transducer, which provides the sensitivity to the BZ projection, is a traditional Hall transducer [29].

To provide the equal sensitivity levels for all three transducers of such 3D sensor, it is necessary to take into account the following: contrary to the traditional Hall transducers, which have the sensitivity as an inverse function of semiconductor active layer thickness d, the sensitivity of the transducer of type #1 is inversely proportional function of the semiconductor area width W. So it is recommended that the active area width W should be minimal and approximately equal to the semiconductor layer thickness d.

There are two other options to build up 3D magnetic thin film sensors. They are based on type #2 and type #3 transducers.

The structure of type #2 transducer is shown in **Figure 15**.

The type #2 transducer is fed by the direct current source, and one pair of mutually opposed current contacts (e.g., 3 and 5) is connected to one output of the power source (e.g., positive), and another pair (4 and 6 correspondently) to the other power source contact (negative, correspondently). In accordance with this connection scheme, the four current flow circuits are formed: I_{34}, I_{36}, I_{54} and I_{56}, where the indices in the marked currents correspond to the numbers of current contacts. These current flow circuits geometrically form the square sides. The matter of principal importance is that the currents that flow in opposed square sides are equal by value and opposed by sign: $\vec{I}_{34} = -\vec{I}_{65}$; $\vec{I}_{36} = -\vec{I}_{45}$. The output signal of transducer is the voltage difference between the voltage contacts. The informative signals about the magnetic field vector projections B_X, B_Y, B_Z are the voltages V_X, V_Y, V_Z, which in first approximation are determined as:

$$V_X = V_8 - V_{10};\ V_Y = V_7 - V_9;\ V_Z = \{(V_8 - V_9) + (V_{10} - V_7)\}/2 \tag{2}$$

The difference of type #2 transducer from its analog [29, 30] is the limitation of the area of current transition between the 2nd outer and 11 inner insulating areas. This provides the increase of the sensor sensitivity to the B_X, B_Y magnetic fields and decreases the cross-impact between the informative signals.

The sensitivity increase is explained by the reduction of the transducer active area size. Contrary to the traditional Hall transducers (sensitive to the field B_Z perpendicular to the transducer plane), where the size that determines the sensitivity is the active area thickness, in case of the B_X, B_Y field transducers, the determinative size is the current scattering region. The smaller area occupies the current scattering region, the higher is the voltage difference between the two corresponding voltage contacts, and therefore the higher sensitivity is.

The structure of type #3 transducer is shown in **Figure 16**.

Arms 2, 4 and contacts 6, 8, 10, 12 form the first vertical Hall transducer, and arms 3, 5 and contacts 7, 9, 11, 13—the second vertical Hall transducer. The first transducer is intended for measuring the B_X magnetic field vector projection, and the second one—for measurement of B_Y projection. The measurement principle of Hall transducers consists in forming the voltage difference at the voltage outputs during the deviation of charge carriers in the semiconductor area under the electromotive force influence.

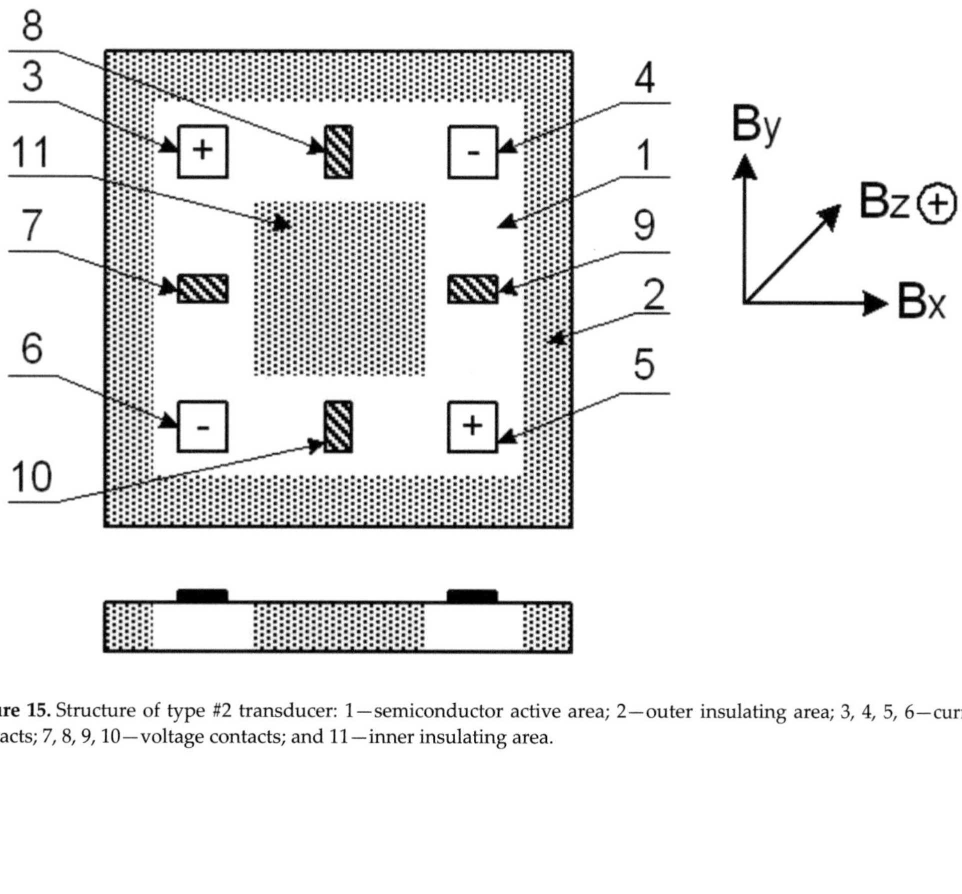

Figure 15. Structure of type #2 transducer: 1—semiconductor active area; 2—outer insulating area; 3, 4, 5, 6—current contacts; 7, 8, 9, 10—voltage contacts; and 11—inner insulating area.

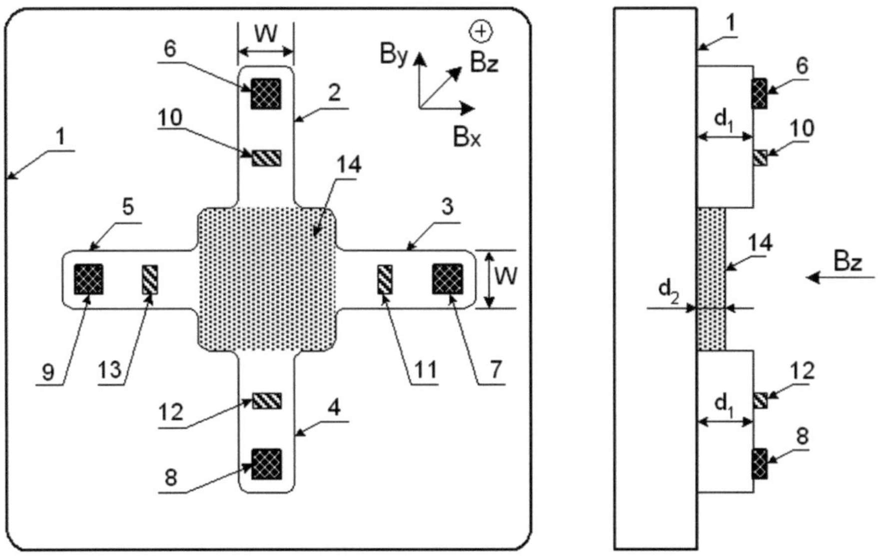

Figure 16. Structure of type #3 transducer: 1—substrate; 2, 3, 4, 5—four arms of cross-shaped figure, formed by the crossing of two semiconductor areas of vertical Hall transducers; 6, 7, 8, 9—current contacts; 10, 11, 12, 13—voltage contacts; and 14—semiconductor area of horizontal Hall transducer.

The operation of type #3 transducer presumes two power supplying modes. The first one provides the operation of the vertical Hall transducers, and the second one for operation of the horizontal one.

The first power-supplying mode presumes the connecting of the current contacts of the vertical Hall transducers, when the currents in their semiconductor areas are flowing in mutually opposed directions. For this, the current contacts of both vertical Hall transducers are connected into one circuit. Namely, contacts 6 and 8 of the first vertical Hall transducer are connected together and to the first, for example, positive output of the power source, and contacts 7 and 9 of the second vertical Hall transducer are also connected together to the second, therefore, negative output of the power source. So, in the first vertical Hall transducer, the currents are flowing from the top to the bottom (current I_2 in arm 2) and from the bottom to the top (current I_4 in arm 4), and in the second—from left to right (current I_3in arm 3) and from right to left (current I_5 in arm 5). In case of ideal symmetry of the transducer structure, the following equality takes place: $\vec{I} = -\vec{I}; \vec{I} = -\vec{I}$.

Output signals of vertical Hall transducers are formed at the voltage outputs as a voltage difference, which is proportional to the multiplication of the power supply current value of the transducer by the corresponding magnetic field vector projection:

$$V_X = V(12) - V(10) = K_X \times I \times B_X / W \text{–for the first transducer and} \quad (3)$$

and

$$V_X = V(12) - V(10) = K_X \times I \times B_X / W \text{–for the second transducer and} \quad (4)$$

where V(11), V(12), V(13) and V(14) are the voltages at the voltage outputs 11, 12, 13 and 14, correspondingly; V_X, V_Y and K_X and K_Y—output signals and transducing coefficients of the first and second transducers, correspondingly; I—operational current, W—the width of semiconductor areas.

From the physical point of view, the appearance of voltage difference at the voltage contacts of the vertical Hall transducers is explained by the fact that due to the opposed current flow directions in both transducer arms, the deviation of current carriers in those arms has also the opposed direction. In particular, in arm 2 of the first transducer, the carriers are deviated in the direction away from the substrate to the surface of the semiconductor area, and then, in arm 4 of this transducer—in the direction from the surface to the substrate.

In a high-gradient magnetic field, the voltage difference formed at the voltage outputs of the vertical Hall transducers is the informative value of the averaged field induction value. Taking into the consideration that all the voltage outputs of the vertical Hall transducers are equidistant from its center (crossing area), the measured averaged induction value corresponds to the spatial point of the transducer center.

The second power-supplying mode presumes the application of only one pair of current contacts, namely 6 and 8 of the first vertical transducer. Then, output 6 is connected to the first power supply output, and output 8 to the second one. This provides the linear trajectory of the charge carriers in semiconductor area 14 of the horizontal Hall transducer.

The output signal of the horizontal Hall transducer, as a voltage difference, is proportional to the multiplication of the transducer power supply current by the B_Z projection of the magnetic field induction vector, formed at the voltage outputs 11 and 13 of the second vertical Hall transducer: $V_Z = V(13)–V(11) = K_Z.I.B_Z/d_2$, where K_Z is the coefficient of transduction and d_2 is the thickness of semiconductor area of the horizontal Hall transducer.

From the physical point of view, the appearance of the voltage difference at the voltage outputs 11 and 13 is explained by the fact that charge carrier deviation in semiconductor area 14 of the horizontal Hall transducer under the influence of B_Z magnetic field induction vector projection is in the direction from arm 5 to arm 3, or backwards.

As in the vertical transducer, the horizontal Hall transducer measures the value of the magnetic field induction value in the transducer center spatial point. Thus, type #3 transducer allows measuring of all three projections B_X, B_Y and B_Z of the magnetic field induction vector in a single spatial point.

The other distinctive feature of the type #3 transducer is the possibility to change the d_2 thickness of the horizontal Hall transducer semiconductor area. This allows the formation of the transducer with equal sensitivity value to all three magnetic field induction vector projections: $V_X/B_X = V_Y/B_Y = V_Z/B_Z$. Taking into consideration that the sensitivity of vertical Hall transducers is inversely proportional to the semiconductor area width W, and for horizontal transducer—to the thickness of its semiconductor area d_2, the equality of the stated values of sensitivity is provided by the correspondent selection of W/d_2 ratio. It is important that contrary to the d_1 thickness of the vertical Hall transducer areas, the d_2 thickness of the horizontal transducer semiconductor area may change after the output formation. In particular, the decrease of d_2 thickness may be realized by partial etching of the semiconductor layer.

In relation to the analogs [29], the type #3 transducer provides the increase of the measurement accuracy and construction simplification.

The increasing accuracy is caused by the fact that horizontal Hall transducer is placed in the center (crossing area) of vertical Hall transducer. This provides the high spatial alignment of all transducers (two vertical and one horizontal), and therefore, all the three projections B_X, B_Y and B_Z of the magnetic field induction vector are measured at a single spatial point. During the measurement of high-gradient fields, this gives the ability to decrease the magnetic field induction vector measurement error in several times.

The third type, magnetoresistors, is mostly bipolar or field-effect transistors whose structures and operating conditions are optimized with respect to the magnetic sensitivity of its output currents. There are three major effects of magnetotransistors, namely the current deflection effect, the injection modulation and the magnetodiode effect.

We have proposed the structure of magnetotransistors adapted to three dimensional sensing of orthogonal B_X, B_Y and B_Z projections of the magnetic field vector (**Figures 17–19**). These are drift-aided lateral double-collector p-n-p transistors. The emitter area E and two collector p-type areas C1, C2 are embedded into plate-like n-type base region. The two base contact n^+ type areas B1, B2 are used to apply the bias voltage needed to establish the lateral accelerating electric field in the region. The emitter injects holes into the base region. Under the influence of the accelerating field in the base, these holes form a minority carrier beam. In the presence of normal B_Z magnetic field (**Figure 17**), the electric field in the base region rotates for the Hall angle of majority carries, and the hole beam tilts with respect to the electric field for the Hall

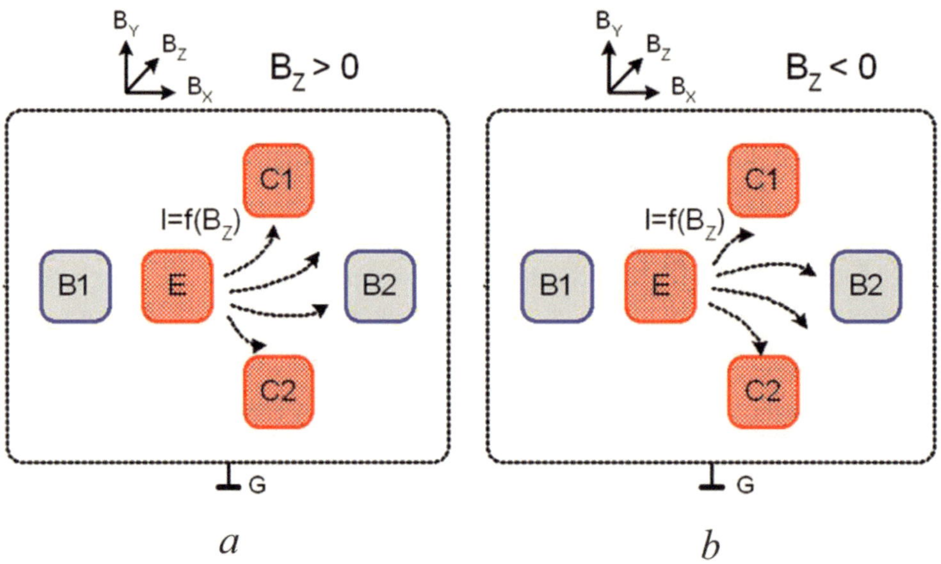

Figure 17. Structure of magnetotransistor (a) and current deflection (b) $I = f(B_Z)$ under B_Z magnetic field vector projection: B1, B2—base contacts; E—emitter area; C1, C2—collector areas; and G–connected to ground a p-n junction wall insulation.

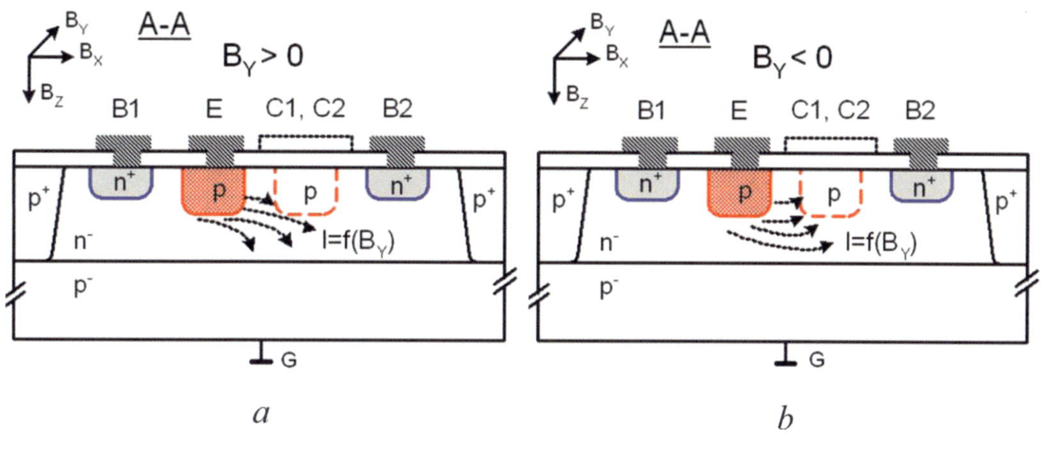

Figure 18. Structure of magnetotransistor (a) and current deflection (b) I = f(B_Z) under B_Z magnetic field vector projection: B1, B2 (n^+ type)—base contacts; E (p type)—emitter area; C1, C2 (p type)—collector areas; G—a substrate connected to ground (p^- type).

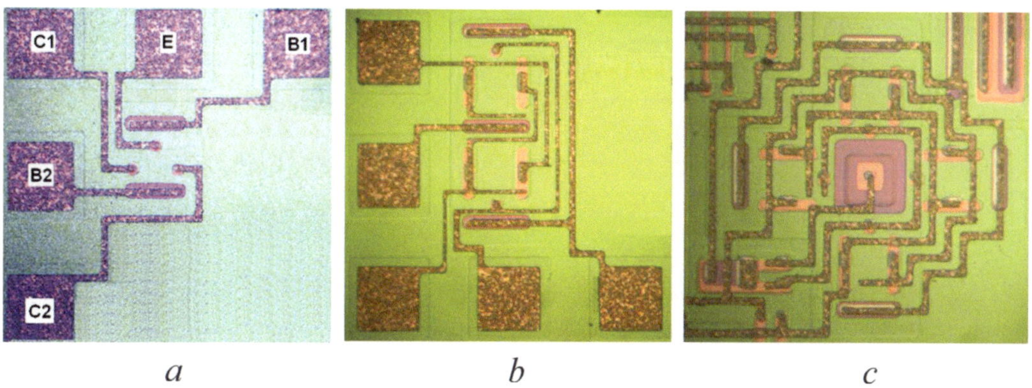

Figure 19. Photos of magnetotransistors: single (a), doubled (b) and quad (c) structures.

angle. The deflection of a beam leads to an imbalance in the two C1, C2 collector currents. In the presence of B_Y magnetic field vector projection (**Figure 18**), the hole beam tilts up to collectors or down toward a p^- type substrate connected to ground. Doubled and quad (**Figure 19**) structures are used to provide a set of signals in correspondence with three projections of magnetic field vector.

Based on the sensors proposed and described above, a set of magnetometers for detection and characterization of magnetic particles has been developed. For illustration only, an example of software windows for control and measurement result visualization, as well as a measurement resolution archived (in this case about 10^{-7}T), are shown in **Figures 20** and **21**. As it was noted, the main advantage of such sensors is a possibility of measuring three orthogonal projections of a magnetic field vector that is quite new in characterization of magnetic particle technique.

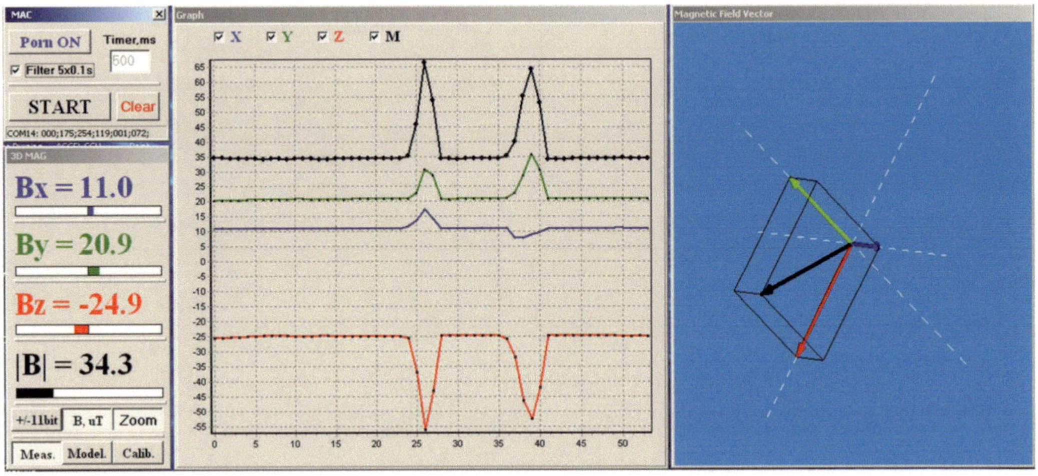

Figure 20. Software of magnetometer based on 3D sensors.

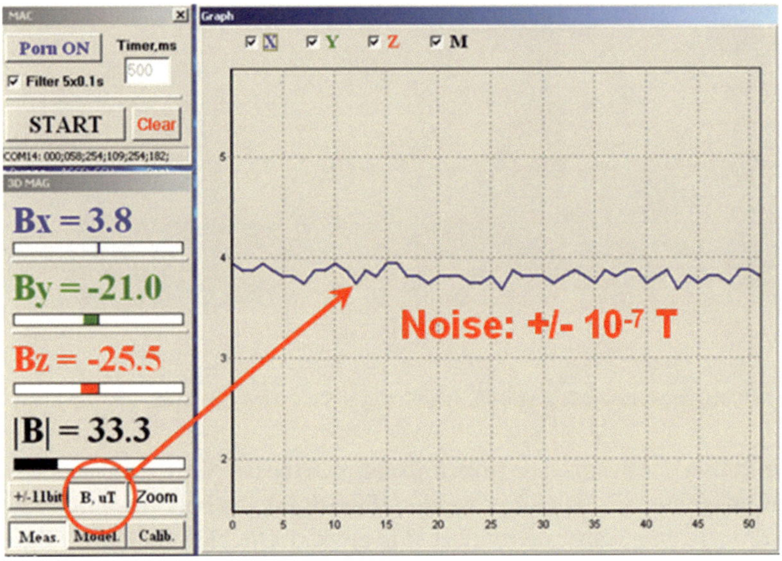

Figure 21. Measurement resolution (noise) of magnetometer based on 3D sensors.

3. Conclusion

It is possible to say that the interest to nanomaterials with magnetic properties increases every year. These materials have found their applications in medicine for diagnostics and treatment of serious diseases, where treatment by other methods is more expensive and durable. However, synthesis of such nanomaterials with stable properties is not completely elaborated, since besides the size factor, the character of their building microgeometry, functional properties and methods of identification should be considered. Special attention should be paid to sizes, topography and biocompatibility of initial nanoparticles. Just these parameters finally determine the functional properties.

It is shown in the paper that identification of synthesized nanoparticles can be realized most effectively by the change of their magnetic properties. This is possible both at the stage of their synthesis and after biofunctionalization. Success of the proposed approach formed the basis of the ideology of the Internet of Things use for monitoring the stability of nanoparticles microstructure and their functional properties in application in medicine.

Acknowledgements

We are infinitely grateful to Mykhailo Gonchar and Tetiana Prokopiv from the Institute of Cell Biology of the National Academy of Sciences of Ukraine, Department of Analytical Biotechnology, for kindly provided samples of Fe3O4 powders for our research. Also we would like to acknowledge financial support of the Ministry of Education and Science of Ukriaine under Grant 0116U004142 and partially by funds of 543994-TEMPUS-1-2013-1-BE-TEMPUS-JPCR MMATENG (Modernization of two cycles (MA, BA) of competence-based curricula in Material Engineering according to the best experience of Bologna Process – EU).

Author details

Zoia Duriagina[1]*, Roman Holyaka[1], Tetiana Tepla[1], Volodymyr Kulyk[1], Peter Arras[2] and Elena Eyngorn[3]

*Address all correspondence to: zduriagina@ukr.net

1 Lviv Polytechnic National University, Lviv, Ukraine

2 KU Leuven, Sint-Katelijne-Waver, Belgium

3 Technische Universität Berlin, Berlin, Germany

References

[1] De Crozals G, Bonnet R, Farre C, Chaix C. Nanoparticles with multiple properties for biomedical applications: A strategic guide. Nano Today. 2016;**11**(4):435-463

[2] Nikiforov VN. Biomedical applications of magnetic nanoparticles. Nauka i tehnologii v promyishlennosti. 2011;**1**:90-99

[3] Filippousi M, Angelakeris M, Katsikini M, Paloura EC, Efthimiopoulos I, Wang Y, Zamboulis D, Van Tendeloo G. Surfactant effects on the structural and on the magnetic properties of iron oxide nanoparticles. Journal of Physical Chemistry C. 2014;**118**(29):16209-16217. DOI: 10.1021/jp5037266

[4] Uvarova IV, Maksymenko VB. Bio-compatible Materials for Medical Products. Kyiv: KiM; 2013. p. 232

[5] Fang C, Zhang M. Multifunctional magnetic nanoparticles for medical imaging applications. Journal of Materials Chemistry. 2009;**19**:6258-6266. DOI: 10.1039/B902182E

[6] Peng S, Wang C, Xie J, Sun S. Synthesis and stabilization of monodisperse Fe nanoparticles. Journal of the American Chemical Society. 2006;**28**:10676-10677

[7] Gillich T, Acikgöz C, Isa L, Schlüter AD, Spencer ND, Textor M. PEG-stabilized core-shell nanoparticles: Impact of linear versus dendritic polymer shell architecture on colloidal properties and the reversibility of temperature-induced aggregation. ACS Nano. 2013;**7**:316-329

[8] Lalatonne Y, Paris C, Serfaty JM, Weinmann P, Lecouvey M, Motte L. Bis-phosphonates-ultra small superparamagnetic iron oxide nanoparticles: A platform towards diagnosis and therapy. Chemical Communications. 2008;**22**:2553-2555

[9] Hosseini AG, Bagheri M, Mohammad-Rezaei R. Synthesis and fluorescence studies of dual-responsive nanoparticles based on amphiphilic azobenzene-contained poly(monomethyl itaconate). Journal of Polymer Research. 2016;**23**(8):12. DOI: 10.1007/s10965-016-1061-y

[10] Rozenfel'd LH, Chekman IS, Tertyshna AI. Nanotechnology in medicine, pharmacy and pharmacology. Farmakolohiya ta likars'ka toksykolohiya. 2008;**1-3**:3-14

[11] Barrera C, Herrera AP, Bezares N, Fachini E, Olayo-Valles R, Hinestroza JP, Rinaldi C. Effect of poly(ethylene oxide)-silane graft molecular weight on the colloidal properties of iron oxide nanoparticles for biomedical applications. Journal of Colloid and Interface Science. 2012;**377**:40-50

[12] Liu Y, Li Y, Li XM, He T. Kinetics of (3-aminopropyl)triethoxylsilane (APTES) silanization of superparamagnetic iron oxide nanoparticle. Langmuir. 2013;**29**:15275-15282

[13] Sun J, Zhou S, Hou P, Yang Y, Weng J, Li X, Li M. Synthesis and characterization of biocompatible Fe_3O_4 nanoparticles. Journal of Biomedical Materials Research Part A. 2006;**80**(2):333-341

[14] Shu Z, Wang S. Synthesis and characterization of magnetic nanosized Fe_3O_4/MnO_2 composite particles. Journal of Nanomaterials. 2009;**2009**:340217. DOI: 10.1155/2009/340217

[15] Ahangaran F, Hassanzadeh A, Nouri S. Surface modification of Fe_3O_4/SiO_2 microsphere by silane coupling agent. International Nano Letters. 2013;**3**:23

[16] Arsalani N, Fattahi H, Nazarpoor M. Synthesis and characterization of PVP-functionalized superparamagnetic Fe_3O_4 nanoparticles as an MRI contrast agent. eXPRESS Polymer Letters. 2010;**4**(6):329-338

[17] Husain Q. Magnetic nanoparticles as a tool for the immobilization/stabilization of hydrolases and their applications: An overview. Biointerface Research in Applied Chemistry. 2016;**6**(6):1585-1606

[18] Huang HY, Lovell JF. Advanced functional nanomaterials for theranostics. Advanced Functional Materials. 2017;**27**(2):1603524. DOI: 10.1002/adfm.201603524

[19] Pankhurst Q, Connolly J. Applications of magnetic nanoparticles in biomedicine. Journal of Physics D: Applied Physics. 2003;**36**:167-181

[20] Xu C, Sun S. New forms of superparamagnetic nanoparticles for biomedical applications. Advanced Drug Delivery Reviews. 2013;**65**:732-743

[21] Long NV, Thi CM, Yong Y, Cao Y, Wu H, Nogami M. Synthesis and characterization of Fe–based metal and oxide based nanoparticles: Discoveries and research highlights of potential applications in biology and medicine. Recent Patents on Nanotechnology. 2014;**8**:52-61

[22] Huang DJ, Lin H-J, Okamoto J, Chao KS, Jeng H-T, Guo GY, Hsu C-H, Huang C-M, Ling DC, Wu WB, Yang CS, Chen CT. Charge-orbital ordering and Verwey transition in magnetite measured by resonant soft X-ray scattering. Physical Review Letters. 2006;**96**(9):096401. DOI: 10.1103/PhysRevLett.96.096401

[23] Radisavljevic I, Kuzmanovic B, Novakovic N, Mahnke HE, Vulicevic LJ, Kurko S, Ivanovic N. Structural stability and local electronic properties of some EC synthesized magnetite nanopowders. Journal of Alloys and Compounds. 2017;**697**:409-416. DOI: 10.1016/j.jallcom.2016.11.090

[24] Nikiforov VN, Ignatenko AN, Irhin VY. The magnetism of the magnetite nanoparticles: Effects of finite size and coating. Izvestiya RAN. Seriya fizicheskaya. 2014;**78**(10):1336-1340. DOI: 10.7868/S0367676514100159

[25] Duriagina ZA, Holyaka RL, Borysyuk AK. Automated widely diapazon magnetometer for magnetic alloys phase analysis: Development and application. Uspehi Fiziki Metallov; 2013;**14**:33-66

[26] Kondyr AI, Borysyuk AK, Pazdriy IP, Shvachko SH. The use of a vibrating magnetometer for phase analysis of special steels and alloys. Vybratsyy v tekhnyke y tekhnolohyyakh. 2004;**34**(2):41

[27] Koh I, Josephson L. Magnetic nanoparticle sensors. Sensors. 2009;**9**:8130-8145. DOI: 10.3390/s91008130

[28] Janssen XJA, van Ijzendoorn LJ, Prins MWJ. On-chip manipulation and detection of magnetic particles for functional biosensors. Biosensors and Bioelectronics. 2008;**23**:833-838. DOI: 10.1016/j.bios.2007.08.023

[29] Popovic RS. Hall Effect Devices. CAT# IPE331. 2nd ed. Bristol and Philadelphia, USA;CRC Press, 2003. p. 420. ISBN: 9780750308557. Available from: https://www.crcpress.com/Hall-Effect-Devices-Second-Edition/Popovic/p/book/9780750308557

[30] Popovic DR, Dimitrijevic S, Blagojevic M, Kejik P, Schurig E, Popovic RS. Three-axis teslameter with integrated Hall probe. IEEE Transactions on Instrumentations and Measurement. 2007;**56**(4):1396-1402. DOI: 10.1109/TIM.2007.900133

[31] wikipedia.org. Magnetoresistance [Internet]. Available from: https://en.wikipedia.org/wiki/Magnetoresistance

[32] honeywell.com. 3-Axis Digital Compass IC HMC5883L [Internet]. 2010. Available from: https://aerocontent.honeywell.com/aero/common/documents/myaerospacecatalog-documents/Defense_Brochures-documents/HMC5843.pdf

[33] Bolshakova I, Holyaka R. Multiposition 3-D Magnetic Field Sensor [Internet]. European Patent Office, Patent No. WO2005029604. 2005. Available from: https://worldwide.espacenet.com/searchResults?ST=singleline&locale=en_EP&submitted=true&DB=&query=WO2005029604

[34] Bolshakova I, Holyaka R. Method for Measuring Magnetic Field [Internet]. European Patent Office, Patent No. WO2012054000. 2012. Available from: https://worldwide.espacenet.com/searchResults?ST=singleline&locale=en_EP&submitted=true&DB=&query=WO2012054000

[35] Bolshakova I, Holyaka R. Magnetic Field Measuring Sensor [Internet]. European Patent Office, Patent No. WO2006028425. 2006. Available from: https://worldwide.espacenet.com/searchResults?ST=singleline&locale=en_EP&submitted=true&DB=&query=WO2006028425

[36] Bolshakova I, Holyaka R. Magnetic Field Measuring Sensor [Internet]. European Patent Office, Patent No. WO2006028427. 2006. Available from: https://worldwide.espacenet.com/searchResults?ST=singleline&locale=en_EP&submitted=true&DB=&query=WO2006028427

[37] Bolshakova I, Holyaka R. Magnetic Field Measuring Sensor [Internet]. European Patent Office, Patent No. WO2006028426. 2006. Available from: https://worldwide.espacenet.com/searchResults?ST=singleline&locale=en_EP&submitted=true&DB=&query=WO2006028426

[38] Bolshakova I, Holyaka R, Murari A. Method for Measuring Magnetic Field [Internet]. European Patent Office, Patent No. EP2630511. 2015. Available from: https://worldwide.espacenet.com/searchResults?ST=singleline&locale=en_EP&submitted=true&DB=&query=EP2630511

[39] Bolshakova I, Holyaka R, Gerasimov S. Magnetic Field Measurement with Continuous Calibration [Internet]. European Patent Office, Patent No. GB2427700. 2009. Available from: https://worldwide.espacenet.com/searchResults?ST=singleline&locale=en_EP&submitted=true&DB=&query=GB2427700

[40] Hotra Z, Holyaka R, Marusenkova T, Ilkanych V. Algorithms of semiconductor magnetic field sensor devices power consumption minimization. In: ELNANO; 2012; Kyiv. pp. 29-30. Available from: http://journals.kpi.ua/publications/text/29_30_2012.pdf

[41] Hotra Z, Holyaka R, Marusenkova T. Optimization of microelectronic magnetic sensors on the splitted Hall structures. In: Prace Instytutu Elektrotechniki, editor. Proceedings of Electrotechnical Institute, Instytut Elektrotechniki. Issue 247. 2010. pp. 13-18. Available from: https://pbn.nauka.gov.pl/polindex-webapp/browse/article/article-4d222aa0-9679-45e1-9467-5c5de8b262f6

[42] Hotra Z, Holyaka R, Bolshakova I, Yurchak I, Marusenkova T. Spatial models of splitted Hall structures. In: Proceedings of VIIth International Conferenceon Perspective Technologies and Methods in MEMS Design; MEMSTECH–2011; IEEE.2011. pp. 5-8. Available from: http://ieeexplore.ieee.org/document/5960249/

[43] Holyaka R, Yurchak I, Marusenkova T, Ilkanych V. Microprocessor noise-immune signal transducer for galvanomagnetic smart sensor devices. In: International Conferenceon Modern Problems of Radio Engineering Telecommunications and Computer Science; TCSET–2012; IEEE.2012. p. 430. Available from: http://ieeexplore.ieee.org/document/6192682/

[44] Bolshakova I, Holyaka R, Hotra Z, Marusenkova T. Methods of modeling and calibrating 3D magnetic sensors based on splitted Hall structures. Electronics and Nanotechnologies. Proceedings of the XXXI International Scientific Conference. 2011; p. 38. Available from: http://www.journals.kpi.ua/publications/text/2011_38.pdf

Chapter 18

Biomaterials for Tissue Engineering Applications in Diabetes Mellitus

Mônica Fernandes Gomes, José Benedito Amorim,
Lilian Chrystiane Giannasi and
Miguel Angel Castillo Salgado

Additional information is available at the end of the chapter

http://dx.doi.org/10.5772/intechopen.69719

Abstract

The search for ideal implants or alternative scaffolds is a challenge for biomedical science researchers, especially in diabetic patients. Many alternative bioactive materials have been used in the regenerative medicine, especially in patient with complex metabolic disorder as diabetes mellitus. Among them, we discussed the following alternative material scaffolds, including amniotic membrane (AM), homogenous demineralized dentin matrix (HDDM), platelet-rich plasma (PRP), and alloplastic materials as porous polyethylene and polyurethane. These biomaterials were applied in the craniomaxillofacial complex and liver injury, resulting in tissue regeneration and microstructural reconstruction due to their effective inductive and conductive properties. Additionally, diabetes disease and its general biophysical mechanism and systemic complications were described in order to improve the comprehension of the physiopathology of this comorbidity and its effects in the tissues. The AM, HDDM, and PRP in implantation sites initiated an inductive cascade as chemotaxis of progenitor cells, mitogenesis, angiogenesis, and differentiation into wide variety of cells. The cell recruitment, division rate, and differentiation of cell lines are under the direct control of several growth factors and stem cells which are present in these biomaterials. Further, some alloplastic materials have triggered satisfactory tissue responses when used in treatments of craniofacial deformities or in anatomical reconstructions.

Keywords: regenerative medicine, amniotic membrane, homogenous dentin demineralized matrix, platelet-rich plasma, alloplastic material, diabetes mellitus

1. Introduction

Diabetes mellitus (DM), a complex metabolic disorder, is a syndrome characterized by abnormalities in carbohydrate, lipid, and protein metabolism, which results either from a partial or an absolute insulin deficiency or from target tissue resistance to its cellular metabolic effects. This disease is characterized by the presence of few or by the absence of functional pancreatic β-cells in the islets of Langerhans and by a substantial reduction or inexistence of insulin secretion. This cellular dysfunction is an important defect in the pathogenesis of type 2 diabetes [1, 29].

Many strategies have been used as a biological dressing, an alternative bioactive material scaffold, and alloplastic material, including amniotic membrane (AM), demineralized dentin matrix (DDM), platelet-rich plasma (PRP), and alloplastic materials as porous polyethylene (porous PTFE) implant, in the degenerative complications in the diabetes disease. The AM is a high-throughput source for multipotent mesenchymal stem cells (MSCs) with the ability to differentiate into wide variety of cells, such as chondroblasts, osteoblasts, adipocytes and fibroblasts, myocytes, endothelial cells, neuronal cells, and hepatocytes, leading to formation of cartilage, bone, connective, muscle, blood vessel, nerve, and liver tissues, respectively. This membrane acts as a barrier prevents the entry of pathogens and toxins, preserves tissue structure, and consequently reduces the levels of local pro-inflammatory cytokines. The DDM is a bioactive tissue able to stimulate the bone repair process, to increase bone mass, and to improve bone microstructure without causing any rejection or infection. It is noteworthy that it acts as a reservoir for various proteins that are embedded within the matrix and are available to local cells, including bone morphogenetic proteins (BMPs), insulin-like growth factors I and II (IGF I and IGF II), platelet-derived growth factor (PDGF), fibroblast growth factor (FGF), transforming growth factor beta (TGF-β), and vascular endothelial growth factor (VEGF).

Based on the literature, we consider that the rich AM and DDM biological properties could be associated with other traditional graft materials, improving their biodynamic effects when implanted in host sites. Some alloplastic materials are well tolerated by the host tissues and present good stability in the implementation site, including porous polyethylene and polyurethane. These materials have been used in orthopedic, articular, orbital, cranial, and maxillofacial reconstruction. The presence of porous allows the ingrowth of the host tissue, which promotes better anchorage of the polymers in the host tissue.

Therefore, these alternative therapies may provide beneficial effects on tissue regeneration and reconstruction, in particular, in diabetes, since one of the main complications is delayed and abnormal repair process by accumulation of advanced glycation endproducts (AGEs) on tissues. These products cause severe widespread damage to tissue through upregulation of inflammation and cross linking of collagen and other proteins.

2. Diabetes mellitus

Diabetes mellitus is known as a metabolic disorder, characterized by deficiency of insulin secretion or action, leading to chronic hyperglycemia and disturbances of carbohydrate, fat, and protein metabolism, may develop in, called diabetes mellitus (DM). According to the International Diabetes Federation, 8.8% of the adult population worldwide has diabetes. Of all individuals with diabetes, only 10–15% have type 1 diabetes mellitus (T1DM); type 2 diabetes mellitus (T2DM) is the most common form. However, T1DM is the most common form of diabetes in children (<15 years of age), and >500,000 children are currently living with this condition globally. The classification of diabetes mellitus disease defines both process and stage of the disease. The processes include type 1, autoimmune and nonautoimmune, with beta-cell destruction; type 2 with varying degrees of insulin resistance and insulin hyposecretion; gestational diabetes mellitus; and other types where the cause is known [1].

3. Development of the diabetes disease

Alterations in the molecular mechanisms of via insulin signaling may provide a deficiency, resistance, and/or decrease of insulin release into bloodstream, leading to development of the diabetes mellitus. To improve the comprehension of the physiopathology of this comorbidity and its effects in the tissues, firstly, it is important to know the biophysical mechanism involved in the glucose transport and the mechanism of insulin release, as well as the relationship of the glycation and diabetic complications.

3.1. Biophysical mechanism involved in the glucose transport

Fasting blood glucose levels are between 70 and 99 mg/dL in healthy individuals. These values can vary depending on food intake, renal function, physical activity, and the method used for testing. The major dietary polysaccharides, also called carbohydrates, for many species are starch and glycogen which are a storage form of glucose arisen from plants and animals, respectively. After food intake, the digestion process of vegetable (starch) and animal (glycogen) polysaccharides initiates from mouth, which are broken down by the salivary amylase (ptyalin), leading to the formation of oligosaccharides. Afterward, pancreatic and digestive enzymes break down these macromolecules to monosaccharides (glucose, fructose, and galactose) from upper small intestine. Posteriorly, these monosaccharides enter into enterocytes by luminal membrane transporters. It is worth highlighting that the pancreatic amylase is produced by the pancreatic acinar tissue, while the maltase, isomaltase, saccharase, and lactase are generated by the enterocytes. Pancreas is known as a glandular organ in the upper abdomen, which serves as two glands in one: a digestive exocrine gland (e.g., production of pancreatic amylase) and a hormone-producing endocrine gland especially composed of islets Langerhans. These islets are constituted of four hormone-producing cell types: insulin-secreting beta cells, glucagon-secreting alpha cells, somatostatin-secreting delta cells, and pancreatic, polypeptide-secreting F cells. Both the glands are vital to the body's survival (**Figure 1**).

Several factors can influence in the insulin secretion, including amino acids, fatty acids, hormones, and drugs; however, the blood glucose concentration levels are considered the main physiological regulators of insulin release. The glucose molecule ($C_6H_{12}O_6$) depends on a transporter protein to enter the cells, seeing that is impermeable to cell membrane. Thus, the glucose uptake from intestinal lumen occurs through ion-coupled membrane cotransporters

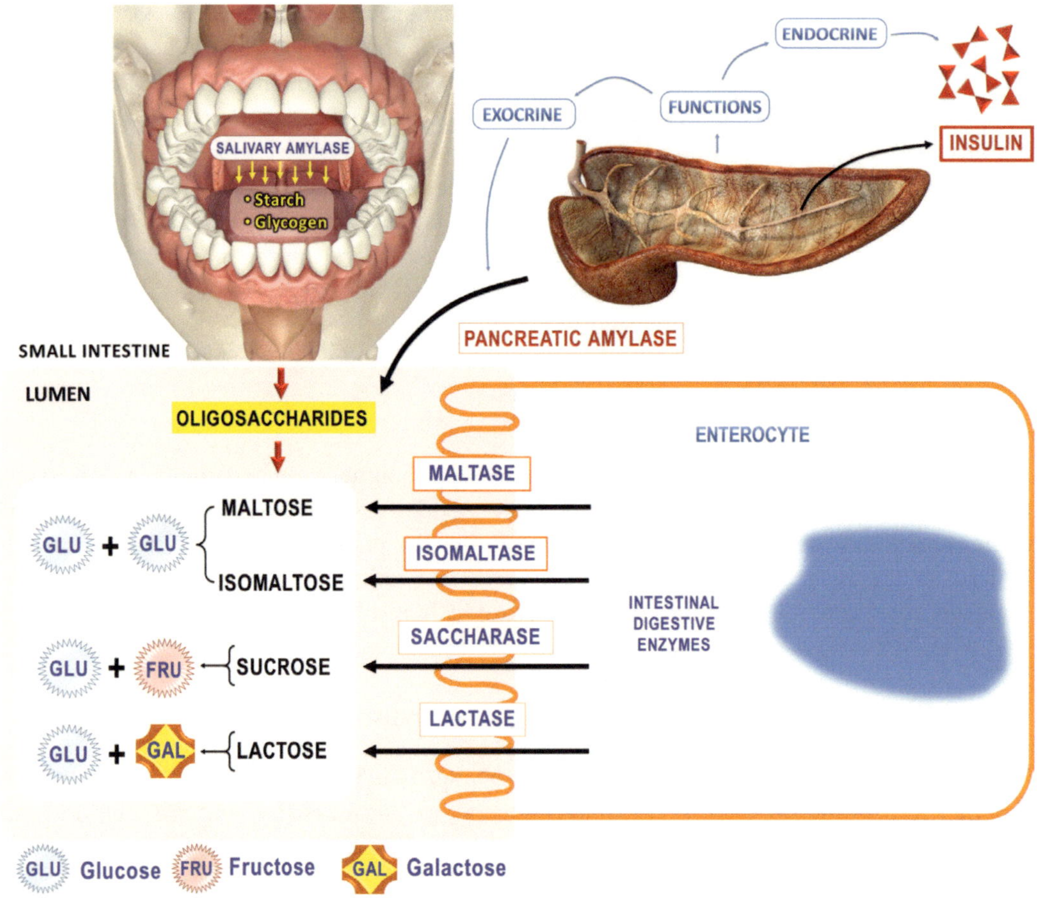

Figure 1. Diagram showing carbohydrate degradation by oral, pancreatic, and intestinal enzymes.

which are molecular machines that use the electrochemical energy of a transmembrane ionic gradient to energize the transport of another solute [37]. Among them, it is known as the type 1 sodium-glucose cotransporter (SGLT-1) that is located at the cell membrane of the apical region of the enterocytes. This cotransporter is coded by the SLC5A1 gene, which was the first mammalian cotransporter carrier protein to be identified, cloned, and sequenced [4].

The present review discusses the members of the sodium-glucose cotransporter (SGLT) gene family (the SLC5A gene family) have been identified in a broad range of tissues, including intestine, kidney, muscle, neurons, and thyroid [21, 36]. The SGLT-1 protein has two allosteric sites that carry a galactose or glucose molecule and sodium ions ($2Na^+$) into the cells, while fructose molecule passes across the luminal membrane through the GLUT-5 (fructose transporter). Both the transport proteins are found at the apical surface of the enterocytes in the small intestine [9]. The glucose transporters (GLUTs) are sodium-independent membrane transport proteins which carry particularly monosaccharides, including glucose fructose and galactose. Various members of solute-carrier (SLC) gene family have been found in literature [28]. None of these transport mechanisms requires cell energy expenditure, characterizing a biological process of passive transport.

Considering the intestinal epithelial cells, the glucose entry depends on the electrochemical gradient of Na^+, as well as its translocation from luminal membrane to basolateral membrane. These differences of extracellular and intracellular concentrations of Na^+ are kept by the sodium-potassium pump. The potassium ions (K^+) into cells pass to extracellular matrix (ECM) through specific channels, located at the basal membrane that maintains pump activity. The sodium-potassium pump is an active transport mechanism which requires energy expenditure using adenosine triphosphate (ATP). The internal cell membrane potential of the enterocytes is negative in relation to the external cell membrane, typically about −70 millivolts. This voltage arises from differences in concentration of the sodium and potassium electrolyte ions, leading to glucose entry into cells. The glucose molecule passes into the extracellular matrix (ECM) through the glucose transporter-2 protein (GLUT-2). Then, it enters in the bloodstream due to the differences of gradient concentration between the ECM and blood capillary (**Figure 2**). Considering the biological property of the blood vessels, intercellular clefts, fenestrations, and cell membrane discontinuity of the endothelial cells also facilitate the gradient diffusion into bloodstream [24].

3.2. Mechanism of insulin release

As previous discussion, the elevated levels of blood glucose generate insulin release from pancreatic β-cells at the islets Langerhans, which plays a pivotal role in the glucose homeostasis and acts in a coordinated fashion on cellular events that regulate the metabolic and growth processes in the human body.

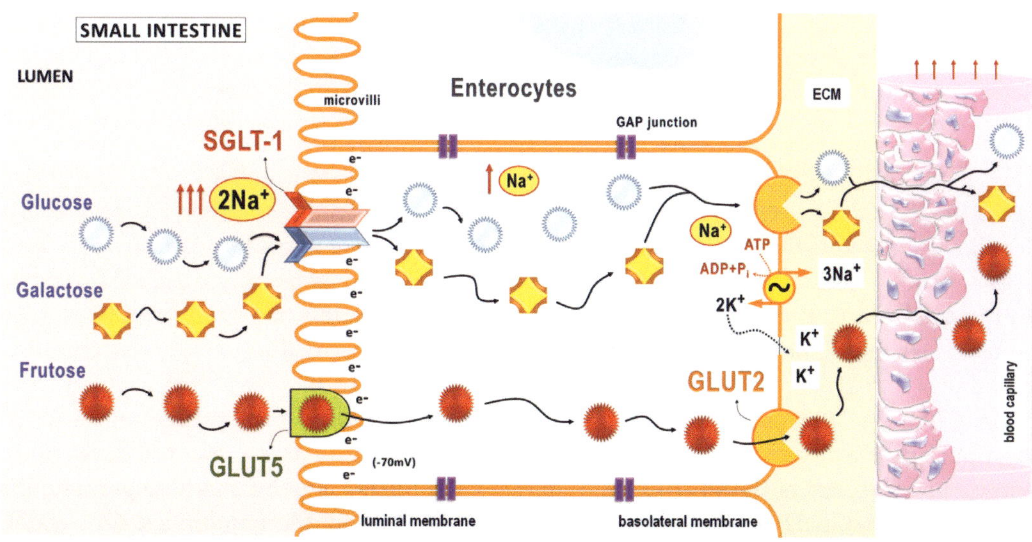

Figure 2. Diagram showing the glucose absorption by the enterocytes.

In humans, the threshold value for glucose-stimulated insulin secretion is about 100 mg/dL (5.6 mmol/L) of plasma glucose concentration. This hormone is secreted in a biphasic manner in response to a marked increase in blood glucose. An initial burst of insulin secretion may last 5 to 15 min, resulting from the secretion of preformed insulin secretory granules. This response is followed by more gradual and sustained insulin secretion that results largely from the synthesis of new insulin molecules.

The primary mechanism is through binding to sulfonylurea receptor (SUR-1) on functioning pancreatic β-cells. Afterward, binding closes the linked ATP-sensitive potassium channels (K_{ATP} channels), which leads to decreased potassium influx and subsequent depolarization of the β-cell membrane. This event provides voltage-dependent T-type calcium (Ca^{2+}) and sodium (Na^{+}) channel opening, resulting in an influx of Ca^{+} channel, as a consequence, causing translocation and exocytosis of insulin-secretory granules to the cell surface. Then, the insulin from the ECM enters into the bloodstream, following toward the target cells [5].

Another mechanism that promotes ATP-sensitive potassium channel closure is increased ATP levels, which are produced into two distinct metabolic pathways: glycolysis (anaerobic process) and mitochondrial respiratory chain (aerobic process). Initially, the glucose molecule passes across the β-cell membrane through the glucose transporter 2 (GLUT-2), and then, it is phosphorylated to form glucose-6-phospate (G-6-P) by two enzymes: hexokinase IV (HK-IV) or glucokinase (GK) of lower affinity for glucose and hexokinase I (HK-I) of higher affinity for glucose. However, the higher affinity enzyme is strongly inhibited by the G-6-P, transferring to the GK a key role in the glucose phosphorylation at the pancreatic β-cells. This process constitutes the first flux determining step for glycolysis. The G6P is broken down into 2-phosphoenolpyruvate by the pyruvate kinase which is transformed in two pyruvates molecules (glycolysis). After that, these molecules enter into mitochondria where two different biochemical processes occur, including the cycle of Krebs and the oxidative phosphorylation cycle. Finally, several ATP molecules are generated. The glycolysis and mitochondrial respiratory chain pathways result in the synthesis of 2 ATPs and 34–36 ATP molecules, respectively, totalizing about 36–38 ATP molecules/mol of glucose [5, 34]. It is noteworthy to highlight the promotor mechanisms to K_{ATP} channels closure lead to the insulin release into the blood stream (**Figure 3**).

After the insulin arrival onto target cells, it links to a specific cell membrane receptor, called insulin receptor substrate (IRS-1) protein, initiating a cascade of phosphorylation events. As a consequence, vesicles containing glucose transporter proteins, especially the GLUT-4, move to the cell surface. The high GLUT-4 level onto cell membrane leads to increase of glucose uptake. Depending on the cell activity, two processes can occur due to high concentration of intracellular glucose: glycolysis or glycogen synthesis. To elucidate better this mechanism, GLUT-4 is an insulin-stimulated glucose transporter that has the primary form of the transporter present in skeletal muscle tissue and adipose tissue. It is present in cells and in intracellular vesicles of the smooth ER. In target cells, the effect of insulin is to promote the translocation of GLUT-4 transporter from intracellular pool to cell membranes. As a result, more transporters are available in the plasma membrane, and glucose uptake by target cells is, thereby, increased.

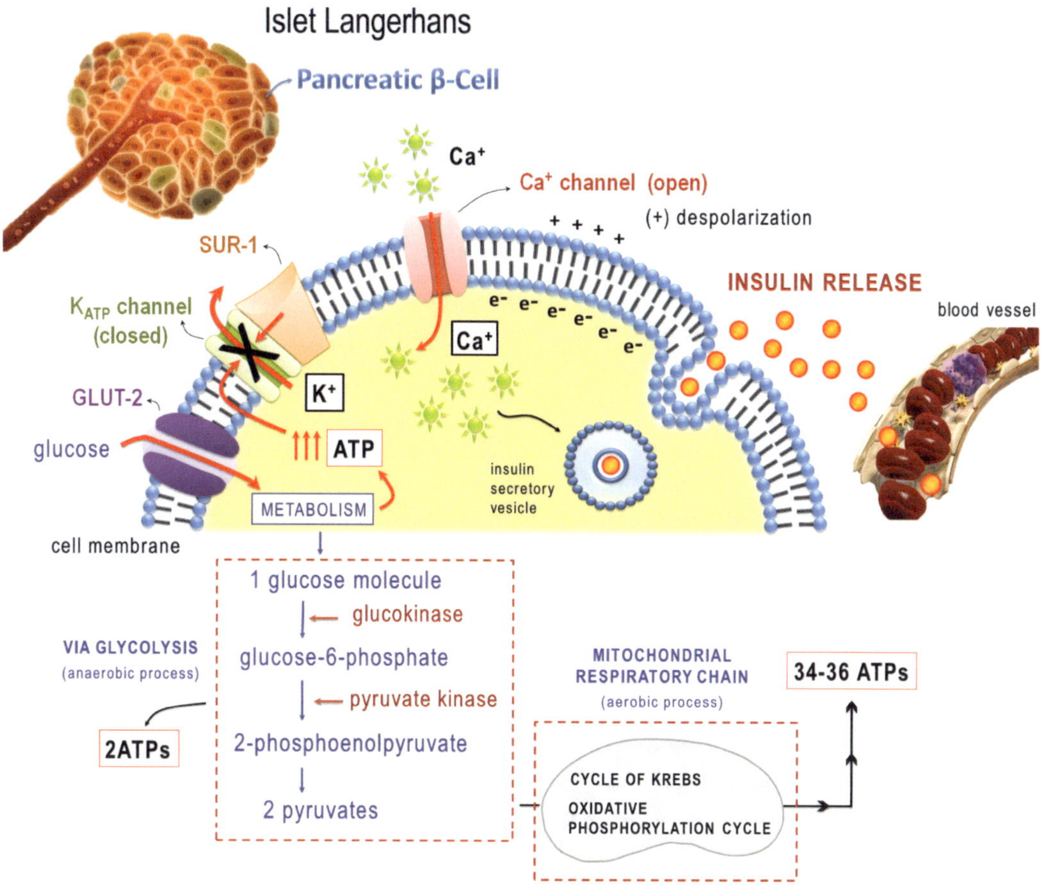

Figure 3. Mechanism of insulin release from pancreatic β-cell to bloodstream.

Based on these basic concepts, we can consider that methods for SGLT and GLUT inhibition can be considered an suitable alternative means of managing glucose control and, as a consequence, a further treatment option to control the diabetes disease.

3.3. Relationship of the glycation process and diabetic complications

Persistently elevated glucose levels during long-standing diabetes induce structural and functional changes in various proteins in the body, including plasma protein (albumin, globulins, and fibrinogen) and collagens, through a glycation or nonenzymatic glycosylation process [3, 23, 39, 44]. This process involves an excessive chemical attachment of sugar molecule (e.g., glucose or fructose) to proteins, lipids, and nucleic acids, without the involvement of enzymes. Various deleterious effects from glycation of plasma protein may be produced, including alteration in drug binding in the plasma, platelet activation, generation of oxygen free radicals, impaired fibrinolysis, and impairment in immune system regulation. On the other hand, the structural impairment in collagen alters the osteoblast differentiation, leading to disorders of bone remodeling process and, as a consequence, skeletal fragility [20, 39]. The glycation and oxidation of lipids and nucleotides can also form a single product from various heterogeneous molecules (heterogeneous adducts) through amino-carbonyl reactions, known as advanced glycation endproducts (AGEs) [23, 27]. As the direct products of diabetic metabolic remodeling, AGEs are an important environmental medium for changes of cells, cytokines, and extracellular matrix components including collagens, proteoglycans, laminin, and vitronectin [11, 20]. These changes in the extracellular matrix cause specific alterations in bone formation and bone remodeling. Bone tissue turnover has been shown to be suppressed; studies have shown decreases in the percentage of osteoclasts and osteoblasts as well as a decrease in osteocalcin synthesis [8, 11, 16]. Moreover, AGEs exhibit the ability to inhibit cellular differentiation, proliferation, and migration, which could delay the repair process [20].

Therefore, the protein glycation, elevated level of AGEs accumulation in tissue collagen, and high plasma AGE concentration lead to major diabetic complications like retinopathy, nephropathy, neuropathy, atherosclerosis, degenerative and cardiovascular diseases, inflammatory arthritis, osteoporosis, susceptibility to infections, delayed tissue repair, and severe periodontal disease [3, 17, 39]. Other generic complications are impaired neovascularization and microvascular complications, resulting in ulcers and chronic nonhealing wounds with the characteristics of high amputation rate [20, 46].

Currently, the major concerns with the use of AGEs inhibitors as therapeutic agents are low effectiveness, poor pharmacokinetics, and undesirable side effects. Several AGEs inhibitors possess potent antiglycation activity and are devoid of undesirable side effects. These small molecules inhibitors can, therefore, serve as scaffolds for the development and designing of new AGEs inhibitors as clinical agents [23].

Studies have described that high levels of AGEs induce increased index values of oxidative stress in bone cells due to the absence of the counterbalancing effects of endogenous antioxidants [17]. Thus, it is considered that bone loss can occur as a result of activated osteoclastogenesis, attenuated osteoblastogenesis, and/or stimulated osteoblast apoptosis, which can be induced by the elevated intracellular levels of reactive oxygen species (ROSs) [13, 17]. ROSs produced either endogenously or exogenously are often associated with the principle of oxidative stress, which can attack lipids, proteins, and nucleic acids simultaneously in living cells [45]. This free radical or byproducts of an aerobic metabolism, including superoxide anion (O_2^-), hydrogen peroxide (H_2O_2), and hydroxyl radicals (OH), can confer reactivity to different biological targets. Increasing evidences have suggested that oxidative stress plays a major role in the pathogenesis of diabetes mellitus due to structural and/or functional damage of the pancreatic β cells [38, 45]. Some studies have described that the persistent hyperglycemia has reduced recruitment of mesenchymal stem cells (MSCs) toward the wounds, compromised their cellular differentiation, and affected the quantity of endothelial progenitor cells, preventing vasculogenesis and delaying the healing process [46]. Accumulating evidence also suggests that the impaired MSCs can compromise mobilization and function of bone cells, resulting in low bone turnover and formation [17].

Recently, alternative material scaffolds have been used to accelerate tissue repair process or to promote implant-based tissue reconstruction, besides the advent of stem cell therapy as a pivotal basis of tissue regeneration or replacing, bringing great hope to diabetic patients.

4. Biological and alloplastic materials for tissue engineering

The goal of tissue engineering is to create grafts that enhance tissue repair following trauma, infection, or neoplasm, and for developmental abnormalities; these procedures represent a challenge in surgery. To obtain effective and suitable tissue regeneration or osseointegration, the following are necessary: osteoprogenitor and osteoconductive cells, which offer potential to differentiate and facilitate the various stages of tissue regeneration, growth factors (GFs), and structural integrity with the absence of any kind of local infection [17].

As a compensatory factor, some biomaterials for tissue engineering with promotive properties, including amniotic membrane (AM), demineralized dentin matrix (DDM), platelet-rich plasma (PRP), and porous polyethylene (porous PTFE) in diabetic and nondiabetics experimental model were described as follows.

4.1. Diabetic experimental models

The main drugs widely used to induce diabetic experimental models are streptozotocin and alloxan. The cytotoxic action of both the diabetogenic agents is mediated by reactive oxygen species (ROS). The streptozotocin drug enters into pancreatic β-cells via a glucose transporter (GLUT-2) and causes alkylation of DNA. DNA damage induces activation of polyADP-ribosylation that leads to depletion of cellular NAD^+ and ATP. Enhanced ATP dephosphorylation supplies a substrate for xanthine oxidase, resulting in the formation of superoxide radicals as hydrogen peroxide and hydroxyl radicals. Furthermore, toxic amounts of nitric oxide are release, causing inhibition of mitochondrial aconitase activity and DNA damage and triggering the destruction and necrosis of these cells. The alloxan drug promotes the superoxide radical formation, which undergoes the dismutation to hydrogen peroxide. Thereafter, highly reactive hydroxyl radicals are formed by the Fenton reaction. The ROS action and the simultaneous massive increase of cytosolic calcium concentration cause a rapid destruction of the pancreatic β-cells [41].

Experimental models for monohydrate alloxan-induced diabetes in rabbits and rats were developed. The diabetes was diagnosed due to the blood glucose levels above 200 mg/dL (**Figure 4**). This drug, in particular, promotes destruction of the pancreatic β-cells and parenchyma disorganization with presence of atypical acinar cells, scarce secretion vesicles, atrophy, and a decreased number of Langerhans islets in some regions, leading to severe pancreas dysfunction and, as a consequence, the development of diabetes type 2. Three months after the diagnosis of diabetes in experimental models, general clinical complications can be found, including weight loss, polyuria, polyphagia, ketoacidosis due to accentuated hypoglycemia, convulsions, diabetic foot, acute dermatitis, erythematous, and ulcerated lesion associated with fungal and bacterial infections located in the submandibular region extending to trunk and neck regions, and intracranial abscess in regions of surgical defect due to impaired bone repair and susceptibility to infections. Severe chronic gingivitis can be also evidenced (**Figure 4**) [11, 13, 42].

It is important to consider that diabetic wounds exhibit impaired angiogenesis, reduced growth factor levels, and reduced chemotactic ability to recruit inflammatory cells to the

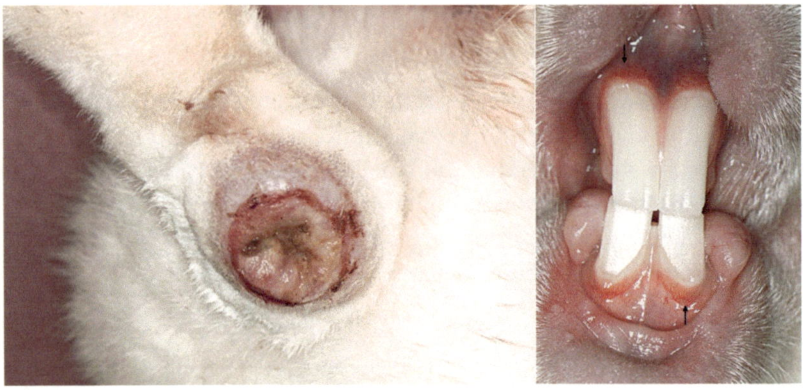

Figure 4. Classical clinical complications in diabetes showing diabetic foot and chronic gingivitis (arrows) in rabbit after 3 months persistent hyperglycemia.

wound, besides that the poor vascularization and maintenance of a chronic inflammatory state limit the healing capacity. Therefore, we discuss in this chapter some conductive and inductive materials for tissue engineering, including AM, DDM, PRP, and porous PTFE, which were applied in diabetic and nondiabetic experimental models or individuals.

4.2. Biological materials

4.2.1. Amniotic membrane (AM)

The AM is a tissue of fetal origin and is composed of three major layers: a single epithelial layer, a thick basement membrane, and an avascular mesenchyme that contain growth factors, cytokines, and other active substances [26]. AM has two types of cells of different embryological origins: amnion epithelial cells derived from embryonic ectoderm and amnion mesenchymal cells from embryonic mesoderm. It is adjacent to the trophoblast cells and lines the amniotic cavity. It can be easily separated from the underlying chorion, with which it never truly fuses at the cellular level. Its nutrition and oxygen are obtained from surrounding chorionic fluid, the amniotic fluid, and the fetal surface vessels. This structure protects the developing embryo against mechanical aggression, provides an environment where the embryo can grow without distortion by pressure from surrounding structures, and also plays an important role during parturition [15].

As a highlight of great potential for clinical application, the amniotic membrane has been widely used in regenerative medicine due to its effective biological properties (biocompatibility, low immunogenicity, anti-inflammatory, and antimicrobial activities), adequate physicomechanical action (permeability, stability, elasticity, flexibility, and resorbability), excellent tissue adhesion, easy delivery of biomodulatory agents including growth factors and genetic materials, and rich MSC source [22, 31, 33]. Current investigations described that pluripotent stem cell lines derived from AM have the ability to differentiate into endothelial and neuronal cells, osteogenic, chondrogenic, adipogenic, skeletal myogenic, hepatic lineages, and able suppress T-cell proliferation [19, 31, 35]. These properties show the greatest prospects to clinical applications in regenerative medicine, which leads to processes of tissue replacing, engineering or regenerating, and preservation of organ functions [12, 26, 43].

Several researches developed in the Center of Biosciences Applied to Persons with Special Care Needs (CEBAPE) in the São Paulo State University (UNESP) has applied the AM as a biologic dressing and a bioactive scaffold material to treat oral mucositis, surgical hepatic resection areas in experiments *in vivo*, and wound areas from incisional and excisional biopsies of oral lesions originated of inflammatory and development processes. This membrane was also applied to stimulate the guided bone regeneration (GBR) in experimental bone defects and to accelerate the tissue repair.

The methods of preparation and storage of the human (h) and homogenous (H) AM were developed by the CEBAPE which are described in the literature [15, 18, 26, 43]. Additionally, the clinical procedures were performed in diabetic and nondiabetic patients or in different animal models, leading to highly favorable results that are reported as follows. We can prove its effects on the tissues through the clinical, biochemical, histological and immunohistochemical outcomes, in particular, in experimental models. As a xenogenic material, a human amniotic membrane (hAM) was used for guided bone regenerative on surgical defects in healthy rabbits. The biocompatibility and low immunogenicity of the hAM were shown since we evidenced newly formed bone tissue in intimate contact with hAM in several surgical defect areas and discrete infiltration of mononuclear inflammatory cells surrounding the hAM in some regions; moreover, its osteogenic and mitogenic effects were very evident (**Figure 5**).

Other application could be in oral biopsy areas to lesion diagnosis, especially in areas of difficult suture as alveolar bone ridge in maxilla and mandible. Studies also showed the use of hAM on wounds produced from incisional and excisional biopsies of alveolar gingiva of the upper premolar region and cheek mucosa, respectively. The diagnosis was plan lichen (inflammatory process) in the alveolar gingiva and intramucosal pigmented nevus (development process). The outcomes were highly satisfactory in both processes, inducing the rapid closure of

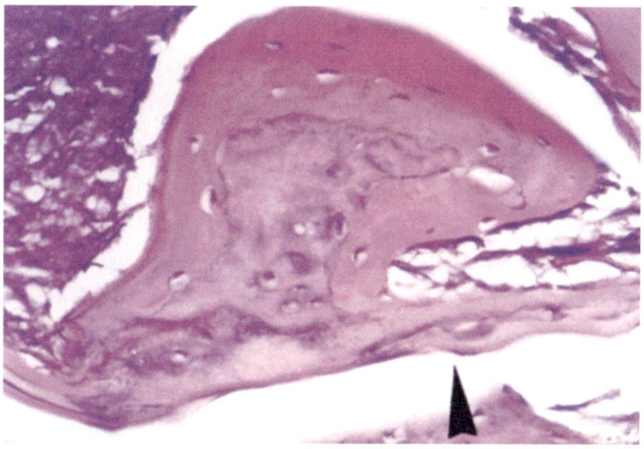

Figure 5. Human amniotic membrane (arrow-head) used as a bioactive xenogenic material which was applied in the guided bone regeneration on the surgical defect in rabbits: newly formed bone formation from the membrane and few infiltrations of mononuclear inflammatory cells, especially lymphocytes, were evidenced (Source: Gomes et al., [15]).

wounds due to stimulation of the tissue repair, relief of pain, and no sign of local infection [18]. Probably, relief of pain occurs due to its strong action of mechanism barrier and good stability on the wound surface, covering nerve endings.

In previous research using homogenous amniotic membrane (HAM) on the surface of surgical liver, resection in rats was performed. The HAM was obtained female rats by cesarean at the 20th day of pregnancy. Fetal amniotic membrane was separated aseptically from the chorionic membrane and rinsed several times in a sterile physiological saline solution and a phosphate buffer at pH 7.4 until the removal of all debris (**Figure 6**). After its laboratory preparation, the HAM was placed on the surface of surgical liver resection. Clinically, this membrane showed an immediate stability and a strong hemostatic action after its implantation on the wound. Histologically, the HAM behavior was observed in 10, 20, 30, and 40 days, showing the following results. Initially, the HAM was biocompatible to hepatic tissue once it intimately adhered to hepatic tissue in regenerative process; subsequently, it underwent metaplasia turning into a highly well-cellularized and vascularized tissue called amino-hepatic tissue. Discrete infiltration of mononuclear inflammatory cells was seen surrounding this biological membrane. This histological characteristic was one of the prominent aspects that can suggest its low immunogenicity in surgically impaired liver tissue. HAM remnants can be found on the injured liver tissue surface in some specimens, after 40-day postsurgery (**Figure 7**). In biochemical analysis of plasma and tissue samples, the aspartate aminotransferase (AST), alanine aminotransferase (ALT), and gamma-glutamyltransferase (GGT) enzymatic levels were significantly low; in contrast, the alkaline phosphatase (ALP) enzymatic levels and the total protein concentrations were strongly high in the rats treated with HAM. Thus, angiogenic, mitogenic, and hepatic-protector effects of the HAM were proved, leading to a better microstructural reconstruction of the injured liver [12].

In our study, other beneficial application of the HAM was as a potent biological dressing in 5-fluorouracil-induced oral mucosites (OM) in rats (**Figure 8**). The pathophysiology of the analyzed OM and its histomorphological features were divided into four phases: inflammation (3 days), cell proliferation (7 days), tissue organization (14 days), and tissue repair (21 days). Among the periods of inflammation and cell proliferation phases, the HAM was incorporated with the wound surface, acting as a protective barrier against microbial infection and preventing any infection signs. Other aspects were discrete to moderate infiltration of inflammatory cells in the connective tissue and early total re-epithelialization in most specimens. The incorporated HAM into connective tissue was degraded and, then resorbed. Regarding the tissue organization and repair phases, the histological features were described the follow. At 14 days, discrete mononuclear inflammatory cells surrounding the connective tissue, muscle layer with normal aspect, a tissue maturation with regularly disposed, remodeled collagen fibers, and rare blood vessels were found. Finally, at 21 days, the oral mucosa showed normal aspects (**Figure 9**). Thus, we can consider that the HAM provides an excellent environment for cell proliferation and neovascularization, stimulating healing process and functioning as a promising biological dressing. Previous studies confirm the growth factors are found into amniotic membrane, including transforming growth factor-β1 and β2 (TGF-β1 and β2), basic fibroblast growth factor (bFGF), epidermal growth factor (EGF), transforming growth factor-α (TGF-α), keratinocyte growth factor (KGF), and hepatocyte growth factor (HGF) [33], reinforcing our findings. Regarding the biodegradable property of the HAM, we hypothesize that the avascular stromal matrix of the HAM is similar

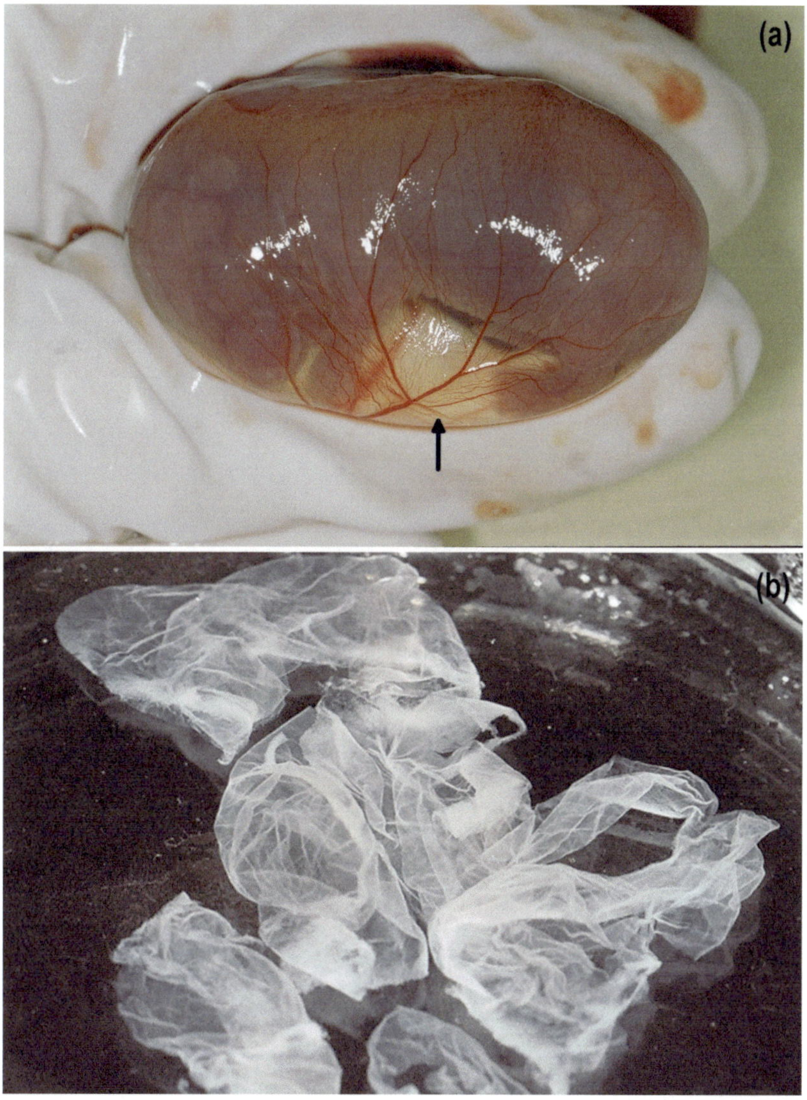

Figure 6. (a) Fetus, amniotic, and chorionic membrane (arrow) and (b) fresh HAM separated from the chorionic membrane and placed in sterile physiological saline solution and phosphate buffer at pH 7.4.

extracellular matrix of the connective tissue in the submucosa region, favoring its degradation during the healing dynamic processes. Investigations report the AM presence of fibronectin, elastin, nidogen, collagen types I, III, IV, V, and VI, elastin, and hyaluronic acid [31, 33].

4.2.2. Homogenous demineralized dentin matrix (HDDM)

Homogenous demineralized dentin matrix (HDDM) arises from tooth structure which has been widely used to stimulate bone regeneration, inducing osteoblastic differentiation and proliferation and chemotaxis. Hence, HDDM accelerates the bone repair process, increases bone mass, and improves bone quality without causing any rejection or infection. HDDM

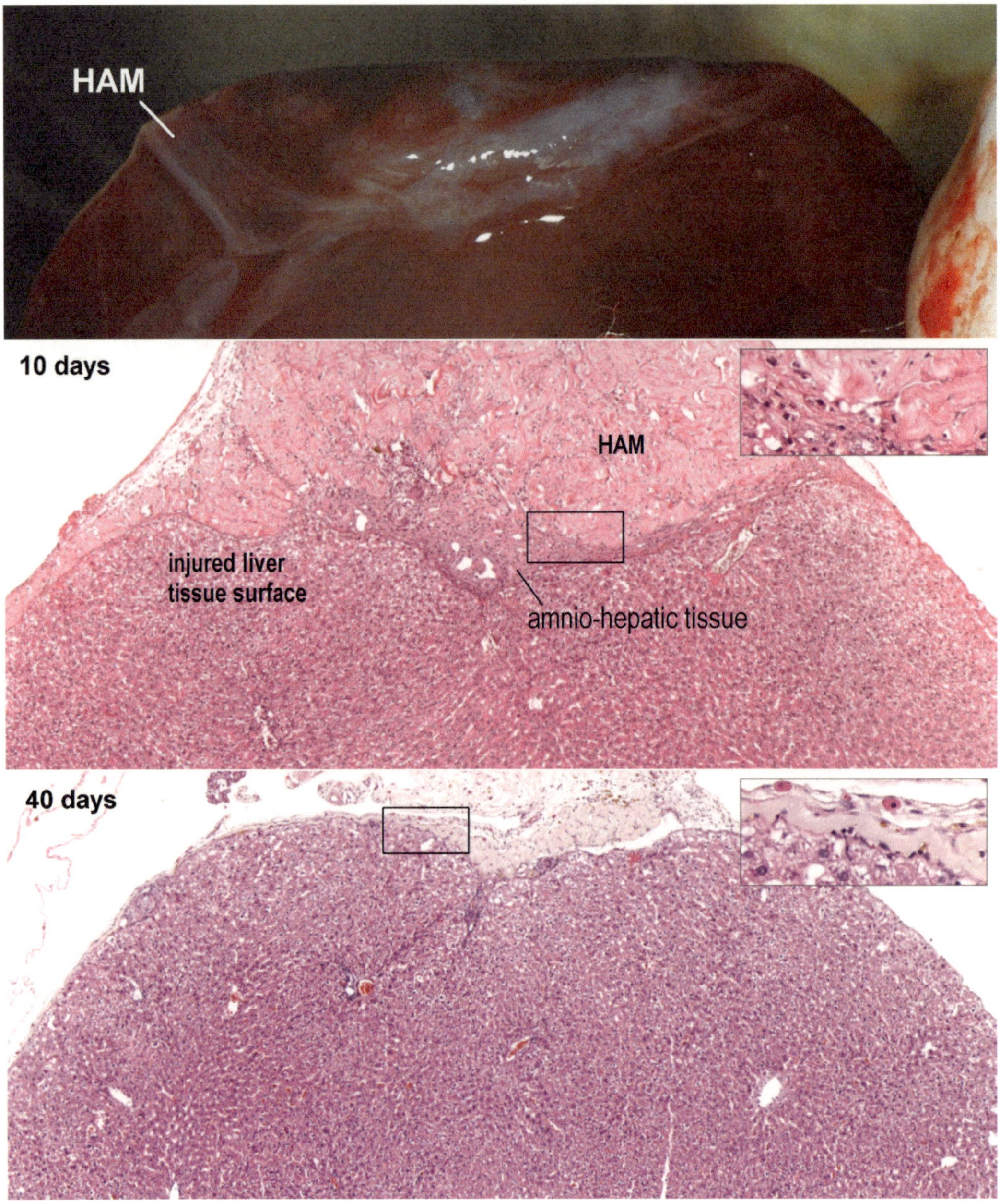

Figure 7. HAM placed on the surgically induced liver wound, HAM intimately adhered to the surface of the injured liver tissue and underwent metaplasia turning into amnion-hepatic tissue in 10 days post-surgery, and remnants of HAM on liver surface in the final step of the liver tissue regeneration in 40 days post-surgery. (hematoxylin-eosin, magnification ×25).

slices comprise an extracellular matrix rich in collagen, without vessels, and act as a reservoir for various proteins that are embedded within the matrix and are available to local cells. Regarding its main property, it is proved that the osteoinductive cascade begins with chemotaxis of bone progenitor cells, mitogenesis, angiogenesis, and bone cell differentiation. The cell recruitment, division rate, and differentiation of these cell lines are under the direct

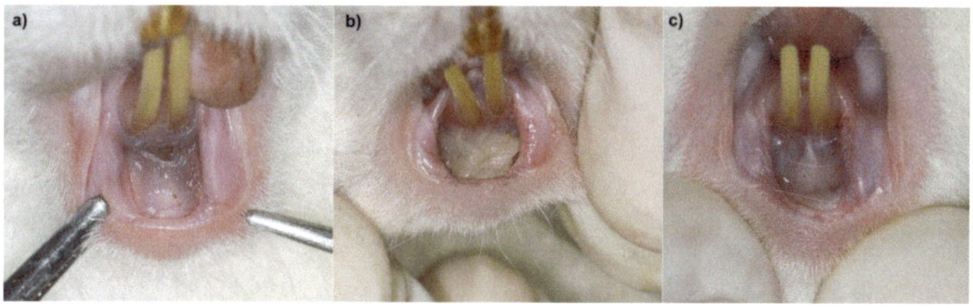

Figure 8. Oral mucositis in labial fornix region of inferior incisors (a), HAM placed on the wound (b), and normal oral mucosa surface after 7 days of treatment with HAM (c).

control of HDDM growth factors, including BMPs [11, 14, 15]. Others growth factors include TGF-β, IGF I and IGF II, PDGF, FGF, TGF-β, and VEGF [17].

HDDM slices strongly favor the chemotaxis of appropriate cells into the surgical defect, the transformation of undifferentiated mesenchymal cells into osteoprogenitor cells, osteoblast proliferation and differentiation, and the synthesis of the extracellular matrix with mineralization of osteoid tissue that leads to accelerated bone maturation and remodeling. It is worth highlighting that the HDDM slices initially yielded hemostatic effects by mechanical action after their implantation in the periphery of the defect. Furthermore, it was demonstrated that the HDDM was incorporated into the newly formed bone matrix and resorbed during the bone remodeling process. Moreover, numerous osteoblasts and osteoprogenitor cells were also located adjacent to the HDDM, confirming its strong chemotactic activity. Studies suggested that HDDM provides bioactive molecules for the host of proteins and growth factors can be released through dentin matrix degradation and exposed dentinal tubules when HDDM is used in the form of slices [6, 11, 14, 15, 17]. Furthermore, when this biomaterial is used in the periphery of surgical defect, it is able to stimulate the reconstruction of the bone architectural microstructure,

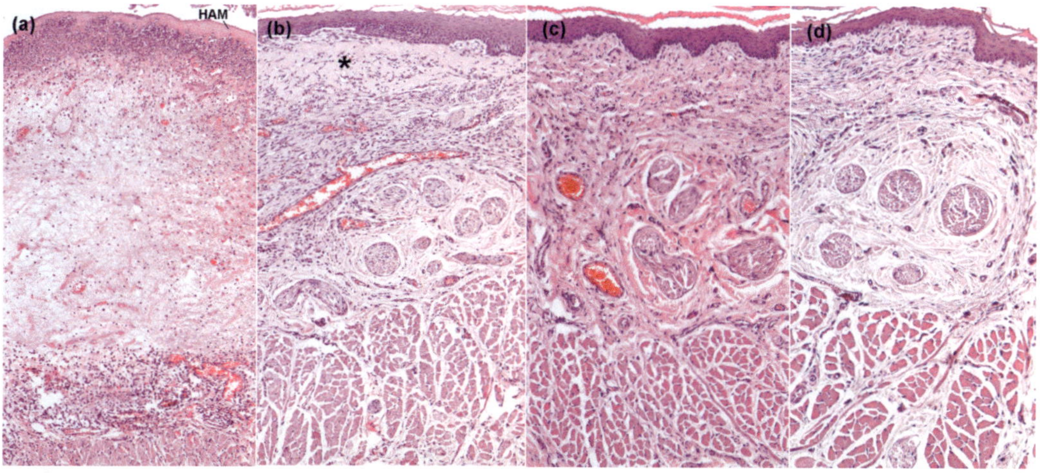

Figure 9. 5-Fluorouracil-induced oral mucosites. Histomorphological features divided into four phases: (a) inflammation (3 days), (b) cell proliferation (7 days), (c) tissue organization (14 days), and (d) tissue repair (21 days). Asterisk: HAM remnants incorporated into connective tissue in the submucosa region (proliferation phase) (hematoxylin-eosin, magnification ×25).

resulting in formation of mature bone trabeculae and bone marrow. Initially, bone trabeculae were mature at the periphery and immature at the centre of the bone defect; thus, centripetal bone growth was seen in the bone repair process (**Figures 10, 11**). Finally, an exophytic bone growth can be evidenced due to the highest chemotactic, mitogenic, and osteogenic potential of the HDDM. Dependent of the animal's age, a physiological conversion from red to yellow bone marrow can be observed, leading to large and regular medullary spaces (**Figure 12**).

Concerning the evidence of HDDM biocompatibility, two hypotheses are suggested by the present study. First, cell surface histocompatibility antigens and sequestered antigens are absent; however, they may be present throughout the extension of the remnant odontoblastic processes within the HDDM. Second, the decalcification process for teeth could cause the denaturation of surface proteins on the plasma membrane of the odontoblastic processes. It was considered that the plasma membrane of the odontoblastic processes could present not only cell surface histocompatibility antigens, but also sequestered antigens that, when not exposed to the cells of the immune system, do not trigger immunological reactions.

The resorption process of the HDDM slices presented the following phases: (1) the dentin matrix was first incorporated into the newly formed bone tissue; (2) next, the dentin was degraded during bone remodeling; and (3) the area in which the dentin matrix was located was replaced by new bone tissue (**Figure 13**).

Therefore, the HDDM could be considered as an ideal bone graft material since it stimulates osteoblast differentiation and proliferation and potentially leads to efficient bone regeneration. We suggested that it could be used as a scaffold for stem cells and bone growth factors, since its biological properties have shown an excellent bioactive material for bone tissue engineering in diabetes.

Autogenous demineralized dentin matrix (ADDM) has been used to accelerate the dental socket wound healing process in humans. The ADDM is considerate biocompatible with surgical human dental sockets. Initially, radiopaque image suggestive of remaining ADDM slices can be found. Moreover, the radiographic bone density of the dental sockets treated with ADDM was similar to that of the surrounding normal bone from 90th day. It is worth highlighting that alveolar bone architecture is improved when the ADDM is used as an engineering tissue (**Figure 14**).

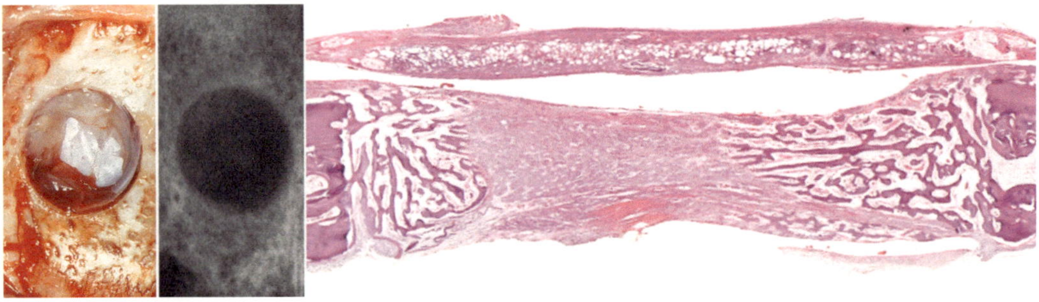

Figure 10. Clinical and radiographic images of surgical defect in the parietal bone of diabetic rabbits taken at 15 days. HDDM slices promoting hemostatic effects by mechanical action after their implantation and their osteogenic activity after 15 days in diabetic rabbit (H.E., original magnification ×25) (Source: Gomes et al. [17]).

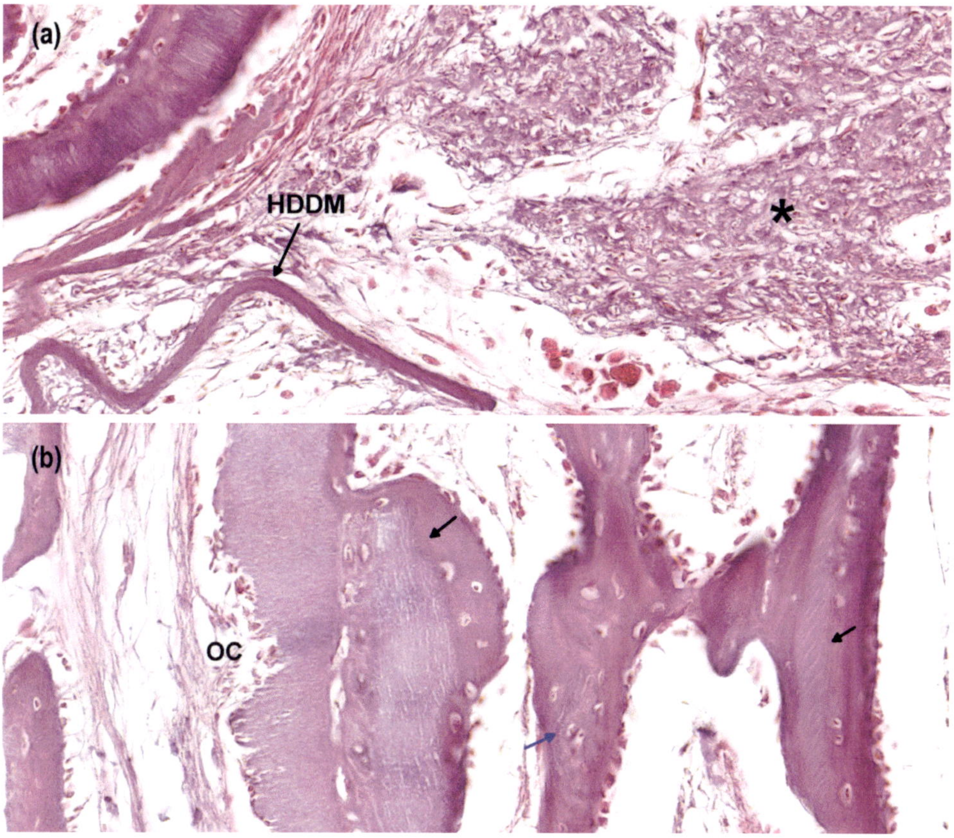

Figure 11. Performance of the HDDM inside the surgical bone defect after 15 days in diabetic rabbit: (a) intense osteogenic connective tissue showing numerous osteoblasts (asterisk) showing chemotactic activity; (b) HDDM slices in resorption process by osteoclasts (OC), resorbed HDDM slices (blue arrow), and HDDM slices incorporated into the newly formed bone tissue (black arrows) (hematoxylin-eosin, magnification ×200).

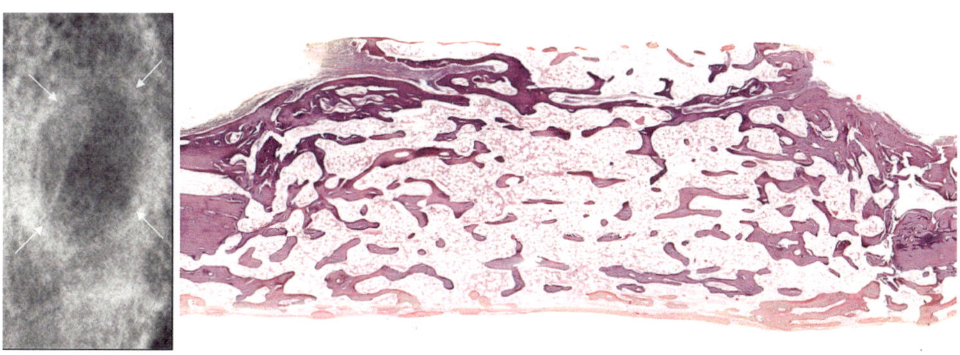

Figure 12. Radiographic image of surgical defect in the parietal bone of diabetic rabbits taken at 90 days and defect region completely filled with mature bone trabeculae a bone marrow (hematoxylin-eosin, magnification ×25) (Source: Gomes et al. [17]).

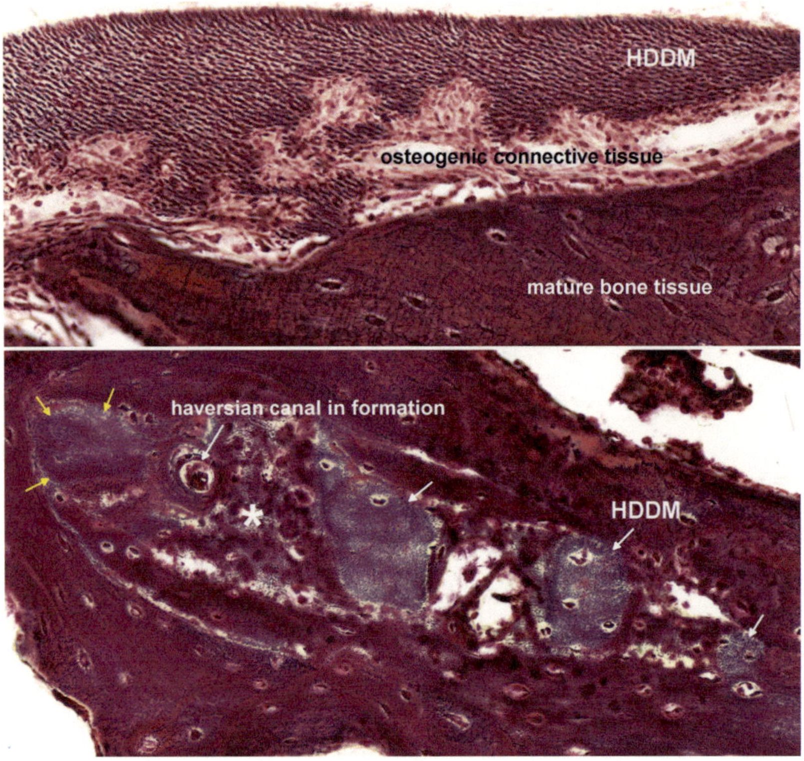

Figure 13. Surgical bone defect region after 90 days in diabetic rabbit. Phases of the HDDM degradation during the bone remodeling process: HDDM in degradation (white arrows), exchange of HDDM by well-cellularized and immature bone tissue with remnants of dentin matrix (asterisk), and replacement of the preexisting HDDM by mature bone tissue (yellow arrows) (Schmorl's stain, magnification ×200).

4.2.3. Platelet-rich plasma (PRP)

Platelet-rich plasma (PRP) has been described as a source of growth factors and a potentially ideal scaffold for tissue engineering in regenerative medicine. PRP is an autologous product that contains highly concentrated number of platelet in a small volume of plasma from whole blood by gradient density centrifugation. It is a proven source of growth factors like platelet-derived growth factor (PGDF) and transforming growth factor β1 and β2 (TGF-β1 and β2), which is obtained by sequestering and concentrating platelets by gradient density centrifugation [25, 47]. This substrate is considered to ameliorate tissue regeneration due to presence of essential various cytokines and growth factors (GFs). Among them, platelet-derived growth factor (PDGF), transforming growth factor β1 (TGF-β1), insulin-like growth factors (IGF), platelet factor 4 (PF-4), fibroblast growth factor 2 (FGF-2), vascular endothelial growth factor (VEGF), and epidermal growth factor (EGF) are considered to be the most important. Subsequently, through stimulation of vascular ingrowth, macrophages arrive and start producing their own cytokines and GFs, some similar to those produced by platelets. This results in a new and continued local tissue repair and regrowth [40]. The cytokines and growth factors also

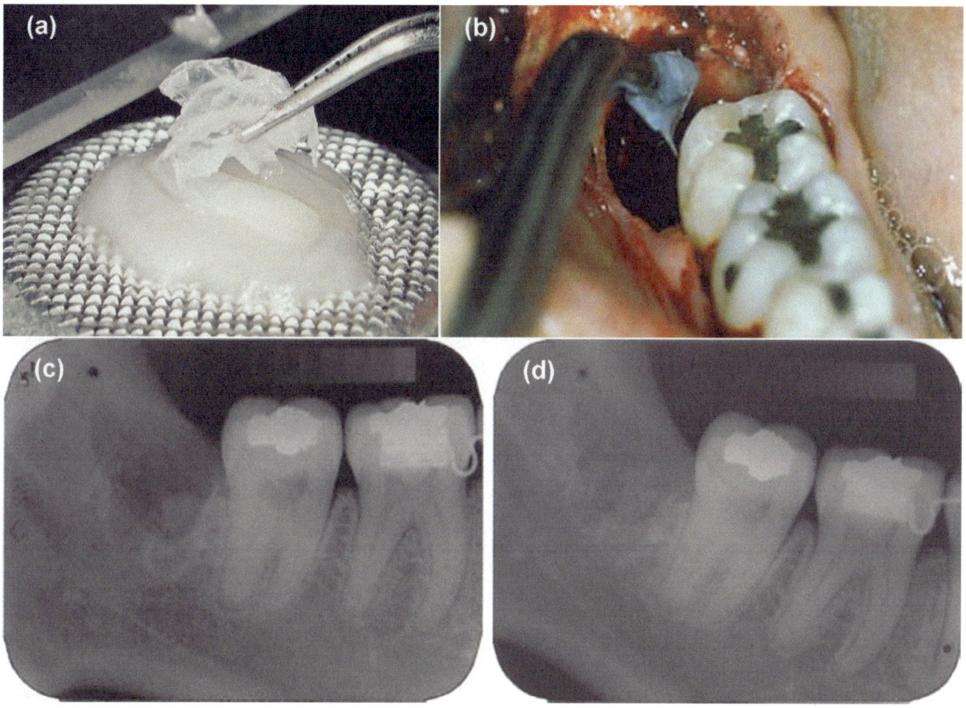

Figure 14. ADDM cut into slices with frozen microtomy (a) and placed into human dental socket covering its surface (b). Periapical radiographs after 15 (c) and 90 days (d) post-surgery. Arrow: radiopaque image suggests remaining of ADDM slices (Source: Gomes et al. [10]).

recruit resident stem cells to the site of injury, where they are stimulated to secrete additional growth factors and anti-inflammatory cytokines, causing more increase in collagen and matrix synthesis. Furthermore, the recruited stem cells will react with the environment to differentiate in parenchymal cells or replace the injured tissue [32].

Some investigators have reported that platelet-rich plasma (PRP) becomes more efficient in tissue repair when associated with bone grafts, bioactive material scaffold, or mesenchymal stem cells (MSCs) which are recruited at the wound site, accelerating tissue repair. This complex could exhibit a synergistic effect. PRP contains numerous proteins including PDGF ($\alpha\alpha$, $\beta\beta$, and $\alpha\beta$ isomers), TGF-b (β1 and β2 isomers), VEGF, angiopoietins, FGF, platelet-derived epidermal growth factor (PDEGF), epidermal growth factor (EGF), IGF, and fibronectin. When it is placed into bone surgical defect, it is quickly filled by a loose connective tissue and newly formed bone trabeculae in a linear arrangement. This connective tissue exhibits myxoid features and numerous newly formed blood vessels, showing intense angiogenic activity and stimulating the formation of bone marrow tissue. The myxoid tissue consisted of very delicate collagen fibers, numerous spindle cells, and a large amount of extracellular amorphous matrix (**Figure 15**). It is possible that the myxoid tissue and intense vascularization have occurred due to the preservation of bioactive angiogenic factors in the PRP, e.g., VEGF and angiopoietins, after its confection in laboratory [17].

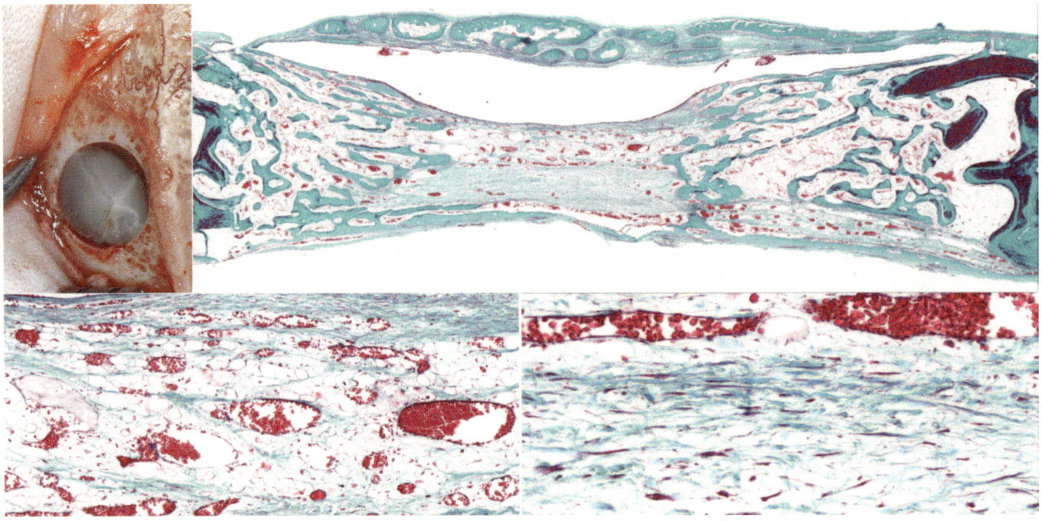

Figure 15. PRP into surgical bone defect stimulating angiogenic activity and myxoit tissue with very delicate collagen fibers, numerous spindle cells, and a large amount of extracellular amorphous matrix, in a period of 30 days (Source: Gomes et al. [17]).

4.3. Alloplastic materials

The ideal implant should be composed of material that is biocompatible, affordable, and nondegradable. Although many materials have proven suitable, alloplastic materials are an attractive implant type, in particular, for the treatment of maxillofacial deformities and for reconstruction [7, 30]. To choose an ideal alloplastic material, first, we must identify the usage needs and their biological and physical properties as biocompatibility, easy of obtaining, non-carcinogenic and nonallergenic material, few production of inflammatory response, essential predeterminant physical-chemical properties to tissue reconstruction and regeneration, no susceptible to infections, sequelae or tissue sequestrum, easy anchorage, handling and sterilization without losing your characteristics and physical structure, high porosity to promote surrounding tissue ingrowth, similar consistency to replaced tissue, stability of contact surface, and good permanence in the implanted region to allow the suitable tissue reconstruction, especially in large defects. Among the several types of available alloplastic materials, porous polyethylene (porous PTFE), hydroxyapatite, and polyurethane resin derived from castor oil implants have been widely indicated in literature. It is worth mentioning the efficient effects of the porous PTFE implant when used in hard and soft tissues. Scientific investigations have shown its application in partial or total ossicular replacement, restorative and esthetic rhinoplasty, orbital implants, ear reconstructions, and augmentation or correction of craniofacial deformities such as temporal and mandibular regions, zygomatic arch, nose, and paranasal areas [7, 8]. Clinical and experimental studies have proven that the PTFE porosity aids surrounding tissue invasion, triggering good adhesion, and stability of the implant in the host site (**Figure 16**). However, this material can be considered poorly biocompatible by tissues, especially in diabetes, due to moderate-intense infiltration of mononuclear inflammation cells as lymphocytes, macrophages, and foreign body-type multinucleated giant cells, especially in uncontrolled diabetes.

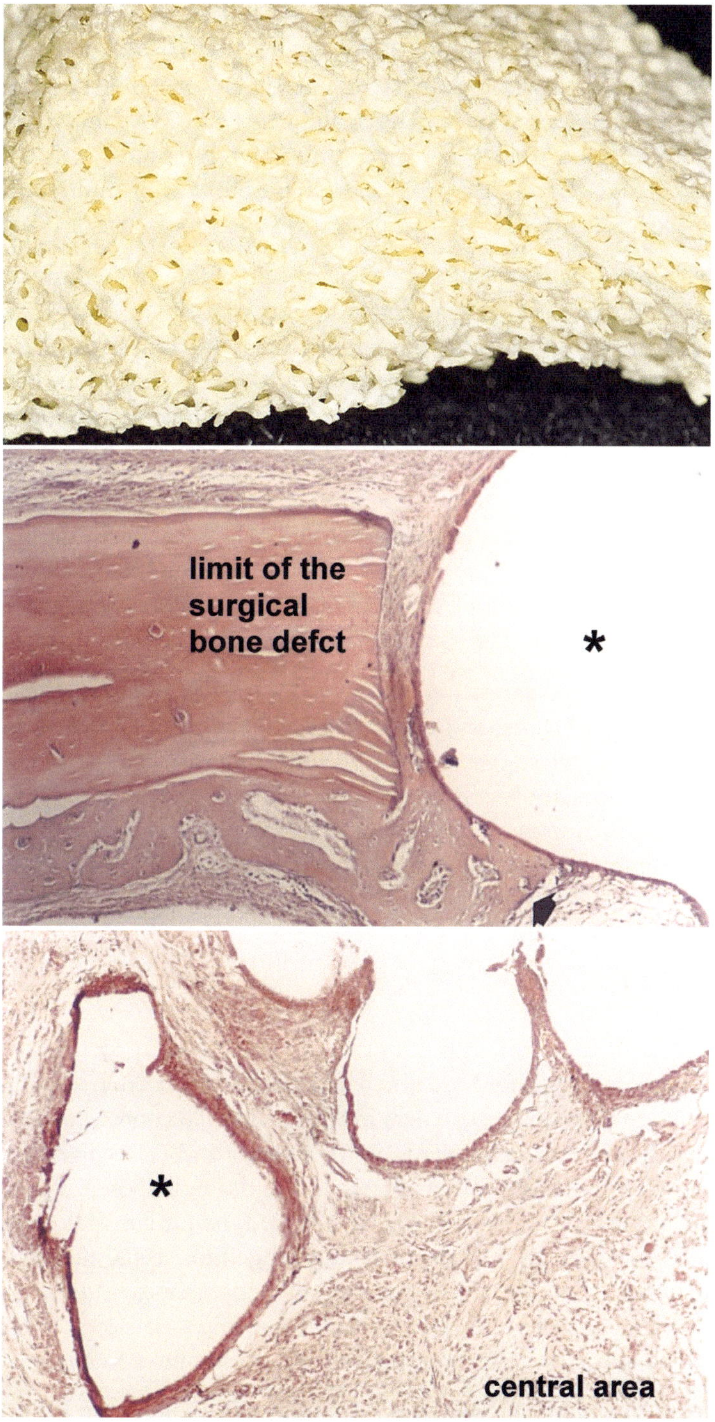

Figure 16. Porous PTFE implanted into surgical bone defect in diabetic rats after 15 days. The defect was filled by fibrous connective tissue with discrete infiltration of mononuclear inflammatory cells and few newly formed bone trabeculae (arrow). The cavities show negative images (asterisk) of the polyethylene, and the soft tissue represents the pore areas (hematoxylin-eosin, magnification ×200).

As a compensatory factor, some therapeutic drugs could be used in association with alloplastic material, such as salmon calcitonin and verapamil. Systemic administration of salmon calcitonin has been widely used in patients with pathologies that affect the bone metabolism, including osteoporosis, osteogenesis imperfect, and Paget's disease, because of its recognized ability to inhibit bone resorption [8]. For support therapy, the salmon calcitonin can be indicated to treat aggressive cherubism, promoting the inhibition of multinucleated giant cell formation and osteoclastic activity, leading to a regressive process of the disease [16].

Regarding to the verapamil, it is used for treatment of high blood pressure, cardiac arrhythmias, and pectoris angina, as well as it is often indicated for diabetic individuals. Diabetic individuals are great users of this drug. The verapamil, a potent calcium channel blocker, promotes the inhibition of voltage-dependent calcium channel, L-type, reducing the Ca^{2+} entry into the cells and, consequently, altering their functions. Osteoclast lineages are affected, resulting in the inhibition of bone resorption and, thus, favoring the new bone tissue formation.

Some researches demonstrated that the verapamil has influenced the calcium balance in the mineralized tissues. It changes the calcium trophic hormone action, altering the endocrine regulator effect in the bone turnover process, since it acts directly in the inhibition of the osteoclasts activity. On the other hand, the increase of intracellular calcium levels of osteoblast stimulates the IGF-I, IGF-II, and TGF-β production, favoring the bone tissue formation [2, 8].

In addition, implant neovascularization is essential to recruit inflammatory and progenitor cells, initiating the wound healing and/or leading to an exponential formation of new tissues [8, 17]. Thus, researches have suggested an additional bioactive material to the porous polyethylene implant, as mesenchymal stem cells (MSCs) or angiogenic and mitogenic factors, so that they can present synergistic effects and improve the environment and, consequently, the development of desirable tissues in a variety of reconstructive and restorative surgeries.

5. Considerations

Considering the exponential growth trend of chronic diseases, in particular, diabetes mellitus, ideal bioactive biomaterials for hard and soft tissue reconstruction and regeneration have been a great challenge for many researchers in worldwide. To choose an ideal biomaterial to be implanted in diabetes, we must first know the cellular and molecular pathophysiological mechanisms following immeasurable damages in the tissue response against any extrinsic factors, including impacts or grafts. In regard to the biodynamic constraints of the repair process, the recruitment of cells, as inflammatory and progenitors cells, and growth factors and cytokine production or action are largely impaired in diabetics individuals, in particular, with uncontrolled hyperglycemia. As undesirable results, these materials could increase the risk for and severity of the diabetic complications instead of favoring the tissue regeneration process.

Based on basic concepts of biomaterial, it is important to highlight that the biocompatibility, material storage without loss of viability, ease of obtaining the material, favorable cost and benefit relationship, and mainly preservation of its inductive properties must be considered as pivotal parameters for tissue-engineered implants which could be used in diabetes. Other concerns of researchers are to identify the reliable material sources that do not pose any risk to the donor or to the individual that will receive the cells, and to accomplish the isolation of cells with high potential of expansion and proliferation.

In this regard, we discussed some property and effects of biomaterials and alloplastic materials that could surely be indicated in cell therapy and tissue engineering, especially in diabetic individuals. As a compensative factor in diabetes, we suggested the clinical application of the AM and DDM in isolation or in association with other materials commonly found in the literature. The highly effective biological properties of these scaffolds favored the development of impaired tissues through the restoring, maintaining, or improving tissue function, corroborating with the basic principles of tissue engineering. Special attention has been paid to AM due to rich amount of growth factor and MSCs in its mesenchymal stroma, playing a key role in the creation of implantable tissue. Moreover, the AM is considered a promising source since it is usually discarded after childbirth (in humans and animals); therefore, it could be used without ethical or religious troubles when collected. The use of this membrane in some centers can be submitted the same rules for the use of organ and tissue transplants; however, it is worth noting the importance to conduct strict control procedures as serological and microbiological screening of this membrane.

Acknowledgements

The funding source of the CEBAPE researches was supported by São Paulo Research Foundation, known as FAPESP (grant numbers: 1996/3203-2, 2002/03952-8, 2003/06017-0, 2004/04447-0, 2004/08656-0, and 2007/54996-9).

Author details

Mônica Fernandes Gomes*, José Benedito Amorim, Lilian Chrystiane Giannasi and Miguel Angel Castillo Salgado

*Address all correspondence to: mfgomes@ict.unesp.br

Center of Biosciences Applied to Persons with Special Care Needs (CEBAPE), Institute of Science and Technology, São Paulo State University (UNESP), São José dos Campos, São Paulo, Brazil

References

[1] American Diabetes Association (ADA). Diagnosis and classification of diabetes mellitus. Diabetes Care. 2014;**37**(Suppl 1):S81–S90

[2] Amorim JBO, Teixeira SS, Negrato GV, Vilela-Goulart MG, Gomes MF. Histomorphometric analysis of verapamil effects on alveolar socket repair on diabetic alloxan-induced rats. Ciência Odontológica Brasileira. 2009;**12**(1):37-48

[3] Aragno M, Mastrocola R. Dietary sugars and endogenous formation of advanced glycation endproducts: Emerging mechanisms of disease. Nutrients. 2017;**9**(4):2-16. pii: E385. DOI: 10.3390/nu9040385

[4] Aronson PS, Boron WF, Boulpaep EL. Physiology of membranes. ln: Boron WF, Boulpaep EL, editors. Medical Physiology. 1st ed. Philadelphia: WB Saunders; 2005

[5] Ashcroft FM, Rorsman P. Diabetes mellitus and the β-cell: The last ten years. Cell. 2012; **148**(6):1160-1171

[6] Carvalho VA, Tosello Dde O, Salgado MA, Gomes MF. Histomorphometric analysis of homogenous demineralized dentin matrix as osteopromotive material in rabbit mandibles. International Journal of Oral & Maxillofacial Implants. 2004;**19**(5):679-686

[7] Choi SY, Shin HI, Kwon TY, Kwon TG. Histopathological and scanning electron microscopy findings of retrieved porous polyethylene implants. International Journal of Oral & Maxillofacial Surgery. 2017;**46**(5):582-585

[8] Claro FA, Lima JR, Salgado MA, Gomes MF. Porous polyethylene for tissue engineering applications in diabetic rats treated with calcitonin: Histomorphometric analysis. International Journal of Oral & Maxillofacial Implants. 2005;**20**(2):211-219

[9] Davidson NO, Hausman AM, Ifkovits CA, Buse JB, Gould GW, Burant CF, Bell GI. Human intestinal glucose transporter expression and localization of GLUT5. American Journal of Physiology. 1992;**262**(3 Pt 1):C795–C800

[10] Gomes MF, Abreu PP, Morosolli ARC, Araújo MM, Goulart Md. Densitometric analysis of the autogenous demineralized dentin matrix on the dental socket wound healing process in humans. Brazilian Oral Research. 2006;**20**(4):324-330

[11] Gomes MF, Banzi EC, Destro MF, Lavinicki V, Goulart Md. Homogenous demineralized dentin matrix for application in cranioplasty of rabbits with alloxan-induced diabetes: Histomorphometric analysis. International Journal of Oral & Maxillofacial Implants. 2007;**22**(6):939-947

[12] Gomes MF. Homogenous amniotic membrane as biological dressing in the liver regeneration and function in rats [livre-docência thesis]. São José dos Campos (SP): Institute of Science and Technology, UNESP – São Paulo State University; 2007

[13] Gomes MF, Destro MFS, Banzi ECF, Vieira EMM, Morosolli AR, Vilela-Goulart MG. Optical density analysis of homogenous demineralized dentin matrix in the bone regeneration of alloxan-induced diabetic rabbits. Brazilian Oral Research. 2008;**22**(3):275-280

[14] Gomes MF, dos Anjos MJ, Nogueira Tde O, Catanzaro Guimarães SA. Autogenous demineralized dentin matrix for tissue engineering applications: Radiographic and histomorphometric studies. International Journal of Oral & Maxillofacial Implants. 2002;**17**(4):488-497

[15] Gomes MF, dos Anjos MJ, Nogueira TO, Guimarães SA. Histologic evaluation of the osteoinductive property of autogenous demineralized dentin matrix on surgical bone defects in rabbit skulls using human amniotic membrane for guided bone regeneration. International Journal of Oral & Maxillofacial Implants. 2001;**16**(4):563-571

[16] Gomes MF, Ferraz de Brito Penna Forte L, Hiraoka CM, Augusto Claro F, Costa Armond M. Clinical and surgical management of an aggressive cherubism treated with autogenous bone graft and calcitonin. ISRN Dentistry. 2011;**2011**:340960

[17] Gomes MF, Valva VN, Vieira EM, Giannasi LC, Salgado MA, Vilela-Goulart MG. Homogenous demineralized dentin matrix and platelet-rich plasma for bone tissue engineering in cranioplasty of diabetic rabbits: Biochemical, radiographic, and histological analysis. International Journal of Oral & Maxillofacial Surgery. 2016;**45**(2):255-266

[18] Gomes MF, Werbicky V, Nogueira TO. Lyophilized human amniotic membrane over wounds in areas of oral biopsy. Revista da Associação Paulista de Cirurgiões Dentistas. 2001a;**55**(5):327-331

[19] Han K, Lee JE, Kwon SJ, Park SY, Shim SH, Kim H, Moon JH, Suh CS, Lim HJ. Human amnion-derived mesenchymal stem cells are a potential source for uterine stem cell therapy. Cell Proliferation. 2008;**41**(5):709-725

[20] Han Y, Sun T, Tao R, Han Y, Liu J. Clinical application prospect of umbilical cord-derived mesenchymal stem cells on clearance of advanced glycation end products through autophagy on diabetic wound. European Journal of Medical Research. 2017;**22**(1):11

[21] Hardman TC, Dubrey SW. Development and potential role of type-2 sodium-glucose transporter inhibitors for management of type 2 diabetes. Diabetes Therapy. 2011; **2**(3):133-145

[22] Hortensius RA, Ebens JH, Harley BA. Immunomodulatory effects of amniotic membrane matrix incorporated into collagens scaffolds. Journal of Biomedical Materials Research Part A. 2016;**104**(6):1332-1342

[23] Jahan H, Choudhary MI. Glycation, carbonyl stress and AGEs inhibitors: A patent review. Expert Opinion on Therapeutic Patents. 2015;**25**(11):1267-1284

[24] Johnson LR. Physiology of the Gastrointestinal Tract. Vol. 2. 3rd ed. New York: Raven Press; 1994

[25] Kaur P, Maria A. Efficacy of platelet rich plasma and hydroxyapatite crystals in bone regeneration after surgical removal of mandibular third molars. Journal of Maxillofacial and Oral Surgery. 2013;**12**(1):51-59

[26] Lima Gde M, Severo MC, Santana-Melo Gde F, Carvalho MA, Vilela-Goulart Md, Salgado MA, Gomes MF. Amniotic membrane as a biological dressing for 5-fluorouracil-induced oral mucositis in rats. International Journal of Oral & Maxillofacial Surgery. 2015;**44**(7):845-851

[27] Lin JA, Wu CH, Lu CC, Hsia SM, Yen GC. Glycative stress from advanced glycation endproducts (AGEs) and dicarbonyls: An emerging biological factor in cancer onset and progression. Molecular Nutrition & Food Research. 2016;**60**(8):1850-1864

[28] Lin L, Yee SW, Kim RB, Giacomini KM. SLC transporters as therapeutic targets: Emerging opportunities. Nature Reviews Drug Discovery. 2015;**14**(8):543-560

[29] Lin Y, Sun Z. Current views on type 2 diabetes. Journal of Endocrinology. 2010;**204**(1):1-11

[30] Mahoney NR, Grant MP, Iliff NT, Merbs SL. Exposure rate of smooth surface tunnel porous polyethylene implants after enucleation. Ophthalmic Plastic and Reconstructive Surgery. 2014;**30**(6):492-498

[31] Mohan R, Bajaj A, Gundappa M. Human amnion membrane: Potential applications in oral and periodontal field. Journal of International Society of Preventive and Community Dentistry. 2017;**7**(1):15-21

[32] Moussa M, Lajeunesse D, Hilal G, El Atat O, Haykal G, Serhal R, Chalhoub A, Khalil C, Alaaeddine N. Platelet rich plasma (PRP) induces chondroprotection via increasing autophagy, anti-inflammatory markers, and decreasing apoptosis in human osteoarthritic cartilage. Experimental Cell Research. 2017;**352**(1):146-156

[33] Niknejad H, Peirovi H, Jorjani M, Ahmadiani A, Ghanavi J, Seifalian AM. Properties of the amniotic membrane for potential use in tissue engineering. European Cells & Materials. 2008;**15**:88-99

[34] Rorsman P, Braun M, Zhang Q. Regulation of calcium in pancreatic α- and β-cells in health and disease. Cell Calcium. 2012;**51**(3-4):300-308

[35] Sanluis-Verdes A, Sanluis-Verdes N, Manso-Revilla MJ, Castro-Castro AM, Pombo-Otero J, Fraga-Mariño M, Sanchez-Ibañez J, Doménech N, Rendal-Vázquez ME. Tissue engineering for neurodegenerative diseases using human amniotic membrane and umbilical cord. Cell and Tissue Banking. 2017;**18**(1):1-15

[36] Sasseville LJ, Longpré JP, Wallendorff B, Lapointe JY. The transport mechanism of the human sodium/myo-inositol transporter 2 (SMIT2/SGLT6), a member of the LeuT structural family. American Journal of Physiology—Cell Physiology. 2014;**307**(5):C431–C441

[37] Sasseville LJ, Morin M, Coady MJ, Blunck R, Lapointe JY. The human sodium-glucose cotransporter (hSGLT1) is a disulfide-bridged homodimer with a re-entrant c-terminal loop. PLoS One. 2016;**11**(5):e0154589

[38] Schieber M, Chandel NS. ROS function in redox signaling and oxidative stress. Current Biology. 2014;**24**(10):R453–R462

[39] Singh VP, Bali A, Singh N, Jaggi AS. Advanced glycation end products and diabetic complications. Korean Journal of Physiology & Pharmacology. 2014;**18**(1):1-14

[40] Sommeling CE, Heyneman A, Hoeksema H, Verbelen J, Stillaert FB, Monstrey S. The use of platelet-rich plasma in plastic surgery: A systematic review. Journal of Plastic, Reconstructive & Aesthetic Surgery. 2013;**66**(3):301-311

[41] Szkudelski T. The mechanism of alloxan and streptozotocin action in B cells of the rat pancreas. Physiological Research. 2001;**50**(6):537-546

[42] Vieira EM, Ueno CS, Valva VN, Goulart Md, Nogueira Tde O, Gomes MF. Bone regeneration in cranioplasty and clinical complications in rabbits with alloxan-induced diabetes. Brazilian Oral Research. 2008;**22**(2):184-191

[43] Vilela-Goulart Md, Teixeira RT, Rangel DC, Niccoli-Filho W, Gomes MF. Homogenous amniotic membrane as a biological dressing for oral mucositis in rats: Histomorphometric analysis. Archives of Oral Biology. 2008;**53**(12):1163-1171

[44] Vlassara H, Brownlee M, Cerami A. Nonenzymatic glycosylation: Role in the pathogenesis of diabetic complications. Clinical Chemistry. 1986;**32**(10 Suppl):B37–B41

[45] Yang H, Jin X, Kei Lam CW, Yan SK. Oxidative stress and diabetes mellitus. Clinical Chemistry and Laboratory Medicine. 2011;**49**(11):1773-1782

[46] Rodrigues M, Wong VW, Rennert RC, Davis CR, Longaker MT, Gurtner GC. Progenitor cell dysfunctions underlie some diabetic complications. American Journal of Pathology. 2015;**185**(10):2607-2618

[47] Kumar KA, Rao JB, Pavan Kumar B, Mohan AP, Patil K, Parimala K. A prospective study involving the use of platelet rich plasma in enhancing the uptake of bone grafts in the oral and maxillofacial region. Journal of Maxillofacial and Oral Surgery. 2013;**12**:387-389